CISM COURSES AND LECTURES

Series Editors:

The Rectors of CISM
Sandor Kaliszky – Budapest
Mahir Sayir – Zurich
Wilhelm Schneider – Wien

The Secretary General of CISM
Giovanni Bianchi – Milan

Executive Editor
Carlo Tasso – Udine

The series presents lecture notes, monographs, edited works and
proceedings in the field of Mechanics, Engineering, Computer Science
and Applied Mathematics.
Purpose of the series is to make known in the international scientific
and technical community results obtained in some of the activities
organized by CISM, the International Centre for Mechanical Sciences.

INTERNATIONAL CENTRE FOR MECHANICAL SCIENCES

COURSES AND LECTURES - No. 422

ROMANSY 13

THEORY AND PRACTICE OF ROBOTS AND MANIPULATORS

PROCEEDINGS OF THE THIRTEENTH
CISM-IFToMM SYMPOSIUM

EDITED BY

ADAM MORECKI
WARSAW UNIVERSITY OF TECHNOLOGY

GIOVANNI BIANCHI
POLYTECHNIC OF MILAN

CEZARY RZYMKOWSKI
WARSAW UNIVERSITY OF TECHNOLOGY

This volume contains 292 illustrations

This work is subject to copyright.
All rights are reserved,
whether the whole or part of the material is concerned
specifically those of translation, reprinting, re-use of illustrations,
broadcasting, reproduction by photocopying machine
or similar means, and storage in data banks.
© 2000 by CISM, Udine
Printed in Italy
SPIN 10796360

In order to make this volume available as economically and as
rapidly as possible the authors' typescripts have been
reproduced in their original forms, This method unfortunately
has its typographical limitations but it is hoped that they in no
way distract the reader.

ISBN 3–211–83333–1 Springer–Verlag Wien New York

PREFACE

The CISM-IFToMM Ro.Man.Sy Symposia have played a dynamic role in the development of the theory and practice of robotics. The proceedings of the thirteenth symposia present world view of the state of the art.

Ro.Man.Sy 2000 held July 3-6, 2000 in Zakopane, Poland was attended by 74 participants from 15 countries.

The proceedings of this thirteenth edition of Ro.Man.Sy focus mainly on problems of mechanical engineering and control.

In his opening lecture B. Roth presented an overview of the theoretical basis for the mechanical aspects of robot design.

In his general lecture M. Vukobratovič discussed theory and practice of new frontiers of robotics.

The 50 regular papers included in this volume illustrate significant contribution in mechanics (13 papers), motion control (7), synthesis and design (8), legged locomotion (11), sensing and machine intelligence (2), applications (5) and biomechanical aspects of robots and manipulators (4). They appear here in the order and form in which they were presented during the symposium.

The next Ro.Man.Sy will be held in Udine, Italy in July 2002.

Adam Morecki
Giovanni Bianchi
Cezary Rzymkowski

CONTENTS

Page

Preface

OPENING LECTURE:

The Theoretical Basis for the Mechanical Aspects of Robot Design
by B. Roth 3

GENERAL LECTURE:

New Frontiers in Robotics
by M. Vukobratovič 15

CHAPTER I: Mechanics

Coordination of Parallel Arrays of Binary Actuators
by P.-H. Yang and K.J. Waldron 43

A Formal-Numerical Approach to Determine the Accuracy of a Parallel Robot
in a 6D Workspace
by J-P. Merlet 51

Symbolic Calculation of Robot's Base Reaction-Force/Torque Equations
with Minimal Parameter Set
by M. Grotjahn and B. Heimann 59

Mechanics of the New UWA Robot
by K. Miller 67

On Kinematic Singularities of Nonholonomic Robotic Systems
by K. Tchoń 75

Experimental Determination of Robot Workspace by Means of CATRASYS
(Cassino Tracking System)
by M. Ceccarelli, E. Ottaviano and M. Toti 85

A Pantograph Mechanism with Large-Deflective Hinges for Miniature
Surface Mount Systems
by M. Horie, T. Uchida and D. Kamiya 93

Adaptive λ- Tracking for Rigid Manipulators
by A. Mazur and C. Schmidt 103

Mobility Analysis of the 3-PSP Mechanism
by R. Di Gregorio and V. Parenti-Castelli 113

Detaching and Grasping Strategy Inspired by Human Behavior
by M. Kaneko, T. Shirai and T. Tsuji 121

Cross-Country Capabilities of a Walking Robot: Geometrical, Kinematic and Dynamic Investigation
by V.E. Pavlovsky and A.K. Platonov 131

Optimization of Robot Gripper Parameters Using Genetic Algorithms
by S. Krenich and A. Osyczka 139

Design of Spatial Fixed-Sequence Manipulator for Precise and Approximate Reproduction of Gripper Predetermined Positions
by V. Arakelian and M. Dahan 147

CHAPTER II: Control of Motion

A Powered-Caster Holonomic Robotic Vehicle for Mobile Manipulation Tasks
by R. Holmberg and O. Khatib 157

Motion Coordination and Hybrid Position/Force Control of a Mobile Micromanipulator Actuated by Direct–Drive Vibromechanisms
by A. Ferreira and P. Minotti 169

Coordination Control of a Human/Manipulator System
by J.H. Chung and S.A. Velinsky 179

Stability of Cooperating Manipulators with Symmetric Position/Force Control and Time Delay
by A. Schneider, I. Zeidis and K. Zimmermann 187

Remote Control of Periodic Robot Motion
by T. Insperger and G. Stépán 197

Dynamic Control of Multiple Joint Manipulators Interacting with Dynamic Environment
by A. Tuneski and M. Vukobratovič 205

A Comparison between PD-Controls in Terms of Normalized and Unnormalized Quasi-Velocities
by K. Kozłowski and P. Herman 215

CHAPTER III: Synthesis and Design

The Modular Design of a Long-Reach, 11-Axis Manipulator
  by J. Angeles, A. Morozov, L. Slutski, O. Navarro and L. Jabre 225

Structure Synthesis of Parallel Manipulators
  by V. Glazunov, A. Kraynev, G. Rashoyan, A. Trifonova and M. Esina 235

Design of Manipulators Under Dynamic and Kinematic Performances
  by S. Guerry and F.B. Ouezdou 241

Influence of Leg Flexibilities on the Trajectory Planning of a 3-DOF Spherical
Parallel Manipulator
  by J. Knapczyk and G. Tora 249

Study on the Specific Characteristics of Various Actuators
  by T. Hayashi 257

Micro-Manipulation and Adhesion Forces
  by D.S. Haliyo, Y. Rollot, S. Regnier and J.C. Guinot 265

Universal Dental Robot — 6-DOF Mouth Opening and Closing Training Robot WY-5
  by H. Takanobu, T. Maruyama, A. Takanishi, K. Ohtsuki and M. Ohnishi 275

Development of the Design of POLYCRANK Manipulator Without Joint Limits
  by K. Nazarczuk, K. Mianowski and S. Łuszczak 285

CHAPTER IV: Legged Locomotion

Emotion-based Walking for a Biped Humanoid Robot
  by Hun-ok Lim, A. Ishii and A. Takanishi 295

Design and Control of the Humanoid Robot ARMAR
  by K. Berns, T. Asfour and R. Dillmann 307

On Dynamics of Movement of Walking Machines with Gears Made on the Basis Cycle
Mechanisms of Walking
  by E.S. Briskin, V.V. Chernyshev, A.V. Maloletov and S.V. Sherstobitov 313

Development of Walking Machines: Novel Leg Drive Design and Control
  by T. Zielińska and J. Heng 323

Autonomous Locomotion of Walking Machines in Rough Terrain
  by M. Frik, A. Buschmann, M. Guddat, M. Karataş and D.C. Losch 331

Design and Control of a Biped Robot
by K. Löffler, M. Gienger and F. Pfeiffer 339

Six Link Mechanisms for the Legs of Walking Machines
by A.P. Bessonov, N.V. Umnov, V.V. Korenovsky, E.E. Silvestrov and S.V. Khoborkov 347

Design, Analysis and Measurements Problems of Mili–Walking Machines
Using Multi–Body System Formulation
by J. Frączek and A. Morecki 355

Modeling, Simulation and Nonlinear Control of a Combined Legged and Wheeled Vehicle
by J. Müler and M. Hiller 363

A New Local Path Planner for a Nonholonomic Wheeled Mobile Robot in Cluttered
Environments
by G. Ramirez and S. Zeghloul 371

Three-Dimensional Simulation of Walk of Anthropomorphic Biped
by F. Gravez, O. Bruneau and F.B. Ouezdou 379

CHAPTER V: Sensing and Machine Intelligence

Cooperative Micro Object Handling by Dual Micromanipulators Under Vision Control
by A. Ferreira and S. Hirai 391

A Comparative Study of Torque Control Using a Wrist or a Base Force/Torque Sensor
by F. Geffard, C. Andriot and G. Morel 401

CHAPTER VI: Applications

Quality Feature Based Adjustment of Robot Programs Exemplified
for the Welding Process – MAGROB
by R.D. Schraft, J. Neugebauer and W. Schaaf 411

Application of the RNT Robot to Milling and Polishing
by K. Mianowski, K. Nazarczuk, M. Wojtyra, W. Szynkiewicz, C. Zieliński
and A. Woźniak 421

Path Planning in Complex Environments for Industrial Robots
with Additional Degrees of Freedom
by F. Valero, V. Mata and M. Ceccarelli 431

Robotic Deburring Using a Fuzzy Force Controller
by R. Bicker and K. Burn 439

A Distributed SMA Actuator System and Associated Self-Guiding Control Strategy for a Scalable Endoscope Steering Device
by Ph. Bidaud, J. Szewczyk, N. Troisfontaine and J.-C. Guinot 447

CHAPTER VII: Biomechanical Aspects

Low Energy Biped Locomotion
by S. Gruber and W. Schiehlen 459

Jumping Motion of an Object Controlled by a Muscle Contraction
by J. Viba, I. Tipans, O. Kononova and J.-G. Fontaine 467

Inverse Simulation Study of Trampoline-Performed Somersaults
by W. Blajer and A. Czaplicki 479

Functional Biomechanics of Human Grasping and Sensory Requirements for Simple Robotic End-Effectors
by R.B. Addis and B. Ravani 489

APPENDIX A

Programme and Organizing Committee 503

APPENDIX B

List of Participants 507

Opening Lecture

The Theoretical Basis for the Mechanical Aspects of Robot Design

Bernard Roth

Department of Mechanical Engineering, Stanford University, USA

Abstract. This paper considers the basic concepts and tools that have been developed for robot design. Many of these concepts have been adapted from classical mechanical design, and some are newly developed for robotic systems. Taken all together, there is now a very large body of results and theory available to the robot designer and analyst. This paper presents an overview of the current state of knowledge regarding the mechanical aspects of robot design. The main objective of this presentation is to develop a framework into which piecemeal and disparate results can combined to provide a theoretical basis which supplements the empirical aspects of the subject with a coherent theory.

1 Introduction

This paper presents a listing of the major results that have been developed in the areas of robotics that are generally covered in the Ro.Man.Sy meetings. The material covered herein has been developed in the preceding thirty-five years, and contains the result of the work of many individuals. In this paper, no single result is identified with any researcher. Instead, these results and methods are treated as though they belong to the entire research community. What is attempted here is the identification of a useful framework into which all the results can be placed in an orderly manner. Such a framework forms the basis of a coherent theory. It is within such a theory that the interrelationships between results can best be revealed, and it also exposes certain gaps which require further research in order to complete the theory.

1.1 The basic question

The basic question we face is: how can the subject of robotics be presented as a coherent theory? Most sciences seem to progress once the material under study is divided into various groupings. So what should be the divisions of our subject? We could proceed, as some authors do, using a division by function. In which case we could have categories such as: Manipulation, Mobility, Vision. For manipulation we could have subcategories such as: arms, hands, fingers; for mobility the subcategories might be: wheels, legs, treads. Alternatively, we might divide by traditional engineering subject areas. In which case an appropriate division would be between Kinematics, Dynamics and Control. These are the categories we will use in this paper.

Sections 2, 3 and 4 contain lists of the main subject areas that have been researched in the kinematics, dynamics and control of robotic systems. As a subsection of each area, there is a list of the problem areas currently receiving attention as "hot topics." Typically the results in these subsections are still not as well developed as the material in the main section of that respective area.

2 Kinematics

In this section we summarize the major results in the kinematics of robots. These results are placed roughly in chronological order:
    Kinematic classification (i.e., identify devices by joint type rather than commercial name)
    Systematic position analysis (i.e., a deterministic way to set up the analysis equations)
    Velocity, force and Jacobians (determined from position equations or directly)
    Orientation of end-effector uncoupled from its position by using three intersecting axes
    Determination of the number of solutions for the inverse kinematics of series manipulators
    Obstacle avoidance techniques
    Determination of conditions for series structures to be solvable in terms of a fourth or lower
        degree polynomial (so called, analytic structures)
    Calibration methodologies and theories developed
    Trajectory paths synthesized
    Static flexibility and backlash studied
    Manipulability and dexterity defined and evaluated for various arm designs
    Workspace types such as dexterous and reachable defined and computed
    Modularity introduced as a practical design approach for manipulators
    The direct kinematics of in-parallel (multi-loop) structures solved
    Structural parameters determined according to workspace design requirements
    All series and single loop structures solved for both direct and inverse kinematics
    Singularities identified and studied — various types, for series and in-parallel mechanisms
    Changes of configurations during a trajectory-generation analyzed
    Singularity "free" designs developed
    Internal forces analyzed in devices with redundancy
    Movement in null space studied
    Motion capabilities for devices with generic vs. special structural parameters (such as
        parallelism and intersection of axes)
    Forces and velocities determined for systems composed of cooperating robots
    All topological variants enumerated
    Distribution of active and passive joints in in-parallel systems studied
    Large-degree-of-freedom systems analyzed and constructed
    Conceptual and virtual devices simulated

2.1   Kinematics — Current

    Micro and submicro devices studied
    Hybrid (combined series and in-parallel) structures studied
    Adaptive configurations studied
    Mems devices studied
    Biomimetics — including humans — becomes a formally recognized area of study

Robots designed to be in the same spaces and physically interacting with humans
Whole body robots are studied (i.e., more than just arms, hands and legs)
More real-world design problems are studied (e.g., surgery, space, underwater)

3 Dynamics

In this section we summarize the major results in the dynamics of robots. The results are placed in very rough chronological order:

Lagrangian formulations developed
Newton-Euler (recursive methods) developed
Kane's method and Appel's method are applied
The "great counting controversy," over which method has shorter computation loads
Operational space method developed
Flexible structures analyzed
Articulated n-bodies and spatial algebra methods developed
Direct-drive dynamics studied
Isotropic characteristics studied
Augmented object models developed for multi-arm manipulation
Equivalent mass and inertia of multi-arm systems found to be additive
Drive system flexibilities modeled
Impact with environment modeled and analyzed
Biped and multi-leg dynamic walking studied
Payloads incorporated into models
Dynamic interactions (e.g. contact impact) modeled and studied
Hopping, hitting, juggling and other dynamic games modeled and studied
Holonomic and non-holonomic systems modeled
Hamiltonian used to study dynamic systems
Real-time simulations of complex dynamics obtained
Closed loop and multi-loop system dynamics modeled
Balancing of structure against gravity loading studied
Decoupling of dynamic effects between members studied
Symbolic computation methods developed
Parallel structures in computational algorithms developed
Error dynamics studied

3.1 Dynamics — Current

Branching mechanisms (such as a torso with arms, legs and head) modeled
Complex dynamic interactions with the environment are modeled
Macro-mini systems optimized
Games with fast dynamic changes are modeled and studied
Fast walking and running studied
Swimming and flying dynamics are studied
Insect dynamics modeled and studied

4 Control

In this section we summarize the major results in the control of robots. The results are placed in rough chronological order:
    PD control - using either finite differences or direct measures of velocity.
    Bang-bang control studied
    Force and position control
    Resolved rate control developed
    Impedance and compliance control developed
    Teleoperation with force feedback studied
    Hybrid position/force control introduced
    Natural and artificial constraints introduced
    Potential function introduced
    Model based control introduced
    Actuator dynamics, backlash, friction and flexibility modeled
    Decoupling of nonlinear aspects
    Stability studied
    System identification techniques used
    Programming control languages developed
    Optimal minimum-time control studied for pick-and-place moves
    Path following with minimum time studied
    Variable structure control (adaptive, sliding modes) studied
    Reciprocal vs. orthogonal screws for combining position and force control
    Null-space motions maintain end-positions while reconfiguring systems
    Simultaneous control of posture and task studied
    Sensor integration studied
    Fine motion control modeled and analyzed
    Dexterous hands and their interactions with the environment are studied
    Controlled, but not explicitly programmed, tasks are implemented
    Simulated hardware used to develop controls and to control virtual systems
    Neural networks, fuzzy control, geometric control, etc., applied to robotic systems
    Control structures for dealing with unstructured environments developed

4.1 Control — Current

    Touch and tactile sensing used to explore environments
    Simulations for large degree-of-freedom systems
    Real-time connection of motion with sensing
    Natural movements studied and replicated
    Variable environmental constraints included
    Control hardware integrated into structure
    Automatic error recovery included
    "Bump but don't bruise" systems developed
    Model-free and autonomous learning

5 Framework for a Theory

The above lists certainly are not complete. Hopefully, they do a fairly decent job of summarizing — in very brief headings — many of the highlights that have been studied by the robotics research community in the areas of central interest to the Ro.Man.Sy community. There are many frameworks these results could be fit into. Probably the most fruitful way to proceed would be to have a separate framework for each of our three subject areas. For kinematics a very straightforward constructive way would be a bottom-up approach, gradually increasing the complexity of the objects being studied. If we started at the very basic level, one would be dealing with links and joints. Then one could determine ways to describe these basic objects, and their relationships to one another. Following this, one could determine how to combine these basic objects so as to form meaningful groups. These groups would the mechanisms of which robotic devices are composed. Finally, one could study how to analyze, design and control the mechanisms. Ultimately one could study specific applications of the theory to complex practical objects.

What has been just outlined is in fact one way of looking at the structures and frameworks of the classical theories of kinematics. So, it seems quite logical to attempt to fit all the results in robotic kinematics into the classical frameworks that have already been developed over the years. Now, however, we need to incorporate more general groups of objects, the so-called robots, into our framework.

Similarly when it comes to dynamics we could also follow along the lines used in most standard texts. Usually classical dynamics is divided into the study of particles and the study of rigid bodies. Clearly, in robotics, particles play a very small part, whereas rigid bodies form the heart of the subject. So, in our taxonomy we will ignore the study of particles, as a subject in its own right, and include it as part of rigid body theory. Also, it is necessary to keep in mind that much of dynamics depends upon the kinematics of the system, and that this is especially true for the complex and time-varying geometry of many robotic systems. So, in fact, the dynamics section here should be thought of as also including much of the material on kinematics in Section 5.1.

Since control depends on large part upon both the kinematics and the dynamics of a system, the material in the controls section should be thought of as also containing much of the material on kinematics and dynamics given in Sections 5.1 and 5.2. Thus we save considerable space by not repeating material already noted in a previous section. In dealing with controls we take a high level view of the classification, since that seems more in keeping with the general approach to the subject.

The remainder of this paper, gives a brief sketch of the general structures that can be used to classify the research results in robotic kinematics, dynamics and control on a very basic level.

5.1 Kinematics

Following the above prescription, we first make a classification of the objects being studied. This is then followed by basic tools to study these objects Then one deals with the combinations of the basic objects Finally, we come to the start of dealing with the composite devices, and their analysis. In this way one gets for example:

Kinematic classification
types of joints
types of chains
special parameters

Systematic analysis
coordinate frames
coordinates (planar, spherical, spatial)
(It should be emphasized that the systematic analysis for robotic devices is exactly the same as for traditional mechanisms)

Type synthesis
open chains (including redundant and hyper redundant chains)
closed chains (including the study of passive joints and active joints)
hybrid chains
adaptive chains (including modularity)

Kinematic analysis (zeroth order)
Orientation uncoupled from position (three intersecting axes)
Number of inverse and direct solutions
    series
    in-parallel
    hybrid
Solvable and analytic structures
Obstacle avoidance (higher order too (?))
Calibration (higher order too (?))

Kinematic analysis (first order)
Velocity, static forces and Jacobians
Principle of virtual work
First-order properties
Duality
Singularities (higher order too (?))
    series
    in-parallel
    hybrid
Internal forces
Force closure (higher order too)

Kinematic analysis (second order)
Curvature
    point-paths
    line-paths (curves)
    planes (surfaces)

Acceleration
- points
- bodies
- geometric

<u>Kinematic analysis (whole space)</u>

Trajectory generation
- pick-and-place (point, rigid body motion)
- continuous path (point, rigid body motion)

Workspace
- boundary and shape analysis
- singularity analysis
- optimization

Manipulability
- ratings
- whole "arm"
- fingers
- legs
- multiple members

Wrist geometry

Stiffness analysis

<u>Kinematic size synthesis (design)</u>

Determine number of meaningful parameters
- series
- in-parallel
- hybrid

Results of parameter specialization
- series
- in-parallel
- hybrid

Parameter families for X (where X is any desired criteria)
- series
- in-parallel
- hybrid

5.2 Dynamics

For the dynamics of robot systems, assuming we already have the kinematic aspects, the basic categories for their classification could be listed as follows:

<u>Dynamic analysis</u>

Equations of motion
- coordinates
- degrees of freedom
- coupled and uncoupled systems

discrete systems
continuous systems
methods
rigid bodies
flexible bodies
Equivalent mass and inertia
multiple systems
single system
isotropic systems
Interaction with environment
impact (high and low energy)
Small oscillations

<u>Design for dynamics</u>
Effective inertia
series (for drive and for environment)
in-parallel (for drive and for environment)
hybrid (for drive and for environment(?))
Acceleration profile
series
in-parallel
hybrid (?)
Flexible structures
series
in-parallel
hybrid (?)
Drives
direct
tendon
gear
harmonic

5.3   Control

This subsection lists a basic structure to use for classifying the types of results reported in Ro.Man.Sy symposia on robot and manipulator control. As stated at the beginning of this section, relevant material previously listed under kinematics and dynamics will not be repeated here, in order to conserve space. In contrast to the previous sections, this material does not seem to benefit from a bottom-up approach. It some sense it seems to be categorized less hierarchically than the material under kinematics and dynamics. This should not be surprising, given the system aspect of the subject matter.

<u>Modeling</u>
Representations and tools
Open-loop and closed loop control (linear and nonlinear)
Feedback and feedforward

Multiple-loop and multiple-input
Errors
- steady state
- transient
- sampling

Sensors, tasks and environment

Simulation
Rigid and flexible systems
- position
- force
- hybrid

Obstacles and environment

Analysis
Mathematical tools
Continuous
Discrete (sampled data)
Nonlinear and time-varying systems

Optimization
Minimum time
- point-to-point
- path following

Minimum effort
- point-to-point
- path following

Margin of stability
Minimum error
- point-to-point
- path following

Adaptive

Design for control (physical systems)
Structures optimized for X (where X is any desired criteria (?))
Sensing
- position
- force
- hybrid
- environment

Fine motion and gross motion
Impact
- compliance
- resonance (?)

Virtual systems
Sensory feedback to user
    position
    force
    hybrid
    virtual environment
    auditory
    visual
Fine motion and gross motion
Evaluation of user's performance

6 Conclusions and a Final Remark

One of the advantages of viewing research results in terms of a comprehensive formalistic structure is that it becomes easy to see missing aspects of the theory. In the foregoing, several missing pieces, in each of the three subject areas, are denoted by the use of question marks. Clearly the reader can expand this very skeletal structure, and produce a lot more question marks. It is the intention of this paper to merely point the way, and provide a conceptual framework into which to place previous and future results. Clearly such an analysis could easily be used to find new areas of study. For this, as they say, "the details are left as an exercise for the reader."

To carry this idea forward, it is suggested that every result in the first part of the paper (Sections 2, 3 and 4) be mapped into one or more of the items in Section 5. If an item cannot be easily mapped, then a new category needs to be added to Section 5. (Before doing this, the readers might wish to review the items in Sections 2, 3 and 4, and add any omissions they feel are important.) Finally, after Section 5 has been expanded to include the known results, it can be further expanded, as has been done with the items noted with the question marks, to include logical subcategories which have not yet been researched. This is best done by simply, and rather mechanically, appending additional subcategories to each of the categories. These subcategories may be new ones, or simply copied from elsewhere on the list. It is best to do this without judging if the subcategories makes any sense or not. Then, after some time, to return to the expanded list and view it with a critical eye.

Whether or not one uses the taxonomy of Section 5 to generate new research questions, it is hoped that the idea of placing the results of robot and manipulator research in the frameworks of the classical subjects of kinematics, dynamics and control will be of value to researchers in conceptualizing the place of a specific research result in the big picture of our shared research enterprise.

During the presentation of this paper at the Ro.Man.Sy Symposium, I requested that the participants add to my lists of results. I have attempted to incorporate all such suggestions, and I want to thank all those who contributed.

General Lecture

New Frontiers in Robotics

Miomir Vukobratovic

"Mihailo Pupin" Institute, Belgrade, Yugoslavia

Abstract. In the paper the usefulness of active systems is considered, whereby their advantages and disadvantages are emphasized. Some characteristic examples of breakdowns and disasters of objects and constructions are presented, such as the catastrophe of the Tacoma bridge, collapse of cooling towers, the breakdown of vertical mine transport and others.
Also, some possible solutions of the mentioned constructions with properties of active response to various external perturbations of high intensity are given. Beside these, basically hypothetical solutions of active systems, in the paper the problems of road vehicles control, active control of aerofoil flutter, control of buildings and structures, variable mechanical structure, are briefly presented.
All these examples of active systems can be classified as future special dedicated robotic systems.

Keywords: active systems, disaster, variable geometry, aerofoil flutter, cooling tower, shaft vehicle, active suspension, variable mechanical structure, safety engineering.

1 Introduction to Active Systems

Regardless of the fact that the category of so called active systems has existed practically for two decades, it is necessary to say something more about the nature of active systems, due to the simple fact that the number of passive (traditional) systems being potential candidates to become active is growing considerably. To date, active systems have been synonymous with active suspension of road, and particularly of railway vehicles. Only in the course of the last decade the idea of active systems has spread from the field of mechanical engineering to the constructions and structures of civil engineering. While active systems in vehicles were aimed at adjusting some of the suspension parameters for maintaining the vehicle dynamic performance, such as motion smoothness, i.e. the ride comfort of road vehicles, or, maintaining the lateral contact force (wheels -- railway tracks) in permitted limits (for motion safety), active systems in civil engineering up to now predominantly had the goal of maintaining the static performance of civil constructions by changing their stiffness according to variable external loads. Bearing in mind these remarks, active systems could be defined as systems which, by automatic varying of their characteristic dynamic parameters in certain ranges, keep their dynamic performance, which enables satisfactory system functioning, in conditions of either some internal or particularly external perturbations which can even have extreme character. While there is no need to speak about the advantages of active systems as compared with the passive, which in some way incorporate themselves into the process of adjusting their characteristics to the changed conditions in which the system (construction) operates, still some of the disadvantages should be emphasized. Let us mention for instance the implementation problem which is technological by its character, so it becomes less important with time, and the second problem: the maintenance costs, which are inherent and can never be neglected. The active systems must be operative and ready to react in subseconds under extreme conditions. However, it can happen that they will never be used under such conditions (e.g. active response of important object (structure) under high ground accelerations, typhoons, or high intensity gusts).

2 Some Characteristic Examples of Disasters of Objects

Due to limited space, only a few examples of disastrous outcomes of objects of special importance will be presented.

2.1 Tacoma Narrows bridge catastrophe

In 1940, after more than one year of functioning, at a wind speed of only about 19 m/s, the suspension bridge of 845 m span in Tacoma (USA) collapsed. Subjected to bending and torsional vibrations of the deck, lasting several hours, the last phase of its structural instability, when the vibrations amplitude became about 75 cm was recorded with a movie camera (Fig. 1), including the collapse itself.

Figure 1. Vibrations of Tacoma bridge before the catastrophe (after [2]).

The Tacoma bridge catastrophe quite later evoked interest within the scientific community. So, based on the assumption that the bridge catastrophe was preceded by the oscillation of its deck which evolved to the so called critical flutter oscillations, the calculations and simulations of the bridge flutter were performed [1,2]. Based on data concerning the design and dynamic-static bridge parameters, the equations of motion of the bridge deck undergoing bending and torsional vibrations in steady airflow gave as the result unstable eigenvalues of the torsional vibration mode and in that way confirmed that the disastrous outcome of the bridge could be classified as flutter of its structure. It should be mentioned that the problem of flutter has been very much discussed in aeronautics between the 50-s and 60-s, when many cases of airplane crashes due to self-excited flutter vibrations at critical flight speeds were recorded. After that period the flutter problem became less critical thanks to the more complete dynamic calculations of the structures and adequate raising of their stiffness. However, during the last years attention was dedicated to the active flutter control [3] which points to the justified interest of the researchers for this problem, the adequate solution of which by active control can contribute to enlarging the margins of aircraft critical speeds.

2.2 Collapse of cooling towers

In 1965, at Ferrybridge, Yorkshire, U.K., three cooling towers collapsed in a gale. They belonged to the group of eight in two rows of four. All three were in the front row, indicating that the interaction between the towers has been important [4] (See Fig. 2).

Figure 2. Collapse of the third tower at Ferrybridge(after [4]).

Elucidation of the mode of failure essentially confirmed that it was a quasi-static failure, i.e. a tensile failure of the shell under the wind loading that would have occurred in a steady wind with the highest speed reached temporarily [4]. Investigation of the cooling towers collapse was carried out in two stages. In the first stage, wind tunnel-tests were carried out based on which pressure distribution over the towers are determined. In the second stage, the results of the first stage were fed into the membrane theory computations of stresses in the shell produced by the combination of the wind pressure and the shell's own weight [4]. The results were very sensitive to the details of pressure distribution.

The redistribution of the forces by the presence of the other towers could lead to failure. In Fig. 3 the wind - tunnel model of Ferrybridge power station is presented [4].

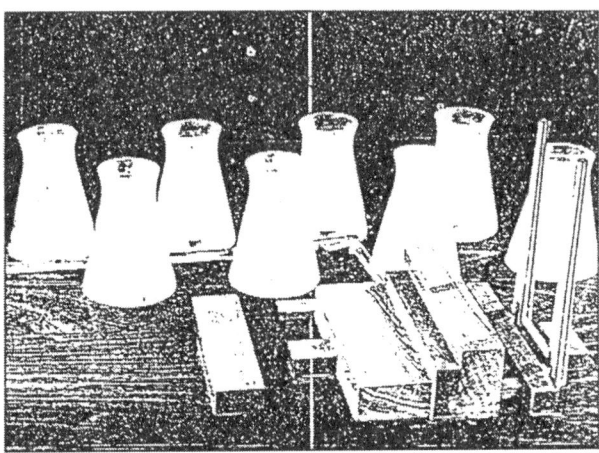

Figure 3. Wind-tunnel model of Ferrybridge power station (after [4]).

2.3 Disaster of vertical shaft vehicle transport.

Nowadays ore transport is performed by shaft vehicles weighting over 10 tons, at speeds attaining 15 m/s, and at a distance of over 1000 m. In Fig.4. the cross-section of vertical mine transport is given. The problem of shaft vehicle motion along vertical guides (railtracks) is a delicate dynamic task which can be investigated as the problem of constrained motion of a rigid body interacting with dynamic environment.

Differing from the traditional approach, it is necessary to take into account not only the shaft vehicle dynamic, but also the dynamics of the railtracks (vertical guides). The simulation results of the new approach to this contact task: motion of shaft vehicle interacting with dynamic environment, have demonstrated that the resonant states of such a system in motion, as compared with the assumption of a nondynamic character, shift towards lower working speeds amounting to about 15 m/s or somewhat lower which already represent the working speeds of shaft vehicles [5].

In Fig.5. the dismounted transport railtracks after the shaft guides damage and destruction are shown.

3 Some Examples of Active Constructions and Structures

It is evident that the examples presented in the previous section can provoke the expected question or dilemma: can the safety of such and similar objects be attained by their "activation" (active responses) or, otherwise, by traditional raising of the safety coefficient?

The traditional way of safety attainment of systems and objects leads to significant waste of constructive materials and energy, which is definitely the weak point of such design procedure of the system and objects activation (robotization) of systems and constructions and demands supplementary maintenance costs of the corresponding equipment (sensors, actuators, controllers). Thus, a very delicate techno-economic problem arises: the decision about entering into system robotization must be made very carefully and selectively.

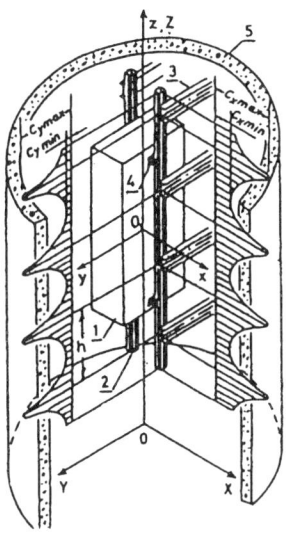

Figure 4. Scheme of the cross-section of vertical mine transport.

Also discussed in this section, are a few examples given of transforming passive constructions into active, and some of the possible ways of their robotization are considered. We are aware of certain risk of such exploits being unpurposeful today, but at the same time believe in a general concept of active systems to which the future belongs.

3.1 Active control of suspension bridge vibrations

Based on the data about the static -- geometrical and dynamic parameters of the suspension bridge at Tacoma, as well as the data about the windspeed on the day of the disaster [1,2], it has been calculated that the open-loop eigenvalues for $V = 20$ m/s in the torsional mode possessed positive real parts. In the case that actuators would be applied, the exerting forces would be located symmetrically on each side of the bridge mid-span, and the closed-loop eigenvalues for the same critical speed would have negative real parts in the torsional mode of oscilllations. In that case, instead of the construction of the structural unstable behavior due to critical flutter vibrations, damped vibrations of the bridge deck are obtained, as presented in Fig. 6 [1,2]. One hypothetical solution of the active control of the suspension bridge with variable load including its variable dynamic load due to seismic ground accelerations is given in Fig. 7.

While the stiffness of the bridge cables are changed in the bridge supports by changing the pulling force, the damping is changed by "hydraulic feet" (active absorbers). In that way the ability to compensate seismic impacts to a certain degree is attained apart from the ability to control the structural vibration of such an object.

In Fig. 8 one of the possible solutions in bridge vibration control is presented. A system of mass blocks may be used as auxiliary elements to produce the control forces. These blocks can be located between the main girders of the bridge and supported by the springs that are able to dissipate energy.

It is also possible to use an auxiliary damper to assist in producing the control force. A system of dampers may be installed in the hangers of a suspension bridge and some control devices produce the required control force, as shown in Fig. 9 [6].

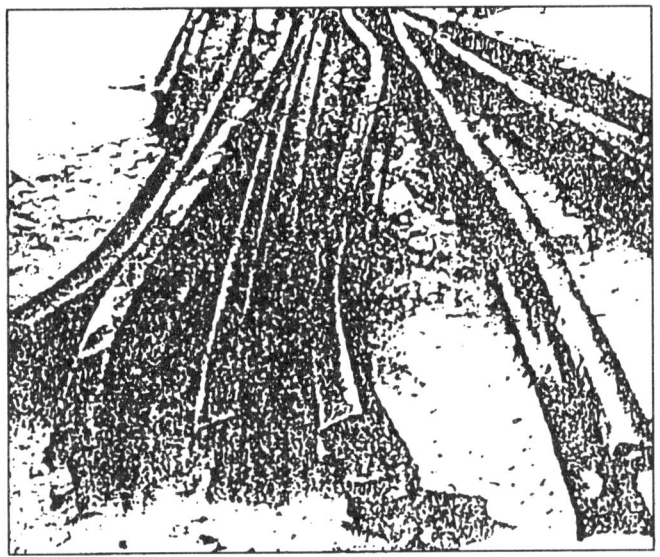

Figure 5. Dismounted transport railtracks after damage.

At the end of 1998, the Akashi-Kaikyo Bridge, the world's longest suspension bridge was opened to traffic [7]. It has a main span of 1991 m, a total length of 3911 m, and main towers of 290 m. (Fig. 10). By its category this bridge belongs to the semi-active construction. Honshu-Shikoku Bridge Authority has conducted various investigations, including wind tunnel tests, for many years, and control methods for the vibration of the main towers were carefully developed. Based on the results of investigation, a cruciform - shaped cross-section was selected for the tower shafts. As a damping device, tuned mass damper (TMD), tuned liquid damper (TLD), friction damper, etc. had been examined, and tuned mass damper as well as hybrid mass damper (combination of passive type and semi-active type) were chosen from their reliability and cost. In order to reduce the amplitude of the vortex induced oscillation caused by wind slower than the design wind speed, tuned mass dampers are installed inside the tower shafts. The dynamical calculations of the bridge construction, under the worst possible dynamical loads, as well as the calculation of the flutter critical speed of its mechanical structure, demonstrated that its safety is jeopardized only above 8.5° on the Richter scale and at wind gusts higher than 280 km/h.

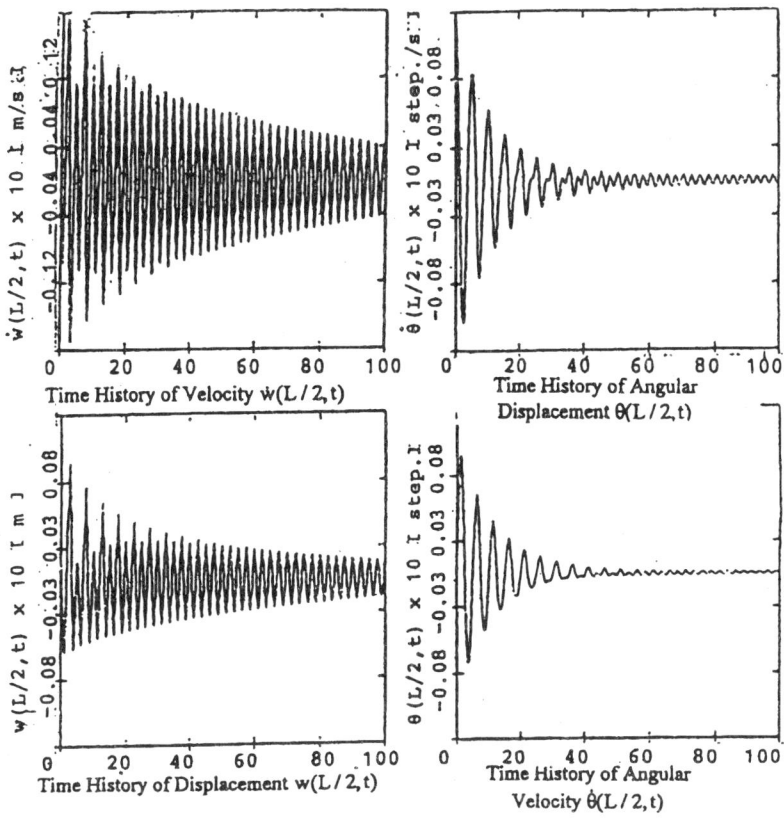

Figure 6. Time histories of some characteristic simulation results of vibrations active control of Tacoma bridge (after [1]).

The simulation experiments concerning the flutter critical speed have been confirmed experimentally on a 1/100 model of the bridge (approximately 40m long) to express the prototype accurately, and a boundary layer wind tunnel of the Goettingen-type, which had the test section of 41x30x4m, was thus constructed.

Relating to dynamic interaction of soil and objects it is necessary to underline extensive dynamic soil-structure interaction research showing that the interaction effect is composed of two effects, i.e. effective seismic motion and dynamic stiffness. Effective seismic motion, sometimes called the kinematic interaction effect, is defined as the response of massless foundations exposed to seismic excitation. The response of massless foundation against any seismic excitation is different from what is obtained from conditions that no foundation exists on a free surface, because the presence of the foundation restricts the seismic excitation.

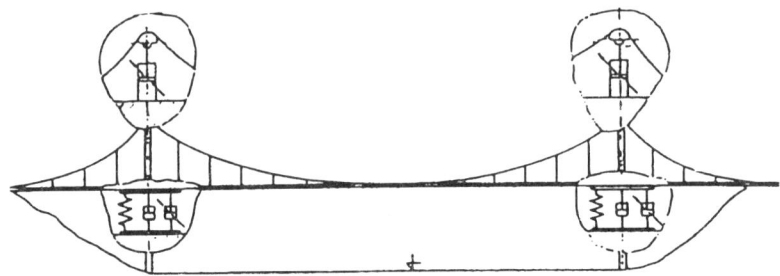

Figure 7. Hypothetical solution of suspension bridge active control.

The response of a large scale foundation which is embedded into the ground is not only affected by vibration properties of the geological layer but depending on the configuration of the contact surface differs between the foundation and the ground because the foundation constrains the vibration of the ground [7]. Dynamic stiffness is introduced to conside the effect of dispersion of vibration energy of the foundation into the ground, instead of conventional stiffness in which the ground spring is supposed to be a static one. In the dynamic stiffness, coupled oscillation between the foundation and the ground is taken into account, and the stiffness becomes variable by the frequency of oscillation.

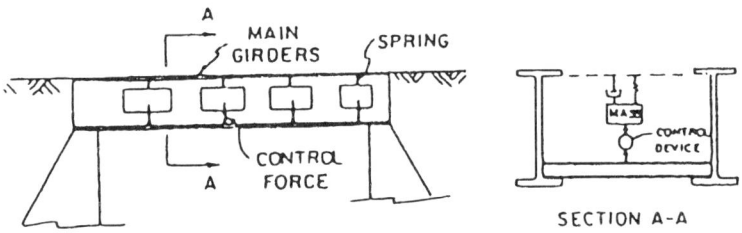

Figure 8. Auxiliary mass for control (after [6]).

3.2 Active response of cooling tower structure

In order to prevent the possible collapses of cooling tower structures, as the ones at Ferrybridge, as well as adequate dynamic calculation of the structure itself, it is possible to foresee some "active solution" to compensate for the perturbations under various condition of dynamic loads by wind of various intensity, and also under the condition of seismic shocks up to a certain level of intensity. Safety of such structures due to seismic shocks has been realized up till now mainly by means of passive absorbers arranged circularly in a ring, built into the foundation of the cooling tower. Active damping of the structure could be ensured by including active absorbers into the foundation ring, by means of which the possibility of variable damping would be realized, which is an exceptionally important parameter, responsible for the dynamic response of the structure.

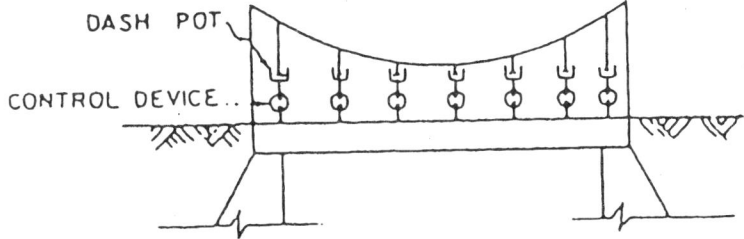

Figure 9. Auxiliary dampers for control.

Figure 10. Akashi-Kaikyo Bridge (after [7]).

Beside controlling the damping characteristics, with such civil structures the possibility of automatically changing both the inertial and stiffness parameters is important, being of exceptional importance for an adequate dynamic response under external loads, such as strong gales and wind gusts. In Fig 11.a one hypothetical solution of variable stiffness and inertia of the cooling tower structure is shown, by means of internal rings (one or several), which are moving along the internal structure according to variable external dynamic load. In Fig. 11b one hypothetical solution of variable structure inertiality of the cooling tower, is also shown schematically. The solution basis consists of one internal fixed ring with lateral guide along which the masses are moving according to some law, changing the inertiality of the whole structure in accordance with variable external dynamic loads. Also with this type of civil structure of great importance is the dynamic approach to the soil- structure contact (interaction). The dynamic nature of this contact imposes an adequate dynamic analysis, which can serve as a basis for active (dynamic) control of the foundation of such and similar types of civil structures.

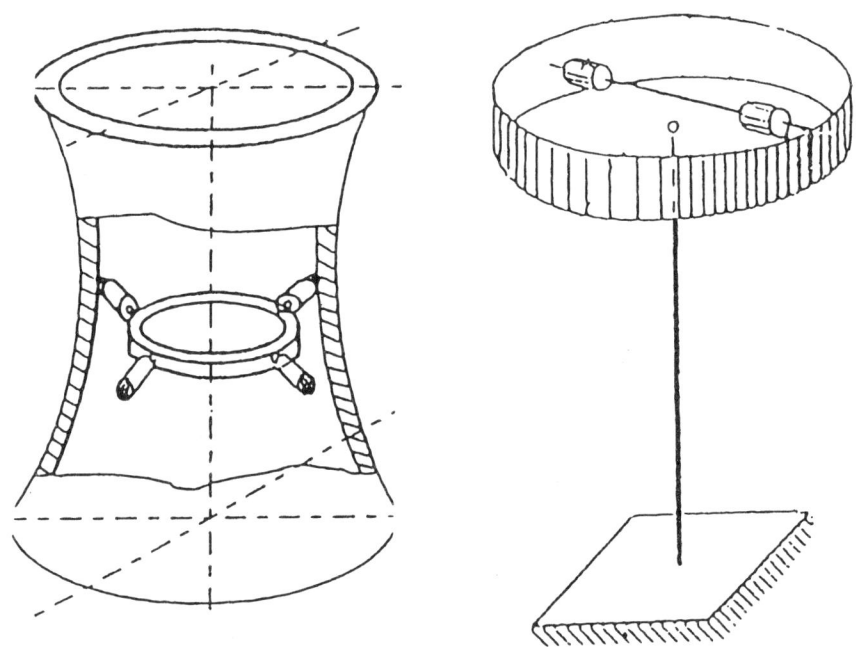

Figure 11. Hypothetical solutions of active cooling tower structure.

3.3 Active control of vertical shaft vehicle transport

Taking into account the environment dynamics (dynamics of the guides), and not only the dynamics of the shaft vehicle, simulation results were obtained [5], demonstrating that in the case of such treatment of this delicate control task the resonant states shift towards shaft vehicle lower motion speeds as compared with the case when this problem is solved by not taking into account the dynamic nature of the guides (Fig. 12). For compensating the perturbations, which by the dynamic state of the shaft vehicle interacting with guides dynamics can lead to resonant regime and to the destruction of the mine vertical transport, in Fig. 13 another hypothetical solution of the active control of the dynamic contact of the shaft vehicle with the guides is given. In a controller, control signals are generated based on information from the sensors, which are used for adjusting the variable damping of the absorbers built in the supports along the guides and shown in Fig. 13.

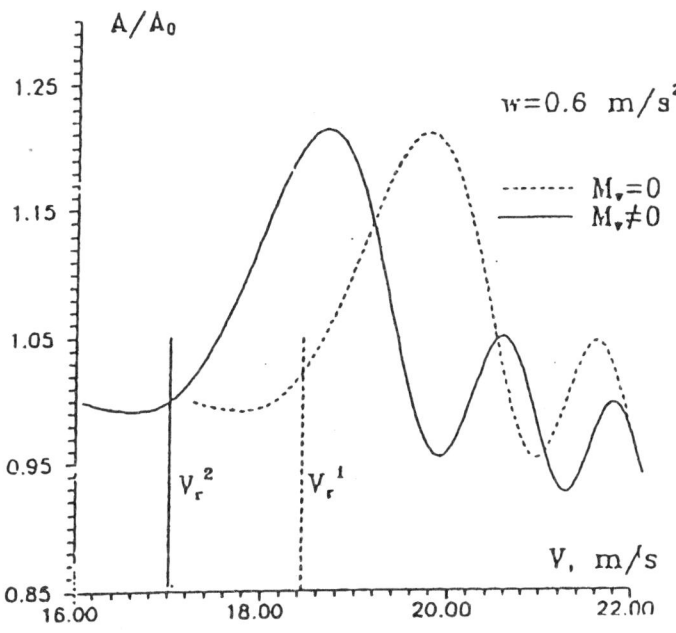

Figure 12. Passing through resonant zones Notations: M_v - equivalent contact mass, V_r - vehicle motion velocity corresponding to the resonant case; A/A_0 - relative amplitude.

3.4 Active control of tall buildings

One of the new application areas for control system design has to do with the protection of buildings from dynamic loadings such as strong earthquakes, high wind and blasts. Buildings have traditionally relied on their strength and ability to dissipate energy to survive under severe dynamic loading.

In recent years, worldwide attention has been directed toward the use of control and automation to mitigate the effects of these dynamic loads on these structures.

Active control of tall buildings can be performed, by which the displacements would be compensated in the case of strong gales, or in the cases of ground acceleration due to seismic perturbations. Mathematical models of the building structure dynamics offer a good basis for the synthesis of a controller which is to control the displacements of the structure making them smaller and reducing them to allow displacement. Control of the building in the case of ground displacement, i.e. its acceleration is shown in Fig. 14 [8]. Displacement of the first story unit and the control force for the first story unit are presented in Fig. 15 [8].

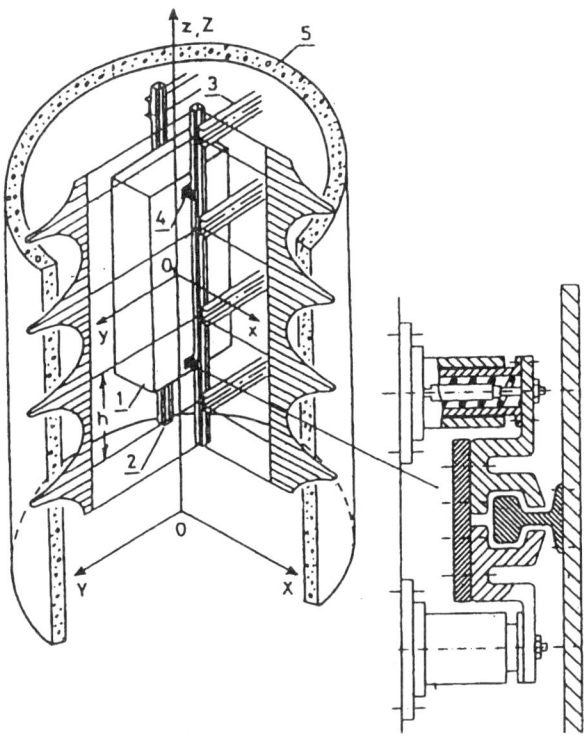

Figure 13. Active control of shaft vehicle interacting with guides dynamics.

Modern tall buildings are known to be subject to wind-induced oscillations. Field studies show that they usually oscillate at the fundamental frequency of the building, coupled sometimes with torsional motion when the torsional and lateral oscillation frequencies are close. One of the most common devices currently in use is in the form of tuned mass damper (TMD) system. TMDs have been installed in several known tall buildings in Canada, USA, Japan and China. These systems are located on the upper floor for suppressing primarily the first fundamental mode in wind-induced motion [6]. Fig. 16 describes the structure - TMD system schematically. The building is modeled in this figure as a simplified, single degree - of - freedom system with a mass m_1, damping constant c_1, and spring constant k_1, which represent, respectively, the first-mode model mass, damping, and stiffness of the building, and m_2, c_2, and k_2 represent the corresponding quantities associated with the TMD. An elementary vibrational analysis shows that, when mass m_2 is "tuned" to the primary mass m_1 by matching the natural frequency of mass m_2 (with mass m_1 fixed) to that of mass m_1, when m_2 is absent, then significant reduction of the primary mass vibration results. Significant reduction of building displacements can be achieved when m_2 is only ~ 1% of the total equivalent mass of the building. For instance, in the case of the Citicorp Center in New York, the TMD weigh approximately 400 tons [6].

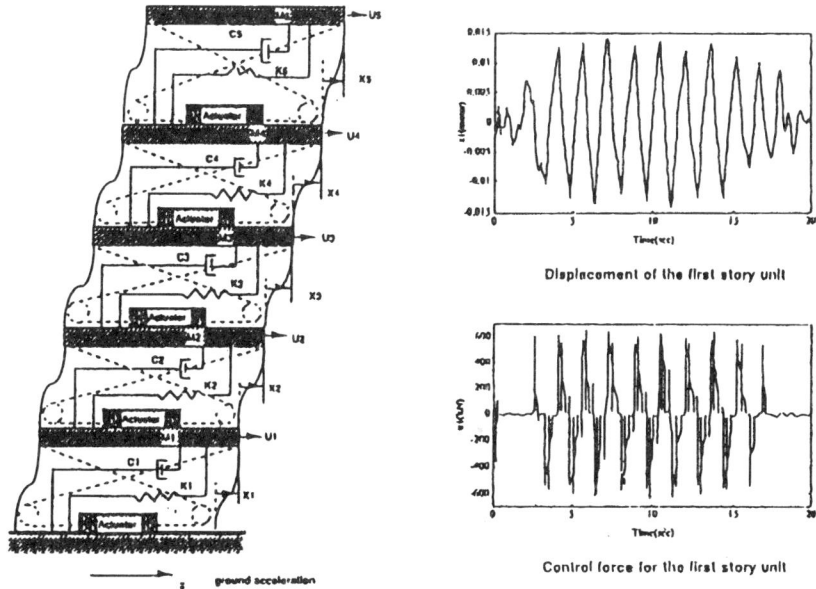

Figure 14. Five-story building structure with active tendon controllers (after [8]).

Figure 15. Control of five-story building structure (after [8]).

Passive systems are capable of reducing the building displacement by approximately 40%. TMD effectiveness significantly increases through the use of an actuator force as a function of the building motion variables. Other possible benefits in using active control include reduction of mass m_2 of the TMD and reduction of stroke length in excursion associated with mass m_2 while keeping building motion within acceptable limits.

In [6] the effectiveness of added active control was examined numerically using approximately the same parameters as those in Citicorp Center. In Fig. 17 displacement and acceleration results are shown for cases (a) without TMD, (b) TMD without active control and (c) TMD with active control u(t).

3.5 Active control of road vehicle

Control of road or railtrack vehicles is already present in the form of practical realization of semi-active or active suspension aimed at satisfying the driving comfort or raising safety [9]. Active suspension of a road vehicle can be schematically presented as in Fig. 18. Synthesis of one solution of active suspension in the scope of complete dynamic control has been performed, based on the complete dynamics of a road vehicle [10]. In Fig. 19 only part of the simulation results are presented referring to load variations on the vehicle wheels in the cases of passive and active control of the suspension characteristics [10].

3.6 Active control of aerofoil flutter

In [3] the procedure for aerofoil flutter control by means of active control was presented. Control is achieved by adding a controlled force of appropriate amplitude and phase to the case of natural flutter, aimed at annulling the natural destabilizing tendency of aerodynamic load on the aerofoil. This control force can be introduced to the aerofoil by active control of the ailerons or flaps. However, in [3] a scheme which can be more efficient at high frequencies and which, hence, could serve as a supplement to the conventional methods, is in question. For carrying out the active control by means of sound, acoustics is used. Although the pressures, generated by the loudspeaker are small, it was demonstrated that they can, in fact, induce significant and useful control forces on the aerofoil. Such procedure offers new flexibility in system design, because the loudspeakers can be installed on suitable locations on the aerofoil or remotely mounted. Description of the theory, indispensable for understanding the details of this idea, is also given.

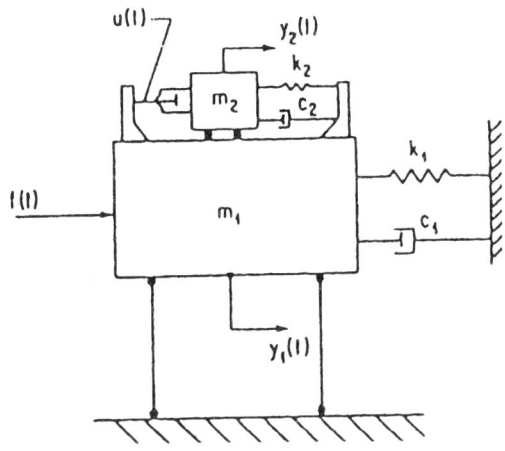

Figure 16. Building - TMD system (after [8]).

According to the approach of Nissim [11], a linear mechanical systems is considered, in which the force exerted by the system on its surrounding at time t depends only on the immediate state of the system. It was shown that the influence of the mechanical system on its surrounding depends on the eigenvalues of the diagonal matrix, determining the total power being transferred to the surrounding. Total power output is of critical importance. If it is positive, it indicates that the system will be stable, as energy is always transferred outwards. If it is negative, then any motion of the system will draw energy from outside, and the disturbance will grow [3]. A simplified mathematical model was developed, demonstrating that a loudspeaker (considered as a piston) mounted on the aerofoil or remotely mounted can supply the necessary control forces. Experimental results are given, too, in which the aerofoil in state of flutter was stabilized by loudspeakers, Figures 20 and 21 [3].

New Frontiers in Robotics

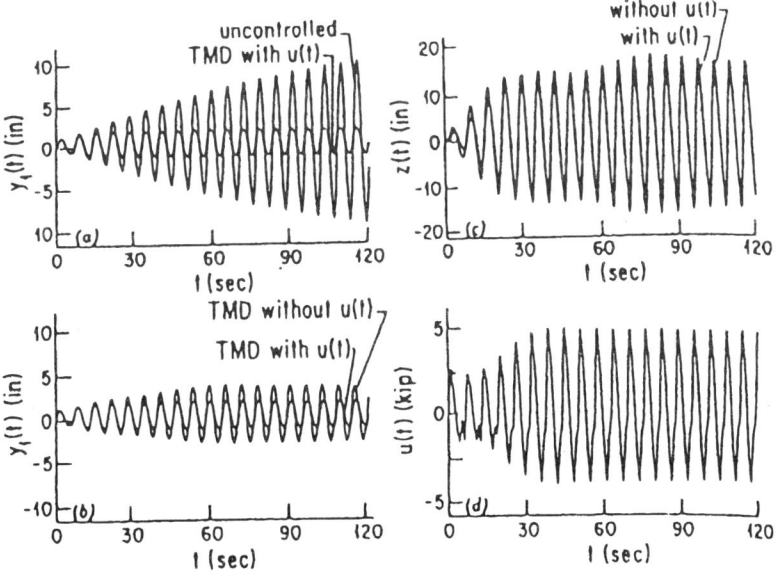

Figure 17. Displacement results for cases a, b, and c. (after [8]).

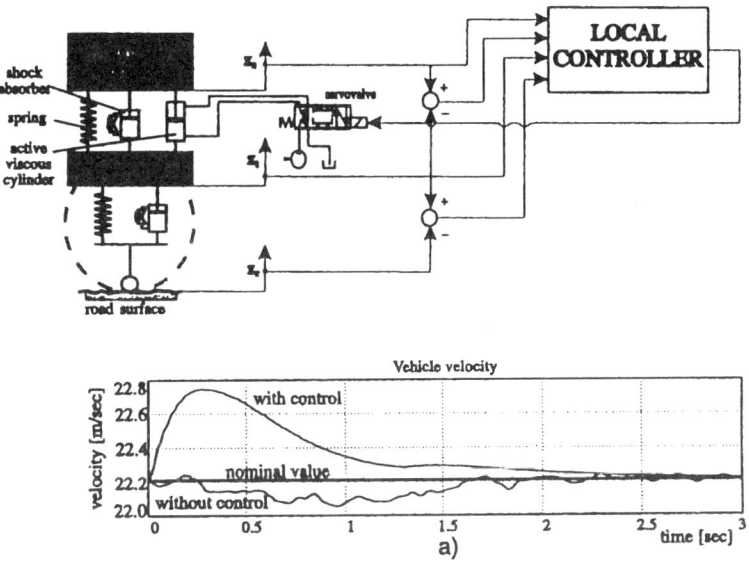

Figure 18. Active suspension systems of road vehicle.

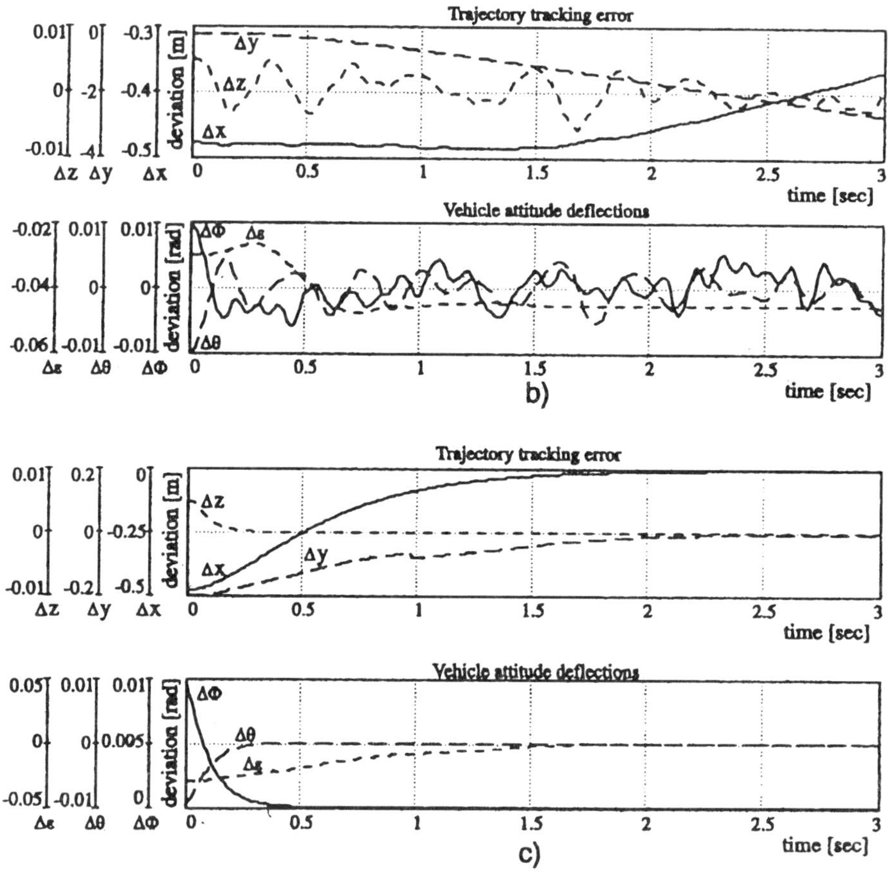

Figure 19. Simulation results of the vehicle model for the case of slipping road as a perturbation type: a) real vehicle speed, b) accuracy indices for programmed tracking trajectory, without and c) with automatic vehicle control
(x,y,z - cartesian coordinates, ϕ, θ, ε - roll, pitch and yaw angles, respectively).

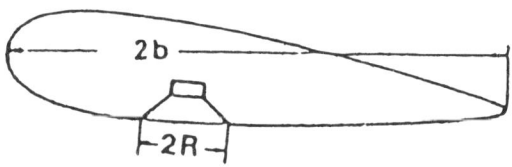

Figure 20. Loudspeaker mounted within the wing.

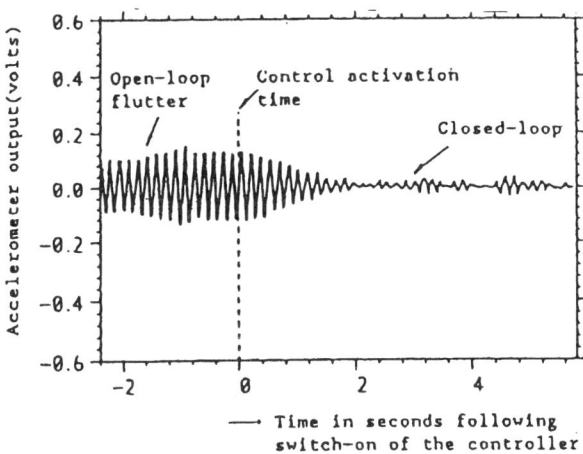

Figure 21. Typical individual time history trace of the bending acceleration of the wing (after [3]).

4 Variable Geometry Structure

We have seen that the real and contemporary need for systems that behave compatibly with variable operating conditions lead the research efforts towards the so called active systems. This understood the activation of system elements which conventional design considered as passive. The natural extension of this idea was to allow some additional motions, thus giving the system the ability to adapt to various tasks. This concept is called *variable geometry*. It is proposed that the systems and constructions should be designed so as to feature variable geometry in appropriate margins. This means the adaptation of geometrical parameters relevant for the change of system properties (like stiffness and distribution of static and dynamic load) [12,13].

4.1 Problem statement

Consider a system of k interconnected rigid bodies. The constructive connections between the bodies determine the number of *degrees of freedom* (DOFs). Let it be n. The set of generalized coordinates that define the system position is $q = (q_1, \dots, q_n)$.

Figure 22 presents one example from the field of robotics. A part of a robot is shown: a joint actuated by a hydraulic cylinder. It has one DOF. It has to be emphasized that this example could also refer to other systems with active linked structure, such as concrete distributors, dredges, cranes, etc.

The quantities that define the constructive connections among the bodies belong to the set of *geometric parameters*. For instance, in the example parameter l defines some constructive connections. With a conventional design geometric parameters are constant. The values of geometric parameters strongly influence the dynamic (or static) performance of the system. Let us shortly discuss the values of geometric parameter l in the examples from Fig. 22. In the example, we are interested in the load capacity and the speed of operation. These requirements are in conflict since the value of parameter l that gives higher joint torque at the same time gives lower joint velocity. In the example we found the conflict when choosing values for geometric parameter.

In such cases, the usual engineering approach suggests choosing numerical value so as to achieve compromise. Here, we shall search for a different solution. Conflict can be solved, if the system is redesigned to allow variation of geometric parameters. Thus, we come to the notion of *systems with variable geometry*. This means that the constructive connections allow some other functional motions beside the ones previously defined (i.e. beside q). In examples from Fig. 22 the variation of geometric parameter l can adapt the system to different requirements. Geometric parameters (like l) now become additional coordinates marked by s (see Fig. 22).

4.2 General form of dynamic model and problems of implementation

Consider a system of k rigid bodies. Let the connections among them be such that beside the motions $q = (q_1, ... , q_n)$, additional displacements $s = (s_1, ... , s_m)$ are possible. In this way, the system has $n+m$ DOFs. The coordinates from the first set are called *joint coordinates* and the additional DOFs are called *variable geometry coordinates*. In this case the dynamic model can be written in the form of two submodels:

$$H_{qq}(q,s)\ddot{q}+H_{qs}(q,s)\ddot{s}+h_q(q,\dot{q},s,\dot{s})=\tau_q+D_q(q,s)w \qquad (1)$$

$$H_{sq}(q,s)\ddot{q}+H_{ss}(q,s)\ddot{s}+h_s(q,\dot{q},s,\dot{s})=\tau_s+D_s(q,s)w \qquad (2)$$

where H_{qq}, H_{qs}, H_{sq} and H_{ss} are the blocks of inertial matrix, while h_q and h_s take care of centrifugal, Coriolis, gravitational, and other position and velocity dependent effects (e.g. springs and dampers), τ_q and τ_s represent the corresponding driving torques (or forces), w is the vector of some external action, matrix D defines the relative position of this action.

When discussing implementation, one should start with the question whether it is possible do decompose the problem or it has to be considered as a whole. The answer is that partition is possible and sometimes necessary, but the general approach offers better prospects. Let us explain this.

Pre-process adaptation. The first approach decomposes the problem into two phases. Geometry adjustment procedure is performed first. It means positioning of s-coordinates to appropriate values s^*. After that, practical operation of the system starts and it involves q-coordinates. Operation is performed with geometry locked. Hence, the approach is called *pre-process adaptation*. To apply it, two conditions have to be satisfied: *the task* must be known in advance, and for a particular task there must exist at least one constant geometry that allows execution.

Pre-programmed adaptation with in-process realization. This approach understands that the change of geometry is performed simultaneously with the practical operation of the system. However, the law $s^*(t)$ for variation of geometry is calculated off-line, in advance, and stored in the computer memory. During the execution, servosystems simply track this reference law. For industrial robots this seems to be a very appropriate approach. For advanced robotics, the request for immediate response to different unexpected events (like obstacles) makes the approach unsuitable. In constructions, it is possible to change the geometry simultaneously with the process but only if the process is slow.

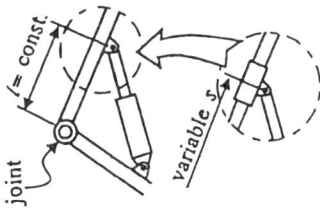

Figure 22. One example of multibody system.

On-line adaptation. This approach is considered as general and it understands that the change of geometry *(s)* is performed in-process and according to the current dynamic requirements. This means that the reference s^* is calculated and executed on-line. The control system must allow the calculation of the reference s^*. This includes: first, solution of the relevant dynamic characteristics (by means of measurement or estimation), and second, calculation of reference based on the dynamic characteristics. The problem of measurement or estimation cannot be discussed in general but only for particular systems. Calculation of the reference $s^*(t)$ can be performed in a heuristic way or systematically, based on optimization technique.

4.3 Robotic mechanism with variable geometry

The discussion on modelling and control is based on the example from Fig. 22. For simulation, configuration from Fig. 23 was adopted. First, we discuss the choice of coordinates q. From Fig. 23 one concludes that coordinates could be chosen as joint motions (angles: φ_1, φ_2) or actuator motions (linear displacement of pistons: p_1, p_2). For the fixed geometry both sets of coordinates are acceptable due to one-to-one correspondence. After introducing variable geometry, the way of choosing coordinates q becomes more important. From the standpoint of kinematics, it is still more convenient to utilize the joint motion ($q = \varphi$), since it directly determines the motion of the end-effector. In dynamics, however, the fact that drives actually act along p, excludes this choice. Joint torques T are not suitable any more to play the role of driving inputs. They depend on the real drive F and the variable geometry s (see Fig. 23). Thus, we utilize the actuator motion and adopt $q = p$. Then, the driving inputs are $\tau_q = F$.

With the robotic example considered, the aim of variable geometry is to achieve adaptation to different requirements regarding load and speed. Thus, the constraints follow from the actuator limitations: force limit F^M and piston velocity limit $\dot{p}^M$. Our goal is to avoid violation of limits and execute the task. Starting with this aim, we remember the three approaches to variation of geometry and discuss the problem of modelling and control.

Pre-process adaptation. In the *phase of geometry adjustment*, coordinates s change towards values s^* that will meet the expected requirements regarding load and speed. Coordinates q are kept constant ($q = C$). Modelling of such system starts from equations (1) and (2). If the condition $q = C$ is introduced into (1) and (2), the later submodel (eq. (2)) decouples from (1) τ_s is now the control input while τ_q represents the reaction. In simulation, the submodels can be solved one after the other. After adjustment, the robot starts its regular operation. In this phase the geometry is constant, i.e. $s = C$, while coordinates q change in order to execute the task. From the control and modelling point of view, the situation is analogous to the one previously described, only that coordinates q and s switch their roles.

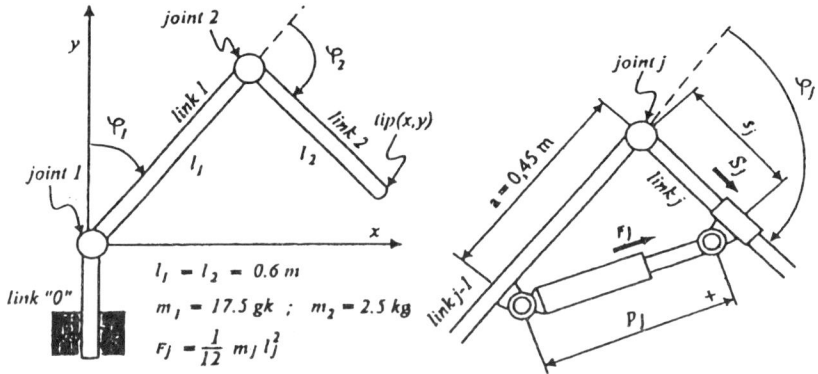

Motion: $A \to B$, the tip moves along straight line with triangular velocity profile
$(x,y)_A = (0.8196, -0.2196)$ $(x,y)_B = (1.1885, -0.1171)$,
execution time $= 0.4$ s

Figure 23. Robot configuration and the motion task prescribed

In-process variation of geometry - simplified model. The adjustment of geometric parameters is made simultaneously with the execution of robot task (according to a pre-programmed law or on-line). So, both coordinates, q and s, change. However, we still do not consider the general problem, but introduce an assumption regarding the drives. The actuators driving the geometry coordinates s are considered to be completely robust to external influence. This means that they will provide some motion law $s(t)$ and this law cannot be disturbed by any effect coming from robot dynamics. Such assumption is justified in many practical cases (e.g., if the powerful hydraulic actuator is applied, or an electrical motor with a gear-box that has large ratio and self-braking in the reverse direction). If it is satisfied, then (2) reduces to

$$H_{ss}(s)\ddot{s}+h_s(s,\dot{s})=\tau_s \qquad (3)$$

where from the control of displacement s can be solved. Some standard concept of control can be used to ensure that coordinates s track the desired law $s^*(t)$. Realized displacement $s(t)$ enters the submodel (1) making it nonstationary. To realize the control of q, some method suitable for nonstationary systems should be applied. For simulation purposes, equation (3) could be first integrated to give $s(t)$. Then, the submodel (1) can be solved.

In-process variation of geometry - general model. Coordinates q and s may change simultaneously and no additional assumption is made. In this case, submodels (1) and (2) are fully coupled. There are $n + m$ DOFs and from the mathematical point of view they could be treated equally, although coordinates q and s still have different purposes and nature. The two submodels, (1) and (2), have to be integrated simultaneously.

Numerical example is selected so as to support the main intention: to promote the concept of variable geometry and indicate its prospects. Modelling will mainly concentrate to show how some constraints of joint actuators could be overcome. Dynamics of *s*-coordinates is kept on the theoretical level (eq. (3)) in order to avoid discussion on specific realizations of the geometry actuation.

For the simulation experiment, the configuration and the motion from Fig. 23 are adopted. In order to perform simulation, the kinematics and the dynamics are described by models:

$$\ddot{X} = J(\varphi)\ddot{\varphi} + A(\varphi,\dot{\varphi}) \;,\; H(\varphi)\ddot{\varphi} + h(\varphi,\dot{\varphi}) = T \tag{4}$$

where $X = [x\; y]^T$, $\varphi = [\varphi_1\; \varphi_2]^T$, J is Jacobian, and A is the the adjoint matrix.

When it appears necessary we pass from coordinates φ to coordinates p and s. The relations between coordinates and their derivatives are

$$\begin{aligned} p_j^2 &= a^2 + s_j^2 + 2as_j \cos\varphi_j, \\ \dot{p}_j &= \frac{2s_j \dot{s}_j + 2a\dot{s}_j \cos\varphi_j - 2as_j \sin\varphi_j \dot{\varphi}_j}{2(a^2 + s_j^2 + 2as_j \cos\varphi_j)^{1/2}}, j=1,2. \end{aligned} \tag{5}$$

The relation between joint torque T_j and piston force F_j is given by

$$F_j = -T_j / d_j, d_j = a(s_j / p_j)\sin\varphi_j \tag{6}$$

For the imposed task, the inverse kinematics gives the time histories of joint angles and velocities: $\varphi(t), \dot{\varphi}(t)$. Calculation of dynamics gives the joint torques $T(t)$. Then we pass from joint coordinates and torques to piston motion p and forces F. For a piston j, motion p_j and velocity $\dot{p}_j$ are found from (5), and force F_j from (6). This calculation includes the parameter s_j. The piston variables $(\dot{p}, F)$ are first calculated for constant values of geometric parameters: $s_1 = s_2 = 0.15m$ (Fig. 24). Velocities and forces are now checked against the prescribed upper limits $\dot{p}_1^M = \dot{p}_2^M = 1m/s$ and $F_1^M = F_2^M = 1200N$. From Fig. 24, one concludes that force F_1 and velocity $\dot{p}_2$ violate their limits while other variables are acceptable. Thus, change of geometry s is needed. The variation of geometry means the solution of law $s^*(t)$ that would prevent violation of any constraint. This can be done off-line (pre-programmed adaptation) or on-line. Here, we show the results of on-line adaptation.

The main idea of on-line adjustment lies in the following. During robot motion, velocities and forces, $\dot{p}_j$ and $F_j, j = 1, 2$, are measured or calculated and the appropriate change of geometry

(s_1, s_2) is found so as to keep them in the allowable regions: $\dot{p}_j < \dot{p}_j^M$, and $F_j < F_j^M$. In order to realize the adjusting, we apply a kind of penalty functions method as explained in Sec.4.2. At the beginning, we adopt that there exist some "home positions" of geometrical coordinates, $s_1^0 = s_2^0 = 0.15m.$.

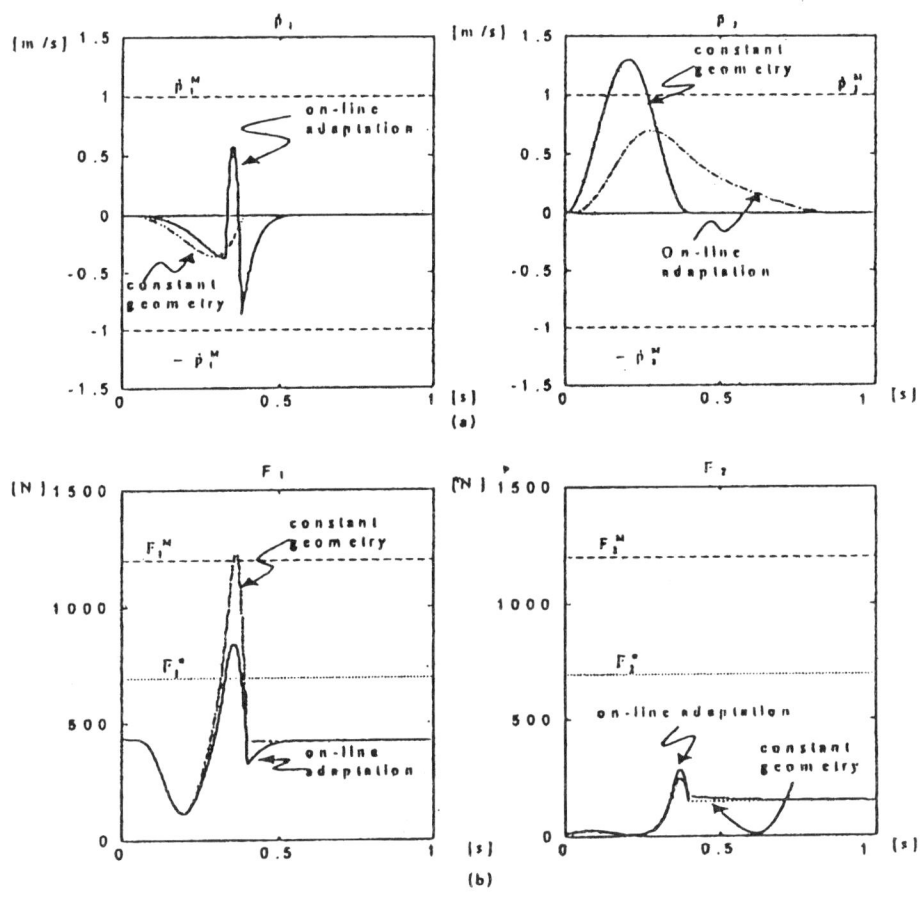

Figure 24. Time histories of piston variables, for constant geometry and for on-line adaptation: (a) velocities $\dot{p}_1$, $\dot{p}_2$, (b) forces F_1, F_2.

In order to distinguish between the critical and the noncritical values of velocities and forces, we define $\dot{p}_j^0 = 0.$ m/s, and $F_j^0 = 700N$, , $j = 1,2$. Change of geometry starts above these values. For the imposed task, the change of geometry (law s^*) is found and shown in Fig. 25.

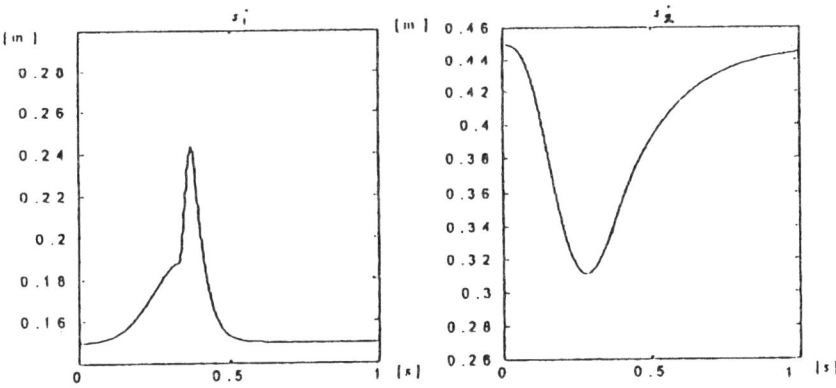

Figure 25. Variation of geometry: time histories of coordinates s_1^*, s_2^*.

5 Concluding Remarks

Real and contemporary needs for systems whose behaviour should be compatible with variable operating conditions oriented the research efforst towards partially or fully active systems.

Based on represented examples of real or hypothetical active systems in different fields, we can speak about very specific robotized constructions, structures, or systems in which the feedback loops are introduced. Thus, we can already talk about the robots of different classes: conventional robots and unconventional robotized structures like suspension bridges, buildings, vehicles, etc.

Active response of technical systems and constructions is already becoming a reality in various system in a broad range of engineering areas (mechanical, electrical and civil engineering). In each of these different engineering fields, there is a necessity to maintain the desired system performance during variable working conditions, as well as during different types of external perturbation that can lead to the so called extreme condition. It is characteristic for active systems that by adjusting the dynamic parameters, their dynamic performance is maintained in different working modes and under variable operating conditions. In the near future it is more realistic to expect a widespread realization of active systems and constructions that could lead to essential improvement of their dynamic performance, particularly under conditions when the capabilities of traditional (passive) systems (constructions, structures) are exhausted.

The techniques of computer-aided design with efficient and reliable user-oriented software have become widely available today. Their application to the automatic presentation of response and the analysis of performance in real time contributes to the development and improvement of active systems and constructions (structures). Important experience in active control such as in the aviation industry, can be utilized in other areas, i.e. in other classes of systems. This primarily concerns practically all kinds of vehicles, from ground transportation (road, railway), sea, to space transportation, then active foundations of civil structures, active buildings, long span suspension bridges, cooling towers, large shell-type roof constructions, etc. Differing from the approach to the synthesis and implementation of active systems where some dynamic parameters (inertial, stiffness, and most often damping) are changed, it is additionally proposed that the systems and constructions should be designed so as to feature variable geometry in purposeful margins [12,13]. In this way, one arrives at a new possibility in solving the problems of active systems and structures. Namely, by changing the relevant dynamic parameters, the maintaining of the requested dynamic performance is influenced, while by changing the geometrical parameters, appropriate distribution of the variable load is adjusted in a most direct way, and indirectly, the stiffness of the constructions and structures.

For the reasons stated, for constructions, objects and systems which can suffer destruction or malfunction, we now arrive at the real necessity to emphasize the importance of a new engineering which is already called Safety and Prevention Engineering. This new field of engineering should combine the efforts and coordinate the activity of scientists in forming a sufficiently general theory and safety concept for technical objects and systems, in developing methods and safety providing means for certain potentially endangered constructions (structures) and manufacture, in implementing and using safety technologies, in better understanding this activity, and as well as in forming a safety culture and promoting it within the society, in which we have to face actively in a new way the various natural (ecological and other) and technical breakdowns and disasters.

6 References

1. Meirovitch L., Gosh D., "Control of Flutter in Bridges", Journal of Engineering Mechanics, Vol. 113, pp.720-737, 1987.
2. Bowers N.A., "Tacoma Narrows Bridge Wrecked by Wind", Engineering News Records, Vol. 14, pp. 647 and 656, 1940.
3. Huang X.Y., "Active Control of Aerofoil Flutter", AIAA Journal, Vol. 25, pp.1126-1132,1987.
4. Tritton D.J., Physical Fluid Dynamics, Van Nostrand Reinhold, pp. 312, 1977.
5. Zajic B., Vukobratovic M., "Influence of the Environment on the Stability of Shaft Vehicle with Parametric Excitation and Small Parameter", Proc. of First ECPD Intern. Conference on Advanced Robotics and Intelligent Automation, Athens, pp. 625-631, 1995.
6. Chang J.C.H., Soong T.T., "Structural Control Using Active Tuned Mass Dampers", Journal of the Engineering Mechanics Division, Proc. of the American Society of Civil Engineers, Vol. 106, No. EM6, Dec., 1980.
7. The Akashi-Kaikyo Bridge, Honshu-Shikoku Bridge Authority, Japan, October 1998.
8. Chuang C.H., "Wu D.N., Wang Q., "LQR for State-Bounded Structural Control", Trans. of ASME, J. of Dyn. Systems, Measurement and Control, Vol. 118, pp. 113-119, 1996.
9. Shladover S.E., Desver, Ch.A., Hedrick, J.K., Tomizuka, M., Alrand, J., Zhang, W., Mc Mahon, D.H., Peng, H., Sheikho leslam, S., Mc Keown, N., "Automatic Vehicle Control Developments in the PATH Program", IEEE Trans. on Vehicular Technology, Vol.40, No 1, pp. 114-130, 1991.

10. Rodic A., Vukobratovic M., "Contribution to the Integrated Control Synthesis of Road Vehicles", IEEE Trans. on Control Systems Technology, Vol. 8, No 1, 1999.
11. E. Nissim, "Flutter Suppression Using Active Control Based on Concept of Aerodynamics Energy", NASA TN-6199, 1971.
12. Vukobratovic M., Potkonjak V., "Systems with Variable Geometry: Concept and Prospects", ASME Journal of Dynamic Systems, Measurement and Control", Vol.121, June issue, 1999.
13. Vukobratovic M., Potkonjak V., Matijevic V., "Internal Redundancy – the Way to Improve Robot Dynamics and Control Performances", Journal of Intelligent and Robotic Systems, Vol. 27, Nos. 1-2, 2000.

Chapter I

MECHANICS

Coordination of Parallel Arrays of Binary Actuators

Po-Hua Yang and K. J. Waldron

Department of Mechanical Engineering
The Ohio State University
Columbus, OH 43210, USA

Abstract. The objective of this study is the examination of the design and control opportunities presented by large arrays of small, simple actuators acting in parallel. An array of two-state actuators, i.e. a bundle of binary actuators acting in parallel, can be regarded as a simplified model of biological muscle. An experimental parallel array system composed of a large number of pneumatic cylinders connected between an output member and base has been constructed. Each cylinder can be either set as on or off. The parallel array is viewed as a programmable force generator, rather than as a motion generator. In this paper, we will discuss the wrench workspace generated by the prototype parallel array and also present an efficient method for controlling the wrench generation.

1 Introduction

In contrast to most biological systems, the use of actuation units in most artificial systems is relatively conservative. In a biological system, muscle fibers are bundled in very large numbers to form muscles (McMahon, 1984). A muscle fiber viewed as a basic actuation unit can be individually excited by electric signals transmitted by nerve filaments. Ideally, we can model the muscle fiber as a two-state actuator which contracts when excited and relaxes when passive. By arranging a large number of fibers in parallel, force can be controlled simply by determining the number and locations of fibers to be excited. Although there is considerable variation. mammals typically have muscle mass of the order of 40% of body mass (Enoka, 1988; McMahon, 1984). This is a larger percentage of system mass than is typical of artificial systems. Body geometry seems to interact with the biomechanical requirements of the system to determine muscle location and mass. In other words, muscles are also used as active structural elements.

Bundling large numbers of active elements into a limited space is becoming feasible with recent developments in smart materials, such as shape memory alloy, piezo-ceramics and ionic-polymer metal composites (Shahinpoor, 1995; Shahinpoor and Mojarrad, 1997). One can envision the design flexibility of being able to add actuators to a machine wherever there is available space, if one is not bound by the convention of minimizing the number of actuators. The study presented here is not directed at the development of the active elements of the actuator array. Our focus is more on the use of these active elements in a robotic system. We propose methods to characterize and control the systems actuated by parallel arrays. The study may also interact with the development of smart materials by providing an attractive application of such elements and identifying desirable characteristics which may be built into them.

2 System Architecture and Formulation of the Problem

A schematic drawing of a parallel array is shown in the Figure 1. Structurally, the mechanism is statically overdetermined. Kinematically, the mechanism is over-actuated. In this work, we focus on the static problem and the behavior of the array in isolation. Ultimately, as is the case with muscles, the actuators must connected between structural members of a mechanical system. The resulting motion depends both on the array and the kinematic structure.

Zero-pitch line-geometry structures defining screw systems (Hunt, 1978) can provide a method of quick examination for a parallel array system. The word "parallel" used here means "acting in parallel" instead of "geometrically parallel". In fact, if all lines of actions are parallel, it will result a rank 3 system which can generate moments in any direction orthogonal to the line of action, or linear forces in the direction parallel to the line of action, but cannot generate more general wrenches. In designing a system with a massive array of actuators, it is very difficult to have a direct sense of how the system will behave. Although line geometries are useful for understanding certain configurations, a more precise metric is necessary for more subtle situations, such as configurations close to singularities. The Jacobian matrix that can be viewed as a collection of all the screw vectors in the parallel array system plays an important role in establishing such precise metrics. The rank of the Jacobian matrix $\mathbf{J}_{(6 \times N)}$ is exactly equivalent to the order of the screw system, where N is the number of actuators. The eigenvalues of the stiffness matrix $k(\mathbf{J} \times \mathbf{J}^T)_{6 \times 6}$ can be recognized as performance indices in the directions of the eigenvectors which can be considered as principle axes of a stiffness system provided by the parallel array.

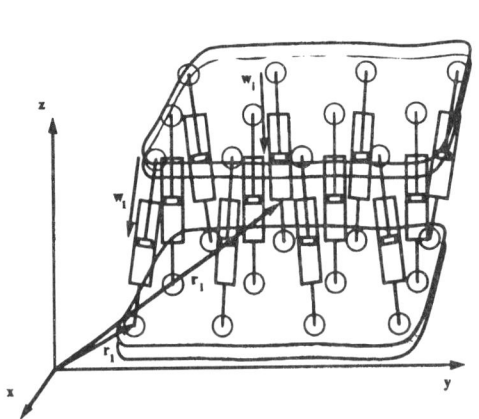

Figure 1. Concept Drawing of a Binary Parallel Array

Figure 2. Photograph of the Experimental Setup

The basic mathematical formulation of the problem to be studied will be introduced here. In the array shown in Fig. 1, the ith actuator is characterized by the direction of the line of action of the force it produces: $\mathbf{w}_i$, together with the position of any point on that line of action: $\mathbf{r}_i$.

We consider the vectors $\mathbf{w}_i$; $i = 1,...,N$ to be unit vectors. Hence, if F is the magnitude of the force produced by each actuator when activated, the resultant wrench $[\mathbf{R}^T \ \mathbf{M}^T]^T$ exerted on the output member can be written in the form:

$$\begin{bmatrix} \mathbf{R} \\ \mathbf{M} \end{bmatrix} = F \begin{bmatrix} \mathbf{w}_1, \ldots, \mathbf{w}_N \\ \lambda_1, \ldots, \lambda_N \end{bmatrix} \mathbf{u}$$
$$= \mathbf{J}_{(6 \times N)} \mathbf{u} \qquad (1)$$

where $\lambda_i = \mathbf{r}_i \times \mathbf{w}_i$

The Jacobian matrix $\mathbf{J}$ represented here is mathematically isomorphic to that used in the inverse rate problem of a serial chain system (Waldron and Hunt, 1991). If actuators are operated in a binary fashion: either fully excited, or off, the elements in the $N \times 1$ vector $\mathbf{u}$ will either be one or zero where one corresponds to an active actuator, and zero corresponds to a inactive actuator.

Given the vector $\mathbf{u}$, Eq. (1) can be used to compute the resultant wrench of the array. This forward computation gives the solution to the direct isometric force synthesis of the system, which might be called the *"workspace problem"*. However, here we are talking about a *"wrench workspace"*. Namely, the set of all wrenches that the system can generate.

The reverse problem of Eq. (1) is formulated by treating the resultant wrench $[\mathbf{R}^T, \mathbf{M}^T]^T$ as known and trying to find the vector $\mathbf{u}$ which is the commanded signal for generating that wrench. The solution of this *inverse problem* or *control problem* gives the necessary information for coordinating the mechanism to perform useful tasks. However, the difficulties imposed by using a discretely actuated parallel array can be better described by the following contradictions. On the one hand, the system is highly redundant, which usually means a larger solution space. On the other hand, the solution space seems very small, since each limb in the system can only be set to one of the two discrete states which the actuator can produce. In fact, most generalized inversion processes are not suitable for our purpose since they yield only continuous numbers instead of a binary vector $\mathbf{u}$.

3 Wrench Workspace and Discrete Convolution

Although a discretely actuated parallel array does not provide a continuous workspace, the total number of possible wrenches that can be produced by the array grows exponentially with the number of actuators employed in the system. The displacement workspace of a discretely actuated hyper-redundant manipulator can be computed by using the Fourier transform and the concept of convolution (Ebert-Uphoff and Chirikjian, 1995). Although we are talking about the wrench workspace here, a similar idea can still be applied. In fact, the relationship between a wrench workspace of a binary parallel array and discrete convolution is more straightforward than the case for the displacement workspace of a discretely actuated hyper-redundant manipulator. A simplified case is used here to illustrate the idea.

For the ease of the explanation, we will assume that only three different wrenches, including a null wrench, can be produced by a planar parallel array unit. This is three different wrenches:

$$\mathbf{f} = \begin{bmatrix} f_x & f_y & m_z \end{bmatrix} \qquad (2)$$

where $[f_x\ f_y\ m_z] = [0\ 0\ 0],\ [1\ 2\ 5],\ or\ [2\ 3\ 4]$.

Since planar wrenches are presented by 3×1 vectors, the wrench workspace can be drawn in a three dimensional space. A space function $\rho(x, y, z)$ is assigned a number that corresponds to the occurrences of the wrench $[x, y, z]$; otherwise, 0 is assigned to that point. We can also view $\rho(x, y, z)$ as a density (or occurrence) function. The wrench workspace for the case represented by Eq. (2) can then be written as

$$\{\rho(x,y,z) = 1\ |\ [x\ y\ z] = [0\ 0\ 0], [1\ 2\ 5], or\ [2\ 3\ 4]\} \tag{3}$$
$$\{\rho(x,y,z) = 0\ |\ elsewhere\} \tag{4}$$

Wrench generators can collaborate if they act in parallel. In other words, the wrench workspace can be expanded by adding more wrench generators acting in parallel. The system of wrench generators acting in parallel is actually a dual of a system of motion generators arranged in series (Waldron and Hunt, 1991). For studying how the workspace is expanded by parallel bundling, we can use the wrench generator defined by Eq. (2) as an element. By stacking only two of them along the z axis, we can find all the possible output wrenches by using a brute force search, since very few units are used here. The result is represented by the density (occurrence) function h as

$$\{h(x,y,z) = 1\ |\ [x\ y\ z] = [0\ 0\ 0], [2\ 4\ 10], or\ [4\ 6\ 8]\} \tag{5}$$
$$\{h(x,y,z) = 2\ |\ [x\ y\ z] = [1\ 2\ 5], [2\ 3\ 4], or\ [3\ 5\ 9]\} \tag{6}$$
$$\{h(x,y,z) = 0\ |\ elsewhere\} \tag{7}$$

In the following, we will show that the same result can be obtained by using discrete convolution. If we perform the discrete convolution for $\{\rho[x, y, z] \otimes \rho[x, y, z]\}$, summation will take the place of integration to give:

$$h[x,y,z] = \sum_{i=1}^{n}\sum_{j=1}^{m} \rho[i,j,k]\ \rho[x-i, y-j, z-k]. \tag{8}$$

By substituting the $\rho[x, y, z]$ in Eq. (3) and (4) into Eq. (8), the equation can be expanded to

$$h[x,y,z] = \rho[0,0,0]\ \rho[x,y,z] + \rho[1,2,5]\ \rho[x-1, y-2, z-5]$$
$$+ \rho[2,3,4]\ \rho[x-2, y-3, z-4]. \tag{9}$$

By substituting x,y, and z with real numbers for the above equation, the corresponding $h[x, y, z]$ can be evaluated. When (x,y,z) is (0,0,0), (2,4,10), or (4,6,8), one of the three terms on the right hand side of Eq. (9) will be one and the other two terms will be zero. This results in $h[0, 0, 0]$, $h[2, 4, 10]$, and $h[4, 6, 8]$ equal to one. When (x,y,z) is (1,2,5), (2,3,4), or (3,5,9), two of the three terms on the right hand side of Eq. (9) will be one and only one term will be zero. This results in $h[1, 2, 5]$, $h[2, 3, 4]$, and $h[3, 5, 9]$ being two. For the other points in x, y, z space, the value of $h[x, y, z]$ will all be zero. These results are identical to the results listed in Eq. (5), (6) and (7).

The relationship between the wrench workspace and discrete convolution is more straightforward than its serial dual. The convolution method is especially useful for when a modularized manipulator is considered, as was shown by Chirikjian's work (Chirikjian and Ebert-Uphoff, 1998).

4 Coordination of Force Generation

The Jacobian matrix $\mathbf{J}$ of the parallel array can be disassembled to six $1 \times N$ row vectors as

$$\mathbf{J}_{6 \times N} = \begin{bmatrix} \mathcal{A}^T & \mathcal{B}^T & \mathcal{C}^T & \mathcal{D}^T & \mathcal{E}^T & \mathcal{F}^T \end{bmatrix}^T \tag{10}$$

The linear forces along the x,y, and z axes can then be written as:

$$R'_x = \mathcal{A}^T \cdot \mathbf{u}; \quad R'_y = \mathcal{B}^T \cdot \mathbf{u}; \quad R'_z = \mathcal{C}^T \cdot \mathbf{u}$$
$$M'_x = \mathcal{D}^T \cdot \mathbf{u}; \quad M'_y = \mathcal{E}^T \cdot \mathbf{u}; \quad M'_z = \mathcal{F}^T \cdot \mathbf{u} \tag{11}$$

where $\mathbf{u}$ is a binary vector.

The wrench produced is not going to be the same as the desired wrench. Based on the error for each component in a wrench system, a performance index can be evaluated by summing the squared "component error".

$$V(\mathbf{u}) = \frac{1}{2}[(R_x - \sum_{i=1}^{N} \mathcal{A}_i\, u_i)^2 + (R_y - \sum_{i=1}^{N} \mathcal{B}_i\, u_i)^2 + (R_z - \sum_{i=1}^{N} \mathcal{C}_i\, u_i)^2]$$
$$+ \frac{1}{2}[(M_x - \sum_{i=1}^{N} \mathcal{D}_i\, u_i)^2 + (M_y - \sum_{i=1}^{N} \mathcal{E}_i\, u_i)^2 + (M_z - \sum_{i=1}^{N} \mathcal{F}_i\, u_i)^2]$$
$$- \frac{1}{2} \sum_{i=1}^{N} u_i\,(u_i - 1)[\mathcal{A}_i^2 + \mathcal{B}_i^2 + \mathcal{C}_i^2 + \mathcal{D}_i^2 + \mathcal{E}_i^2 + \mathcal{F}_i^2] \tag{12}$$

In order to fit the above equation into a quadratic form, it is convenient to have a $\frac{1}{2}$ multiplier. The last line of the equation forces u_i to take on values of 0 or 1. It also cancels the diagonal terms of the weighting matrix $\mathbf{W}$ when we convert the above equation into a quadratic form. The zero eigenvalue can be avoided by this cancellation. We will show that later.

Eq. (12) can be fitted into a quadratic form:

$$V(\mathbf{u}) = \frac{1}{2} \mathbf{u}^T \mathbf{W} \mathbf{u} + \mathbf{b}^T \mathbf{u} \tag{13}$$

where the weighting matrix $\mathbf{W}$ and the bias vector $\mathbf{b}$ are

$$\mathbf{W} = \begin{bmatrix} 0 & \mathcal{A}_1 \mathcal{A}_2 & \cdots & \mathcal{A}_1 \mathcal{A}_N \\ \mathcal{A}_2 \mathcal{A}_1 & 0 & \cdots & \mathcal{A}_2 \mathcal{A}_N \\ \vdots & \vdots & \vdots & \vdots \\ \mathcal{A}_N \mathcal{A}_1 & \mathcal{A}_N \mathcal{A}_2 & \cdots & 0 \end{bmatrix} + \cdots + \begin{bmatrix} 0 & \mathcal{F}_1 \mathcal{F}_2 & \cdots & \mathcal{F}_1 \mathcal{F}_N \\ \mathcal{F}_2 \mathcal{F}_1 & 0 & \cdots & \mathcal{F}_2 \mathcal{F}_N \\ \vdots & \vdots & \vdots & \vdots \\ \mathcal{F}_N \mathcal{F}_1 & \mathcal{F}_N \mathcal{F}_2 & \cdots & 0 \end{bmatrix} \tag{14}$$

$$\mathbf{b} = \begin{bmatrix} \frac{1}{2}(\mathcal{A}_1^2 + \mathcal{B}_1^2 \cdots + \mathcal{F}_1^2) - \mathcal{A}_1 R_x \cdots - \mathcal{F}_1 M_z \\ \frac{1}{2}(\mathcal{A}_2^2 + \mathcal{B}_2^2 \cdots + \mathcal{F}_2^2) - \mathcal{A}_2 R_x \cdots - \mathcal{F}_2 M_z \\ \vdots \\ \frac{1}{2}(\mathcal{A}_N^2 + \mathcal{B}_N^2 \cdots + \mathcal{F}_N^2) - \mathcal{A}_N R_x \cdots - \mathcal{F}_N M_z \end{bmatrix} \tag{15}$$

The performance index $V(\mathbf{u})$ in Eq. (13) can be minimized by choosing an appropriate value of $\mathbf{u}$. There are several ways of implementing quadratic programming. The optimization toolbox provided by MATLAB is one of many good candidates. It is also found that the Hopfield network in the field of neural networks can provide the same functionality with a relatively straightforward implementation (Hopfield and Tank, 1985; Tank and Hopfield, 1986). The process of the Hopfield network is shown in Figure 3. The binary numbers required by coordinating the discretely actuated parallel array naturally suggest a neural net implementation.

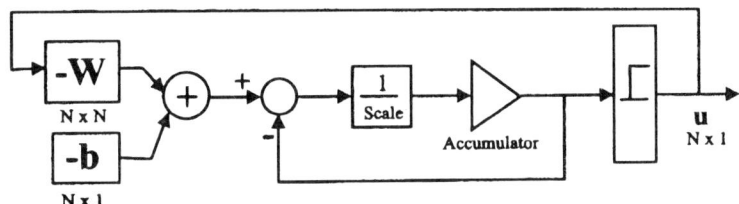

Figure 3. High-gain Hopfield Network

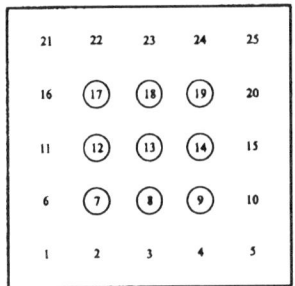

Figure 4. Group A

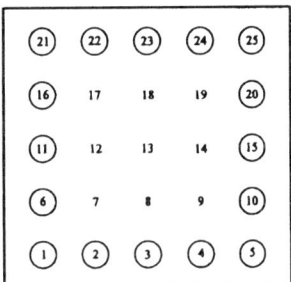

Figure 5. Group B

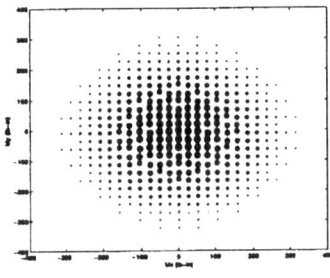

Figure 6. Workspace

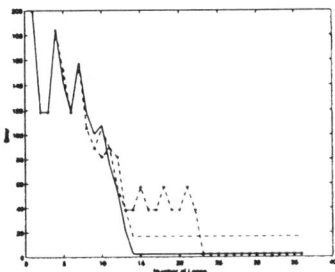

Figure 7. Convergence of error

5 Experiment and Results

A simplified binary parallel array is realized by an experimental model (Yang and Waldron, 1999). A relatively small array of actuators, consisting of twenty-five pneumatic cylinders, is implemented in hardware as shown in Figure 2. Each actuator is connected to the base plate and the upper plate by universal joints. The universal joints are arranged as a 5×5 grid on the base plate and spaced with 5.1 cm of distance between rows and columns. A similar pattern is used for the upper plate except that the space between rows and columns is reduced to 3.2 cm. Each pneumatic cylinder is controlled by a 3-way valve, which allows the actuator to operate in a binary fashion: either fully excited, or off. The force output of each cylinder is set at 50 newtons by regulating the compressed air at a constant pressure of 413 kpa.

With 25 actuators, there are about 32 millions (2^{25}) possible wrenches can be created. The task of finding all these wrenches by a brute force calculation will be time-consuming. However, the complexity of workspace calculation can be greatly reduced by partitioning the actuators into two groups as is shown in Figure 4 and 5 and using the convolution method presented in Section 3. Since the experimental array is mainly configured for producing the moments along the x and y axes and forces along the z axis, the workspace presented by Fig. 6 is simplified by only showing the x- and y- components of the generated moments.

In Section 4, the inverse static problem of the parallel array was solved by an optimization approach, which can be realized by using the Hopfield network with two major advantages: the ease of implementation and a natural binary solution. It will be desirable to find the efficiency of the network by tracking the result of each iteration of the feedback loop as illustrated in Fig. 3. The error of the result can be evaluated by using the mean square error as

$$e_i = \sqrt{(R_x - R'_x)^2 + (R_y - R'_y)^2 + \cdots + (M_z - M'_z)^2} \tag{16}$$

As we plot the error against each iteration as is seen in Fig. 7, we can visualize the speed of convergence. Three different lines in the figure is obtained by using different values for "SCALE" in the feedback loop (see Fig. 3). When this value is properly chosen, the solution can be optimized.

6 Conclusion

Research in robotic systems is often motivated by imitation of biological systems. However, these imitations are usually focused on the structural level instead of lower level actuation such as muscle systems. In this paper, we suggest a mechanical model of a biological muscle system by using massive arrays of binary actuators acting in parallel.

For exploration of such a complicated model, we choose a simpler problem, static force analysis, as our starting point of this research. The major discovery of this work is that by a proper formulation, an optimization approach can be used for solving the inverse static problem of the parallel array. We also find that the formulated optimization problem can be realized by a Hopfield network with two major advantages: ease of implementation and a natural binary solution. The static force study provided by this paper will be valuable for the future development of the parallel array and also helpful for understanding the mechanics of biological muscles.

References

McMahon, T. A. (1984). *Muscles, Reflexes, and Locomotion*. Princeton, NJ: Princeton University Press.

Enoka, R.M. (1988). *Neuromechanical Basis of Kinesiology*. Champaign, IL: Human Kinetics Books.

Shahinpoor, M. (1995). Design, Modeling and Fabrication of Micro-Robotic Actuators with Ionic Polymeric Gel and SMA Micro-Muscles. In *Proc. 1995 ASME Design Engineering Technical Conference*. Boston, MA.

Shahinpoor, M., and Mojarrad, M. (1997). Biomimetic Robotic Propulsion Using Ion-Exchange Membrane Metal Composite Artificial Muscles. In *Proceedings of 1997 IEEE Robotic and Automation Conference*. Albuquerque, NM.

Hunt, K. H. (1978). *Kinematic Geometry of Mechanisms*. Oxford: Clarendon Press.

Waldron, K. J., Hunt, K. H. (1991). Series-Parallel Dualities in Actively Coordinated Mechanisms. *International Journal of Robotics Research*, Vol. 10, No. 5, pp. 473-480.

Ebert-Uphoff, I. and Chirikjian, G. (1995). Efficient Workspace Generation for Binary Manipulators with Many Actuators. *Journal of Robotic Systems*, Vol. 12, No. 6, pp. 383-400.

Chirikjian, G., and Ebert-Uphoff, I. (1998). Numerical Convolution on the Euclidean gruop with applications to workspace generation. *IEEE Trans. on Robotics and Automation*, 14(1):123-136.

Hopfield J. J., and Tank, D. W. (1985). Neural Computation of Decision in Optimization Problems. *Biological Cybernetics*, Vol.52, pp. 141-154.

Tank, D. W., and Hopfield J. J. (1986). Simple 'Neural' Optimization Networks: an A/D Converter, Signal Decision Circuit and a Linear Programing Circuit. *IEEE Transactions on Circuits and Systems*, Vol. 33, No. 5, pp. 533-541.

Yang, P., Waldron, K. J. (1999). Design of a Parallel Mechanism with a Massive Number of Binary Actuators for Wrench Generation. In Soni, A. H., ed.,*Proceedings of the Sixth Applied Mechanisms and Robotics Conference*, Cincinnati, OH. pp. AMR 99-080:01-08.

A Formal-Numerical Approach to Determine the Accuracy of a Parallel Robot in a 6D Workspace

J-P. Merlet

INRIA, France

Abstract. The positioning error of a parallel robot is conditioned by the measurement errors on the leg lengths, these two quantities being linearly related through the pose-dependent jacobian matrix of the robot. An important design problem is to determine the extremum of the positioning errors over a prescribed 6D workspace. This is a difficult problem as the jacobian matrix has a complex formulation, involving thousands of terms. We present the preliminary result for an algorithm that estimate the positioning error with an arbitrary accuracy.

1 Introduction

Parallel robots have been extensively studied this recent years and are now starting to appear as commercial products for various applications, hence the interest for their optimal design. Among other criterion checking the positioning accuracy is an important part of the design process. Surprisingly this issue has been largely ignored in the literature: Patel [4] has studied the positioning error on a trajectory, Ropponen [5] these errors for a given pose, while Masory [1] uses a specific hardware to evaluate the accuracy of a parallel robot. But to the best of the author knowledge, the problem of determining the maximal errors has not been addressed in the literature. In this paper we consider a Gough-type 6 d.o.f. parallel manipulator (Figure 1) constituted of a fixed planar base plate and a planar mobile plate connected by 6 articulated and extensible links.

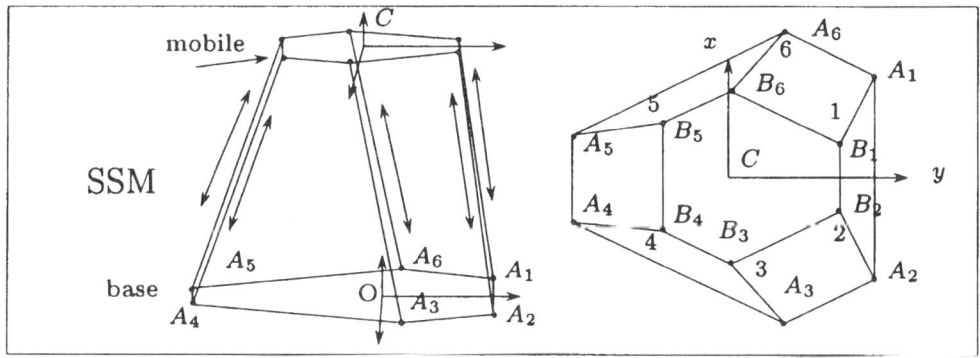

Figure 1. Gough platform.

The pose of the platform may be adjusted by changing the length of the six legs. A reference frame $(O; x; y; z)$ is attached to the base and a mobile frame $(C; x_r; y_r; z_r)$ is attached to the moving platform. The leg i is attached to the base with a ball-and-socket joint whose centre is A_i, while it is attached to the moving platform with an universal joint whose centre is B_i. Let ρ_i be the leg lengths (the distance between A_i and B_i), X a 6-dimensional vector defining the pose of the end-effector: the three first components of X are the coordinates of C in the reference frame, while the three last components are three parameters describing the orientation of the end-effector. In this paper we will use the Euler angles ψ, θ, ϕ. The errors on the leg length measurements $\Delta \rho$ and the positioning errors on the end-effector ΔX are related as follows:

$$\Delta \rho = J^{-1} \Delta X \quad \Delta X = J \Delta \rho \quad (1)$$

where J is the jacobian matrix of the robot and J^{-1} its inverse. In our problem the measurement errors $\Delta \rho$ are known, i.e. we have $-\epsilon_i \leq \Delta \rho_i \leq \epsilon_i$, and we are interested in finding the extremal values of each component of ΔX. Hence the later equation is the one we are interested in. Unfortunately we have a convenient analytical formulation only for J^{-1}; a row of this matrix may be written as:

$$J_i^{-1} = \frac{A_i B_i}{\rho_i} \quad CB_i \times \frac{A_i B_i}{\rho_i} \quad (2)$$

Although this matrix may be easily expressed in terms of the pose parameters, its inverse has a complex formulation [2]. The maximal value of the positioning error on the i-th pose parameters may be written as:

$$\Delta X_i = \sum_{j=1}^{j=6} |J_{ij}| \epsilon_j$$

Even if we assume that we have been able to compute the jacobian matrix, we will have to determine the maximal value of ΔX_i for all the poses in the prescribed workspace. Clearly this is a difficult optimisation problem.

Now we must examine why we must compute the maximum of the positioning errors. Two main reasons may justify this computation:

- prove that a given robot design has a maximal error less than a given threshold ϵ
- compare the positioning accuracy of two robots with different dimensions

It must be note that in both cases it is not necessary to determine the exact value of the maximal positioning error ΔX. This may be shown easily in the former case: assume that we are able to determine a number G such that we can guarantee that $G - \mu |\Delta X| \leq G + \mu$, where μ is an arbitrary error margin. Now suppose that $G + \mu < \epsilon$ or $G - \mu > \epsilon$ then we have determined that in the first case the robot verify the desired accuracy property while in the second case that it fails. Otherwise we cannot answer the question and we will have to decrease μ and start again. Note however that if we are free to fix the value of μ at will, then we may get a very good approximation of the exact value of the maximal positioning error.

Thus our main goal will not to compute the error *exactly* but up to the given accuracy μ. A direct result is that the computation time will be quite low for relatively large μ: in the two problems presented above we will start with large μ and reduce its value only if the result is indeterminate. This approach will be, in general, quite efficient.

2 Positioning Errors and Algorithm Principle

Let S_i be the matrix obtained from J^{-1} by substituting the i-th column of J^{-1} by the vector $\Delta \rho$. By solving the linear system (1) we get the positioning error on X_i as:

$$\Delta X_i = \frac{|S_i|}{|J^{-1}|}$$

Let S_i^j be the minor of S_i obtained by removing the i-th column and j-th row of this matrix We have:

$$\Delta X_i = \frac{\sum_{j=1}^{j=6}(-1)^{i+j}\Delta \rho_j |S_i^j|}{|J^{-1}|} \tag{3}$$

We define now the *semi inverse jacobian* matrix M by its i-th row:

$$M_i = (A_i B_i \quad CB_i \times A_i B_i) \tag{4}$$

A direct consequence of this definition is that:

$$|J^{-1}| = \frac{|M|}{\prod_{k=1}^{k=6}\rho_k} \quad |S_i^j| = \frac{|M_i^j|}{\prod_{k=1,6}^{k \neq j}\rho_k} \tag{5}$$

It must be noted that contrary to $|J^{-1}|$, $|S_i^j|$ the expressions $|M|$, $|M_i^j|$ have no denominator. Reporting these results in equation (3) we get:

$$\Delta X_i = \frac{\sum_{j=1}^{j=6}(-1)^{i+j}\Delta \rho_j |M_i^j|\rho_j}{|M|} \tag{6}$$

Consider now that the pose parameters should lie in some workspace W and let U_{ij}^M be the maximal absolute value of $(-1)^{i+j}|M_i^j|\rho_j/|M|$ for all the possible poses in the workspace. A consequence of equation (6) is that:

$$T_{ij}^M = \sum_{j=1}^{j=6} -\varepsilon_j U_{ij}^M \leq \Delta X_i \leq T_{ij}^m = \sum_{j=1}^{j=6} \varepsilon_j U_{ij}^M \qquad (7)$$

Assume now that we want to determine the extreme values $\Delta^m X_i$, $\Delta^M X_i$ of ΔX_i with a given accuracy μ, i.e. we want to find 2 quantities V_i^m; V_i^M such that:

$$V_i^m - \mu \leq \Delta^m X_i \quad \Delta^M X_i \leq V_i^M + \mu$$

Initial values for V_i^m, V_i^M may be obtained by taking random poses in the workspace. If we are able to compute bounds on T_{ij}^m, T_{ij}^M, for all poses in the workspace and update V_i^m, V_i^M in such way that

$$V_i^m - \mu \leq T_{ij}^m \quad T_{ij}^m \leq V_i^M + \mu \qquad (8)$$

then V_i^m, V_i^M may be considered as extremal values of the positioning error, up to the accuracy μ. So the basic principle of our algorithm for computing the extremal values of ΔX_i is

1. initialize V_i^m, V_i^M by computing the positioning error at poses randomly selected in W
2. compute bounds on T_{ij}^m, T_{ij}^M for all poses in W
3. if these bounds satisfy the inequalities (8) return V_i^m, V_i^M as extremal values of the positioning error
4. otherwise split W in smaller subset. In each subset compute the positioning error at some randomly selected poses and update V_i^m, V_i^M if necessary. Then repeat the process for each subset and return the largest extremal values obtained for all the subsets.

We have now to explain how to perform step 2 of this algorithm and how to split W in smaller subsets.

3 Finding Bounds on the Positioning Errors

We have to determine bounds on the extremal values of a sum of terms, each of them having the generic form $\rho |M_m|/|M|$, where M_m is a minor of M. This sum has a very complex formulation, so we will consider independently each term of the sum, find bounds on the extremal values of each term and consider that the bounds on the sum is the sum of the bounds on each term. Each term is still a complex function of the pose parameters and of the geometry of the robot, as defined by the location of the joints of the robot. So we have decided to use the symbolic computation software MAPLE to compute a generic form for the determinants $|M_m|$, $|M|$. According to the structure of the elements of the matrix J^{-1} these determinants may be written generically as:

$$|\ldots| = \sum_{a=1}^{a=n} F_a$$

$$F_a = C_a x^i y^j z^k \cos^l(\psi)\sin^m(\psi)\cos^n(\theta)\sin^o(\theta)\cos^p(\varphi)\sin^q(\varphi)$$

where the C_js are numerical constants. When running our algorithm a MAPLE program will calculate each determinant and determine all its basic components F_a, that it will then write in a result file, each line representing one F_a as a set of float number (for C_a) and 9 integers representing i; j; k; l; m; n; o, p; q. We will then use a method called interval analysis [3] to evaluate lower and upper bounds for the F_a. Interval analysis is similar to real analysis except that real variables are replaced by intervals and that specific rules are used for each basic arithmetic operations.

More generally if $X_1 = [\underline{x_1}, \overline{x_1}]$ with $\underline{x_1} \leq \overline{x_1}$ denotes an interval it is possible to define all the arithmetic operators (and more complex functions) for intervals. For example the "+" operator for intervals will be defined as:

$$X_1 + X_2 = [\underline{x_1} + \underline{x_2}, \overline{x_1} + \overline{x_2}]$$

With this method we are able to compute lower and upper bounds on any function for given ranges for the input parameters. For example being given 6 ranges for the variables x, y, z, ψ, θ, ϕ we are able to compute two real number F_a^-, F_a^+ such that we can guarantee that for any value of the variable in their ranges we have $F_a^- \leq F_a(x, y, z, \psi, theta, \phi) \leq F_a^+$ (note that some implementation of interval analysis operators take into account round-of errors to get *guaranteed* bounds).

A drawback of interval analysis is that the resulting bounds may be largely over estimated (e.g. the bounds for the function $x^2 + x$, when $x \in [-1, 0]$ are computed as [-1,1], while the real bounds are [-1/4,0]), but they will get closer to the exact values as soon as the size of the input ranges decrease. Thus for given ranges on the pose parameters we are able to compute bounds on $\rho, |M_m|$ and $|M|$ and consequently bounds on $\rho, |M_m|/|M|$ which in turn enable to compute bounds on the positioning errors.

4 The Bisection Process

Suppose that the workspace W is defined as a set of ranges, one for each pose parameters. We will call this type of workspace an *extended box* or EB for short. To create the subsets of W that are necessary in our algorithm we will bisect each range of the workspace: if the initial range is $[x_1, x_2]$, the bisection process will lead to the two ranges $[x_1, (x_1 + x_2)/2]$, $[(x_1 + x_2)/2, x_2]$. Thus, if none of the pose parameters has a constant value, we will get $2^6 = 64$ new EBs as a result of the bisection process. These EBs will be stored in a list and the algorithm will be used for each EB in the list until all the EBs in the list have been processed.

5 Dealing with Other Types of Workspace

We have described in the previous section an algorithm to compute the extremal values of the positioning errors if the workspace is an EB. This algorithm allows us to deal with other type of workspace. Assume for example that the workspace is defined as three ranges for the orientation parameters and a geometric object (e.g. a sphere) as the possible location of the origin of the end-effector. We will assume that we are able to define an EB which include the workspace W so defined and that we may define a test T for an EB which return 0 if the EB is outside W, 1 if the EB is fully inside W and 2 if only a part of the EB is inside W. The previous algorithm will be modified at the bisection level where the test T will be applied on each EB resulting from the bisection. If the test returns 0 we just discard the EB, if it returns 1 we compute the positioning errors for this EB and update if necessary the current extremal values of the positioning errors. If the test returns 2 and the positioning errors for this EB may be larger than the current extremal values, then we store the EB in the list without updating the current extremal errors. A similar approach enable to deal with a workspace defined in terms of bounds on the articular coordinates i.e. we are able to compute the extremal positioning errors *for all* reachable poses of the end-effector.

6 Experimental Results

In this section we will assume that the error ϵ on the measurement of the articular coordinates is 1/100. We will assume that the accuracy μ with which we want to calculate the extremal values of the positioning errors are $\mu_{\Delta x} = 0.0291$, $\mu_{\Delta y} = 0.1329$, $\mu_{\Delta z} = 0.0189$, $\mu_{\Delta \theta_x} = 0.0018$rd, $\mu_{\Delta \theta_y} = 0.00086$rd, $\mu_{\Delta \theta_z} = 0.005$rd (these values have been selected as 10% of the extreme positioning errors computed for a few random poses in the workspace). The computation time has been established on a SUN Ultra 1 workstation.

The origin of the end-effector is located in a workspace defined by $x, y \in [-5, 5]$, $z \in [45, 50]$.

First we assume that the orientation of the end-effector is constant and defined by $\psi = \theta = \phi = 0$. We get the maximal positioning errors as, $\Delta x = \pm 0,145762$, $\Delta y = \pm 0,664774$, $\Delta z = \pm 0,0946431$, $\Delta \theta_x = \pm 0,0092133$rd, $\Delta \theta_y = \pm 0,004327$rd, $\Delta \theta_z = \pm 0,025138$rd in a computation time of 0.1s. If we decrease by half the value of the μs or divide them by 10 we get the same result for the positioning errors in 0.16s and 8.95s respectively. We assume now that the angle may vary in the range [-5,5] degree. The computation time with the same accuracy than above becomes 31.6 seconds.

Then we assume that both angles ψ, θ may vary in the range [-5,5] degrees. The computation time drastically increase and become about 11h.

7 Improving the efficiency

As may be noted in the previous section the efficiency of the algorithm is pretty good up to a 4D workspace, but drastically decrease as soon as we move to a 5D workspace. We will now suggest three possibilities to improve the efficiency.

7.1 Improving the Bounds on the Determinants

Using the nested form. We have seen that we expand all the determinant involved in the calculation as an expression involving a sum of generic terms F_a. But it is well known in the field of interval arithmetics that the Horner (or "nested" form) of an expression leads, in general, to better bounds. For example we have considered previously the expression $x^2 + x$ with x in the range [-1; 0] and we haven that the corresponding interval evaluation was [-1, 1]. If we consider now the nested form $x(x + 1)$ of this expression we get the interval evaluation [-1; 0] and we have reduced by the two the width of the interval evaluation.

The nested form of the determinants can be computed by MAPLE but the main difficulty is to use it in a program. To deal with this problem we will use a parser that has been developed in our laboratory.

This parser is able to read the analytical expression of a function written in a file and, being given ranges for the unknowns, will compute the interval evaluation of the function. Using this parser is somewhat slower than a direction evaluation of the expression but it is expected that the decrease in the width of the interval evaluation will drastically increase the efficiency of the algorithm.

Evaluation of the trigonometric expressions. In the determinants appear the sine and cosine of the orientation angles. These quantities are clearly not independent but are considered as such in term of interval analysis. For example if at some point appears the quantity $\sin\psi + \cos\psi$, the interval evaluation of this quantity for ψ in $[0, 2\pi]$ will be [-2, 2], while the exact value is $[-\sqrt{2}, \sqrt{2}]$. We have developed a program that is able to detect such expression and substitute it by a procedure that will compute better bounds.

7.2 Decreasing the Number of EB

For a 6D workspace the bisection process will lead to up to 64 new EB's each of which in turn may lead to 64 new EB's. Consequently the number of EB stored in the list is quickly growing. To avoid this effect we may use a well known method in the field of interval analysis: instead of bisecting the EB in all its 6 dimensions, we may bisect only one of them. This may have a very large impact on the efficiency. Indeed the bisection of an EB along one dimension leads to 2 new EB's. Imagine now that the algorithm is able to calculate that the maximal positioning error cannot be obtained for these 2 EB's, and consequently that they have not to be added to the list. We will evidently get a similar result by using the algorithm described in section 4 but at the cost of examining 64 EB's, instead of only 2. As bisecting along one dimension will lead in the worst case to the same number of new EB's as bisecting along all the dimensions, it may be seen that bisecting along one dimension may be a better approach.

It remains to determine which of the 6 dimensions should be selected as the bisected dimension. A classical approach for this problem is to use the "smear" function. Let $G(X)$ be a function of $X = \{x_1, x_2, \ldots, x_n\}$ and we define s_j as

$$s_j = Max\left\{\left\|\frac{\partial G(X)}{\partial x_k \partial x_j}\right\|(\overline{x_j} - \underline{x_j})\right\}$$

where the jacobian $\partial G(X)/\partial x_k \partial x_j$ is evaluated using interval analysis. In some sense s_j is a measure of the influence of the variable x_j on the value of G, which is weighted by the width of range on x_j. As a consequence the direction which will be bisected is the one which has the larger s_j.

8 Conclusion

We have presented here a preliminary approach to deal with one of the difficult problems related to parallel robots: determining the extreme values for the positioning errors for given ranges on the errors on the measurements of the articular coordinates. The experimental results show that our algorithm perform pretty well for workspace of dimension up to 4, while being quite slow for larger dimension. We have suggested ways to improve the efficiency of the algorithm, which are currently being implemented.

9 References

1. Masory O. and Jihua Y. Measurement of pose repetability of Stewart platforms. J. of Robotic Systems, 12(12):821-832, 1995.
2. Mayer St-Onge B. and Gosselin C. Singularity analysis and representation of spatial six-dof parallel manipulators. In J. Lenarčič V. Parenti-Castelli, editor, Recent Advances in Robot Kinematics, pages 389-398. Kluwer, 1996.
3. Moore R.E. Methods and Applications of Interval Analysis. SIAM Studies in Applied Mathematics, 1979.
4. Patel A.J. and Ehmann K.F. Volumetric error analysis of a Stewart platform based machine tool. Annals of the CIRP, 46/1/1997:287-290, 1997.
5. Ropponen T. and Arai T. Accuracy analysis of a modified Stewart platform manipulator. In IEEE Int. Conf. on Robotics and Automation, pages 521-525, Nagoya, May, 25-27, 1995.

Symbolic Calculation of Robots' Base Reaction-Force/Torque Equations with Minimal Parameter Set

Martin Grotjahn and Bodo Heimann

Institute of Mechanics, University of Hannover, D-30167 Hannover

Abstract. In this paper a new recursive formulation of the NEWTON-EULER method for tree-structured kinematic topologies is presented. The algorithm leads to a formulation which is linear with respect to the dynamic parameters. Starting from the recursive algorithm general simplification and regrouping rules are formulated for the reduction of the parameter vector to minimal dimension. The rules are only dependent on the kinematic parameters of each link. The algorithm is efficiently implementable in standard computer algebra programs. It is applied to the calculation of the base reaction force/torque equations which can be used for identification and control of robotic manipulators. The application to a 6-d.o.f. standard industrial robot shows the efficiency of the proposed method.

1 Introduction

Joint friction is a major problem in control of industrial robots. It can usually not be accurately predicted and compensated which is necessary for most advanced control algorithms. Hence, in Morel and Dubowksy (1996) the elimination of the friction influence is proposed by estimating the joint torques from measurements of the forces and torques exerted by the robot on its base. To implement this control scheme numerical values of the gravitational and inertial parameters have to be identified. In order to eliminate the perturbing friction influence the base force/torque sensor can also be used for identification Raucent et al.; Liu et al. (1992; 1998).

Common identification schemes, like the LEAST-SQUARES-algorithm, require a representation of the equations which is linear with respect to a parameter set of minimal dimension. Furthermore, for real time applications the computational burden of the equations has to be reduced as far as possible. Especially for robots with more than three d.o.f. the base force/torque equations become complex and the determination of the base parameter set is not trivial Liu et al. (1998).

Therefore, in this paper a new efficient algorithm is presented for symbolic calculation of the base force/torque equations and simultaneous determination of the base parameters. It is based on the NEWTON-EULER approach. Each link's contribution to the equations is calculated as a function of the contribution of the carrying link and the modified DENAVIT-HARTENBERG parameters. Linear dependencies are exploited to formulate general rules for the combination of the parameters of links i and $i - 1$. This can be done recursively link by link to determine the base parameters. The algorithm allows a more efficient analytic and numerical calculation of the equations than the standard NEWTON-EULER approach since determination and exploitation of possible simplifications is much easier.

2 Kinematics

The kinematics of a serial robot can be specified by the modified DENAVIT-HARTENBERG (MDH) notation Khalil and Kleinfinger (1986). Coordinate frame i is fixed to link i. The z_i-axis is the axis of joint i and the x_i-axis is the normal of z_i and z_{i+1}. Coordinate frame i is defined with respect to frame $i-1$ by the homogenous transformation matrix

$$T_i^{i-1} = \begin{bmatrix} R_i^{i-1} & {}_{i-1}r_i^{i-1} \\ 0\ 0\ 0 & 1 \end{bmatrix} = \begin{bmatrix} c_{\theta i} & s_{\theta i} & 0 & a_i \\ s_{\theta i}c_{\alpha i} & c_{\theta i}c_{\alpha i} & -s_{\alpha i} & -d_i s_{\alpha i} \\ s_{\theta i}s_{\alpha i} & c_{\theta i}s_{\alpha i} & c_{\alpha i} & d_i c_{\alpha i} \\ 0 & 0 & 0 & 1 \end{bmatrix} \quad (1)$$

which is a function of the MDH-parameters θ_i, d_i, α_i and a_i. The abbreviations s_x and c_x denote $\sin(x)$ and $\cos(x)$ respectively. The matrix R_i^{i-1} and the vector ${}_{i-1}r_i^{i-1}$ define orientation and position of frame i with respect to frame $i-1$.

Thus, velocity ${}_iv_i$ and angular velocity ${}_i\omega_i$ of link i and the corresponding accelerations can recursively be calculated by the following equations:

$$_iv_i = {}_iv_{i-1} + {}_i\tilde{\omega}_{i-1}r_i^{i-1} + \bar{\sigma}_i e_z \dot{q}_i \quad (2)$$

$$_i\dot{v}_i = {}_i\dot{v}_{i-1} + {}_i\dot{\tilde{\omega}}_{i-1}r_i^{i-1} + {}_i\tilde{\omega}_{i-1}({}_i\tilde{\omega}_{i-1}r_i^{i-1}) + \bar{\sigma}_i\left(\ddot{q}_i e_z + 2\dot{q}_i {}_i\tilde{\omega}_{i-1}e_z\right) \quad (3)$$

$$_i\omega_i = {}_i\omega_{i-1} + \sigma_i \dot{q}_i e_z \quad (4)$$

$$_i\dot{\omega}_i = {}_i\dot{\omega}_{i-1} + \sigma_i \dot{q}_i {}_i\tilde{\omega}_{i-1}e_z + \sigma_i \ddot{q}_i e_z \quad (5)$$

where $e_z = [0\ 0\ 1]^T$ and ${}_ia_{i-1} = R_{i-1}^i {}_{i-1}a_{i-1}$ with $R_{i-1}^i = (R_i^{i-1})^T$. The operator $\tilde{}$ denotes the crossproduct $\tilde{a}b = a \times b$. The joint variable q_i is given by $q_i = \bar{\sigma}_i d_i + \sigma_i \theta_i$ where $\sigma_i = 1$ for rotational, $\sigma_i = 0$ for translational joints and $\bar{\sigma}_i = 1 - \sigma_i$. If frame 0 is fixed the recursion is initialized by $v_0 = \omega_0 = \dot{\omega}_0 = 0$ and $\dot{v}_0 = -g$ where g is the vector of gravity.

3 Dynamic parameters

The dynamic parameters of each link i consist of its inertia tensor about the corresponding coordinate frame ${}_iI_i^i$, its first moment $s_i := [s_{xi}\ s_{yi}\ s_{zi}]^T = m_{ii}r_{Ci}^i$ (r_{Ci}^i: vector from coordinate frame to centre of mass) and its mass m_i. The combination of the independent scalar components in a vector leads to

$$p_i = [I_{xxi}\ I_{xyi}\ I_{xzi}\ I_{yyi}\ I_{yzi}\ I_{zzi}\ s_{xi}\ s_{yi}\ s_{zi}\ m_i]^T. \quad (6)$$

which results in the over-all parameter vector $p = \begin{bmatrix} p_1^T & p_2^T & \cdots & p_n^T \end{bmatrix}^T$ for a n d.o.f. robot.

4 NEWTON-EULER equations

The forces and torques exerted from link j to link $j+1$ can be obtained as a sum of the dynamic contributions of all following links:

$$_j f_{j,j+1} = \sum_{i=j+1}^{n} R_i^j \underbrace{\left[m_{ii} \dot{v}_i + {}_i\tilde{\dot{\omega}}_i s_i + {}_i\tilde{\omega}_i \left({}_i\tilde{\omega}_i s_i \right) \right]}_{= {}_i f_{j,i}},$$

$$_j m_{j,j+1} = \sum_{i=j+1}^{n} R_i^j \left[{}_i I_{ii}^i \dot{\omega}_i + {}_i\tilde{\omega}_i \left({}_i I_{ii}^i \omega_i \right) + \tilde{s}_{ii} \dot{v}_i + \tilde{r}_{ii}^j f_{j,i} \right]. \quad (7)$$

5 Recursive parameter linear form

Without loss of generality, all equations are derived with respect to the base wrenches w since they are used for identification and control of serial robots by base wrench measurements. The adaptation to the calculation of forces and torques at other positions is easily obtained by substituting R_i^0 and $_i r_i^0$.

5.1 Parameter linear formulation

Parameter linearity means that a matrix H exists such that

$$w = \begin{bmatrix} {}_0 f_{0,1} \\ {}_0 m_{0,1} \end{bmatrix} = Hp = \sum_{i=1}^{n} H_i p_i = \sum_{i=1}^{n} \frac{\partial w}{\partial p_i} p_i. \quad (8)$$

We define two new operators $()^*$ and $()^\circ$ with

$$\omega_i^* I_i^\circ := I_i \omega_i \quad (9)$$

where

$$\omega_i^* := \begin{bmatrix} \omega_{xi} & \omega_{yi} & \omega_{zi} & 0 & 0 & 0 \\ 0 & \omega_{xi} & 0 & \omega_{yi} & \omega_{zi} & 0 \\ 0 & 0 & \omega_{xi} & 0 & \omega_{yi} & \omega_{zi} \end{bmatrix} \quad (10)$$

and

$$I_i^\circ - [I_{xxi} \ I_{xyi} \ I_{xzi} \ I_{yyi} \ I_{yzi} \ I_{zzi}]^T. \quad (11)$$

With (7), (9) and $\tilde{a}b = -\tilde{b}a$, the base wrenches can be obtained by

$$w = \begin{bmatrix} {}_0 f_{0,1} \\ {}_0 m_{0,1} \end{bmatrix} = \begin{bmatrix} \sum_{i=1}^{n} R_i^0 \left[\left({}_i\tilde{\dot{\omega}}_i + {}_i\tilde{\omega}_{ii} \tilde{\omega}_i \right) s_i + {}_i \dot{v}_i m_i \right] \\ \sum_{i=1}^{n} R_i^0 \left[\left({}_i \dot{\omega}_i^* + {}_i\tilde{\omega}_{ii} \omega_i^* \right) ({}_i I_i^i)^\circ - {}_i\tilde{\dot{v}}_i s_i + {}_i\tilde{r}_{ii}^0 f_{0,i} \right] \end{bmatrix}. \quad (12)$$

Partial differentiation leads to

$$H_i = \begin{bmatrix} R_i^0 & 0 \\ R_i^0 \tilde{r}_i^0 & R_i^0 \end{bmatrix} \begin{bmatrix} 0 & {}_i\tilde{\dot{\omega}}_i + {}_i\tilde{\omega}_{ii} \tilde{\omega}_i & {}_i \dot{v}_i \\ {}_i \dot{\omega}_i^* + {}_i\tilde{\omega}_{ii} \omega_i^* & -{}_i\tilde{\dot{v}}_i & 0 \end{bmatrix}. \quad (13)$$

5.2 Dependency of H_i on H_{i-1}

Exploiting (2)-(5) H_i can be obtained from H_{i-1} in general by the following recursion:

$$H_i = H_{i-1} L_i + K_i. \tag{14}$$

Matrices L_i and K_i are given in the appendix. Although we use the NEWTON-EULER formulation, the matrix L_i is the transposed of the matrix depicted by Gautier and Khalil in Gautier and Khalil (1988) for the recursive formulation of LAGRANGIAN equations.

6 Determination of base parameters

For the determination of the minimal parameter set initially those parameters are set to zero which do not dynamically contribute to the base wrenches. Subsequently, the remaining parameters are regrouped to eliminate all linear dependencies.

6.1 Parameters without influence

If the contribution of the j-th parameter p_j to the base wrenches h_j is zero — where h_j is the j-th column of H — the parameter has no influence on the equations. This condition can usually be extended to

$$h_j = \text{constant} \tag{15}$$

since static influences are not measurable due to sensor offset and drift. Such parameters occur especially for lower links which do not have six d.o.f.

6.2 Regrouping conditions

If the contribution of a parameter p_j depends linearly on the contributions of some other parameters $p_{1j}, \ldots, p_{kj}$,

$$h_j = \sum_{l=1}^{k} a_{lj} h_{lj} \tag{16}$$

then p_j can be set to zero and the regrouped parameters p_{lj}^r can be obtained by

$$p_{lj}^r = p_{lj} + a_{lj} p_j . \tag{17}$$

6.3 Closed-form regrouping rules

General regrouping rules can be formulated by exploitation of the recursive calculation of H_i (14). If one column or a linear combination of columns of L_i is constant and the corresponding columns of K_i are zero columns, condition (16) is fulfilled and the parameters can be regrouped. This leads to the following rules which were formulated in Gautier and Khalil (1988) for LAGRANGIAN equations. Since the recursion for NEWTON-EULER derived in this paper has a similar structure the same regrouping rules can be formulated.

Rotational joints For rotational joints the joint variable is θ_i and the other MDH-parameters d_i, α_i and a_i are constant. This means that the ninth, the tenth and the sum of the first and fourth columns of L_i and K_i comply with the above conditions. Thus, the corresponding parameters I_{yyi}, s_{zi} and m_i can be eliminated by

$$\begin{bmatrix} I_{xx,i}^r \\ p_{i-1}^r \end{bmatrix} = \begin{bmatrix} I_{xx,i} \\ p_{i-1} \end{bmatrix} + \begin{bmatrix} -1 & 0 & 0 \\ L_{i,1}+L_{i,4} & L_{i,9} & L_{i,10} \end{bmatrix} \begin{bmatrix} I_{yyi} \\ s_{zi} \\ m_i \end{bmatrix} \quad (18)$$

where $L_{i,j}$ denotes the j-th column of L_i.

Translational joints For translational joints the joint variable is d_i which means that $L_{I,i}$ is constant and the first six columns of K_i are zero columns. Therefore, the moments of inertia can be added to those of the carrying link:

$$(I_{i-1}^r)^\circ = I_{i-1}^\circ + L_{I,i} I_i^\circ \iff p_{i-1}^r = p_{i-1} + \begin{bmatrix} L_{I,i} & 0 \\ 0 & 0 \end{bmatrix} p_i \quad . \quad (19)$$

This corresponds to $_{i-1}I_{i-1}^r = {_{i-1}}I_{i-1} + {_{i-1}}I_i$ since the orientation of link i is constant with respect to link $i-1$ for translational joints. In the case of $\omega_{i-1} = \|\omega_{i-1}\| e_{z,i}$ the ninth column of K_i becomes a zero column so that a further regrouping is possible by

$$p_{i-1}^r = p_{i-1} + L_{i,9} s_{zi} \quad . \quad (20)$$

Minimal formulation The regrouping is done link by link beginning with the n-th link. This leads to the formulation which is linear with respect to the base parameter vector p_{min} with minimal dimension m:

$$w = A(q, \dot{q}, \ddot{q}) p_{min} = \sum_{j=1}^{m} h_j p_j^r. \quad (21)$$

7 Application

The algorithm has been numerically verified for different kinematic topologies. Here, the application to the *manutec-r15*, a standard industrial robot with six rotational joints, is shown. The MDH parameters are given in tabular 1. For a more detailed description of the *manutec-r15* see Daemi and Heimann (1998). Due to the symmetric structure of the links the robot has only 31 non-zero dynamic parameters (see tabular 2). Since the last link is rotationally symmetrical $J_{6,yy}$ is equal to $J_{6,xx}$. Starting from this the described regrouping rules lead to the 15 base parameters depicted in tabular 3.

The analytic calculation of the base force/torque equations is implemented by using the computer algebra program MAPLE. Optimized C-code can be exported by MAPLE for use in real-time implementations. The presented algorithm leads to a representation consisting of 1301 additions and 1775 multiplications. The classical NEWTON-EULER approach results in 2756 additions and 4647 multiplications. The premature elimination of the redundant dynamic parameters

i	θ_i	d_i	a_i	α_i
1	$-q_1$	l_0	0	0
2	q_2	0	0	$-\frac{\pi}{2}$
3	q_3	0	l_1	0
4	$-q_4$	l_2	0	$\frac{\pi}{2}$
5	q_5	0	0	$-\frac{\pi}{2}$
6	q_6	0	0	$\frac{\pi}{2}$

Tab.1. MDH-parameters of the manutec-r15

p_1	p_2	p_3	p_4	p_5	p_6
$J_{1,xx}$	$J_{2,xx}$	$J_{3,xx}$	$J_{4,xx}$	$J_{5,xx}$	$J_{6,xx}$
$J_{1,yy}$	$J_{2,xz}$	$J_{3,yy}$	$J_{4,yy}$	$J_{5,yy}$	$J_{6,yy}$
$J_{1,zz}$	$J_{2,yy}$	$J_{3,zz}$	$J_{4,zz}$	$J_{5,zz}$	$J_{6,zz}$
s_{z1}	$J_{2,zz}$	s_{y3}	s_{z4}	s_{y5}	s_{z6}
m_1	s_{x2}	m_3	m_4	m_5	m_6
	s_{z2}				
	m_2				

Tab.2. Non-zero parameters according to (6) of the manutec-r15

$p_{min,r15}$
$I_{zz1} + I_{yy2} + I_{yy3} + l_1^2(m_3 + m_4 + m_5 + m_6)$
s_{z2}
$I_{xx2} - I_{yy2} - l_1^2(m_3 + m_4 + m_5 + m_6)$
I_{xz2}
$I_{zz2} + l_1^2(m_3 + m_4 + m_5 + m_6)$
$s_{x2} + l_1(m_3 + m_4 + m_5 + m_6)$
$I_{xx3} + I_{yy4} + 2l_2 s_{z4} + l_2^2(m_4 + m_5 + m_6) - I_{yy3}$
$I_{zz3} + I_{yy4} + 2l_2 s_{z4} + l_2^2(m_4 + m_5 + m_6)$
$s_{y3} - s_{z4} - l_2(m_4 + m_5 + m_6)$
$I_{xx4} + I_{yy5} - I_{yy4}$
$I_{zz4} + I_{yy5}$
$I_{xx5} + I_{xx6} - I_{yy5}$
$I_{zz5} + I_{xx6}$
$s_{y5} - s_{z6}$
I_{zz6}

Tab.3. Resulting base parameter set of the *manutec-r15*.

reduces the numbers of operations to 1332 additions and 2478 multiplications. These three approaches should theoretically yield the same results. But due the complexity of the equations MAPLE is not able to execute all possible simplifications automatically. Therefore, it is important to reduce the complexity as much as possible in advance. The presented algorithm offers the opportunity to simplify each link's contribution H_i before using it for calculation of the next link's contribution H_{i+1} according to (14). This leads to higher efficiency in comparison to the classical NEWTON-EULER approach.

8 Conclusions

In this paper a new recursive formulation of the NEWTON-EULER method for tree-structured kinematic topologies is presented. It leads to a formulation which is linear with respect to a dynamic parameter set of minimal dimension. It is applied to the calculation of the base reaction force/torque equations of standard industrial robots. The application to a six d.o.f. standard industrial robot shows the computational efficiency of the proposed method.

References

Daemi, M., and Heimann, B. (1998). Separation of friction and rigid body identification for industrial robots. In *Proc. of the 12th CISM-IFToMM Symp. on the Theory and Practice of Robots and Manipulators*, 35–42.

Gautier, M., and Khalil, W. (1988). A direct determination of minimum inertial parameters of robots. In *Proc. of the IEEE Int. Conf. on Robotics and Automation*, 1682–1687.

Khalil, W., and Kleinfinger, J. F. (1986). A new geometric notation for open and closed-loop robots. In *Proc. of the IEEE Int. Conf. on Robotics and Automation*, volume 2, 1174–1179.

Liu, G., Iagnemma, K., Dubowsky, S., and Morel, G. (1998). A base force/torque sensor approach to robot manipulator inertial parameter estimation. In *Proc. of the IEEE International Conference on Robotics and Automation*, 3316–3321.

Morel, G., and Dubowksy, S. (1996). The precise control of manipulators with joint friction: A base force/torque sensor method. In *Proc. of the IEEE Int. Conf. on Robotics and Automation*, volume 1, 360–365.

Raucent, B., Campion, G., Bastin, G., Samin, J., and Willems, P. (1992). Identification of the barycentric parameters of robot manipulators from external measurements. *Automatica* 28(5):1011–1016.

Appendix

$$L_i = \begin{bmatrix} L_{I,i} & L_{s,i} & L_{m,i} \\ 0 & T_i^{i-1} & \end{bmatrix} \tag{22}$$

$$K_i = \begin{bmatrix} R_i^0 & [0 & \sigma_i K_s & \bar{\sigma}_i K_m] \\ R_i^0 & [\sigma_i K_I & \sigma_{i_i} r_i^0 K_s - \bar{\sigma}_i \tilde{K}_m & \bar{\sigma}_{i_i} \tilde{r}_i^0 K_m] \end{bmatrix} \tag{23}$$

In the following the index i is omitted for simplicity, if it is not necessary for understanding.

$$L_I = \begin{bmatrix} c_\theta^2 & -2c_\theta s_\theta & 0 & s_\theta^2 & 0 & 0 \\ c_\alpha c_\theta s_\theta & c_\theta^2 c_\alpha - s_\theta^2 c_\alpha & -s_\alpha c_\theta & -c_\alpha c_\theta s_\theta & s_\theta s_\alpha & 0 \\ s_\alpha c_\theta s_\theta & c_\theta^2 s_\alpha - s_\theta^2 s_\alpha & c_\alpha c_\theta & -s_\alpha c_\theta s_\theta & -s_\theta c_\alpha & 0 \\ s_\theta^2 c_\alpha^2 & 2c_\theta s_\theta c_\alpha^2 & -2s_\theta c_\alpha s_\alpha & c_\alpha^2 c_\theta^2 & -2c_\theta c_\alpha s_\alpha & s_\alpha^2 \\ s_\theta^2 c_\alpha s_\alpha & 2c_\theta s_\theta c_\alpha s_\alpha & s_\theta c_\alpha^2 - s_\theta s_\alpha^2 & c_\alpha s_\alpha c_\theta^2 & c_\theta c_\alpha^2 - c_\theta s_\alpha^2 & -c_\alpha s_\alpha \\ s_\theta^2 s_\alpha^2 & 2c_\theta s_\theta s_\alpha^2 & 2s_\theta c_\alpha s_\alpha & s_\alpha^2 c_\theta^2 & 2c_\theta c_\alpha s_\alpha & c_\alpha^2 \end{bmatrix},$$

$$L_s = \begin{bmatrix} 0 & 0 & 2d & d^2 \\ dc_\theta s_\alpha - as_\theta c_\alpha & -ac_\theta c_\alpha - ds_\theta s_\alpha & as_\alpha & ads_\alpha \\ -dc_\theta c_\alpha - as_\theta s_\alpha & -ac_\theta s_\alpha + ds_\theta c_\alpha & -ac_\alpha & -adc_\alpha \\ 2ac_\theta + 2ds_\theta c_\alpha s_\alpha & -2as_\theta + 2dc_\theta c_\alpha s_\alpha & 2dc_\alpha^2 & d^2 c_\alpha^2 + a^2 \\ ds_\theta \left(s_\alpha^2 - c_\alpha^2\right) & dc_\theta \left(s_\alpha^2 - c_\alpha^2\right) & 2dc_\alpha s_\alpha & d^2 c_\alpha s_\alpha \\ 2ac_\theta - 2ds_\theta c_\alpha s_\alpha & -2as_\theta - 2dc_\theta c_\alpha s_\alpha & 2ds_\alpha^2 & d^2 s_\alpha^2 + a^2 \end{bmatrix}, \quad L_m = \begin{bmatrix} d^2 \\ ads_\alpha \\ -adc_\alpha \\ d^2 c_\alpha^2 + a^2 \\ d^2 c_\alpha s_\alpha \\ d^2 s_\alpha^2 + a^2 \end{bmatrix}.$$

$$K_I = \begin{bmatrix} \dot{q}_i\,_i\omega_{i-1,y} & -2\dot{q}_i\,_i\omega_{i-1,x} & \ddot{q}_i & -\dot{q}_i\,_i\omega_{i-1,y} & -\dot{q}_i^2 - 2\dot{q}_i\,_i\omega_{i-1,z} & \dot{q}_i\,_i\omega_{i-1,y} \\ \dot{q}_i\,_i\omega_{i-1,x} & 2\dot{q}_i\,_i\omega_{i-1,y} & \dot{q}_i^2 + 2\dot{q}_i\,_i\omega_{i-1,z} & -\dot{q}_i\,_i\omega_{i-1,x} & \ddot{q}_i & -\dot{q}_i\,_i\omega_{i-1,x} \\ 0 & 0 & 0 & 0 & 0 & \ddot{q}_i \end{bmatrix}$$

$$K_s = \begin{bmatrix} -\dot{q}_i^2 - 2\dot{q}_i\,_i\omega_{i-1,z} & -\ddot{q}_i & 0 \\ \ddot{q}_i & -\dot{q}_i^2 - 2\dot{q}_i\,_i\omega_{i-1,z} & 0 \\ 2\dot{q}_i\,_i\omega_{i-1,x} & 2\dot{q}_i\,_i\omega_{i-1,y} & 0 \end{bmatrix}, \quad K_m = \begin{bmatrix} 2\dot{q}_i\,_i\omega_{i-1,y} \\ -2\dot{q}_i\,_i\omega_{i-1,x} \\ \ddot{q}_i \end{bmatrix},$$

where $_i\omega_{i-1} = \begin{bmatrix} _i\omega_{i-1,x} \\ _i\omega_{i-1,y} \\ _i\omega_{i-1,z} \end{bmatrix} = R_{i-1}^i\,_{i-1}\omega_{i-1}$.

Mechanics of the New UWA Robot

Karol Miller

Department of Mechanical and Materials Engineering, The University of Western Australia

Abstract. New University of Western Australia Robot is a variation of the well-known Delta parallel robot. Parallel manipulators possess a number of advantages when compared to traditional serial arms. They offer generally much higher rigidity and smaller mobile mass than their serial counterparts. These features allow much faster and more precise manipulations. The main disadvantage of parallel robots is their small workspace in comparison to serial arms of similar size. Our previous work investigated the influence of motor axes orientation on the workspace volume of Delta type manipulators. This research has shown that the Delta configuration is not optimal and that the configuration characterized by $\alpha = ArcTan(1/\sqrt{2}) \approx 35.26$ degrees - the inclination of each motor axis to the horizontal plane (for Delta robot $\alpha = 0$ degrees), and $\beta=60$ degrees - rotation of each motor with respect to the vertical axis (for Delta robot $\beta=0$ degrees), known as the New University of Western Australia Robot – NUWAR – may be advantageous. This contribution presents NUWAR's workspace volume and shape as well as results in kinematics of NUWAR. It was shown, that like for the Delta, there existed analytical closed form solutions of inverse as well as forward kinematics problems. Simulations of trajectories of various shapes (straight line, ellipse, sheared ellipse, cycloid, etc.) confirmed the appropriateness of methods used. The model of kinematics was implemented in the comprehensive Matlab simulation suite, which enables a user to specify trajectories in Cartesian or joint space, and conduct kinematic analyses of the manipulator. The NUWAR direct drive prototype was constructed and is now capable of achieving accelerations of 600 m/s^2.

1 Introduction

Parallel robots, unlike serial, traditional ones, have the end effector connected to the base by several kinematic chains in parallel. Research into the field of parallel robots documented in the literature dates back to the year 1938, when Pollard patented his mechanism for car painting (Pollard, 1938). In 1947 Mc Cough proposed a 6-degree-of-freedom platform, which was later used by Stewart in his flight simulator (Stewart, 1966). Parallel manipulators are particularly suited to a number of typical industrial applications and have presented a lot of interest to various researchers over the years. In recent years several new structures and mechanisms have been proposed, developed for a variety of both established and novel applications, such as packaging, assembly, haptic interfaces, etc. (Clavel, 1988; Pierrot et al., 1991; Badano et al., 1993; Arai and Tanikawa, 1996; Tsumaki et al. 1998).

Parallel manipulators possess a number of advantages when compared to traditional serial arms. They offer generally a much higher rigidity and smaller mobile mass than their serial counterparts. These features allow much faster and more precise manipulations. A catalogue of a large variety of parallel configurations can be found in the book by Merlet (1990) and at his Internet page (Merlet, 2000).

The New University of Western Australia robot (Miller, 1998) (Fig. 1) is an original design that arose from the need for a manipulator, which, while retaining the qualities of the Delta Robot, would have a larger workspace and better torsional stiffness. As seen from Figure 1, the three closed kinematic chains consisting of arms and parallel rods, are identical and actuated by three revolute electric motors rigidly mounted to the top (robot base), and closed below at the common tool-base by revolute joints. The combination of the constrained motion of these three chains results in three translatory degrees of freedom for the robot tool-base. The structure is very rigid (approx. 120Hz lowest natural frequency). Capable of achieving **600 m/s²**, the direct drive NUWAR robot is one of the fastest robots in the world.

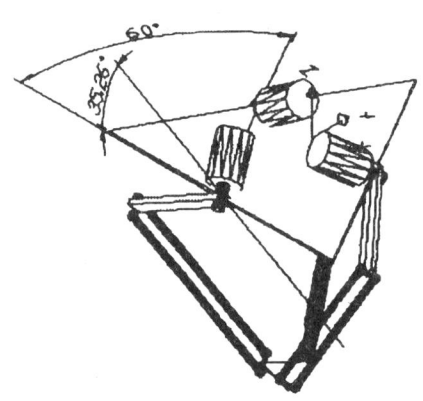 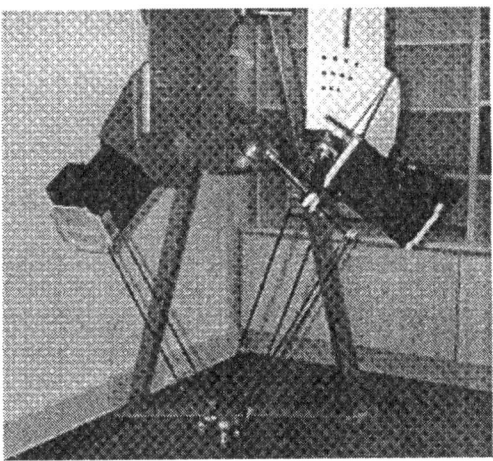

Figure 1. a) Layout of New University of Western Australia Robot; b) NUWAR prototype

As can be seen, NUWAR design is markedly different from previously proposed three-degree-of-freedom parallel manipulators by having the motor axes out of the horizontal plane. The choice of angles $\alpha = ArcTan(1/\sqrt{2}) \approx 35.26$ degrees makes the motor axes orthogonal in three-dimensional space, and $\beta = 60$ – parallel to opposite sides of the base triangle.

2 Computer Models of NUWAR Robot

2.1 Workspace Analysis

NUWAR's arm/forearm assemblies are identical, and located symmetrically on sides of an equilateral triangle. In all models the Z-axis is perpendicular to the base plane and points vertically up.

Mechanics of the New UWA Robot

The X and Y axes are located in the baseplane, with the X axis pointing towards motor #1. The origin is located at the point equidistant to the three motors.

The computer models of the robots were written parametrically so that different configurations can easily be analyzed. Mechanical simulation package ADAMS (1993) and AutoCad/AutoLisp (1997) were used. In addition to the dimensions of each component, two design variables - the angles α and β - were introduced. These two angles determine the orientation of each motor axis (i.e. the controlled revolute joint axes), which connect the arms to the base. As the arm is always perpendicular to it's motor axis, the angles α and β control the orientation of the three arm/forearm assemblies. Varying α and β angles leaves the locations of the three motors unchanged.

All dimensions have been chosen as in DELTA-740 (Guglielmetti, 1994):
La= length of the arm from joint to joint - 0.26m
Lfa= length of the forearm from joint to joint - 0.48m
Rb= radial distance of each motor from the centerline - 0.194m
Rtp= displacement in radial direction of the midpoint of each pair of spherical joints on the traveling plate - 0.03m
spfa= spacing of the forearms - 0.05m
phi= angular limit of lateral rotation of the revolute joints which join the arms to the forearms - 60 degrees, and
$\alpha = ArcTan(1/\sqrt{2}) \approx 35.26$, $\beta=60$ degrees.

The robot's workspace was derived using AutoCAD. A generalized program was written using AutoLISP in terms of the design variables mentioned above. Additionally a parametric model of the robots was constructed using ADAMS. The ADAMS models contain eleven rigid parts. These are: ground (i.e. the base of the robot), three arms, six forearms, and the traveling plate. ADAMS constraints were used to define the relationship between various parts so that the model is not overconstrained and it does not possess additional "internal" degrees of freedom. The constraints used are: revolute joints (1 DOF, to define the relationship between the base and each arm), universal joints (2 DOF, to represent the relationship between each forearm and its parent arm), spherical joints (3 DOF rotation) are used to define the relationship between the forearms and the traveling plate.

The volumes of the workspace on one side of the baseplane were calculated for the Delta ($\alpha=0$, $\beta=0$) and NUWAR configurations, and the shapes of the workspace were observed and analyzed. The workspace of NUWAR is 9.4% larger than the workspace of the Delta, assuming all dimensions being the same. Close examination of Figures 2 and 3 reveals that the NUWAR's workspace has a larger cross section in the region closest to the baseplane ($z<0.35$ m), but a smaller cross section in the other half of the workspace ($z>0.35$ m). Hence this robot would be more suitable for applications where a larger workspace is required close to the baseplane.

2.2 Kinematics

Forward kinematics problem
The solution of the forward kinematics problem for a three-degree-of-freedom parallel manipulator amounts to finding the intersection of three spheres (e.g. Pierrot, et al. 1991;

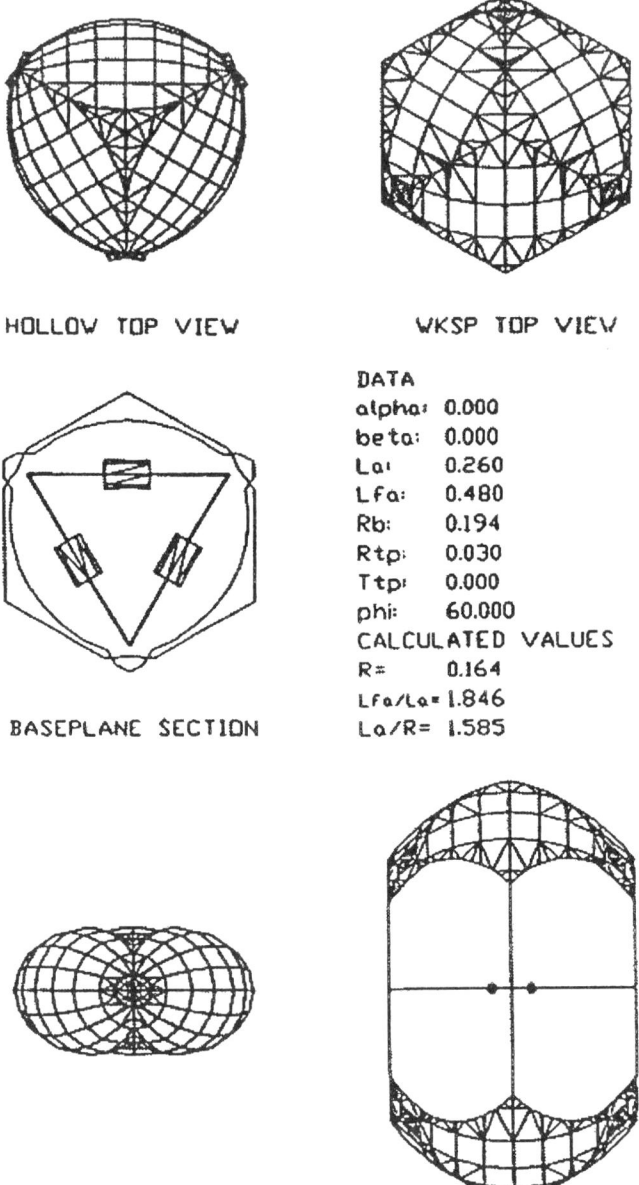

Figure 2. Workspace of Delta ($\alpha=0$, $\beta=0$).

Mechanics of the New UWA Robot

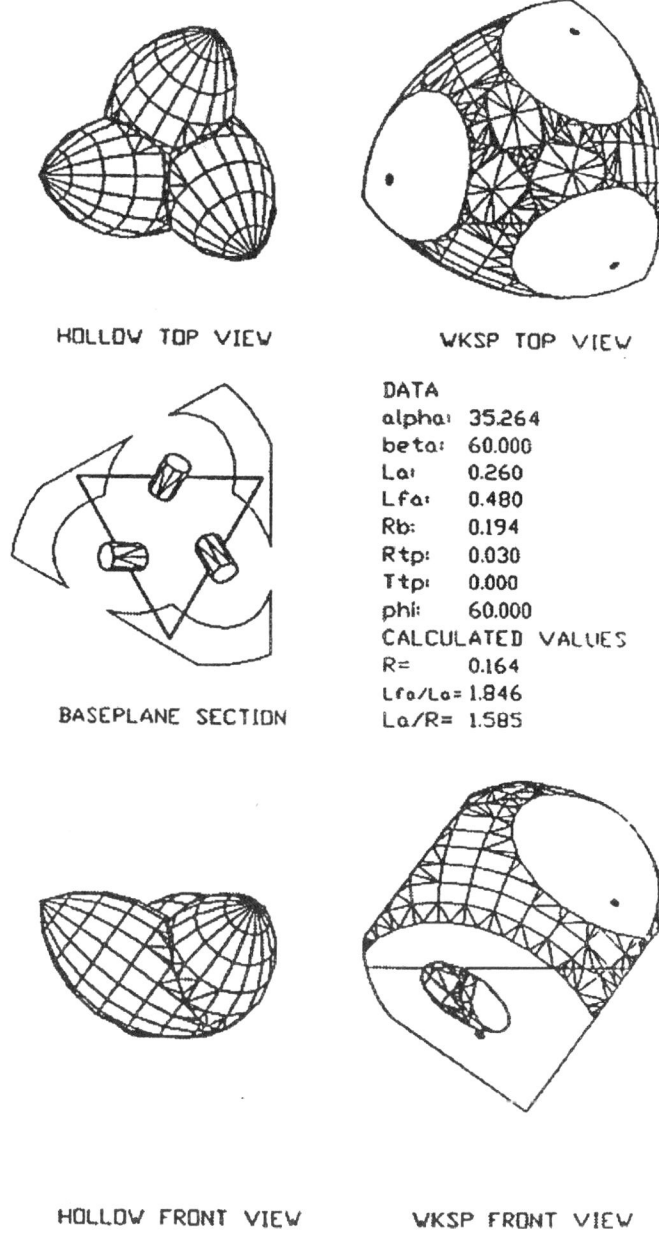

Figure 3. Workspaces of NUWAR ($\alpha=35.26°$, $\beta=60°$).

Tsai, et al. 1996). This requires the solution of a quadratic. The method used in (Sternheim, 1987) allows for the desired solution to be selected in its process, bypassing the need for choosing the appropriate root of the polynomial equation. Therefore, we decided to adapt this method to NUWAR. The details of the simple derivation are omitted here. The reader is referred to the original paper (Sternheim, 1987). The application of the method to the NUWAR involves only accounting for motor axes rotation by angles α and β.

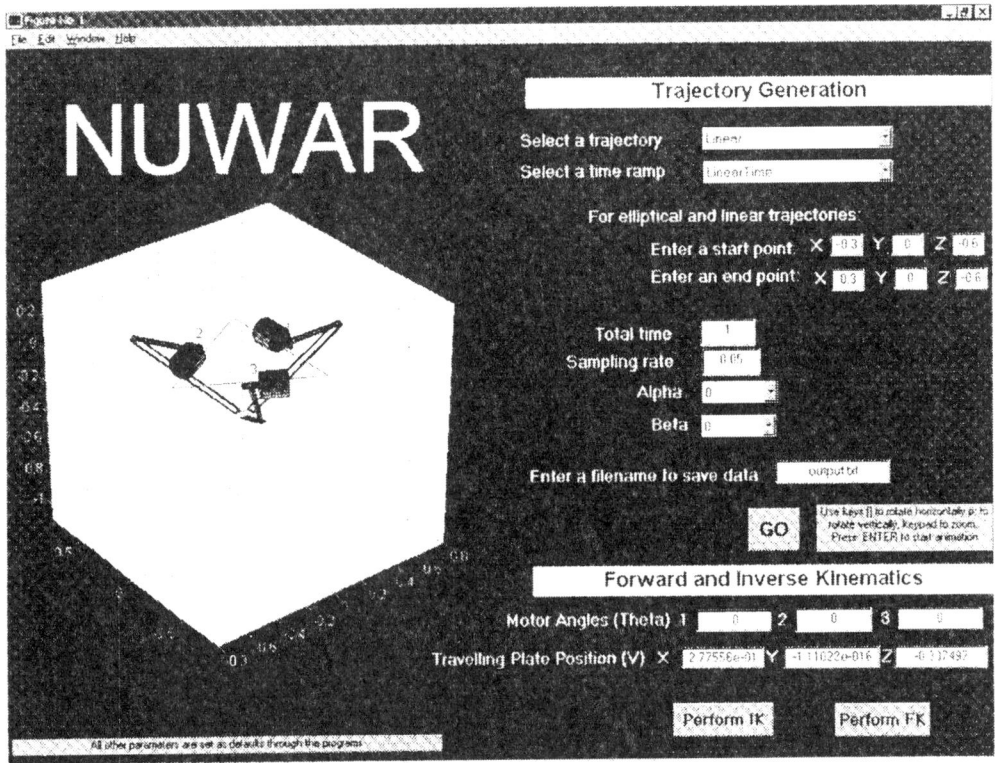

Figure 4. New University of Western Australia Robot graphical user interface

Inverse kinematics problem

There are two solutions for each controlled arm angle for a given position of the travelling plate. In the case of NUWAR it is desirable to keep the controlled arm angles between –90 and 90 degrees. As for the forward kinematics, the algebraic method leads to quadratic equations (Pierrot, et al. 1991, Tsai, et al. 1996). Again, we chose the method of (Sternheim, 1987), which takes a direct geometric approach. With this algorithm one can easily choose the solution, which minimises the possibility of the damage to the robot. The kinematics algorithms were programmed in Matlab and are accessible through a user-friendly graphical interface, Figure 4. The simulation suite allows for calculating motor angles for given position of the end-effector and the position of the end-effector given the motor angles, as well as simulation of linear, elliptical and cycloidal trajectories with various time profiles.

3 Conclusions and Discussion

This paper presents the results of numerical investigation of the New University of Western Australia Robot – an optimally structured Delta-type parallel robot. Our results show that the configuration of the New University of Western Australia Robot - NUWAR is advantageous over the Delta configuration in terms of workspace volume. The shape of the workspace of NUWAR is beneficial for applications involving manipulations relatively close to the baseplane.

The direct as well as inverse kinematics problems were solved and incorporated into a comprehensive simulation suite accessible through a convenient graphical user interface.

The results of numerical simulations have formed a basis for a construction of a prototype, which is now capable of achieving accelerations of 600 m/s^2 !

4 Acknowledgments

The author would like to thank final and third year students who helped in the construction of the prototype, in particular Lyndsey White who programmed the graphical user interface for NUWAR. The financial support of the Australian Research Council is gratefully acknowledged.

References

ADAMS Reference Manual (1993). Version 7.0, Mechanical Dynamics, Ann Arbor, Michigan, USA.
Arai, T. and Tanikawa, T. (1996). Development of a new parallel manipulator with fixed linear actuator. *Proceedings of Japan/USA Symposium on Flexible Automation.* Vol.1, ASME, pp.145-149.
AutoCad (1997) Release 14 Users Guide, Autodesk, Inc., USA.
Badano, F., Betempts, M, Burckhardt, C.,W., Clavel, R., and Jutard, A. (1993). Assembly of Chamferless Parts using a Fast Robot. *Proceedings of 24th International Symposium on Industrial Robots,* Tokyo, pp. 89-96.
Clavel, R. (1988). Delta, a fast robot with parallel geometry. *Proceedings of 18th International Symposium on Industrial Robots,* Lausanne, Switzerland, pp.91-100.
Guglielmetti, P. (1994). *Model-Based Control of Fast Parallel Robots: a Global Approach in Operational Space.* PhD Thesis no. 1228, Swiss Federal Institute of Technology, Lausanne.
Merlet, J.P. (1990). *Les robots paralleles.* Hermes, Paris.
Merlet, J.P. (2000). http://www.inria.fr/saga/personnel/merlet/merlet.html
Miller, K. (1998). *Parallel Robot.* Australian Provisional Patent Application.
Pierrot, F., Dauchez, P., and Fournier, A. (1991). Hexa, a Fast 6-Degree of Freedom Fully Parallel Robot. *Proceedings of International Conference on Advanced Robotics.* Pisa, Vol. 2/2, pp.1158-1163.
Pierrot, F., Dauchez, P., and Fournier, A. (1991). Fast Parallel Robots. *Journal of Robotic Systems,* 8(6), pp.829-840.
Pollard, W.L.V. (1938). *Position Controlling Apparatus.* US patent.
Sternheim, F. (1987). Computation of the direct and inverse geometric models of the Delta-4 parallel robot. *Robotersysteme* 3, pp. 199-203.
Stewart, D. (1966). A Platform with six degrees of freedom. *Proceedings of the Institution of Mechanical Engineers,* vol. 180, part I, no. 15, pp. 371-386.

Tsai, L.-W., Walsh, G., C., Stamper, R., E. (1996). Kinematics of a Novel Three DOF Translational Platform. *Proceedings of IEEE International Conference on Robotics and Automation,* pp. 3446-3451

Tsumaki, Y., Narusew, H., Nenchev, D., N., and Uchiyama, M. (1998) Design of compact 6-dof interface. *Proceedings of IEEE Conference on Robotics and Automation,* pp. 2581-2585

On Kinematic Singularities of Nonholonomic Robotic Systems

Krzysztof Tchoń

Institute of Engineering Cybernetics,
Wrocław University of Technology, Wrocław, Poland

Abstract. Using a control system representation of kinematics we define and investigate posture and configuration singularities of nonholonomic robotic systems. A significance of these singularities and their interdependence are illustrated with an example of a car pulling two trailers.

1 Introduction

The knowledge of kinematic singularities has fundamental significance for motion planning and control of robotic manipulators (holonomic systems). In recent years this knowledge has increased considerably [1]. The problem of kinematic singularities of nonholonomic robotic systems seems to be equally important for motion planning and control of these systems [2–5]. A viewpoint on kinematic singularities usually assumed in robotics focuses on posture singularities of nonholonomic systems, [2,4,6], that correspond to a nonuniform structure of the growth vector of the system distribution [3,7]. On the other hand, the fundamental invariants of this distribution are its so-called abnormal curves [8,9], determined by singular controls of a nonholonomic system. Recently, it has been revealed that both these types of singularities are connected [10].

In this paper we intend to undertake a systematic study of kinematic singularities of nonholonomic robotic systems from a control theory perspective. We begin with representing the kinematics of a nonholonomic system as a driftless control system. Next, relying on geometric properties of a distribution associated with this system, we define posture singularities. Eventually, by interpreting the kinematics of a nonholonomic system in terms of the reachability map of this control system, we propose a natural concept of singular configurations that correspond to singular controls of the system. Both types of singularities are illustrated with an example of a car pulling two trailers. In this example, besides "trivial" singular configurations making the system immobile, there exist "non-trivial" singular configurations in which a motion of the system preserves singular postures.

This paper has been composed as follows. In Section 2 we introduce basic concepts. Section 3 contains definitions of singular postures and configurations. Section 4 is devoted to examples. The paper is concluded with Section 5.

2 Basic Concepts

Consider a robotic system described by generalized coordinates $\mathbf{q} \in \mathbb{R}^n$ and velocities $\dot{\mathbf{q}} \in \mathbb{R}^n$, subject to $l \leq n$ independent phase constraints in the Pfaffian form

$$\mathcal{A}(\mathbf{q})\dot{\mathbf{q}} = \mathbf{0} \ . \tag{1}$$

The constraint matrix $\mathcal{A}(\mathbf{q})$ of size $l \times n$ has full rank and depends on $\mathbf{q}$ analytically. Geometrically, the constraints (1) mean that at every point $\mathbf{q} \in \mathbb{R}^n$ admissible velocities of the system should belong to the null space of $\mathcal{A}(\mathbf{q})$, i.e. $\dot{\mathbf{q}} \in Ker\,\mathcal{A}(\mathbf{q})$. Let $\mathbf{g}_1(\mathbf{q}), \ldots, \mathbf{g}_m(\mathbf{q})$, $m = n - l$, denote analytic vector fields that span $Ker\,\mathcal{A}(\mathbf{q})$. Then, the property $\dot{\mathbf{q}} \in Ker\,\mathcal{A}(\mathbf{q})$ can be rephrased equivalently as that a trajectory of the robotic system is defined by the following driftless control system

$$\dot{\mathbf{q}} = G(\mathbf{q})\mathbf{u} = \sum_{i=1}^m \mathbf{g}_i(\mathbf{q}) u_i \ . \tag{2}$$

The control functions $\mathbf{u}(\cdot)$, driving this system, will be assumed Lebesgue square integrable on a time interval $[0, T]$, $\mathbf{u}(\cdot) \in L_m^2[0, T]$.

We recall that the phase constraints (1) are called *holonomic*, if there exists an analytic map $h : \mathbb{R}^n \longrightarrow \mathbb{R}^k$, $k \leq l$, such that

$$\mathcal{A}(\mathbf{q})\dot{\mathbf{q}} = \mathbf{0} \iff h(\mathbf{q}) = \mathbf{0} \ . \tag{3}$$

If no such a map exists, the constraints are *nonholonomic*. A sufficient condition for nonholonomicity can be expressed in terms of a Lie algebra $\mathcal{L}$ associated with the system (2), defined as the smallest linear space of vector fields containing the system vector fields $\mathbf{g}_1, \ldots, \mathbf{g}_m$, and closed with respect to taking the Lie bracket of vector fields,

$$[\mathbf{X}, \mathbf{Y}](\mathbf{q}) = \frac{\partial \mathbf{Y}}{\partial \mathbf{q}} \mathbf{X}(\mathbf{q}) - \frac{\partial \mathbf{X}}{\partial \mathbf{q}} \mathbf{Y}(\mathbf{q}) \ .$$

Given this Lie algebra $\mathcal{L}$, the constraints are nonholonomic, if at any $\mathbf{q} \in \mathbb{R}^n$, $\mathcal{L}$ satisfies the *rank condition*

$$\dim \mathcal{L}(\mathbf{q}) = n \ . \tag{4}$$

A fundamental geometric characterization of the control system (2) is provided by a system distribution

$$\mathcal{G} = span_{C^\omega} \{\mathbf{g}_1, \ldots, \mathbf{g}_m\}, \tag{5}$$

and an associated sequence of distributions $\mathcal{G}^0 \subset \mathcal{G}^1 \subset \cdots \subset \mathcal{G}^j \subset \cdots$, defined as

$$\mathcal{G}^0 = \mathcal{G}, \quad \mathcal{G}^{j+1} = \mathcal{G}^j + [\mathcal{G}, \mathcal{G}^j], \quad j = 0, 1, \ldots, \tag{6}$$

that is called a filtration of the distribution $\mathcal{G}$ [3]. Let us choose a point $\mathbf{q} \in \mathbb{R}^n$, and compute dimensions $r_j(\mathbf{q}) = \dim \mathcal{G}^j(\mathbf{q})$, $j = 0, 1, \ldots$, of linear subspaces assigned to the point $\mathbf{q}$ by distributions constituting the filtration. It is clear that the sequence $r_0(\mathbf{q}) \leq r_1(\mathbf{q}) \leq \ldots \leq r_j(\mathbf{q}) \leq \ldots \leq n$ will stabilize after a number $p(\mathbf{q})$ of steps at a certain value $r_{p(\mathbf{q})}(\mathbf{q})$. The vector

$$\mathbf{r}(\mathbf{q}) = (r_0(\mathbf{q}), \ldots, r_{p(\mathbf{q})}(\mathbf{q})) \tag{7}$$

is called a growth vector of the distribution $\mathcal{G}$ at $\mathbf{q}$. The distribution $\mathcal{G}$ is regular, if its growth vector is the same at every point, i.e. $r_j(\mathbf{q}) = r_j$ and $p(\mathbf{q}) = p$. In this case the vector

$$\mathbf{r} = (r_0, r_1, \ldots, r_p) \tag{8}$$

is named a growth vector of the distribution $\mathcal{G}$. Clearly, if $r_p = n$, the constraints (1) are nonholonomic.

From now on we shall assume that the system (2) satisfies the rank condition (4), so the constraints (1) are nonholonomic. The corresponding robotic system will also be referred to as nonholonomic. For a nonholonomic system the number $p(\mathbf{q})$ such that $r_{p(\mathbf{q})} = n$ is called a *degree of nonholonomy* at the point $\mathbf{q}$. If the degree of nonholonomy is independent of $\mathbf{q}$, we call it a degree of nonholonomy of the system (2).

The driftless control system (2) represents the kinematics of a nonholonomic robotic system. Its state variables $\mathbf{q} \in \mathbb{R}^n$ play the role of *postures* of the system. For every fixed initial posture $\mathbf{q}_0$ and a chosen control function $\mathbf{u}(\cdot) \in L^2_m[0, T]$, the posture assumed by the system at the time instant T can be computed through the reachability map of the system (2),

$$\mathbf{q}(T) = \varphi_{\mathbf{q}_0, T}(\mathbf{u}(\cdot)) = \mathbf{k}_{\mathbf{q}_0, T}(\mathbf{u}(\cdot)) \ . \tag{9}$$

The map

$$\mathbf{k}_{\mathbf{q}_0, T} : L^2_m[0, T] \longrightarrow \mathbb{R}^n, \tag{10}$$

defined by the expression (9), corresponds to the forward kinematic map of a robotic manipulator, and therefore may be called the *kinematics* of the nonholonomic system. By the rank condition, it follows that the system (2) is completely controllable and small time locally controllable. The first property means that any desired posture $\mathbf{q}$ is reachable from any initial $\mathbf{q}_0$, by applying appropriate control functions, within an arbitrarily small time. Mathematically, this implies that the kinematic map (10) is surjective. The second property asserts that at any posture the system (2) can be driven in all basic directions in $\mathbb{R}^n$ for any $T > 0$.

Respecting a formal analogy between the kinematics (10) and the kinematics of a stationary manipulator, it is justified to regard control functions $\mathbf{u}(\cdot)$ as *configurations* of the nonholonomic robotic system. Pursuing further this analogy we shall introduce the analytic Jacobian

$$J_{\mathbf{q}_0, T}(\mathbf{u}(\cdot)) : L^2_m[0, T] \longrightarrow \mathbb{R}^n \tag{11}$$

of the nonholonomic system as the Gâteaux differential of the kinematics $\mathbf{k}_{\mathbf{q}_0,T}$,

$$J_{\mathbf{q}_0,T}(\mathbf{u}(\cdot))\mathbf{v}(\cdot) = \frac{d}{ds}|_{s=0}\mathbf{k}_{\mathbf{q}_0,T}(\mathbf{u}(\cdot) + s\mathbf{v}(\cdot)) = \int_0^T \Phi(T,t)B(t)\mathbf{v}(t)dt \ . \quad (12)$$

In control theoretic terminology, the analytic Jacobian determines the state reachable at T in the linear time-dependent control system

$$\dot{\boldsymbol{\xi}} = A(t)\boldsymbol{\xi} + B(t)\mathbf{v}(t) \quad (13)$$

associated with the system (2) (a variational system), initialized at $\boldsymbol{\xi}(0) = 0$. This linear system represents a linear approximation to (2) along a given pair $(\mathbf{u}(t), \mathbf{q}(t))$, so

$$A(t) = \frac{\partial\left(G(\mathbf{q}(t))\mathbf{u}(t)\right)}{\partial \mathbf{q}}, \quad B(t) = G(\mathbf{q}(t)), \quad (14)$$

and the matrix $\Phi(t,s)$ appearing under the integral in (12) satisfies a differential equation

$$\frac{d}{dt}\Phi(t,s) = A(t)\Phi(t,s), \quad \Phi(s,s) = \mathbb{1}_n \ . \quad (15)$$

Let us invoke once again the robotic manipulator analogy. It is well known that a basic tool of solving the inverse kinematic problem for robotic manipulators is a Jacobian pseudoinverse enabling one to find out a solution of an underdetermined Jacobian equation $J_{\mathbf{q}_0,T}(\mathbf{q})\mathbf{v} = \mathbf{w}$. Solving an analogous Jacobian equation for the nonholonomic system, $J_{\mathbf{q}_0,T}(\mathbf{u}(\cdot))\mathbf{v}(\cdot) = \mathbf{w}$, that contains the analytic Jacobian defined by (12), amounts to computing a control function $\mathbf{v}(\cdot)$ in the system (13) with zero initial condition, such that $\boldsymbol{\xi}(T) = \mathbf{w}$. A solution of this problem, obtained within the control theory, takes the form

$$\mathbf{v}(t) = B^T(t)\Phi^T(T,t)M_{\mathbf{q}_0,T}^{-1}(\mathbf{u}(\cdot))\mathbf{w} = \left(J_{\mathbf{q}_0,T}^{\#}(\mathbf{u}(\cdot))\mathbf{w}\right)(t), \quad (16)$$

and actually defines a Jacobian pseudoinverse for the nonholonomic system. The matrix

$$M_{\mathbf{q}_0,T}(\mathbf{u}(\cdot)) = \int_0^T \Phi(T,t)B(t)B^T(t)\Phi^T(T,t)dt \quad (17)$$

present in the pseudoinverse definition is known as the Gram matrix of the variational system (13). Its invertibility, being a necessary and sufficient condition for controllability of the system (13), guarantees the existence of the pseudoinverse (16). This existence can be given the following interpretation: if a posture $\mathbf{q}(T)$ is reachable from $\mathbf{q}_0$ using a control function $\mathbf{u}(\cdot)$ then every posture in some neighbourhood of $\mathbf{q}(T)$ may be reached from $\mathbf{q}_0$ by employing a small variation $\mathbf{v}(\cdot)$ of the control $\mathbf{u}(\cdot)$. With a reference to the concept of mobility for nonholonomic systems introduced in compliance with the manipulability matrix and the manipulability of robotic manipulators, the Gram matrix (17) has been called the *mobility matrix* of a nonholonomic system [1].

3 Singularities of Nonholonomic Systems

An overview of basic concepts referring to the kinematics of a nonholonomic robotic system that has been accomplished in the previous section allows us to distinguish two types of singularities: *singular postures* and *singular configurations*. They will be characterized below.

Definition 1. A posture $\mathbf{q} \in \mathbb{R}^n$ of a nonholonomic system is regular, if the growth vector $\mathbf{r}(\mathbf{q})$ of the system distribution remains constant in a neighbourhood of $\mathbf{q}$; otherwise the posture $\mathbf{q}$ is singular.

The posture singularity refers to the generation procedure of available directions of motion of the system (2) by piecewise-constant control, in accordance with the Campbell-Baker-Hausdorff-Dynkin formula [5]. It turns out that the growth vector equal to $\mathbf{r}(\mathbf{q}) = (r_0(\mathbf{q}), \ldots, r_{p(\mathbf{q})})$, implies that in order to move from $\mathbf{q}$ in every direction in $\mathbb{R}^n$, it suffices to generate vector fields spanning the distribution $\mathcal{G}^{p(\mathbf{q})}$, where $p(\mathbf{q})$ denotes the degree of nonholonomy at $\mathbf{q}$. If the posture $\mathbf{q}$ is singular then there exists another posture $\mathbf{q}'$, arbitrarily close to $\mathbf{q}$, such that in order to obtain every direction of motion from $\mathbf{q}'$ we need a different distribution $\mathcal{G}^{p(\mathbf{q}')}$, perhaps containing Lie brackets of lower degrees, so the controls able to move the posture from $\mathbf{q}'$ may become less complicated then from $\mathbf{q}$.. Thus, the posture singularity manifests itself in a non-uniform structure of controls that may change rapidly when passing from one system posture to another.

In this paper, more than in the posture singularities, we are interested in the singular configurations of a nonholonomic system. Relying on the analogy with regular and singular configurations of the manipulator kinematics we shall propose the following definition.

Definition 2. A configuration $\mathbf{u}(\cdot)$ of the kinematics (10) is regular, if the analytic Jacobian (11) is surjective, i.e.

$$J_{\mathbf{q}_0,T}(\mathbf{u}(\cdot)) \left(L_m^2[0,T]\right) = \mathbb{R}^n;$$

otherwise $\mathbf{u}(\cdot)$ is singular.

By the pseudoinverse formula (16), a configuration $\mathbf{u}(\cdot)$ is regular if and only if the system (13) is controllable, i.e. the mobility matrix (17) has full rank

$$rank\, M_{\mathbf{q}_0,T}(\mathbf{u}(\cdot)) = n \ . \tag{18}$$

As we have already noted, at singular configurations $\mathbf{u}(\cdot)$ the Jacobian pseudoinverse does not exist. For this reason, the Newton algorithm of solving the inverse kinematic problem for nonholonomic systems becomes ill posed [1].

In the general case of square integrable control functions, checking regularity of configurations is a difficult task as its prerequisite is computing the fundamental matrix (15). However, within the very special class of analytic control functions, there is a necessary and sufficient condition for regularity more accessible computationally.

Theorem 3. *For the control system (13), define the following sequence of matrices*

$$B_0(t) = B(t), \quad B_{i+1}(t) = A(t)B_i(t) - \frac{d}{dt}B_i(t), \quad i \geq 1 \ .$$

Then, a configuration $\mathbf{u}(\cdot)$ *is regular if and only if there exists a natural number k and a time instant $t \in [0,T]$ such that*

$$rank\ [B_0(t), B_1(t), \ldots, B_k(t)] = n \ . \tag{19}$$

An equivalent characterization of singular configurations of nonholonomic systems can be made in terms of singular extremals of optimal control theory. To this aim, let us examine an optimal control problem of the system (2) with Lagrangian $L(\mathbf{q}, \mathbf{u})$, that amounts to finding out a control function $\mathbf{u}(\cdot) \in L^2_m[0,T]$ such that

$$\int_0^T L(\mathbf{q}(t), \mathbf{u}(t))dt \to \min \ . \tag{20}$$

Necessary conditions for optimality, provided by the Pontryagin's Maximum Principle, require to define a Hamiltonian

$$H(\mathbf{q}, \boldsymbol{\psi}, \psi_0, \mathbf{u}) = \boldsymbol{\psi}^T G(\mathbf{q})\mathbf{u} - \psi_0 L(\mathbf{q}, \mathbf{u}), \tag{21}$$

and to introduce canonical Hamiltonian equations

$$\dot{\mathbf{q}} = \frac{\partial H(\mathbf{q}, \boldsymbol{\psi}, \psi_0, \mathbf{u})}{\partial \boldsymbol{\psi}}, \quad \dot{\boldsymbol{\psi}} = -\frac{\partial H(\mathbf{q}, \boldsymbol{\psi}, \psi_0, \mathbf{u})}{\partial \mathbf{q}} \ . \tag{22}$$

Then, the optimal control function $\mathbf{u}(\cdot)$ should satisfy a maximum condition

$$H(\mathbf{q}(t), \boldsymbol{\psi}(t), \psi_0, \mathbf{u}(t)) = \max_{\mathbf{u} \in \mathbb{R}^m} H(\mathbf{q}(t), \boldsymbol{\psi}(t), \psi_0, \mathbf{u}) \ . \tag{23}$$

A triple $(\mathbf{u}(\cdot), \mathbf{q}(\cdot), \boldsymbol{\psi}(\cdot))$ is called an extremal of the optimal control problem. The extremal is singular, whenever $\psi_0 = 0$ and the Hamiltonian does not depend on $\mathbf{u}$, i.e. when

$$\boldsymbol{\psi}^T G(\mathbf{q}(t)) = \boldsymbol{\psi}^T B(t) \equiv 0, \tag{24}$$

where $B(t)$ has been already defined by (14). Furthermore, in the singular case the canonical equations take the form

$$\dot{\mathbf{q}} = G(\mathbf{q})\mathbf{u}, \quad \dot{\boldsymbol{\psi}}^T = -\boldsymbol{\psi}^T \frac{\partial (G(\mathbf{q})\mathbf{u})}{\partial \mathbf{q}} = -\boldsymbol{\psi}^T A(t), \tag{25}$$

with $A(t)$ also defined by (14). Assuming that control functions are sufficiently smooth (e.g. analytic), a necessary condition for $\mathbf{u}(\cdot)$ being singular takes the form

$$\boldsymbol{\psi}^T B(t) \equiv 0, \quad \dot{\boldsymbol{\psi}}^T B(t) + \boldsymbol{\psi}^T \dot{B}(t) = -\boldsymbol{\psi}^T B_1(t) \equiv 0, \quad \ldots, \quad \boldsymbol{\psi}^T B_k(t) \equiv 0,$$

where matrices $B_k(t)$, $k \geq 0$ have been defined in Theorem 3. Eventually, since the adjoint variable does not vanish, $\psi(t) \neq 0$, we derive a condition

$$rank\ [B_0(t), B_1(t), \ldots, B_k(t), \ldots] < n \qquad (26)$$

that in view of the Theorem 3 is necessary and sufficient for a configuration $\mathbf{u}(\cdot)$ to be singular in the analytic case.

4 Examples

In order to illustrate the concepts of singular postures and singular configurations of nonholonomic systems we shall examine a well known example of a mobile robot composed of a car pulling two trailers [3]. Denote by $\mathbf{q} = (x, y, \varphi_1, \varphi_2, \theta) \in \mathbb{R}^5$ a vector of generalized coordinates of this system whose meaning has been explained in Figure 1. Assume for simplicity that the lengths of all trailers are

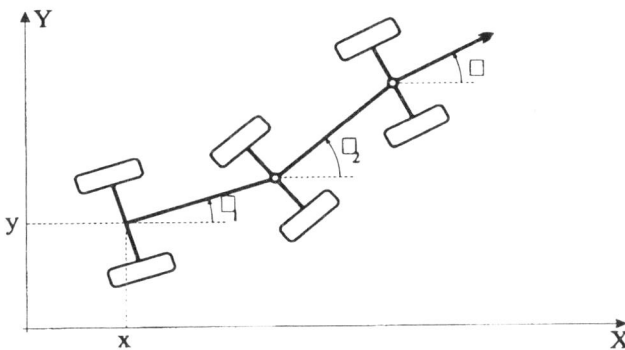

Fig. 1. The car with two trailers

equal to 1. The phase constraints that express a requirement of non-sliding rolling of all wheels have the form

$$A(\mathbf{q}) = \begin{bmatrix} \sin\varphi_1 & -\cos\varphi_1 & 0 & 0 & 0 \\ \sin\varphi_2 & -\cos\varphi_2 & -\cos(\varphi_1 - \varphi_2) & 0 & 0 \\ \sin\theta & -\cos\theta & -\cos(\theta - \varphi_1) & -\cos(\theta - \varphi_2) & 0 \end{bmatrix}. \qquad (27)$$

It is easily checked that the control system (2) representing the constraints (27) may be defined as

$$\begin{pmatrix} \dot{x} \\ \dot{y} \\ \dot{\varphi}_1 \\ \dot{\varphi}_2 \\ \dot{\theta} \end{pmatrix} = \begin{pmatrix} \cos(\theta - \varphi_2)\cos(\varphi_2 - \varphi_1)\cos\varphi_1 \\ \cos(\theta - \varphi_2)\cos(\varphi_2 - \varphi_1)\sin\varphi_1 \\ \cos(\theta - \varphi_2)\sin(\varphi_2 - \varphi_1) \\ \sin(\theta - \varphi_2) \\ 0 \end{pmatrix} u_1 + \begin{pmatrix} 0 \\ 0 \\ 0 \\ 0 \\ 1 \end{pmatrix} u_2 = \mathbf{g}_1(\mathbf{q})u_1 + \mathbf{g}_2(\mathbf{q})u_2 .$$

$$(28)$$

The meaning of the controls is: u_1 = a velocity of the car, u_2 = a velocity of changing the orientation of the car (e.g. by turning its steering wheel). It has been proved in [6] that the control system (28) satisfies the rank condition (4), so the constraints (27) are indeed nonholonomic. Obviously, the distribution (5) associated with this system is equal to $\mathcal{G} = span_{C^\omega}\{\mathbf{g}_1, \mathbf{g}_2\}$, its dimension $r_0(\mathbf{q}) = 2$ being constant. The filtration of $\mathcal{G}$ consists of the following distributions

$$\mathcal{G}^1 = span_{C^\omega}\left\{\begin{pmatrix}\cos(\varphi_2-\varphi_1)\cos\varphi_1\\ \cos(\varphi_2-\varphi_1)\sin\varphi_1\\ \sin(\varphi_2-\varphi_1)\\ 0\\ 0\end{pmatrix}, \mathbf{e}_4, \mathbf{e}_5\right\},$$

$$\mathcal{G}^2 = span_{C^\omega}\left\{\begin{pmatrix}\cos(\varphi_2-\varphi_1)\cos\varphi_1\\ \cos(\varphi_2-\varphi_1)\sin\varphi_1\\ \sin(\varphi_2-\varphi_1)\\ 0\\ 0\end{pmatrix}, \begin{pmatrix}\sin(\varphi_2-\varphi_1)\cos\varphi_1\\ \sin(\varphi_2-\varphi_1)\sin\varphi_1\\ -\cos(\varphi_2-\varphi_1)\\ 0\\ 0\end{pmatrix}\cos(\theta-\varphi_2), \mathbf{e}_4, \mathbf{e}_5\right\},$$

$\mathcal{G}^3 = span_{C^\omega}\{\mathbf{e}_1\cos\varphi_1 + \mathbf{e}_2\sin\varphi_1, (\mathbf{e}_2\cos\varphi_1 - \mathbf{e}_1\sin\varphi_1)\cos(\theta-\varphi_2), \mathbf{e}_3, \mathbf{e}_4, \mathbf{e}_5\}$,

$\mathcal{G}^4 = span_{C^\omega}\{\mathbf{e}_1, \mathbf{e}_2, \mathbf{e}_3, \mathbf{e}_4, \mathbf{e}_5\}$,

where $\mathbf{e}_i$ stands for the i-th unit vector in $\bm{r}^5$. From the structure of this filtration we deduce that the postures of the car pulling two trailers at which $\theta - \varphi_2 = \pm\frac{\pi}{2}$ are singular,

$$S_p = \left\{\mathbf{q} = (x, y, \varphi_1, \varphi_2, \theta) \in \mathbb{R}^5 \mid \theta - \varphi_2 = \pm\frac{\pi}{2}\right\},$$

what means that the wheels of the car are perpendicular to the shaft of the first trailer [6]. Furthermore, it follows that at any regular posture $\mathbf{q} \notin S_p$ the distributions $\mathcal{G}^2$ and $\mathcal{G}^3$ simplify to the form

$$\mathcal{G}^2 = span_{C^\omega}\{\mathbf{e}_1\cos\varphi_1 + \mathbf{e}_2\sin\varphi_1, \mathbf{e}_3, \mathbf{e}_4, \mathbf{e}_5\},$$

$$\mathcal{G}^3 = span_{C^\omega}\{\mathbf{e}_1, \mathbf{e}_2, \mathbf{e}_3, \mathbf{e}_4, \mathbf{e}_5\}.$$

In consequence, at regular postures the growth vector $\mathbf{r}(\mathbf{q}) = (2,3,4,5)$, while at singular postures $\mathbf{r}(\mathbf{q}) = (2,3,3,4,5)$. Respectively, the degrees of nonholonomy equal $p(\mathbf{q}) = 3$ and $p(\mathbf{q}) = 4$.

Now, let us look at singular configurations of the car pulling two trailers. Assuming analytic control functions, it is found immediately from Theorem 3 that a configuration $\mathbf{u}(\cdot) = (0(\cdot), u_2(\cdot))$, with arbitrary second component, is singular. This conclusion is quite intuitive and simply means that the only the steering wheel is turned while the system stands still. In order to discover other singular configurations, we shall introduce the system (28) into a singular posture $\mathbf{q} \in S_p$, and try to make it invariant. To this aim we need to have $(\theta - \varphi_2)(0) =$

$\pm\frac{\pi}{2}$ and $\frac{d}{dt}(\theta - \varphi_2) = 0$. Therefore, a singular configuration of this system will be $\mathbf{u}(\cdot) = (u_1(\cdot), u_2(\cdot))$, with $u_1(\cdot) = u_2(\cdot)$. The maneuvre of the car with two trailers corresponding to this singular configuration has been shown in Figure 2. Executing this maneuvre lies in tracing by the middle point of the car axle a

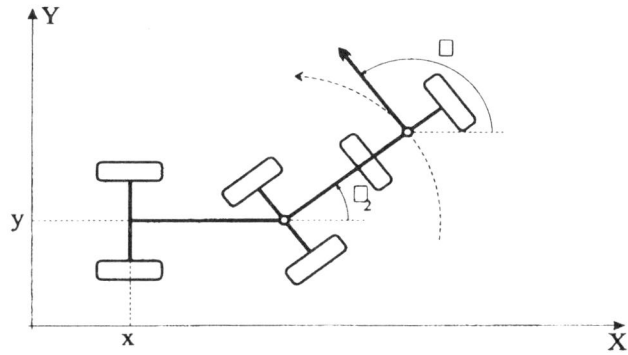

Fig. 2. The car with two trailers in a singular configuration

circle centered at the the middle point of the first trailer axle, while axle middle points of both trailers remaining at rest.

5 Conclusion

We have defined and examined two types of singularities of nonholonomic systems: the singular postures and the singular configurations. Both types of singularities have been characterized in control theoretic terms, with regard to the control system defining the kinematics of a nonholonomic system. Additionally, the configuration singularities have been introduced on the basis of an analogy with the singular configurations of stationary robotic manipulators. We have learnt from our example that, although both types of singularities are conceptually independent, the configurations preserving posture singularities might be singular. As we have already said, the knowledge of singular configurations of a nonholonomic system is necessary in solving the inverse kinematic problem [1]. Analogously to the normal forms of the manipulator kinematics, one can define normal forms of the kinematics of a nonholonomic system. A well known example of such a normal form in a neighbourhood of a regular posture is the chained form, [3], to which one can transform a.o. a car with n trailers. Normal forms of the car with n trailers around singular postures have been obtained recently in [10]. A recognition of a robotic significance of normal forms of nonholonomic systems will be a challenging subject of future research, parallelizing the normal form approach for robotic manipulators [12].

6 Acknowledgments

This research has been done within a research project "Modelling, motion planning, and control of mobile manipulators", supported by the Polish State Committee of Scientific Research.

References

1. Tchoń, K., Mazur, A., Dulęba, I., Hossa, R., Muszyński, R.: Manipulators and Mobile Robots: Modelling, Motion Planning, and Control. Academic Publishing House, Warsaw (2000) (in Polish)
2. Laumond, J.-P.: Singularities and topological aspects in nonholonomic motion planning. In: Nonholonomic Motion Planning, ed. by Z. Li and J. F. Canny, Kluwer Academic Publ., Boston (1992) 755–763
3. Murray, R. M., Li, Z., Sastry, S. S.: A Mathematical Introduction to Robotic Manipulation. CRC Press, Boca Raton (1994)
4. Laumond, J.-P., Sekhvat, S., Lamiraux, F.: Guidelines in nonholonomic motion planning for mobile robots. In: Robot Motion Planning and Control, ed. by J.-P. Laumond, Springer–Verlag, London (1998) 1–54
5. Dulęba, I.: Algorithms of Motion Planning for Nonholonomic Robots. Wrocław University of Technology Publishers, Wrocław (1998)
6. Bellaiche, A., Jean, F., Risler, J.-J.: Geometry of nonholonomic systems. In: Robot Motion Planning and Control, ed. by J.-P. Laumond, Springer–Verlag, London (1998) 55–91
7. Vershik, A. M., Gershkovich, V. Ya.: Nonholonomic dynamical systems, geometry of distributions and variational problems. In: Dynamical Systems VII, ed by V. I. Arnold and S. P. Novikov, Springer–Verlag, Berlin (1994) 1–81
8. Montgomery, R.: A survey of singular curves in sub-Riemannian geometry. J. Dyn. Contr. Syst. **1** (1995) 49–90
9. Jakubczyk, B.: Characteristic varieties of distributions and abnormal curves. Math. Inst. Polish Academy of Sci. (1999) (preprint)
10. Pasillas-Lépine, W., Respondek, W.: On the geometry of Goursat structures. Inst. Nat. Sci. Appl. de Rouen (1999) (preprint)
11. Sontag, E. D.: A general approach to path planning for systems without drift. In: Essays on Mathematical Robotics, ed. by J. Baillieul. S. S. Sastry and H. J. Sussmann, Springer–Verlag, New York (1998)
12. Tchoń, K., Muszyński, R.: Singular inverse kinematic problem for robotic manipulators: A normal form approach. IEEE Trans. Robotics Automat. **14** (1998) 93–104

Experimental Determination of Robot Workspace by Means of CATRASYS (Cassino Tracking System)

Marco Ceccarelli, Erika Ottaviano and Maria Toti

Laboratory of Robotics and Mechatronics, DiMSAT, University of Cassino, Italy

Abstract. CATRASYS (Cassino Tracking System) is a new measuring system that can easily evaluate robot workspace performances. A prototype has been built at the Laboratory of Robotics and Mechatronics in Cassino. The trilateration technique has been used to evaluate position and orientation of robot end-effector. In this paper we present the system and its operation through experimental determination of robot workspace.

1 Introduction

For most robots it is very difficult to know accurately the workspace, especially when orientation of end-effector is to be taken into account. Several instruments for monitoring the position or trajectory of the robot end-effector can be used, see Van Brussel (1990) and Dorf (1988) for arguments. They can be grouped according to the measuring operation and specific technology that is applied. Several instruments use triangulation or trilateration techniques such as cameras, theodolites, laser tracking systems and wire systems. Trilateration and triangulation determine the relative position between points by using the geometry of triangles. Triangulation uses measurements of both distances and angles, whereas trilateration uses only distance measurements.

Laser tracking systems use triangulation technique. The model can be rather complicated. The system can reach high accuracy if it is well calibrated, for example it has been reported the value of 0.01 mm in Dorf (1988). Without a proper compensation of the triangulation error the system will only work for some unpredicted range in the measurement field, as pointed out through previous experiences in Ceccarelli et al. (1998). In addition the system is strongly affected by the relative positions of the measuring units.

In this paper we present CATRASYS, which performs measurements of robot end-effector position and orientation during robot motion by using trilateration technique. It has been conceived at Laboratory of Robotics and Mechatronics in Cassino and a prototype has been built for experimental activity at the Laboratory. Preliminary work has been presented in Gabriele (1994) and first results on experimental validations have been presented in Ceccarelli et al. (1999), Toti (1999) and Ceccarelli et al. (1999).

CATRASYS (Cassino Tracking System) is a new measuring system that can evaluate easily large displacements and it has been successfully applied for evaluation of robot workspace.

2 Trilateration Formulation

A general scheme for trilateration is sketched in Fig.1. Distances from known points O_1, O_2, and O_3 to the robot end-effector are evaluated by using wire devices to measure distances.

The measured distances d_1, d_2 and d_3 are used as radii of arcs from each known point O_1, O_2 and O_3. The position of a reference point H on the robot end-effector is defined as the position of the point at which three arcs intersect as a function of coordinates x_i, y_i and z_i of the points O_i, (i=1,2,3). This can be formulated as

$$
\begin{aligned}
(x_H - x_1)^2 + (y_H - y_1)^2 + (z_H - z_1)^2 &= d_1^2 \\
(x_H - x_2)^2 + (y_H - y_2)^2 + (z_H - z_2)^2 &= d_2^2 \\
(x_H - x_3)^2 + (y_H - y_3)^2 + (z_H - z_3)^2 &= d_3^2
\end{aligned}
\qquad (1)
$$

After some algebraic manipulation, the position of H can be computed through components x_H, y_H and z_H as

$$
\begin{aligned}
x_H &= H_x \frac{-B_2 - \sqrt{B_2^2 - 4B_1 B_3}}{2B_1} + E_x \\
y_H &= H_y \frac{-B_2 - \sqrt{B_2^2 - 4B_1 B_3}}{2B_1} + E_y \\
z_H &= \frac{-B_2 - \sqrt{B_2^2 - 4B_1 B_3}}{2B_1}
\end{aligned}
\qquad (2)
$$

where

$$
B_1 = H_x^2 + H_y^2 + 1, \qquad B_2 = 2H_x(E_x - x_3) + 2H_y(E_y - y_3) - 2z_3,
$$

$$
B_3 = E_x^2 + E_y^2 + A_3^2 - 2E_x x_3 - 2E_y y_3 B_1, \qquad H_x = -\frac{Z_{21} + H_y Y_{21}}{X_{21}}, \qquad (3)
$$

$$
H_y = \frac{Z_{21} X_{32} - Z_{32} X_{21}}{Y_{32} X_{21} - Y_{21} X_{32}}, \qquad E_x = \frac{A_1 + A_2 - E_y Y_{21}}{X_{21}}, \qquad E_y = \frac{-A_1 + A_2}{Y_{32} X_{21}}
$$

with (i, j=1,2,3), (i≠j)

$$
A_i = d_i^2 - x_i^2 - y_i^2 - z_i^2, \quad X_{ij} = x_i - x_j, \quad Y_{ij} = y_i - y_j, \quad Z_{ij} = z_i - z_j \qquad (4)
$$

In order to evaluate position and orientation of robot end-effector it is necessary to consider six parameters or determine the position of two different points H and F of the robot end-effector, as shown in Fig.1. Using three measurements d_1, d_2 and d_3 it is possible to determine

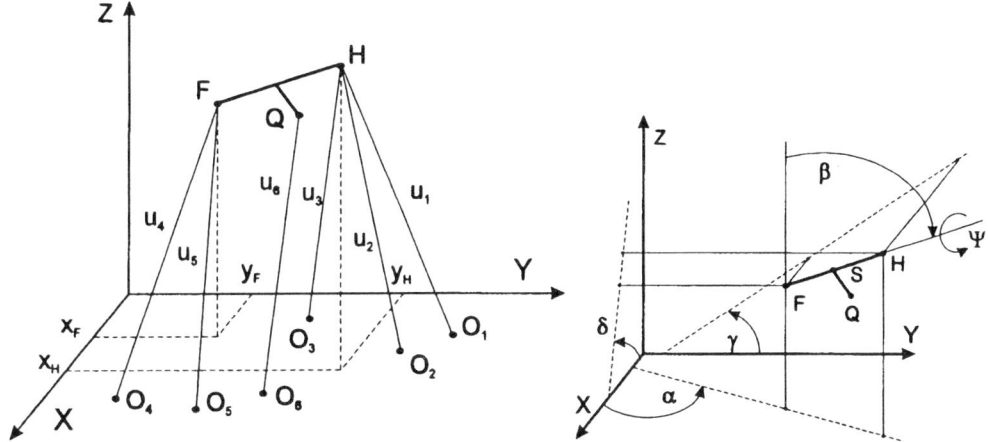

Figure 1. A trilateration scheme for position and orientation measures of robot end-effector.

the position of H. By using two more measures, d_4 and d_5, the position of point F can be determined by measuring the lengths of three edges of a tetrahedron, whose three vertices of the triangular base are O_4, O_5 and H, and the fourth vertex is F, Fig.1. Using similarly Eqs. (1-4) for the measured lengths d_4, d_5 and HF, whose length is fixed and known, it is possible to evaluate the position of F through its components (x_F, y_F, z_F).

Thus it is possible to evaluate the robot end-effector orientation through the orientation angles α and β, or α, β, δ and γ, as shown in Fig.1. In fact the orientation of the robot end-effector, identified by the segment HF, can be described by the angles computed as, Fig.1,

$$\alpha = \tan^{-1}\left[\frac{y_H - y_F}{x_H - x_F}\right] \quad , \quad \beta = \tan^{-1}\left[\frac{(z_H - z_F)}{\sqrt{(y_H - y_F)^2 + (x_H - x_F)^2}}\right] \quad (5)$$

Referring to Fig.1, also angles δ and γ can describe the orientation of end-effector HF by using the expressions

$$\delta = \tan^{-1}\left[\frac{z_H - z_F}{x_H - x_F}\right] \quad , \quad \gamma = \tan^{-1}\left[\frac{z_H - z_F}{y_H - y_F}\right] \quad (6)$$

In addition, the rolling angle ψ can be computed as

$$\psi = \tan^{-1}\left[\frac{z_Q - z_S}{\sqrt{(y_Q - y_S)^2 + (x_Q - x_S)^2}}\right] \quad (7)$$

when a sixth wire transducer is available and position of middle point S is computed from position of H and F, as shown in Fig.1.

Equations (1) to (7) express analytically a general formulation for the trilateration scheme of Fig.1 when the origin of the reference frame OXYZ does not coincide with any point O_i, (i=1,2,3). An evaluation of the accuracy of CATRASYS is rather complicated since several variables and geometrical parameters, including the relative position among the points O_1, O_2, O_3 and H, greatly affect the operation efficiency. An error of 0.3 mm has been evaluated for a measured position of H at a distance of 1m from O by using accuracy data of the transducers, see Toti (1999). However, an evaluation of the accuracy error will be formulated and proved by experimental validation, which will be carried out in a near future. Nevertheless, CATRASYS structure does not require more than one calibration measurement when the relative positions of O_i, (i=1,2,3), are known and kept fixed. At the moment a known distance is used to calibrate easily CATRASYS.

3 CATRASYS (Cassino Tracking System)

CATRASYS system is designed as a mechanical part, an electronic interface unit and a Pesonal Computer, as shown in Fig.2, see Ceccarelli et al. (1999), Toti (1999) and (Ceccarelli et al. (1999).

The core of the mechanical part is composed by a robot end-effector for CATRASYS and five wire transducers, which are located on a Trilateral Platform. Fig.3 shows the prototype. The used wire transducers, which are denoted by T in Fig.2, have a working range of 2500 mm and continuous resolution, as reported in Celesco (1994). The interface unit is an acquisition card AT-MIO-16F5, see National Instruments (1995). Signals from wire transducers are fed to the electronic interface unit, which is connected to a Personal Computer for data analysis. CATRASYS permits to measure continuously both position and orientation during robot motion.

The robot end-effector for CATRASYS is a coupling device: it connects the wires of the five transducers with robot end-effector. Two equal units have been designed ad hoc and the

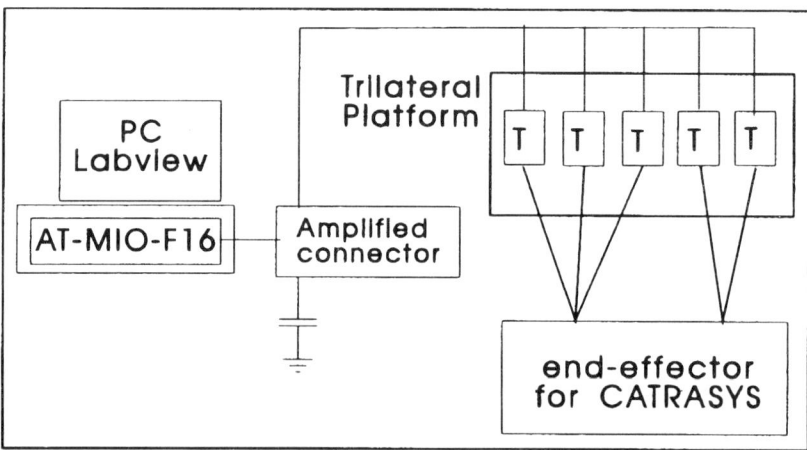

Figure 2. A scheme of the experimental system CATRASYS, (T stands for the wire traducer).

Figure 3. Prototype of Trilateral Platform and robot end-effector for CATRASYS.

prototype is reported in Fig.3. The mechanical design is simple and has fully rotating shaft, which permits the wires to track the robot end-effector without wrapping each other.

A PC IBM 486 with 32 MB of RAM has been used for programming and monitoring the system performance. A virtual instrument has been developed with LabView software and the results are plotted without any other user interface.

Position of robot end-effector for CATRASYS is determined by using the signals of the wire transducers giving d_1, d_2 and d_3 to calculate Eqs.(1) to (4). Orientation is determined when F position is measured and Eqs. (5), (6) and (7) are calculated by also using d_4 and d_5.

Before experimental determinations a calibration of the wire transducers is necessary and it is performed by means of given lengths for each wire transducer.

4 Experimental Determinations

Validation of CATRASYS has been carried out at the Laboratory of Robotics and Mechatronics in Cassino using a PUMA 562 robot, Unimation (1990), and a SCORBOT-ER-V robot, Eshed Robotec (1992), Fig.4.

The workspace boundaries of the manipulators have been determined by an automatic continuous operation, which is based on a suitable programming of a robot so that H traces the workspace boundary. The movement of the reference point H has been programmed along cross-section boundary of workspace by using two basic motion strategies: the external boundary is traced by rotating the first joint of the fully extended part of the manipulator arm; in other parts the motion is performed by means of a zigzag trajectory that goes to and departs from the boundary with a priori distance in order to avoid off-run situations. It is only necessary to specify values of reach parameters delimiting the workspace areas, which are related to each strategy.

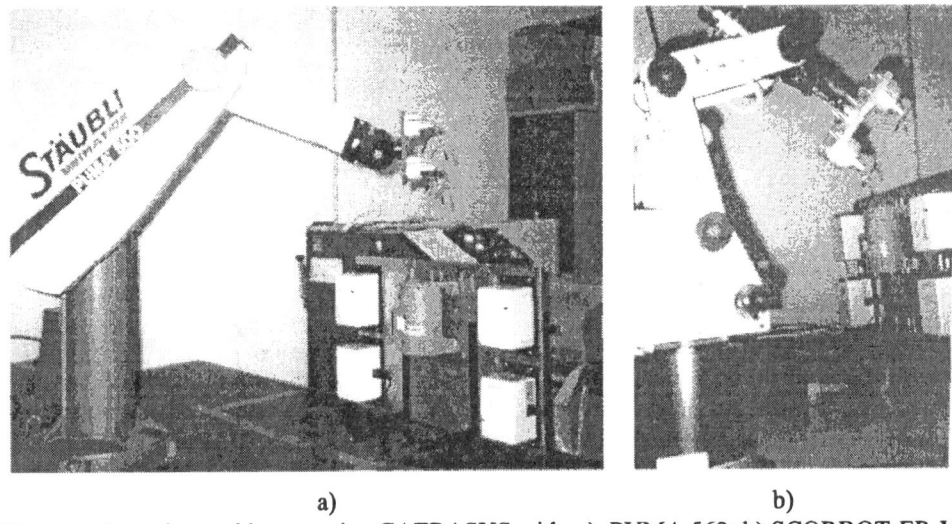

a)   b)

Figure 4. Experimental layout using CATRASYS with: a) PUMA 562; b) SCORBOT-ER-V.

A specific program has been developed with VAL II software for PUMA robot and however, it can be easily applied to any manipulator, as the case of the SCORBOT-ER-V.

Figure 5 shows a representation of workspace boundary of PUMA 562 robot and SCORBOT-ER-V robot in Zr frame through measured positions. The workspace boundary of Fig.5 has been expressed in term of $r = (x_H^2 + y_H^2)^{1/2}$ and $z = z_H$, which are the radial and axial reaches.

Using CATRASYS it is possible to evaluate robot end-effector orientation along any path when a robot is used for trajectory planning. Fig.6 shows an illustrative example, which consists of simultaneous measurements of the orientation of the robot end-effector through angles

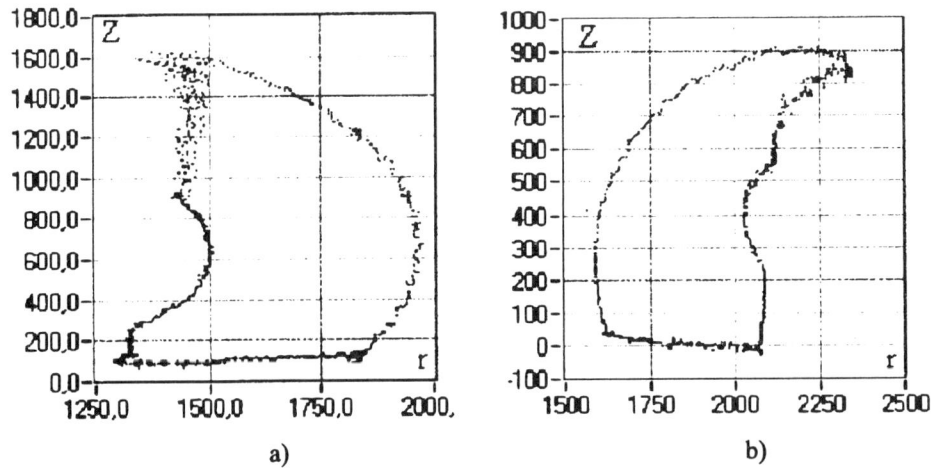

a)   b)

Figure 5. Measured workspace boundary with respect to Zr for:
a) PUMA 562 robot; b) SCORBOT-ER-V robot.

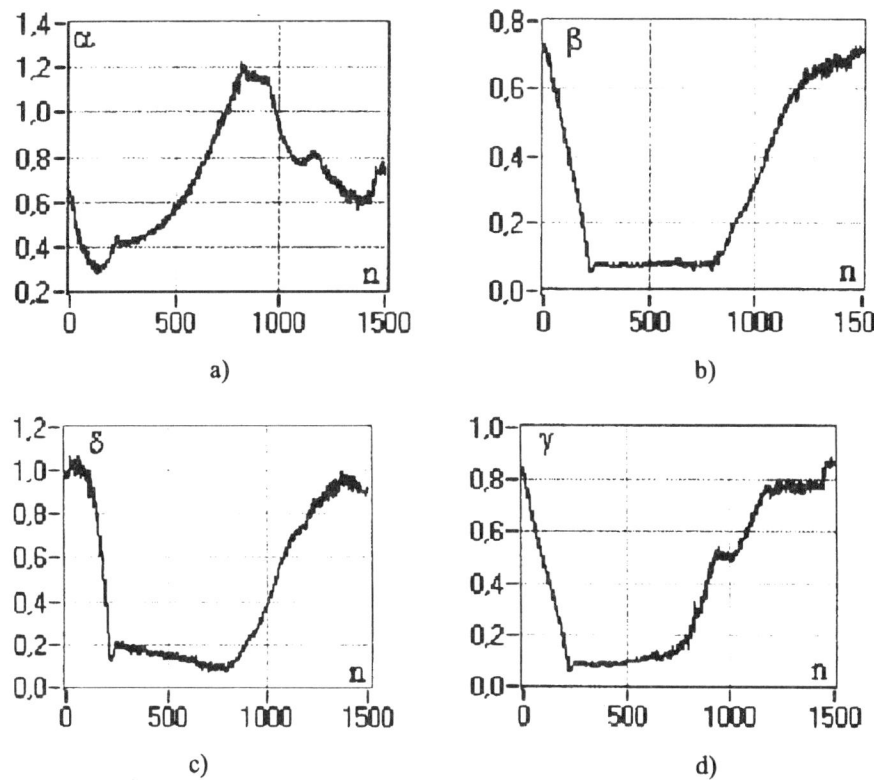

Figure 6. Measured orientation angles of the PUMA robot end-effector during the motion of H along the workspace boundary of Fig.5 a) versus acquisition number n:
a) α angle; b) β angle; b) δ angle; d) γ angle.

α, β, δ and γ during the movement of H for tracing the workspace boundary of Fig.5 a). It is worth to note that CATRASYS can perform position and orientation measures with a continuous data acquisition, which can be related with a not very slow motion of the robot.

5 Conclusion

In this paper we have presented the system CATRASYS (Cassino Tracking System) for large displacement measurement and its application to experimental determination of robot workspace. Experimental validations of CATRASYS have been carried out at the Laboratory of Robotics and Mechatronics in Cassino by using industrial robots like a PUMA 562 robot, Unimation (1990), and a SCORBOT-ER-V robot, Eshed Robotec (1982).
CATRASYS (Cassino Tracking System) exhibits the following features:
- it is portable, low-cost and easy in use;
- the position and orientation can be measured continuously and the motion of a robot can be tracked automatically;

- the working volume depends on the measuring range of the wire transducers only.

The workspace boundaries of the manipulators have been determined by an automatic continuous operation, which is based on a suitable programming of a robot so that H traces the workspace boundary.

References

Ceccarelli, M., Ottaviano, E. and Besa, A. (1998). Experimental Determination of Robot Workspace by Using a Laser System. *Proceedings of the 7th International Workshop on Robotics in Alpe-Adria-Danube Region RAAD'98*, Smolenice. 203-208.

Ceccarelli, M., Toti, M.E. and Ottaviano, E. (1999). CATRASYS (Cassino Tracking System): A New Measuring System for Workspace Evaluation of Robots. *Proceedings of the 8th International Workshop on Robotics in Alpe-Adria-Danube Region RAAD'99*, Munich. 19-24.

Ceccarelli, M., Ottaviano, E., Toti, M. and Lanni, C. (1999). Design and Experimental Activity of CaTraSys (Cassino Tracking System). *Proceedings of the XIV Italian National Congress AIMETA*, Como. CD Proceedings Paper 17. (in Italian).

Celesco Inc. (1994). *User's Manual for Wire Transducers mod. PT101*.

Dorf, C. (Ed.) (1988). International Encyclopedia of Robotics: Applications and Automation", New York: Wiley. Vol.3.

Eshed Robotec (1982). *ACL Reference Guide*.

Gabriele, E. (1994). Numerical and Experimental determination of a robot workspace. *Master Thesis, University of Cassino*. (in Italian).

National Instruments (1995). AT-MIO-16F5 Acquisition Card. *N.I. Catalog*.

Toti, M.E. (1999). Design and Experimental Validation of a Measuring System for Evaluation of Kinematic Characteristics of Robots. *Master Thesis, University of Cassino*. (in Italian).

Unimation (1990). *Val II User's Manual*.

Van Brussel, H. (1990). Evaluations and Testing of Robots, *Annals of CIRP*. 32: 657-664.

A Pantograph Mechanism With Large-Deflective Hinges for Miniature Surface Mount Systems

Mikio Horie[1], Toru Uchida[2], and Daiki Kamiya[1]

[1] Precision and Intelligence Laboratory (P & I Lab.), Tokyo Institute of Technology, Japan

[2] Graduate Student of Tokyo Institute of Technology, Japan

Abstract. In this paper, a new surface mount system with parallel arrangement miniature manipulators is proposed for use in system downsizing. The miniature manipulator consists of a molded pantograph mechanism, which is composed of large deflective hinges and links, both made of the same materials. In order to create such systems, first, durability of the pantograph mechanism is to be confirmed by fatigue tests. Next, the input and output displacement characteristics of the pantograph mechanism are to be experimentally discussed. Finally, propriety of the proposed system should be confirmed.

1 Introduction

Nowadays, it is difficult to meet the needs and demands for producing handy electric devices such as cellular phones due to their rapid, widespread popularity, along with the development of their minimization. Therefore, in the substrate surface mount field, there has been a strong need to speed up present surface mount systems while minimizing the size of the devices. The present surface mount system has been developed by mainly focusing on the improvement of speed functions, however resulting in enlargement of the devices. It has become almost impossible to improve the speed of the system. Therefore, we should consider a productivity increase that would be made possible by minimizing each device, while filling designated spaces to the maximum capacity.

In this study, we propose a new surface mount system which consists of groups of the manipulators that have been minimized by an integral molded pantograph mechanism with hinges and links. In addition, durability against repeated input displacement is to be confirmed with large deflective hinges, even the small size of which can obtain large angular displacements. Moreover, this paper discusses minimization possibilities of a new system by a model-devised surface mount system, made possible after the experiments have clarified the input-output displacement characteristics of a model-devised integral molded pantograph mechanism.

2 Suggestions on a Surface Mount System Consisting of Groups of Miniature Manipulators

The present mount system needs high structural stiffness in order to speed up functions of mounting work, which has a tendency to increase its overall size. Therefore, in this study, the new surface mount system shown in Fig. 1, is proposed based on the idea that placing as many miniature manipulators as possible within a working space should make it possible to minimize the size of devices while maintaining the present productivity. With this new surface mount system, a great number of miniature manipulators that are arranged in a parallel way, can assemble machine parts onto the substrates which simultaneously move on the belt conveyor. An integral molded pantograph mechanism [Horie, 1998] made of high polymer (Polypropylene) materials is used for minimizing these manipulators. Since this type of mechanism is integrally molded with both link and hinge parts in an injection molding method, it does not require

---

This work is partially supported by the Scientific Research Grant-in-Aid of the Mitutoyo Association for Science and Technology(MAST).

assembling, therefore making it easy to be devised even in such cases when minimizing is required. In comparison, the presently used manipulators have difficulties in minimization because of their assemblage in which actuators are placed onto the moving links. In new integral molded pantograph mechanisms, however, actuators are placed onto the fixed links, this making it possible to minimize and lighten arm parts. In this way, arm parts can avoid interference among one another in the working space where large numbers of manipulators operate simultaneously.

The integral molded pantograph mechanism uses spring plate type hinges for revolute pairs, and angular displacements can be obtained by utilizing deformation of the materials used. On the other hand, the presently used pantograph mechanism uses the deformation produced on the hinge surface by operations within a limited elastic dimension. Due to this, this mechanism faces many problems, such as that angular displacements obtained at hinge parts are relatively small, and also that a larger moving region can not be obtained in comparison to the size of its mechanism. In order to obtain large angular displacements at hinge parts, the hypothesis is made that making the length of hinges longer than the thickness should make the aspect ratios of hinges larger. However, these hinges can also lower the structural stiffness of a pantograph mechanism, then worsening the relation between input and output displacements which are regarded as a displacement enlargement mechanism. Therefore, in this study, large deflective hinges, which can obtain large angular displacements by producing larger deformation than the elastic region, are model-devised. Moreover, the durability against repeated input displacements is tested in such cases where large deflective hinges are used as the hinge parts of an integral molded pantograph mechanism. Figure 2 shows the sizes of both the presently used large deflective elastic hinges [Horie, 1997],[Kamiya, 1999],[Kim, 1999],[Kim, 1998] and the large deflective hinges model-devised in this experiment. The subjective working of manipulators is determined as the substrate surface mount of a miniature electric device. From this substrate size, the moving space of the mechanism is determined as 50mm×40mm. In this way, the size of the mechanism is determined where the relative angular displacements between links around the moving space area can be set at 45°. The displacement enlargement ratios of a pantograph mechanism are set to be four times more than the input in the X direction, and five times more than that of the Z direction. The length of hinge parts is set at 200 μm, and the thickness can be adjusted between 30 μm and 300 μm using a metal mold. In addition, the thickness of link parts and the width of the mechanism are both set at 5 mm. The shape and size of the model-devised integral molded pantograph mechanism are shown in Fig. 3. Polypropylene is used as it can both obtain a relatively larger deformation among high polymer materials and be easily molded.

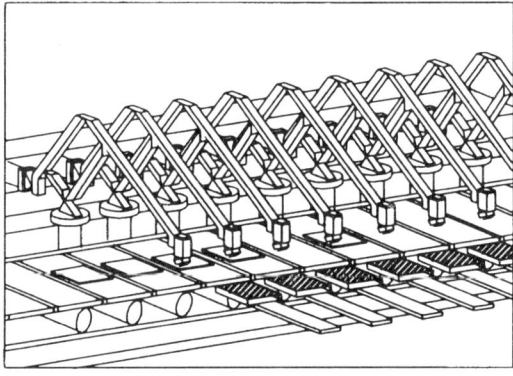

Figure 1. Proposed surface mount system

3 Fatigue Tests of an Integral Molded Pantograph Mechanism with Large Deflective Hinges

Large deflective hinges are known to be able to obtain large angular displacements by producing larger deformations on the materials than the elastic regions. Therefore, when they are used as hinge parts of a pantograph mechanism, it is thought that there might be some possibility that the materials could be destroyed by fatigue under the condition that repeated use creates large deformations. Consequently, these results led to conducting the fatigue test of the model-devised pantograph mechanism using the apparatus shown in Fig. 4. The experimental apparatus consists of three parts: the vibrator based on the slider crank mechanism, the laser displacement sensors, and the oscilloscope. The test is conducted as follows: Using a vibrator, the sinusoidal displacement input at 7.5-mm vibration is given to the input part in the X direction of a pantograph mechanism, and then the displacements of input-output ending points are measured by a laser displacement sensor. The relation between the input vibratory frequencies and the repeatable numbers until hinge parts has fractured is determined after the integral molded pantographs using hinges of different thickness lengths have been examined. The vibration frequencies are then determined by observing waves of input displacements, and also by regulating the input voltage to the motor. However, the experiment is found to be difficult in measuring repeatable numbers because the test using

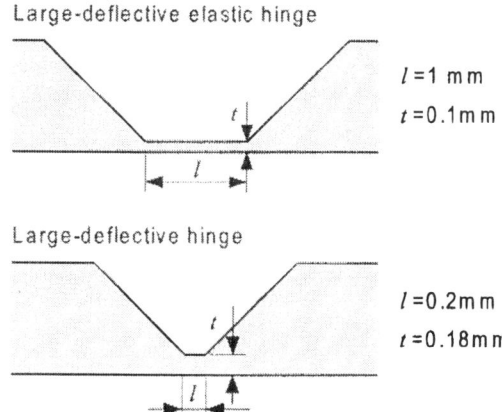

Figure 2. Dimension of large-deflective elastic hinge and large-deflective hinge

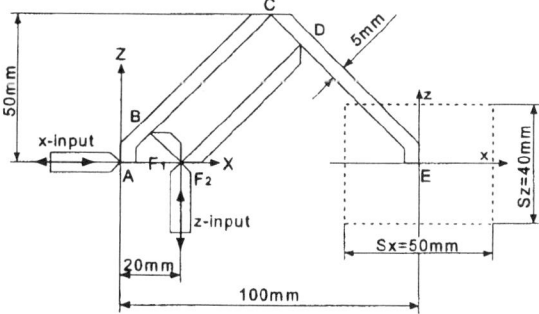

Figure 3. Dimension of pantograph mechanism

high frequencies causes fractures in a relatively short period of time. Therefore, the input vibration frequency is made to gradually rise at about 1 Hz every second from 5 Hz, and the frequency is measured until the hinge parts have fractured. This frequency can then be determined that which has fractured by being repeatedly used a hundred times. The hinge thickness, the vibration frequency, the number of cycles, and failed hinge parts of the integral molded pantograph mechanisms in each conducted experiment, are shown in Table 1. A hyphen mark in Table 1 shows that a pantograph mechanism does not fracture even after a one million-time repetition. The names of the hinge parts are shown in Fig. 2. The details of the experiment are as follows: First, the response waves of input-output displacements are shown in Fig. 5. The pantograph mechanism shown in Fig. 5 (a) did not fracture even after being repeated one million times, resulting in no wave changes. The mechanism in Fig. 5 (b), however, did fracture after being repeated 30600 times, which can be seen as roughness in the wave of output displacements occurring with every repetition. This is thought to be because the hinge parts are made to stretch by the repetition. The roughness of these waves can be seen in all the fractured pantograph mechanisms. Next, the pantograph

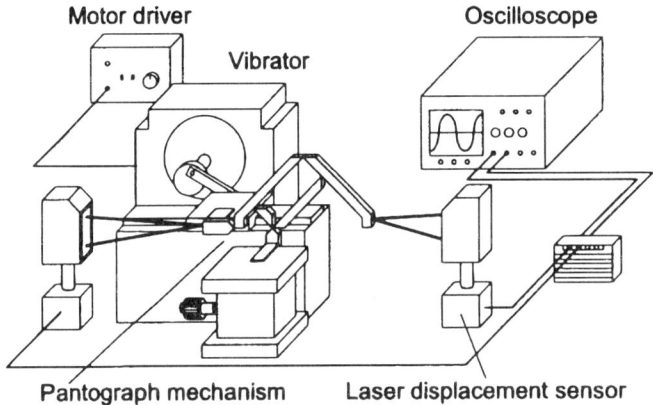

Figure 4. Experimental apparatus

Table 1. Result of fatigue test

Group	A					B			
Exp. No.	1	2	3	4	5	6	7	8	9
Hinge thickness [μm]	60	90	90	90	100	110	110	115	120
Input frequency [Hz]	19.4	15	15	20	22.2	5	20	18	15
Number of cycles	100	-	153900	3140	100	-	3900	30600	-
Broken hinge	D	-	C, D	B,C,D	F_2, D	-	C, D	D	-

Group	B				C			
Exp. No.	10	11	12	13	14	15	16	17
Hinge thickness [μm]	125	125	130	135	150	165	225	260
Input frequency [Hz]	17	20	26	20	26.8	17	10	20
Number of cycles	-	6040	100	8600	100	-	-	-
Broken hinge	-	F_2, D	F_2, D	A	F_2, D	-	-	-

mechanisms model-devised in this experiment were divided into three groups according to the different lengths of their hinge thickness, as shown in Table 1. Then, based on the relation between the vibration frequency and the fracture repeatable number, each of the frequencies enduring after the three pantograph mechanisms were repeatedly used one million times, was determined. The result of this is shown in Fig. 6. The frequencies were 15 Hz in Group A of the pantograph mechanisms, 17.5 Hz in Group B and 20 Hz in Group C. Within the region used in this experiment, it is found out that the greater the thickness of the hinge, the more durable it can be when used at high speeds. Therefore, we deduce that the fracture of pantograph mechanisms in this study can be considered as the fatigue fracture done under repeated stress, made toappear on the hinge parts by the link inertial force. Moreover, as shown in Table 1, it is obvious that the fracture of all the pantograph mechanisms except for that in Experiment No.13 is seen at hinge part D. Therefore, regarding each pantographmechanism used in these experiments, the relation between the maximum repetition and the maximum equivalent inertial force factor [Inertial force / Hinge section area] at hinge part D caused by the inertial force of the output link, was calculated. This result is shown in Fig. 7. Each sign in Fig. 7 corresponds to the groups divided in Fig. 6 (Group A, Group B, and Group C). The

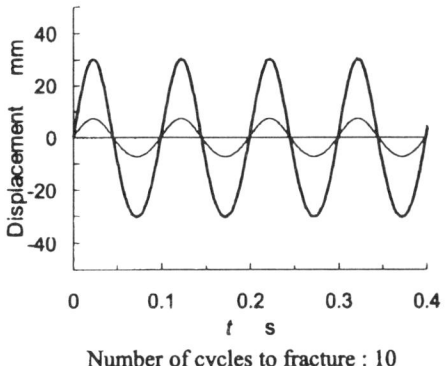
Number of cycles to fracture : 10

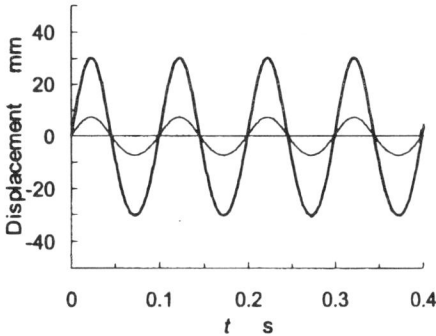
Number of cycles to fracture : 1,000,000

(a) Exp. No. 16

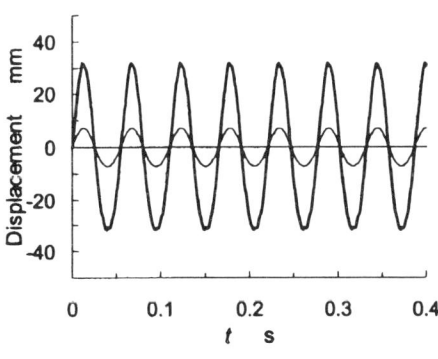
Number of cycles to fracture : 10

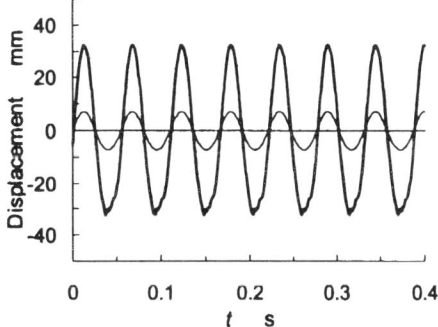
Number of cycles to fracture : 30,000

(b) Exp. No. 8

Figure 5. Waves of input and output displacements (——— : Output displacement, —— : Input displacement)

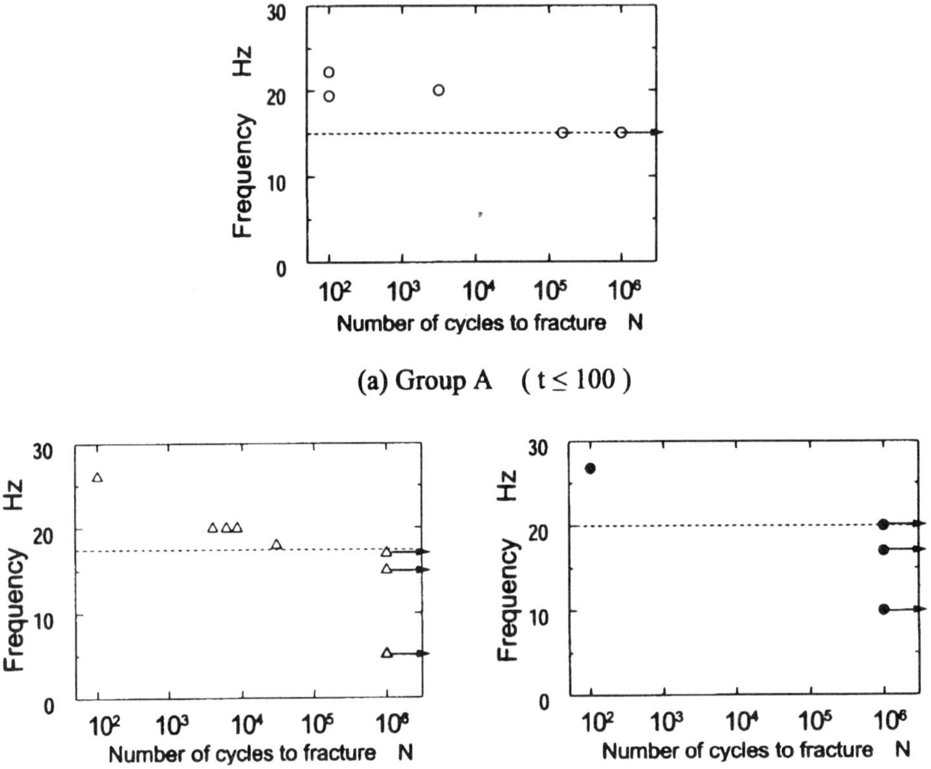

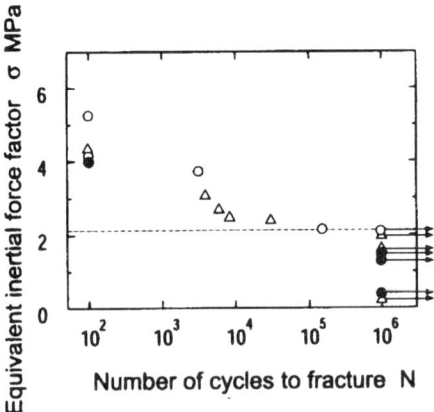

(a) Group A (t ≤ 100)

(b) Group B (100 < t < 150)

(c) Group C (150 ≤ t)

Figure 6. Results of fatigue test

Figure 7. S-N curve

stress on the hinges due to the inertial force of links was calculated by dividing the force transmitted on the parts of revolute pairs by the cross-section area of the hinges. This was done when the pantograph mechanisms with revolute pairs were operated under the same conditions as those of the integral molded pantograph mechanism. Sign ⟶ in the figure shows that the pantograph mechanism did not fracture and also the conclusion time of the experiment. The figure shows a similar tendency with an S-N curve line of a standard fatigue test. The fatigue limit obtained in this experience was 2.1 MPa. When the stretch strength is set at 3.3 MPa the fatigue limit ratios shows 6.4 %. Normally, the fatigue limit ratio of a high polymer material is 0~50 % and the resulting hinge fracture of this fatigue test is thought to appear because repeatable stress has an effect on the hinges cracked by bending.

4 Input-Output Displacement Characteristics of an Integral Molded Pantograph Mechanism

In order to evaluate a model-devised integral molded pantograph mechanism used as a miniature manipulator of a surface mount system, the needed input-output displacement characteristics were obtained by the experimental apparatus shown in Fig. 8. The hinge length of the pantograph mechanism used in this experiment was determined to be 200 μm, with the thickness being 180 μm. In this experiment, displacement inputs were given in both the X direction and Z direction by the linear stepping motor. Then, pictures of output ending points were taken by a CCD camera, and the displacements were then calculated by an image processor. For the stepping motor used for input, the positioning accuracy was set at 10 μm and the step resolution was at 1.02 μm. The measurement resolution of output displacements in this experiment was determined to be 10 μm. The input displacement was set at a 1 mm interval to go along the X-axis, Z-axis as well as along the border around moving regions as shown in Fig. 9. First, errors of output displacements when input parts in the Z direction were fixed and also when input was given only in the X direction, are shown in Fig. 10. Errors were determined as the difference calculated from output displacements when a pantograph mechanism was considered to be an ideal displacement enlarging mechanism. In Fig. 10, Sign "White circle" represents displacement errors in the X direction, Sign "White triangle" represents those in the Z direction, and Sign "Black circle" represents the absolute rates of errors, that is, the square root of the quantity "White circle" squared plus "White triangle" squared. The graph shows the results of measurements when the pantograph mechanism was moved on the X-axis as well as on the border along the

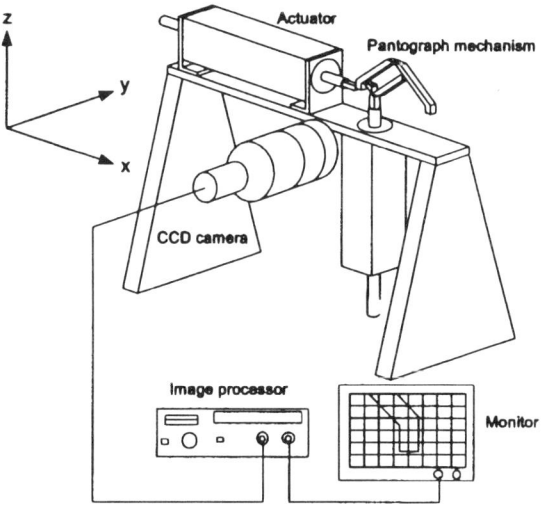

Figure 8. Experimental apparatus

upper and lower moving regions. Next, Figure 11 shows errors of output displacements in the case where input was fixed in the X direction and was given in only the Z direction. In the same way, the errors were measured where the pantograph mechanism was moved on the Z-axis and also on the border of working space. The movement in both plus and minus directions on one line was measured. The differences on the Y-axis at the two measuring points represent hysteresis errors during this time. This result clarified that the displacement enlarging ratios in both the X and Z directions were slightly lower than the estimated values. Also, the displacement errors were seen to be larger around the border of moving space where the angular displacement at hinge parts became larger. The maximum errors of output displacements of the integral molded pantograph was 340 μm, while the maximum hysteresis errors were 130 μm. Moreover, the accuracy ratios of positioning repeatability at ending points E_1 and E_2 were measured after having the locus of the supposed surface mounting job moved back and forth between E_1 and E_2. Measuring the accuracy 15 times showed that the accuracy of positioning repeatability at E_1 was ±9μm, while that of E_2 was ±11μm.

In considering that the accuracy of positioning repeatability is quite high when this pantograph mechanism is used as the surface mount system, and also considering that the positioning accuracy of the present

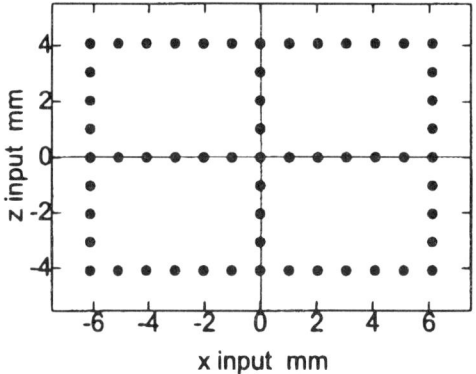

Figure 9. Input displacement

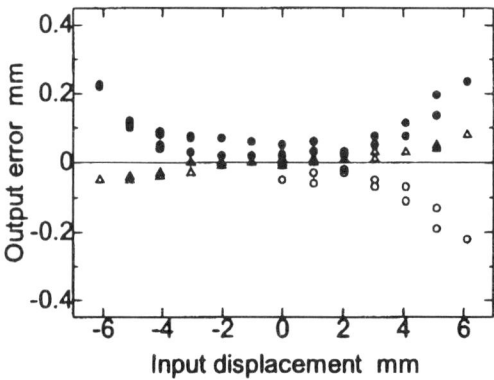

Figure 10. Output displacement error (Zin=0; Moving direction is horizontal)

surface mount system is ±100 μm, the open loop control is therefore proven to be useful. Moreover, if the output ending point can hold some electric parts up to the maximum of 0.5 gf, the time of carrying the parts onto the substrate is then calculated to be 0.1 second based on the fatigue limit of the hinge parts. Furthermore, if each of the detecting, recognizing, maintaining, revolving, mounting of machine parts, as well as eliminating inferior parts requires 0.1 second, as does the present system, the mounting time for miniature manipulators will be 0.8 seconds. Since this enables over 8 manipulators to do the same work, a new system will be able to produce more machine parts than the presently used system. It is also found that a model-devised pantograph mechanism is quite small and only needs less than approximately 0.1 % of the space that rotary head parts of the presently used system require. These above-mentioned facts indicate that the presently used surface mounting system can be minimized even in cases where a great number of manipulators are placed in a certain working space with appropriate actuators.

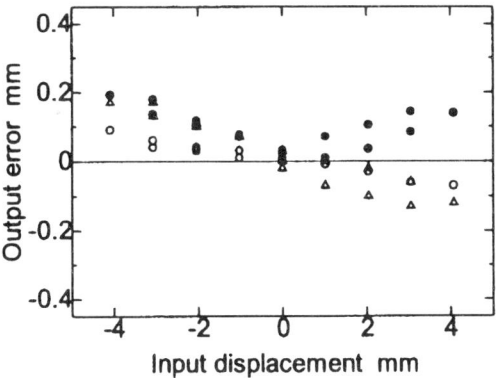

Figure 11. Output displacement error (Xin=0; Moving direction is vertical)

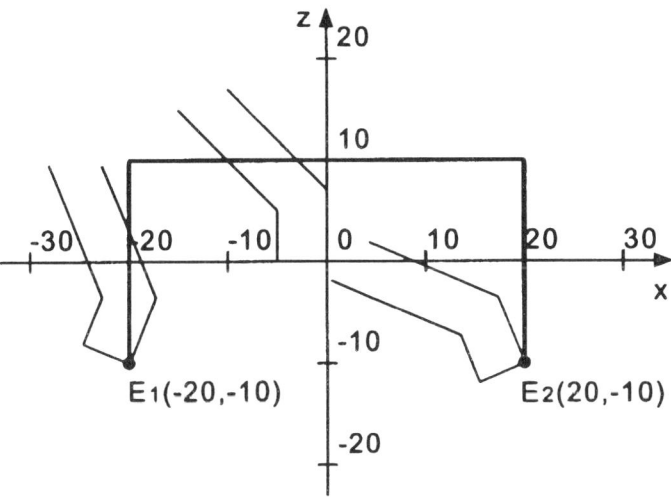

Figure 12. Experimental trajectory (Surface mount job is supposed)

5 Conclusion

In order to minimize the presently used system, a new surface mount system consisting of miniature manipulators was proposed, and integral molded pantograph mechanisms were then model-devised and researched. This research clarifies the durability as well as input-output displacement characteristics of the mechanisms. The results obtained from this research are as follows.

(1) An integral molded pantograph mechanism consisting of large deflective hinges with 200 μm of the length, 180 μm of the thickness and 5 mm of the width did not fracture in displacement input fatigue tests on the mechanism even after repeatedly

used 1 million times. In this test, the maximum relative angular displacement between links was set at 45°.

(2) This research also clarifies that the integral molded pantograph mechanism, with above-mentioned hinge parts, had 340μm of output displacement errors within the working space of 50mm × 40mm, while the hysteresis errors were determined to be 130μm.

(3) The accuracy of positioning repeatability of this integral molded pantograph mechanism was clari fied to be ±11μm in the positioning test where the practical mount work had been estimated. Based on this, the mechanism was evaluated when used as a surface mount system. This indicates that there is some possibility that the proposed system can minimize the present surface mount system.

References

Horie M. , Takayama M. , Kamiya D. and Ikegami, K. (1998). Development of Pantograph Mechanisms Uniformly Molded with Large-Deflective Elastic Hinges and Links made from the Same Materials. *Proceeding of Movic'98* (The Fourth International Conference on Motion and Vibration Control), Vol.3, 913-918.

Horie, M., Nozaki, Ikegami, K., and Kobayashi, F.(1997). Design System of Super Elastic Hinges and Its Application to Micro-Manipulators. *JSME International Journal Ser.C*, Vol.40, No.2, 323-328.

Kamiya, D., Hayama, T., and Horie, M. (1999). Electrostatic Comb-drive Actuators made of Polyimide for Actuating Micromotion Convert Mechanisms. *J.of Microsystem Technologies*, 5, 161-165.

Kim, K., and Horie, M. (1999). Displacement Analysis of Micro Parallel Spring Mechanisms with Large-Deflective Elastic Hinges. *Proceeding of 1999 International Conference on Advanced Manufacturing Technology (ICAMT'99)*, 197-201.

Kim, K., and Horie, M., and Sugihara, T. (1998). Displacement Characteristics of a Parallel Leaf Spring Mechanism with Large-DeflectiveElastic Hinges for Optical Mount. *Proceeding of the 13th. Korea Automatic Control Conference(KACC'98, International Session Papers)*, 484-489.

Adaptive λ-Tracking for Rigid Manipulators

Alicja Mazur[1] and Carsten Schmidt[2]*

[1] Institute of Engineering Cybernetics, Wrocław University of Technology, Poland
[2] Institute for Robotics and Process Control, Technical University of Braunschweig, Germany

1 Introduction

This paper deals with a modification of a simple trajectory tracking algorithm for rigid manipulators, namely a modification of the classical PD controller with static gain, which does not employ any specific knowledge of the robot dynamics. In (Qu and Dorsey, 1991) it has been shown that

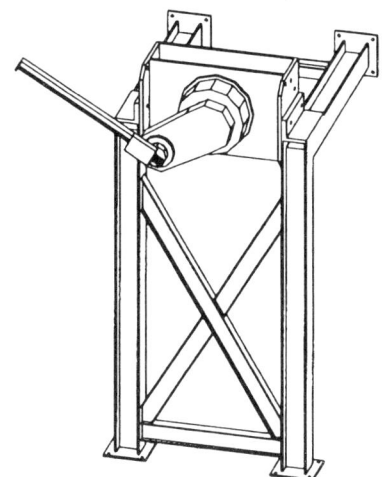

Figure 1. The Experimental Direct Drive Arm (EDDA), which served as a testbed.

the classical PD controller is able to keep position errors within certain bounds. It is, however, impossible to give a simple evaluation for those bounds. Here we introduce a dynamical PD controller, which allows to predefine some error bounds even in the absence of the knowledge of the robot dynamics. The region which the tracking error is converging to, depends only on the design parameters of the PD controller. The algorithm is in fact a universal adaptive control system with a dead zone of width $\lambda > 0$. This application is in the spirit of λ-tracking introduced by (Mazur, 1996, Mazur and Hossa, 1997). An interesting part of the paper is the practical evaluation of the

---

* The authors are grateful to Dr. Achim Ilchmann, Department of Mathematics & Centre for Systems and Control Engineering, University of Exeter, for his help with completing the proof and for the revision of the mathematical parts of the paper.

proposed control algorithm using the rigid manipulator *EDDA* (Experimental Direct Drive Arm) at the Institute for Robotics and Process Control in Braunschweig, see Figure 1. The experiments will demonstrate the successful application of the algorithm in practice. They will further serve to present some relationship between the choice of control parameters and the behaviour of the position tracking error.

The paper is organized as follows. In Section 2 the system class is described. In Section 3 we state the control objective in the main theorem and discuss the theoretical results. The experimental results are given and discussed in Section 4. Section 5 contains some conclusions, and the proof of the main theorem is given in the Appendix.

2 System Class

We consider the motion of a rigid manipulator described by the following nonlinear differential equation:

$$M(x)\ddot{x} + C(x,\dot{x})\dot{x} + G(x) + T(\dot{x}) = u, \tag{1}$$

where $x(t)$ is a $n \times 1$ vector of joint variables, $M(x)$ is an $n \times n$ inertia matrix (symmetric and positive definite), $C(x,\dot{x})$ is an $n \times n$ matrix of Coriolis and centripetal terms, $G(x)$ is an $n \times 1$ vector of gravity terms and $T(\dot{x})$ is an $n \times 1$ vector of friction terms.
All of these functions are assumed to be continuous and to have the following properties:

Property 1. *(Craig)*
The inertia matrix $M(x)$ satisfies, for some (unknown) $m_1, m_2 > 0$,

$$m_1 I \leq M(x) \leq m_2 I, \quad \text{for all} \quad x \in R^n.$$

Property 2. *(Craig)*
The matrix of Coriolis and centripetal terms has the special form

$$C(x,\dot{x}) = \begin{bmatrix} \dot{x}^T N_1(x) \\ \vdots \\ \dot{x}^T N_n(x) \end{bmatrix}$$

where the $n \times n$ matrices $N_i(x)$, $i = 1,\ldots,n$ are symmetric and bounded. (In fact, x appears only as the argument of sine and cosine functions.)

Property 3. *(Berghuis)*
For manipulators with only revolute joints we have, for some $G_M > 0$,

$$\|G(x)\| \leq G_M, \quad \text{for all} \quad x \in R^n.$$

3 Main Theorem

In this section we state and explain the main theorem. We do not assume any knowledge of the system parameters, only the structural Properties 1-3 are required to hold. Our goal is to present a simple adaptive algorithm so that $x(t)$ asymptotically tracks (within a prespecified fixed error) any desired reference trajectory $x_d(t)$ with bounded first and second derivatives.

To this end we use the following notation for the ball of radius $\eta > 0$ centered at 0 in the n-dimensional Euclidean space:

$$\overline{B}_\eta(0) := \{e \in R^n \mid \|e\| \leq \eta\}.$$

Theorem 4. *Let* $\lambda > 0$, $P = diag\{P_1,\ldots,P_n\}$, *and* $D = diag\{D_1,\ldots,D_n\}$ *for some* $P_i, D_i > 0$, *with* $i = 1,\ldots,n$. *We define*

$$e(t) = x(t) - x_d(t), \quad \text{position tracking error}$$
$$E(t) = Pe(t) + D\dot{e}(t), \quad \text{combined tracking error}$$

for some desired reference trajectory $x_d(\cdot)$ *with* $\dot{x}_d(\cdot), \ddot{x}_d(\cdot) \in L_\infty$. *If the time-varying proportional feedback*

$$u(t) = -k(t)E(t), \tag{2}$$

with gain adaptation

$$\dot{k}(t) = \begin{cases} (\|E(t)\| - \lambda) \cdot \|E(t)\|, & \text{if } \|E(t)\| > \lambda, \\ 0, & \text{if } \|E(t)\| \leq \lambda, \end{cases} \tag{3}$$

for arbitrary $k(0) \in R$ *is applied to (1) with arbitrary* $x(0) \in R^n$, *then the closed-loop system (1),(2),(3) has an absolutely continuous unique solution* $(e(\cdot), k(\cdot))$ *on a maximally extended interval* $[0, \omega)$, $\omega \in (0, \infty]$. *If* $\dot{x}(\cdot)$ *is bounded, then we have*

(i) $\omega = \infty$, *i.e. no finite escape time*,
(ii) $\lim_{t \to \infty} k(t)$ *exists and is finite*,
(iii) $E(t) = Pe(t) + D\dot{e}(t) \to \overline{B}_\lambda(0)$ *as* $t \to \infty$,
(iv) $e_i(t) \to \overline{B}_{\frac{\lambda}{P_i}}(0)$ *as* $t \to \infty$,
(v) $\dot{e}_i(t) \to \overline{B}_{\frac{\lambda}{D_i}}(0)$ *as* $t \to \infty$.

The proof of the theorem is delegated to the Appendix.

Note that the reference trajectory might have jumps since we do only assume that it is essentially bounded and its first and second order derivatives exist almost everywhere. Note also that if there is measurement noise which is known to be smaller than λ, then the controller copes with it due to the dead zone.

We have introduced the matrices P and D in order to scale independently the size of the region which the i-th element of the position tracking error is converging to. Note that the i-th element of the combined tracking error E_i is a signal from the local P_iD_i-controller with static gain designed for the i-th joint of a rigid manipulator. The time-varying control u, equation (2), is an output signal of a PD-controller with dynamical gain. The universal adaptive controller with dead zone produces the global gain $k(t)$ which is used to scale the gain of all local P_iD_i-controllers. Theorem 4 states that the proposed control law u without any information about robot dynamics ensures the convergence of the combined tracking error E, a linear combination of the position error e and the velocity error $\dot{e}$, to the predefined region. A simple consequence of this is the convergence of the position error e_i of the i-th joint and of the velocity error $\dot{e}_i$ to prespecified regions.

If the robot is perturbed, then the adaptive controller drives it back again and this is the main reason for applying an adaptive controller instead of running experiments and estimating the gain. If the gain becomes very large and the perturbation has settled down, then the process engineer might reset the gain.

4 Experimental Results

According to Theorem 4 (iii), the combined tracking error $E(t)$ converges to a ball of radius λ centered at 0, and the controller gain $k(t)$ converges to a finite value, see (ii). This will be illustrated by experiments with different values for λ, showing the behaviour of $E(t)$ and $k(t)$. Furthermore, a relationship for the position tracking error has been stated in (iv). It relates the i-th element of the position tracking error e_i to the ratio $\frac{\lambda}{P_i}$. These results will be illustrated by experiments with different values for P_i.

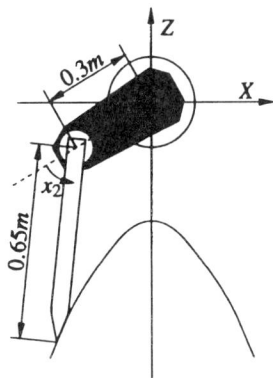

Figure 2. Test trajectory for EDDA

The experiments were conducted using the experimental two link arm EDDA (Experimental Direct Drive Arm) at the Institute for Robotics and Process Control in Braunschweig (see Figure 1). The joints are directly driven by the electromagnetic forces of the motors, and thus no gears inertia and friction will obscure the typical nonlinear coupling effects of robot dynamics. This makes EDDA an ideal testbed for analysing the behaviour of λ-tracking with robot dynamics.

The following Cartesian test trajectory has been used (t in seconds and X and Z in meters):

$$X(t) = 0.35 \cdot \sin(\frac{\pi}{2}t - \frac{\pi}{2}), \qquad Z(t) = -0.625 - 0.225 \cdot \sin(\pi t + \frac{\pi}{2}). \tag{4}$$

It is shown in Figure 2 that the movement starts at the left turning point with no position and velocity error. Since we set the controller gain k initially to zero, the robot starts its movement by dropping out of the tracking error convergence zone $\bar{B}_\lambda(0)$ defined by λ. This fact can be observed in Figure 3, which shows the first joint component of the combined tracking error $E(t)$.

As a result, the controller gain k increases quickly up to a value of about 28. By then the combined tracking error E has returned to the dead zone and k stops changing. It starts increasing again, when the movement changes direction at the right turning point of the trajectory. The acceleration in conjunction with gravitation causes the tracking error E to leave the dead zone for a short period, which leads to a further increase of k. After several cycles, however, the departures from $\bar{B}_\lambda(0)$ become ever shorter and the increase of k is slowing down, as can be seen in Figure 5.

Figures 4 and 6 are demonstrating the convergence to different regions $\bar{B}_\lambda(0)$ for different values of λ. The corresponding position tracking errors can be seen in Figure 7. It should be mentioned, that a choice of a large value for λ usually causes k to reach its final value at once.

The experiments in Figure 8 are showing the position tracking error $e_1(t)$ for the first joint. According to Theorem 4 (iv) it should converge to the $\frac{\lambda}{P_1}$-strip. The diagrams are showing this for a succession of different parameters P_1. The position tracking error of the second joint is much smaller than the corresponding limit $\frac{\lambda}{P_2}$, since the test movement requires much less torque for moving the second joint. The value of k is thus determined by the combined tracking error of the first joint, which just reaches the convergence zone.

5 Conclusion

A new control scheme for rigid manipulators namely a dynamical version of the classical PD controller has been presented. With the classical PD controller it shares the simplicity and no knowledge of the robot dynamics is necessary. Yet unlike the classical PD controller the range of the tracking error is predictable and a function of the controller parameters only. A model of the robot dynamics in particular is not necessary to determine the convergence region for the tracking error.

6 Appendix

PROOF OF THEOREM 4

Using the notation

$$d_\lambda(E) := \begin{cases} (\|E\| - \lambda), & \text{if } \|E\| > \lambda, \\ 0, & \text{if } \|E\| \leq \lambda, \end{cases}$$

the closed-loop system (1)-(3) may be written as

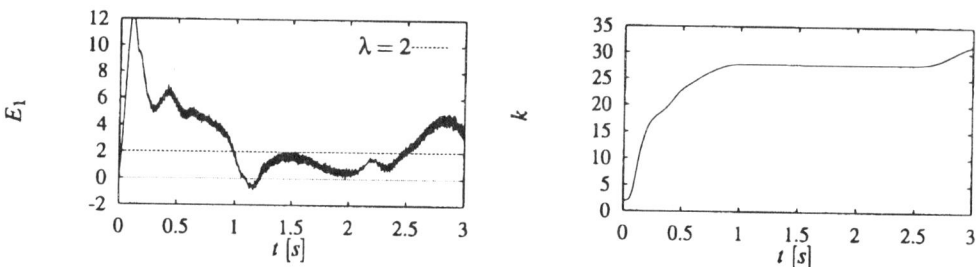

Figure 3. The first joint component of the combined tracking error $E(t)$ and the corresponding controller gain $k(t)$. $\lambda = 2$, $P_1 = 120$, $D_1 = 4.4$.

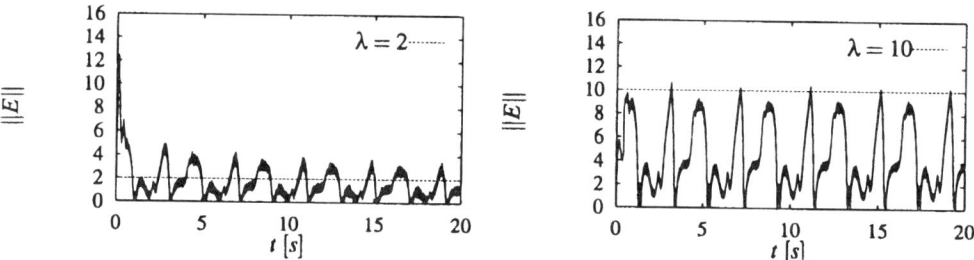

Figure 4. $\|E(t)\|$ for different values of λ. $P_1 = 120$, $P_2 = 27.28$, $D_1 = 4.4$, $D_2 = 1$.

$$\left.\begin{array}{l} M(x)\ddot{e} = -k[Pe + D\dot{e}] - C(x,\dot{x})\dot{e} - C(x,\dot{x})\dot{x}_d - G(x) - T(\dot{x}) - M(x)\ddot{x}_d, \\ \dot{k} = d_\lambda(E) \cdot \|E\|. \end{array}\right\} \quad (5)$$

By the Properties 1-3 it follows from the classical theory of ordinary differential equations that there exists a unique solution $(e(\cdot), k(\cdot))$ which is absolutely continuous and can be maximally extended over some interval $[0, \omega)$, $\omega \in (0, \infty]$. We now proceed in six steps.

STEP 1: We prove boundedness of $k(\cdot)$ on $[0, \omega)$.

Seeking a contradiction suppose that $\lim_{t \to \omega} k(t) = \infty$. It is easily seen that the first equation in (5) is equivalent to

$$\dot{E} = -kDM(x)^{-1}E + P[\dot{x} - \dot{x}_d] - D\ddot{x}_d - DM(x)^{-1}[C(x,\dot{x})\dot{x} + G(x) + T(\dot{x})]. \quad (6)$$

Set, for some $\rho > 0$ to be specified later,

$$\Delta_\rho(E) := \begin{cases} \|E\|_{D^{-1}} - \rho, & \text{if } \|E\|_{D^{-1}} > \rho, \\ 0, & \text{if } \|E\|_{D^{-1}} \le \rho, \end{cases}$$

where $\|E\|_{D^{-1}} := \sqrt{E^T D^{-1} E}$. Now differentiating the Lyapunov like function

$$V_\rho(E) := \frac{1}{2}\Delta_\rho(E)^2$$

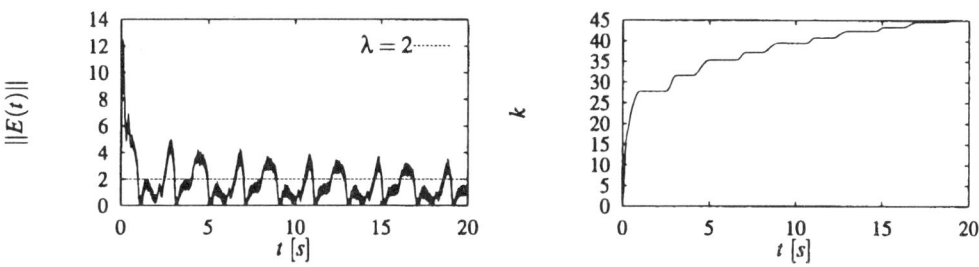

Figure 5. $\|E(t)\|$ and corresponding $k(t)$ with $\lambda = 2$, $P_1 = 120$, $P_2 = 27.28$, $D_1 = 4.4$, $D_2 = 1$.

along the solution of (6) yields, for almost all $t \in [0, \omega)$,

$$\frac{d}{dt}V_\rho(E(t)) = \Delta_\rho(E)\|E\|_{D^{-1}}^{-1} E^T D^{-1} \dot{E}$$
$$= -k\Delta_\rho(E)\|E\|_{D^{-1}}^{-1} E^T M(x)^{-1} E + \Delta_\rho(E)\|E\|_{D^{-1}}^{-1} E^T \xi, \quad (7)$$

where

$$\xi := D^{-1} P[\dot{x} - \dot{x}_d] - \ddot{x}_d - M(x)^{-1}[C(x,\dot{x})\dot{x} + G(x) + T(\dot{x})].$$

Since for some $\hat{m}_1, \hat{m}_2 > 0$ we have $\hat{m}_1 \leq M(x)^{-1} \leq \hat{m}_2$ for all $x \in R^n$ and by assumption $\dot{x}, \dot{x}_d, \ddot{x}_d \in L_\infty$, it follows that $\xi \in L_\infty$. Therefore applying the inequality

$$p_1 \|E\| \leq \|E\|_{D^{-1}} \leq p_2 \|E\|,$$

which holds for $p_1 := \min_i D_i^{-1/2}$ and $p_2 := \min_i P_i^{-1/2}$, and choosing $t_1 \in [0, \omega)$ sufficiently large such that $k(t) \geq 0$ for all $t \in [t_1, \omega)$, yields, for some $A_1 > 0$ and almost all $t \in [t_1, \omega)$,

$$\frac{d}{dt}V_\rho(E(t)) \leq -\frac{\hat{m}_1}{p_2} k(t) \Delta_\rho(E)\|E\| + A_1 \Delta_\rho(E).$$

Now $\Delta_\rho(E) \geq 0$ implies

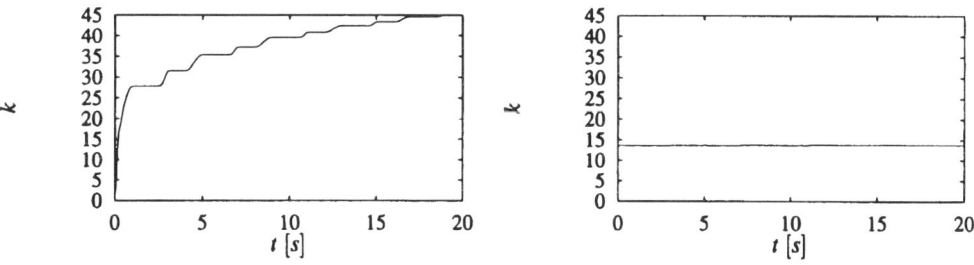

Figure 6. The controller gain $k(t)$ for different values of λ. $P_1 = 120$, $P_2 = 27.28$, $D_1 = 4.4$, $D_2 = 1$.

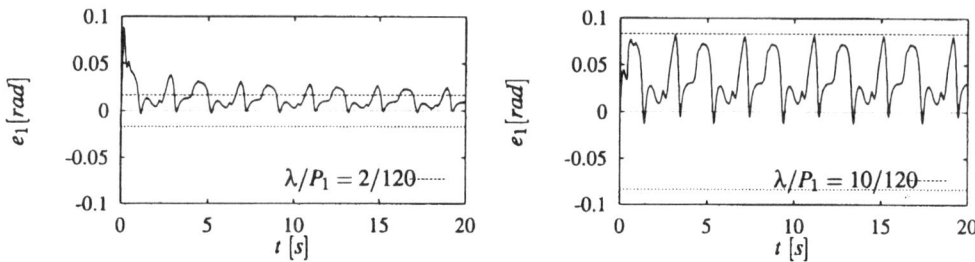

Figure 7. The tracking error $e_1(t)$ of the first joint for different values of λ. $P_1 = 120$, $P_2 = 27.28$, $D_1 = 4.4$, $D_2 = 1$.

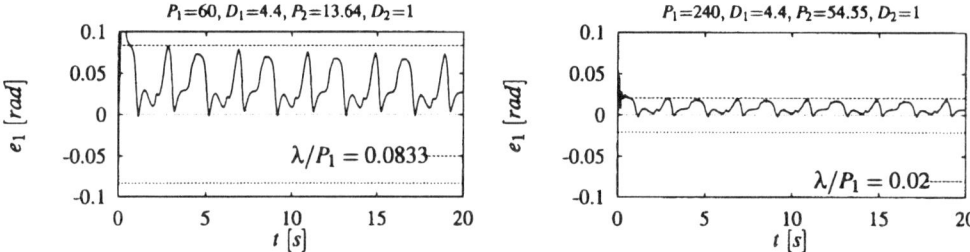

Figure 8. Position tracking error $e_1(t)$ of the first joint for different values of P. $\lambda = 5$, D is left unchanged.

$$1 \leq \frac{p_2}{\rho}\|E\| \qquad (8)$$

and hence

$$\frac{d}{dt} V_\rho(E(t)) \leq -\left(\frac{\hat{m}_1}{p_2} k(t) - \frac{p_2}{\rho} A_1\right) \Delta_\rho(E)\|E\|.$$

If we choose $t_2 \in [t_1, \omega)$ such that

$$\frac{\hat{m}_1}{p_2} k(t) - \frac{p_2}{\rho} A_1 \geq 0 \quad \text{for all} \quad t \in [t_2, \omega)$$

and use (see Lemma 2 in (Ilchmann and Ryan, 1994))

$$\Delta_{p_1\lambda}(E) \geq p_1 d_\lambda(E),$$

then we obtain, for almost all $t \in [t_2, \omega)$,

$$\frac{d}{dt} V_{p_1\lambda}(E(t)) \leq -\left(\frac{\hat{m}_1}{p_2} k(t) - \frac{p_2}{\rho} A_1\right) p_1 d_\lambda(E)\|E\| = -\left(\frac{\hat{m}_1}{p_2} k(t) - \frac{p_2}{\rho} A_1\right) p_1 \dot{k}. \qquad (9)$$

Now integration and substitution yields, for all $t \in [t_2, \omega)$,

$$0 \leq V_{p_1\lambda}(E(t)) \leq V_{p_1\lambda}(E(t_2)) - \frac{\hat{m}_1 p_1}{2p_2}\left(k(t)^2 - k(t_2)^2\right) + \frac{p_2 A_1 p_1}{p_1 \lambda}\left(k(t) - k(t_2)\right). \tag{10}$$

This contradicts unboundedness of $k(\cdot)$ and hence the statement is proved.
STEP 2: Boundedness of $k(\cdot)$ on $[0,\omega)$ applied to (6) yields $\omega = \infty$.
STEP 3: We prove boundednes of $E(\cdot)$.
Since $k \in L_\infty$ and (8) holds true, there exists some A_2 such that (7) yields

$$V_\lambda(E(t)) \leq V_\lambda(E(0)) + A_2 \int_0^t \Delta_\rho(E(\tau))\|E(\tau)\|\, d\tau = V_\lambda(E(0)) + A_2[k(t) - k(0)]. \tag{11}$$

Applying boundedness of k once more proves the claim.
STEP 4: We prove $\lim_{t \to \infty} d_\lambda(E(t)) = 0$, which yields (iii).
Since k is bounded and

$$k(t) \geq k(0) + \int_0^t d_\lambda(E(\tau))\|E(\tau)\|\, d\tau \geq k(0) + \int_0^t d_\lambda(E(\tau))^2\, d\tau$$

we have $d_\lambda(E)^2 \in L_1$. Applying boundedness of $k, E, \dot{x}, \dot{x}_d, \ddot{x}_d$ to (7) yields $\frac{d}{dt} \frac{1}{2} d_\lambda(E)^2 \in L_\infty$. Now by Lemma 2.1.7 in (Ilchmann, 1993) the claim follows.
STEP 5: We prove (iv).
Let $\varepsilon > 0$ be arbitrary. Since

$$0 \leq |P_i e_i(t) + D_i \dot{e}_i(t)| \leq \|E(t)\|,$$

by (iii) there exists some $t' > 0$ such that, for almost all $t \geq t'$,

$$-(\lambda + \varepsilon) \leq P_i e_i(t) + D_i \dot{e}_i(t) \leq \lambda + \varepsilon \tag{12}$$

and hence we conclude by Variation-of-Constants

$$-\frac{\lambda + \varepsilon}{P_i} \leq e_i(t) - \exp^{-\frac{P_i}{D_i}(t-t')} e_i(t') \leq \frac{\lambda + \varepsilon}{P_i},$$

whence

$$e_i(t) \to \overline{B}_{\frac{\lambda+\varepsilon}{P_i}}(0) \quad \text{as} \quad t \to \infty.$$

Since $\varepsilon > 0$ was chosen arbitrarily, the claim follows.
STEP 6: Statement (v) is a consequence of (12) and (iv).
This completes the proof.

References

Berghuis, H. (1993). *Model-Based Robot Control: from Theory to Practice*. Ph.D. Dissertation, Twente University, Enschede.

Craig, J. (1988). *Adaptive Control of Mechanical Manipulators*. New York: Addison-Wesley.

Ilchmann, A., and Ryan, E. (1994). Universal λ-tracking for nonlinearly-perturbed systems in the presence of noise. *Automatica* 30(2):337–346.

Ilchmann, A. (1993). *Non-Identifier-Based High-Gain Adaptive Control*. London: Springer.

Mazur, A., and Hossa, R. (1997). Universal adaptive λ-tracking controller for wheeled mobile robots. In *Proceedings of the Fifth IFAC Symposium on Robot Control SYROCO'97*, 33–38.

Mazur, A. (1996). *Robot Control Algorithms Based on the Principle of Universal Adaptive Control*. PhD dissertation, (in polish), Wrocław University of Technology, Wrocław.

Qu, Z., and Dorsey, J. (1991). Robust tracking control of robots by linear feedback law. *IEEE Transactions on Automatic Control* 36(9):1081–1084.

Mobility Analysis of the 3-PSP Mechanism

Raffaele Di Gregorio[1] and Vincenzo Parenti-Castelli[2*]

[1] Department of Engineering, University of Ferrara, Italy
[2] DIEM, University of Bologna, Italy

Abstract. The mobility analysis of a mechanism aims to find the singularity configurations of mechanism itself and is of basic importance for control of motion. This paper presents the mobility analysis of the three degrees of freedom 3-PSP mechanism. The conditions for the occurrence of architecture singularities and configuration singularities are found. Finally, several singular configurations are also interpreted geometrically.

1 Introduction

Spatial parallel mechanisms are single or multiple loop mechanisms that feature a fixed link (base) connected to a movable link (platform) by a number of independent chains (legs). A certain number of kinematic pairs is actuated and provide the platform with a corresponding number of degrees of freedom (dofs) with respect to the base.

Six-dof parallel mechanisms have been extensively studied in the last decade. Many applications, however, require less than six dofs. In particular, many tasks require three dofs. Three-dof mechanisms have recently been studied, which provide the platform with a motion of pure translation, pure rotation, or with a combination of rotation and translation (see Di Gregorio and Parenti-Castelli, 1999a, for references).

The mobility analysis of a mechanism, that is to find the singularity configurations, i.e., the configurations for which the linear mapping between the independent and dependent variables at the rate level is singular, is of fundamental importance for the control of motion. Indeed, in general, in a singular configuration the motion of the platform is completely out of control. The singularity analysis of mechanisms has been extensively studied in the literature. In particular, in (Gosselin and Angeles, 1990; Ma and Angeles, 1991) a classification of the types of singularities has recently been reported, and architecture, configuration, and representation singularities have been pointed out. In particular, architecture singularities depend on the geometric parameters of the mechanism and make the mechanism singular in all configurations. In architecture singular configurations the mechanism can undergo finite motions when the actuators are locked or a finite motion of the inputs produces no motion of the outputs (Gosselin and Angeles, 1990).

---

[*] The authors would like to thank Dr. C. Innocenti for the fruitful discussions.
The funding of the Italian MURST is gratefully acknowledged.

In this paper the mobility analysis of a rotation-translation 3-dof parallel mechanism is presented. The base and the platform of the mechanisms are connected to each other by three serial legs of type PSP, P and S being prismatic and spherical pairs respectively (see Fig. 1). Three prismatic pairs in the base are actuated thus providing the platform with three dofs. The mechanism, henceforth called 3-PSP mechanism, is topologically symmetric, i.e., the role of the base and the platform can be interchanged. Applications of this mechanism are reported in (Di Gregorio and Parenti-Castelli, 1998) where the ability to locate a plane fixed to the platform in a given position is exploited in biomedical devices.

This paper presents in analytic form the conditions that provide architecture and configuration singularities as functions of the geometric and configuration parameters of the 3-PSP mechanism. A geometric interpretation of several singularities is also shown.

2 Mobility Analysis of the 3-PSP Mechanism

Figure 2 shows the i-th leg of the mechanism. S_b and S_p are arbitrary reference systems fixed to the base and to the platform respectively, points O_b and O_p are their origins. With reference to Figures 1 and 2, the following closure equations in matrix form can be written

$$\mathbf{p} - \mathbf{a}_i + \mathbf{R}_{bp}{}^P\mathbf{b}_i = s_i\mathbf{v}_i - t_i \mathbf{R}_{bp}{}^P\mathbf{w}_i \qquad i=1,2,3 \tag{1}$$

where $\mathbf{p}=(O_p-O_b)$, $\mathbf{a}_i=(A_i-O_b)$, $\mathbf{b}_i=(B_i-O_p)$, $\mathbf{v}_i$ and $\mathbf{w}_i$ are unit vectors of the sliding directions of the prismatic pairs of the base and of the platform respectively. Points A_i (B_i) are arbitrarily

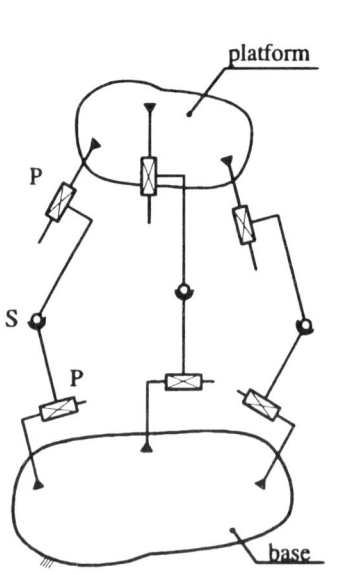

Figure 1. The 3-PSP mechanism

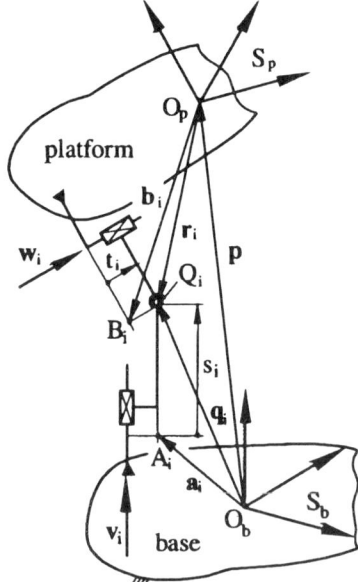

Figure 2. The i-th leg of the 3-PSP mechanism

chosen on the line through point Q_i (center of the spherical pair) and parallel to (unit vector) v_i (w_i), $s_i = A_i Q_i$ ($t_i = B_i Q_i$) are the variables of motion of the prismatic pairs in the base (platform). All vectors are measured in system S_b unless specified by a left hand side superscript (p) that indicates the reference system (S_p) the vector is measured in. Rotation matrix R_{bp} defines the orientation of system S_p with respect to system S_b.

System (1) is the closure equation system of the 3-PSP mechanism. It represents a system of nine scalar equations in twelve variables: the six joint variables s_i and t_i, i=1,2,3, the three components of vector p, and the three orientation parameters that define the matrix R_{bp}. Three variables can be arbitrarily assigned, i.e., the mechanism has 3-dofs, and the remaining six variables can be found by solving system (1).

2.1 Singularity Conditions in the Large (for finite displacements)

It is worth noting that if three platform orientation parameters are given, i.e., if matrix R_{bp} is given, system (1) becomes a linear system in the remaining nine variables s_i, t_i, i=1,2,3, and the three components of vector p. Thus, a one to one correspondence exists between the orientation of the platform and the remaining nine variables. For this reason, the three orientation parameters of the platform are taken as reference parameters to characterize the configuration of the 3-PSP mechanism. This correspondence fails when for a given matrix R_{bp} system (1) is singular. The platform orientations that make system (1) singular but not impossible identify singular configurations in which the 3-PSP mechanism gains one or more dofs in the large, i.e., some links have finite displacements.

Inspection of system (1) reveals that vector p can be linearly eliminated. In fact, subtracting, in order, the three scalar equations for i=2 and respectively the three scalar equations for i=3 from the three scalar equations for i=1 and rearranging them yields

$$\begin{bmatrix} v_1 & -v_2 & 0 & -R_{bp}{}^P w_1 & R_{bp}{}^P w_2 & 0 \\ v_1 & 0 & -v_3 & -R_{bp}{}^P w_1 & 0 & R_{bp}{}^P w_3 \end{bmatrix} \begin{Bmatrix} s \\ t \end{Bmatrix} = \begin{Bmatrix} a_2 - a_1 + R_{bp}({}^P b_1 - {}^P b_2) \\ a_3 - a_1 + R_{bp}({}^P b_1 - {}^P b_3) \end{Bmatrix} \quad (2)$$

where $s = (s_1, s_2, s_3)^T$ and $t = (t_1, t_2, t_3)^T$. T is for transpose.

System (2) is singular when the determinant of its coefficient matrix, M, is zero, i.e., when the following condition holds

$$\det[M] = 0 \quad (3)$$

where

$$M = \begin{bmatrix} v_1 & -v_2 & 0 & -R_{bp}{}^P w_1 & R_{bp}{}^P w_2 & 0 \\ v_1 & 0 & -v_3 & -R_{bp}{}^P w_1 & 0 & R_{bp}{}^P w_3 \end{bmatrix} \quad (4)$$

By using an algebraic manipulator it can be demonstrated (Di Gregorio and Parenti-Castelli, 1999b) that the following relationship holds

$$\det[\mathbf{M}] = \mathbf{n}_1 \cdot \mathbf{n}_2 \times \mathbf{n}_3 \tag{5}$$

where the position $\mathbf{n}_i = \mathbf{v}_i \times \mathbf{w}_i$, i=1,2,3, has been adopted.
By taking into account the expression (5) Eq. (3) becomes

$$\mathbf{n}_1 \cdot \mathbf{n}_2 \times \mathbf{n}_3 = 0 \tag{6}$$

The mechanism configurations which satisfy equation (6) (vectors $\mathbf{n}_i$, i=1,2,3, linearly dependent) make system (2) singular and, as a consequence, make system (1) singular too.

Equation (6) is a scalar equation in the three orientation parameters of the platform and contains all the geometric parameters which identify the geometry of the 3-PSP mechanism. Thus, the values of the orientation parameters which satisfy equation (6) identify a surface in the three dimensional space of the orientation parameters of the platform (for a 3-PSP mechanism whose geometry is given). This surface is the geometric locus of the configurations of the mechanism which are singular in the large.

When system (1) is singular for any feasible R_{bp}, some relations exist among the geometric parameters of the mechanism (architecture singularity). Under this condition, a motion in the large of the platform may occur for special geometry of the mechanism.

2.2 Singularity Conditions in the Small (for infinitesimal displacements)

Differentiating system (1) with respect to time yields

$$\dot{\mathbf{p}} + \omega \times \mathbf{b}_i = \dot{s}_i \mathbf{v}_i - \dot{t}_i \mathbf{w}_i - t_i \, \omega \times \mathbf{w}_i \qquad i=1,2,3 \tag{7}$$

where ω is the angular velocity of the platform with respect to the base. System (7) is a linear system of nine equations in the twelve unknowns $\dot{s}_i$, $\dot{t}_i$, i=1,2,3, $\dot{\mathbf{p}}$, and ω.

Vector $\dot{\mathbf{p}}$ can be eliminated from system (7) in a similar way that vector $\mathbf{p}$ was eliminated from system (1) to obtain system (2). After rearranging, the following system is obtained

$$\omega \times (\mathbf{q}_1 - \mathbf{q}_2) = \dot{s}_1 \mathbf{v}_1 - \dot{s}_2 \mathbf{v}_2 - \dot{t}_1 \mathbf{w}_1 + \dot{t}_2 \mathbf{w}_2 \tag{8.1}$$

$$\omega \times (\mathbf{q}_1 - \mathbf{q}_3) = \dot{s}_1 \mathbf{v}_1 - \dot{s}_3 \mathbf{v}_3 - \dot{t}_1 \mathbf{w}_1 + \dot{t}_3 \mathbf{w}_3 \tag{8.2}$$

where (see Figures 1 and 2) the relationships $\mathbf{r}_i = \mathbf{b}_i + t_i \mathbf{w}_i$, and $\mathbf{r}_i - \mathbf{r}_j = \mathbf{q}_i - \mathbf{q}_j$, i≠j, have been used.

System (8) is a linear system of six equations in the nine variables $\dot{s}_i$, $\dot{t}_i$, i=1,2,3, and ω. For any three arbitrarily given variables, system (8) provides the remaining six. The component of vector $\dot{\mathbf{p}}$ can then be found from any three i-th equations of system (7).

Direct problem. If vector $\dot{\mathbf{s}}$ is taken as input, system (8) can be written as

$$\begin{bmatrix} (\mathbf{q}_2-\mathbf{q}_1)\times \mathbf{R}_{bp}{}^P\mathbf{w}_1 & -\mathbf{R}_{bp}{}^P\mathbf{w}_2 & 0 \\ (\mathbf{q}_3-\mathbf{q}_1)\times \mathbf{R}_{bp}{}^P\mathbf{w}_1 & 0 & -\mathbf{R}_{bp}{}^P\mathbf{w}_3 \end{bmatrix} \begin{Bmatrix} {}^b\omega \\ \dot{t}_1 \\ \dot{t}_2 \\ \dot{t}_3 \end{Bmatrix} = \begin{bmatrix} \mathbf{v}_1 & -\mathbf{v}_2 & 0 \\ \mathbf{v}_1 & 0 & -\mathbf{v}_3 \end{bmatrix} \begin{Bmatrix} \dot{s}_1 \\ \dot{s}_2 \\ \dot{s}_3 \end{Bmatrix} \quad (9)$$

and solved for $\dot{t}$ and ω. Here $(.)\times$ is the skew symmetric matrix related with the vector $(.)$. System (9) is singular when the following condition holds

$$\det[\mathbf{N}] = 0 \quad (10)$$

where

$$\mathbf{N} = \begin{bmatrix} (\mathbf{q}_2-\mathbf{q}_1)\times \mathbf{R}_{bp}{}^P\mathbf{w}_1 & -\mathbf{R}_{bp}{}^P\mathbf{w}_2 & 0 \\ (\mathbf{q}_3-\mathbf{q}_1)\times \mathbf{R}_{bp}{}^P\mathbf{w}_1 & 0 & -\mathbf{R}_{bp}{}^P\mathbf{w}_3 \end{bmatrix} \quad (11)$$

The position vectors q_i, i=1,2,3, that appear in the entries of the matrix **N** can be written as functions of the actuated joint variables, s_i, i=1,2,3. The actuated joint variables, provided the platform orientation parameters which satisfy Eq. (6) are not considered, can be written in explicit form by system (2) as functions of the platform orientation parameters (which define $\mathbf{R}_{bp}$). Thus the position vectors q_i, i=1,2,3, can be written as functions of the orientation parameters of the platform. Moreover, the unit vectors $\mathbf{R}_{bp}\,^P\mathbf{w}_i$, i=1,2,3, which appear in the expression (11), depend only on the platform orientation parameters, because w_i, i=1,2,3, are vectors fixed in the platform. As a consequence, the matrix **N** is a function only of the platform orientation parameters and of the geometry of the 3-PSP mechanism. Hence, for a given geometry of the 3-PSP mechanism, condition (10) represents a surface in the three dimensional space of the three orientation parameters of the platform. This surface is the geometric locus of the configurations of the mechanism which are singular in the small.

Inverse problem. If vector ω is taken as input of system (9) a linear system is obtained in the unknowns $\dot{s}$ and $\dot{t}$, which, as expected, has the same coefficient matrix **M** of system (2), which is given by the expression (4). As a consequence the inverse problem becomes singular under the same conditions given by equation (6) discussed in the previous section.

3 Discussion of the Singularity Conditions

The singularity conditions are given by the Eqs. (6) (det[**M**]=0) and (10) (det[**N**]=0).
Architecture singularities are those that satisfy Eq. (6) and/or Eq. (10) for any feasible values of matrix $\mathbf{R}_{bp}$. Some of these singularities occur if:
  a) the three unit vectors w_i, i=1,2,3, are parallel (see Fig 3) (**M** and **N** singular);
  b) the three unit vectors v_i, i=1,2,3, are parallel (**M** singular);
  c) the three lines for points Q_i, i=1,2,3, parallel to w_i, i=1,2,3, intersect at a point D and the three lines for points Q_i, i=1,2,3, parallel to v_i, i=1,2,3, intersect at a point C (see Fig. 4). In this

case the three vectors n_i, i=1,2,3, are linearly dependent since they are orthogonal to CD and span a two dimensional space instead of a three dimensional one (**M** singular).

d) any two unit vectors v_i and v_j are mutually parallel and the corresponding two vectors w_i, and w_j are also mutually parallel. In this case n_i is parallel to n_j, thus they span the same space and are not independent (**M** singular) (see Fig.5).

Configuration singularities are those that occur for certain values of R_{bp} and satisfy conditions (6) and/or (10). Some of these occur if:

i) any two unit vectors (v_i, w_i), i=1,2,3, are parallel (s_i and t_i not determined) (**M** singular);

ii) any pair of unit vectors n_i, i=1,2,3, are parallel (**M** singular) (see Fig. 6);

iii) any two points Q_i, i=1,2,3, coincide (**N** singular). However, this singularity has no practical relevance since it cannot occur in practice because of link interference.

iv) the three points Q_i, i=1,2,3, are aligned (see Fig. 7) (**N** singular).

4 Conclusions

The mobility analysis of the 3-dof 3-PSP mechanism is presented. Analytical conditions (loci) for the occurrence of architecture and configuration singularities have been reported. The singularity loci are surfaces in the orientation space of the platform. Finally, several singularities have been shown and geometrically interpreted.

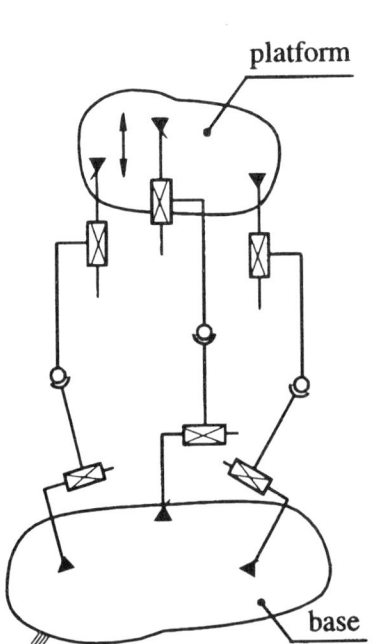

Figure 3. Architecture singularity: case (a)

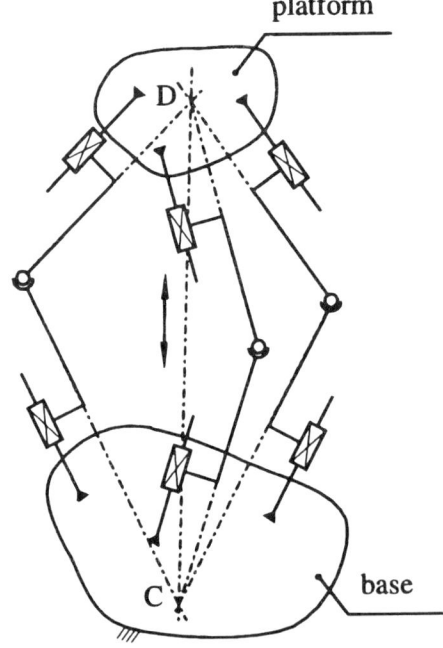

Figure 4. Architecture singularity: case (c)

Mobility Analysis of the 3-PSP Mechanism

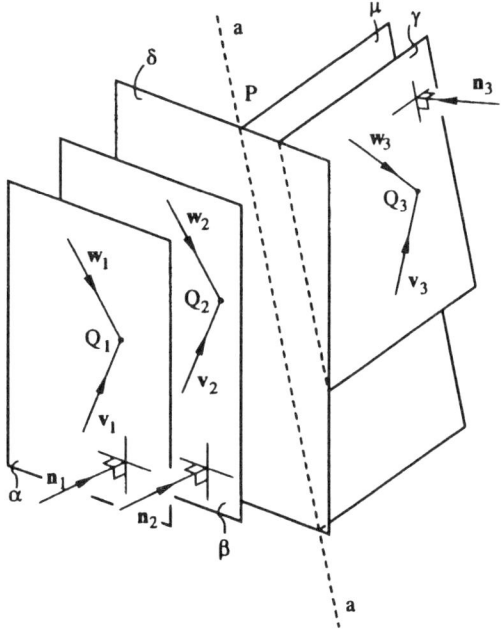

Figure 5. Architecture singularity: case (d)

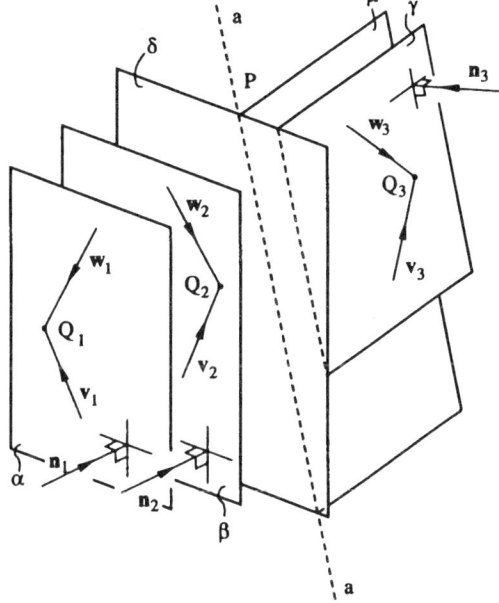

Figure 6. Configuration singularity: case (ii)

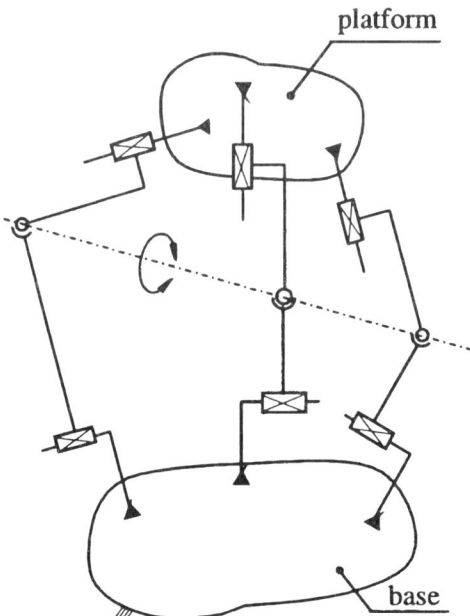

Figure 7. Configuration singularity: case (iv)

References

Di Gregorio, R. and Parenti-Castelli, V. (1998). Kinematics of a 6-dof fixation device for long bone fracture reduction. accepted for *J. of Intelligent and Robotic Systems* (JIRS).

Di Gregorio, R. and Parenti-Castelli, V. (1999a). Position analysis in analytical form of the 3-PSP Mechanism. *Proceedings of the 1999 ASME Design Engineering Technical Conferences, DETC99*, DETC99/DAC-8689, Sept. 12-15, 1999, Las Vegas, Nevada (USA).

Di Gregorio, R. and Parenti-Castelli, V. (1999b). Mobility analysis of the 3-UPU mechanism assembled for a pure translational motion. *Proceedings of the 1999 IEEE/ASME Int. Conf. on Advanced Intelligent Mechatronics*, Sept. 19-23, 1999, Atlanta, Georgia (USA), 520-525.

Gosselin, C., and Angeles, J. (1990). Singularity analysis of closed-loop kinematic chains. *IEEE Transactions on Robotics and Automation*, Vol. 6, No. 3, 1990, 281-290.

Ma, O., and Angeles, J. (1991). Architecture singularities of platform Manipulators. *Proceedings of the 1991 IEEE Intl Conference on Robotics and Automation*, Sacramento, CA, 1991, 1542-1547.

Detaching and Grasping Strategy Inspired by Human Behavior

Makoto Kaneko[1], Tatsuya Shirai[1] and Toshio Tsuji[1]

Department of Industrial and Systems Engineering,
Hiroshima University, Higashi-Hiroshima, 739-8527, Japan.

Abstract. Through grasp experiments by human achieving an enveloping grasp for a small cylindrical object placed on a table, we found an interesting grasping motion, where a human changes the finger posture from upright to curved ones after each finger makes contact with the object. During this motion, the object is automatically lifted up through either rolling or sliding motion between the fingertip and the object. A series of this motion is called as *Detaching Assist Motion (DAM)*. An advantage of *DAM* is that most of grasping motions can be done on the table instead of in the air. Therefore, we can avoid the worst scenario where the object falls down to the table. We first discuss the basic mechanism of *DAM* by human experiments. We then apply the *DAM* to a grasping motion by a multi-fingered robot hand. We show that the *DAM* can be explained by using Self-Posture Changing Motion. We also show some simulation and experimental results to confirm that a small object can be grasped easily by applying the *DAM*.

1 Introduction

Multi-fingered robot hands have a potential advantage to perform various skillful tasks like human hands. For considering the grasp strategy of robot hand, human motion often provides us with a good hint[1]-[4]. Cutkosky[1] has analyzed manufacturing grips and correlation with the design of robot hands by examining grasps used by humans working with tools and metal parts. Bekey et al.[2] have presented the automatic grasp planner which generates an order set of grasp according to task description, heuristics, and geometry of an object. Kang and Ikeuchi[3] have proposed the *contact web* and the *grasp cohesive index* for automatic classification of human grasping. Shimizu et al.[4] have developed the sensor glove *MK III* that can measure the grasping force and its distribution. While these works[1]-[4] discussed the classification of either final grasp patterns or grasp postures, we are particularly interested in considering the whole grasping procedure where the hand approaches an object placed on a table and finally achieves an enveloping grasp. Through the observation of human grasping, we found that human changes his (her) grasping strategy according to the size of objects, even though they have similar geometry. We called the grasp planning *Scale-Dependent Grasp*[5]-[6]. In this work, we focus on cylindrical objects whose diameter is small as shown in Figure.1. For such objects, two characteristic patterns are observed. One is that human first picks up the object from the table, and then finally achieves the target grasp through a grasp transition from the fingertip to the enveloping grasps as shown in Figure.1(a). The other one is that human first approaches the object, and then the finger

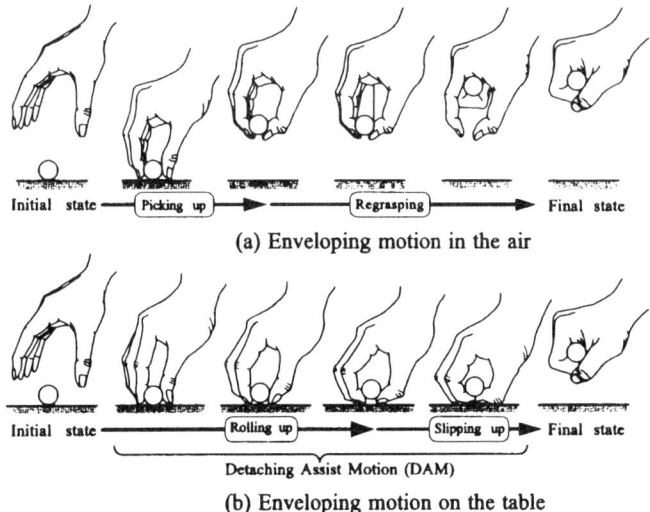

Figure 1. Two grasp strategies enveloping for an object placed on a table

posture is changed from upright to curved ones gradually as shown in Figure.1(b). It is generally difficult for the robot hand to achieve the motion as shown in Figure.1(a), and it may often fail in changing the phase from fingertip to enveloping grasps. If the robot hand fails in manipulating the object in the air, the object will fall down on the table, which is a worst scenario. On the other hand, in the way as shown in Figure.1(b), the robot hand can manipulate the object on the table. Therefore, it is not necessary to worry about falling down the object, even if the robot hand fails in detaching the object from the table. In this paper, we call the series of motion as shown in Figure.1(b) *Detaching Assist Motion (DAM)*. The most attractive feature of *DAM* is that it can be achieved by extremely simple finger motion. Therefore, the *DAM* is easily applicable for the robot hands. Our goal is to analyze the *DAM* and implement it into the grasping strategy of a multi-fingered robot hand.

2 Analysis of the Detaching Assist Motion by Human

2.1 What is Detaching Assist Motion?

An enveloping grasp can be achieved by the following three fundamental tasks: detaching an object from a table, lifting it up toward the palm, and firmly grasping. For detaching the object whose size is larger than fingertip, a human often utilizes the *wedge–effect* where a simple pushing motion of the bottom part of object makes the object detach from the table as shown in Figure.2(a). Due to its simple motion planning, we can easily implement it into the grasping procedure of a multi-fingered robot hand. Either under significant friction or for an object with small diameter, we can not detach the object since the finger forces balance within the object

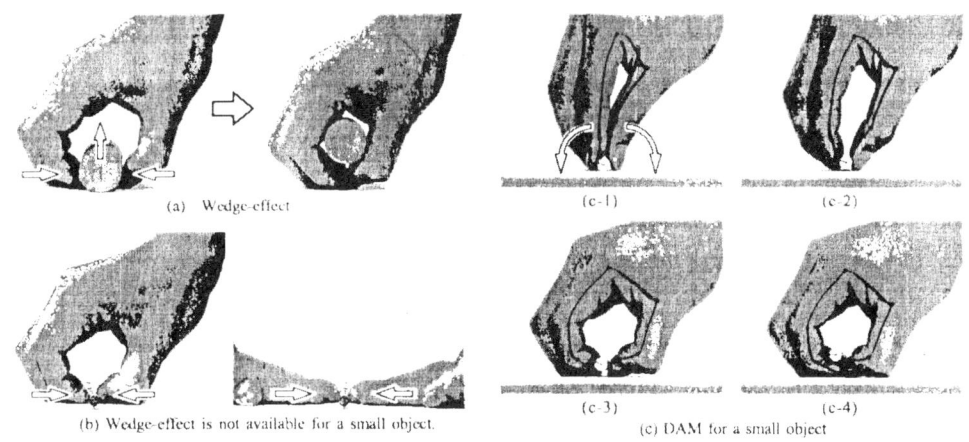

Figure 2. Grasping motion by human

and do not produce a lifting force any more as shown in Figure.2(b). Under such a situation, human detaches such a small object without the *wedge–effect*. By changing finger posture from upright to crooked ones as shown in Figure.2(c), the object is automatically lifted up from the table through either a rolling motion or a sliding motion between the object and the fingertip. We call this grasping motion *DAM*.

2.2 Analysis of the Change of Finger Posture

Why does the *DAM* work effectively for detaching the object from a table? What kind of principle exists behind it? In this subsection, to clarify the basic working mechanism of *DAM*, we examine both finger posture and object position while human purposely applies the *DAM* as shown in Figure.2(c). The seven markers are attached at the side of object and each joint of index finger and thumb as shown in Figure.3(a). We measure the coordinates of markers from the video image sequences recorded by video camera system, where the sampling time is $1/30[sec]$. The absolute angle of index finger θ_i (positive for CCW) and thumb θ_i (positive for CW), the center of object $\boldsymbol{P}_B$ $(= [P_{Bx}, P_{By}]^t)$ and the rotation of object θ_B (positive for CCW) can be obtained from the image sequences.

Figures.3(b) through (d) show experimental results for a cylindrical object with the diameter of $8[mm]$, while human utilizes the *DAM* from the initial posture (Figure.2(c-1)) to the final posture (Figure.2(c-4)), where Figures.3(b),(c) and (d) show the changes of $\Delta\theta_i$ and $\Delta\theta_t$, the trajectory of $\boldsymbol{P}_B$ and the change of ΔP_{By} and $\Delta\theta_B$, respectively. $\Delta\theta_i$, $\Delta\theta_t$, ΔP_{By} and $\Delta\theta_B$ are $\Delta\theta_i = \theta_i - \theta_{i0}$, $\Delta\theta_t = \theta_t - \theta_{t0}$, $\Delta P_{By} = P_{By} - P_{By0}$ and $\Delta\theta_B = \theta_B - \theta_{B0}$, respectively, where subscript 0 denotes the value at initial posture ($0[sec]$).

From Figure.3(b), it can be seen that both fingertips rotate uniformly with respect to time and finally keep constant in posture. An interesting behavior appears for the object motion when

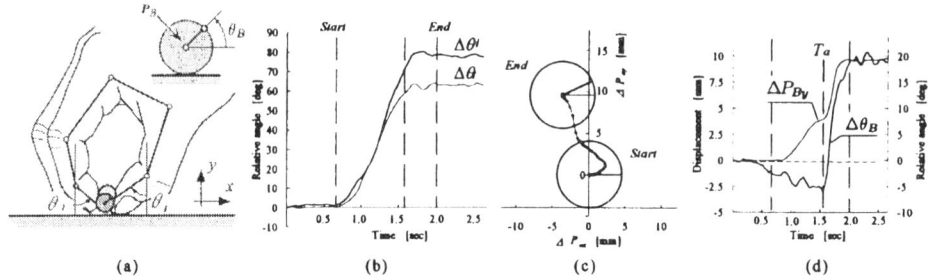

Figure 3. Visual observation during *DAM*

$t = 1.57[sec] (= T_a)$. At the moment of T_a, the object suddenly starts to move up with rotating motion as shown in Figure.3(d), while it slowly moves before T_a.

2.3 Basic Working Mechanism of DAM

Let us discuss what is really happening during the *DAM* by using the fingertip model as shown in Figure.4. We assume that the object is small enough to ensure that a simple pushing motion in the horizontal direction can not lift up the object as shown in Figure.4(a). Now, for simplifying the discussion, let us simplify the fingertip model as shown in Figure.4(b). Before T_a, we can observe from video image that the object and the fingertip keep the rolling contact. If we can assume that each fingertip does not slip on the surface of object, both fingertips will rotate from the initial to the final postures according to the geometrical restriction between finger and object as shown in Figure.4(b) and (c). We call this phase *Rolling–up phase*. As the object is lifted up, the normal direction of friction cone gradually changes upwards while the contact point moves towards the bottom of object. Finally, the moment the contact force is away from the friction cone, the object

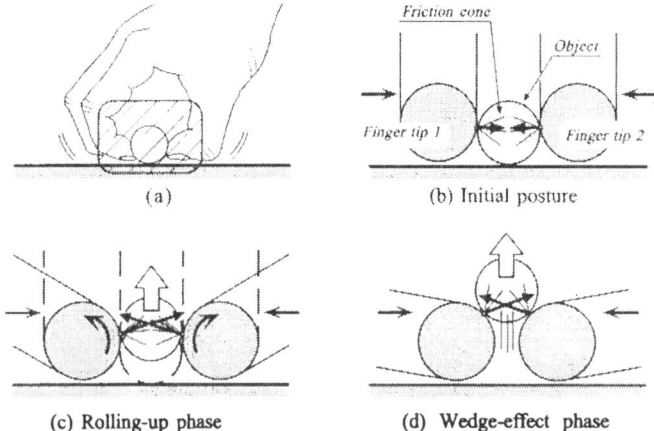

Figure 4. The basic mechanism of DAM

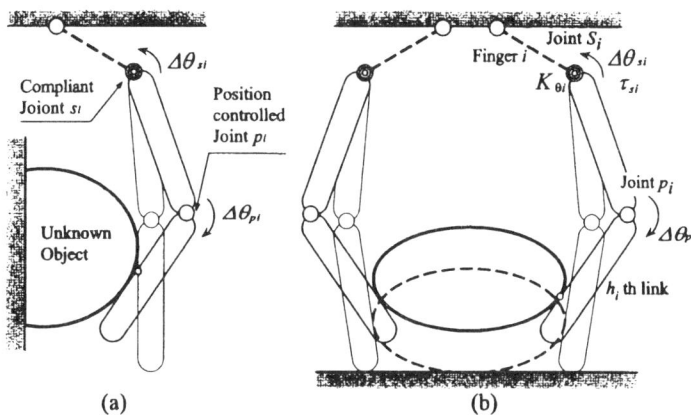

Figure 5. Self-Posture Changing Motion

slips on the surface of fingertips, and the *wedge–effect* occurs as shown in Figure.4(d). Once the contact force is away from the boundary, the *wedge–effect* can be continuously expected. We can also observe that the fingertips of index finger and thumb slip simultaneously the moment of T_a. This is basic mechanism why the object suddenly moves upwards after T_a. We call the final phase *Wedge–effect phase*. These are the outline of the working mechanism of *DAM*. The phase from *Rolling–up* to *Wedge–effect* is automatically switched depending upon the finger rotating motion.

3 Application for Robot Hands

While several strategies which are equivalent to the human *DAM* can be considered, we utilize compliant motion of link system having at least one compliant joint (s_i-th) and one position–controlled joint (p_i-th) as shown in Figure.5(a). Now, suppose that we impart an arbitrary angular displacement $\Delta\theta_{pi}$ at the position–controlled joint p_i for such a link system contacting with an environment. Under such a situation, the link system will automatically change its posture while keeping contact between the environment and the link system, if $\Delta\theta_{pi}$ is given appropriately. This series of motion is termed as *Self–Posture Changing Motion (SPCM)*[11], [12]. *SPCM* is conveniently utilized for detecting an approximate contact point between a link system and an unknown object under the assumption that the object does not move during sensing. In this work, we remove this assumption and allow the object to move according to the contact force imparted by the link as shown in Figure.5(b). We express the series of motions as $SPCM\{\boldsymbol{K}_\theta, \Delta\boldsymbol{\theta}_p\}$, where $\boldsymbol{K}_\theta = diag[k_{\theta 1}, \ldots, k_{\theta n}]^t \in R^{n \times n}$ and $\Delta\boldsymbol{\theta}_p = [\Delta\theta_{p1}, \ldots, \Delta\theta_{pn}]^t \in R^{n \times 1}$ denote the stiffness matrix of compliant controlled joints and the angular displacement matrix of position controlled joints, respectively.

In *SPCM*, h_i-th link keeps making contact with the object during the change of finger posture. From the basic behavior, we can see that it is almost equivalent to the *DAM* by human.

Now, we discuss the condition that the robot hand can lift up the object by utilizing the *SPCM*. Suppose that the robot hand utilizes $SPCM\{K_\theta, \Delta\theta_p\}$ for the object whose mass is m_B. Let $f_c = [f_{c1}^t, \ldots, f_{cn}^t]^t \in R^{3n \times 1}$ and $W_{ext} \in R^{6 \times 1}$ be the contact force vector at each contact point and the load wrench, respectively. The equation of the force and the moment balancing on the object can be expressed as

$$W_{ext} = -G^t f_c, \quad (1)$$

where $G^t \in R^{6 \times 3n}$ is the grasp matrix and given by

$$G^t = \begin{bmatrix} I_3 & \cdots & I_3 \\ (R_B{}^B P_{CB1} \times) & \cdots & (R_B{}^B P_{CBn} \times) \end{bmatrix}.$$

Suppose that the load wrench is $W_{ext} = [0, 0, -(m_B g + f_{ez}), 0, 0, 0]^t$, where g and f_{ez} are the acceleration of gravity and the virtual force in the gravitational direction at the center of gravity of object, respectively. Any component of f_c can not exist outside of the friction cone at each contact point. If all components of f_c exist inside of the friction cone with $f_{ez} = 0$, the object does not move since the resultant force acting on the object balances. If the resultant force does not balance without pushing down towards gravitational direction ($f_{ez} > 0$), the object is necessarily lifted up from the table when such a virtual external force f_{ez} is removed. Based on this consideration, we discuss following issue.

[Problem formulation]

Search $SPCM\{K_\theta, \Delta\theta_p\}$ where the contact force f_c balances with W_{ext} by utilizing a virtual external force $f_{ez} \geq 0$.

Let $\tau_s = [\tau_{s1}, \ldots \tau_{sn}]^t \in R^{n \times 1}$ be the torque at compliant joints. All compliant joints rotate $\Delta\theta_s$ according to the angular displacement $\Delta\theta_p$ under the assumption that the object does not move by $SPCM\{K_\theta, \Delta\theta_p\}$. Under $SPCM\{K_\theta, \Delta\theta_p\}$, torque in compliant joints is given $\tau_s = -K_\theta \Delta\theta_s (= \tau_{bias})$, where we assume $\tau_s = 0$ initially. The relationship between f_c and τ_s is expressed as follows,

$$\tau_s = J^t f_c, \quad (2)$$

where $J^t \in R^{n \times 3n}$ denotes the Jacobian matrix mapping from f_c to τ_s. Therefore, the relationship among f_c, W_{ext} and τ_s can be expressed as eq.(3).

$$\begin{bmatrix} W_{ext} \\ \tau_s \end{bmatrix} = \begin{bmatrix} -G^t \\ J^t \end{bmatrix} f_c \quad (3)$$

Bicchi[7], [8], Zhang, Gao and Gruver[9], Omata and Nagata[10] pointed out that eq.(3) contains the indeterminate contact force which neither affects τ_s nor appears in W_{ext}. To cope with this problem, we assume an extremely small compliance at each contact point between each finger and the object. This assumption releases us from such an indeterminate contact force, since the contact force f_c is uniquely determined under the contact stiffness $K_P = diag[K_{P1}, \ldots, K_{Pn}] \in R^{3n \times 3n}$ at the contact point as shown in Figure.6(a), where $K_{Pi} = diag[k_{Pxi}, k_{Pyi}, k_{Pzi}] \in R^{3 \times 3}$. We assume the displacement $\delta x \in R^{6 \times 1}$ and $\delta\theta_s \in R^{n \times 1}$ under w_{ext}, where δx and $\delta\theta_s$ denote the displacement and the rotation of

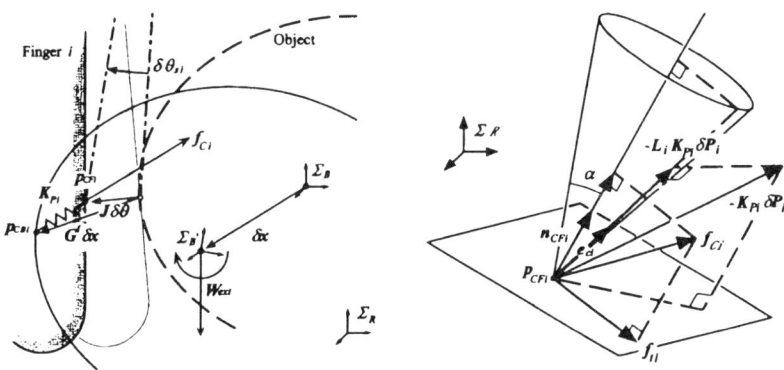

(a) Relationship between displacement and restoring force (b) Contact force outside of friction cone

Figure 6. Notation of parameters

object and the angular displacement of compliant joints. Now, the relationship among displacement vectors δP, δx, and $\delta \theta_s$ is given by

$$\delta P = G\delta x - J\delta \theta_s, \qquad (4)$$

where $\delta P = [\delta P_1^t, \ldots, \delta P_n^t]^t$ and $\delta P_i \in R^{3 \times 1}$ is the position vector from P_{CFi} to P_{CBi}. The relationship among f_c, δP, $\delta \theta_s$ and τ_s can be expressed as follows,

$$f_c = -K_P \delta P, \qquad (5)$$

$$\tau_s = \tau_{bias} - K_\theta \delta \theta_s. \qquad (6)$$

From eqs.(1), (2), (4), (5) and (6), we derive the contact force vector f_c in the following,

$$f_c = -K_{Pe}G(G^t K_{Pe} G)^{-1}(W_{ext} + G^t f_{bias}) + f_{bias}, \qquad (7)$$

where $K_{Pe} = K_P(I - JK_{\theta e}^{-1}J^t K_P)$, $K_{\theta e} = K_\theta + J^t K_P J$, and $f_{bias} = K_P J K_{\theta e}^{-1} \tau_{bias}$.

However, it does not guarantee whether each contact force f_{ci} obtained from eq.(7) exists within the friction cone or not. Because eq.(7) does not contain any constraint with respect to the friction cone. Now, suppose that f_{ci} is computed outside of the friction cone as shown in Figure.6(b). Since actual f_{ci} never exists outside of the friction cone, f_{ci} as shown in Figure.6(b) implies a local slip at the contact point. This means that f_{ci} exists on the friction boundary. Based on this, we change the orientation of f_{ci} into the nearest friction boundary, where the direction e_{ci} is given by $e_{ci} = \cos \alpha n_{CFi} + (\sin \alpha)/(\|f_{ti}\|)f_{ti}$, where $f_{ti} = f_{ci} - (n_{CFi}^t f_{ci})n_{CFi}$. For restricting the direction of restoring force $-K_{Pi}\delta P_i$ towards the direction of e_{ci}, we introduce the linear transformation matrix $L_i \in R^{3 \times 3}$ defined by $L_i = e_{ci}e_{ci}^t$. On the other hand, $L_j = diag[1, 1, 1]$ when f_{cj} exists within the friction cone. From eq.(5), we can obtain:

$$\hat{f}_c = -LK_P \delta P, \qquad (8)$$

where $L = diag[L_1, \ldots, L_n] \in R^{3n \times 3n}$, and "$\hat{*}$" denotes the value when contact force is projected on the friction boundary. From eqs.(1), (2), (4), (6) and (8), the contact force vector $\hat{f}_c$

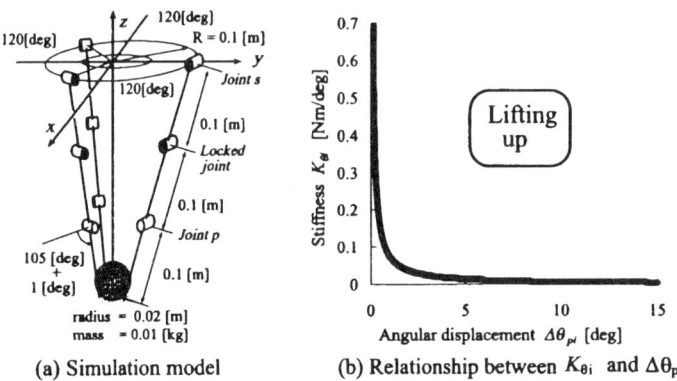

Figure 7. Simulation result

can be obtained as follows,

$$\hat{f}_c = -\hat{K}_{Pe}G(G^t\hat{K}_{Pe}G)^{-1}(W_{ext} + G^t\hat{f}_{bias}) + \hat{f}_{bias}, \qquad (9)$$

where $\hat{K}_{Pe} = LK_P(I - J\hat{K}_{\theta e}^{-1}J^tLK_P)$, $\hat{K}_{\theta e} = K_\theta + J^tLK_PJ$, and $\hat{f}_{bias} = LK_PJ\hat{K}_{\theta e}^{-1}\tau_{bias}$. By using these equations, we can judge whether the robot hand can lift up the object by utilizing $SPCM\{K_\theta, \Delta\theta_p\}$ or not.

Figure.7 shows a simulation result for a three–fingered robot hand, where each finger has three joints. Each finger consists of the compliant controlled, locked and position–controlled joint. The object is the sphere whose mass is $m_B = 0.01[kg]$, radius is $0.02[m]$ and the friction angle $\alpha = 10[deg]$. Figure.7(b) shows the $k_{\theta i}$–$\Delta\theta_{pi}$ map where the robot hand can lift up the object by $SPCM\{K_\theta, \Delta\theta_p\}$. In our simulation, we set $k_{\theta 1} = k_{\theta 2} = k_{\theta 3} = k_\theta$, $\Delta\theta_{p1} = \Delta\theta_{p2} = \Delta\theta_{p3} = \Delta\theta_p$, and $K_{Pi} = diag[k_P, k_P, k_P]$ where $k_P = 1000[N/mm]$ for simplicity. When the shape, the mass of object and the friction coefficient are given, we can roughly design the joint compliance with respect to $\Delta\theta_{pi}$ by using this graph.

4 Experimental Results

We implement the *SPCM* into the grasping procedure and execute the grasping experiment for an object placed on a table. The robot hand consists of three finger units and each finger has three links. The length of each link is $40[mm]$, $25[mm]$, and $30[mm]$ in order from the base, respectively, and the radius of each fingertip is about $5[mm]$.

Figures.8(a) and (b) show the experimental results where the robot hand utilizes the *wedge–effect* in Figure.8(a) and the *SPCM* for detaching a cylindrical object from the table in Figure.8(b). The robot hand grasps the cylindrical object ($\phi 12[mm]$) covered with rubber sheet in order to increase the surface friction. The robot hand can not lift up the object by utilizing the *wedge–effect* as shown in Figure.8(a) since the contact forces balance within the object. On the other hand, it

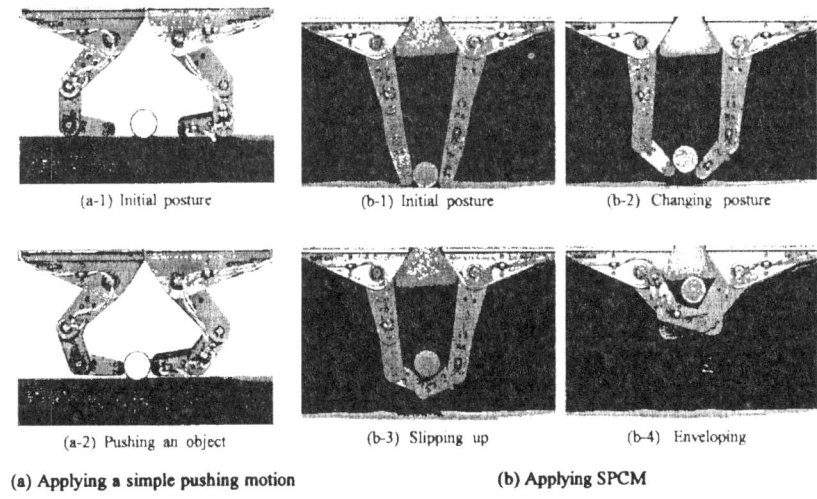

(a-1) Initial posture (b-1) Initial posture (b-2) Changing posture

(a-2) Pushing an object (b-3) Slipping up (b-4) Enveloping

(a) Applying a simple pushing motion   (b) Applying SPCM

Figure 8. Experimental result

can lift up easily the object by utilizing the *SPCM* as shown in Figure.8(b). Each finger consists of compliant controlled, locked, and position–controlled joint. As each position–controlled joint rotates from $15[deg]$ to $80[deg]$, the robot hand lifts up the object from the table (*Rolling–up phase*) as shown in Figure.8(b-2). Finally, the contact condition between the finger link and the object results in sliding contact (*Wedge–effect phase*) as shown in Figure.8(b-3). After every finger–tip link rotates $80[deg]$, the constant torque control is applied for achieving an enveloping grasp as shown in Figure.8(b-4).

5 Conclusion

We newly found the *DAM* through the observation of human grasping motion. We analyzed the finger posture during the *DAM* and explained the basic working mechanism of the *DAM*. We also examined the condition leading to the *DAM* by using *SPCM* which is easily implemented for robot hand and equivalent to the *DAM*. We also implemented the *SPCM* into the grasp procedure of a multi–fingered robot hand and verified its effectiveness experimentally.

This work has been supported by CREST of JST (Japan Science and Technology).

References

[1] M. Cutkosky: "On Grasp Choice, Grasp Models, and the Design of Hands for Manufacturing Tasks," IEEE Trans. on Robotics and Automation, Vol. 5, No. 3, JUNE, pp. 269–279, 1989.
[2] G.A. Bekey, H. Liu, R. Tomovic and W. Karplus: "Knowledge-Based Control of Grasping in Robot Hands Using heuristics from human motor skills," IEEE Trans. on Robotics and Automation, Vol. 9, No. 6, DECEMBER, pp. 709–722, 1993.

[3] S.B. Kang and K. Ikeuchi: "Toward Automatic Robot Instruction from Perception—Recognizing a Grasp from Observation," IEEE Trans. on Robotics and Automation, Vol. 9, No. 4, AUGUST, pp. 432–443, 1993.

[4] S. Shimizu, M. Shimojo, S. Sato, Y. Seki, A. Takahashi, Y. Inukai and M. Yoshioka: "The Relation between Human Grip Types and Force Distribution Pattern in Grasping," Proc. of the IEEE Int. Workshop on Robot and Human Communication (ROMAN'96), pp. 286–291, 1996.

[5] M. Kaneko, Y. Tanaka and T. Tsuji: "Scale-dependent Grasp," Proc. of the IEEE Int. Conf. on Robotics and Automation, pp. 2131–2136, 1996.

[6] T. Shirai, M. Kaneko, K. Harada and T. Tsuji: "Scale-Dependent Grasps," Proc. of the Int. Conf. on Advanced Mechatronics, pp. 197–202, 1998.

[7] A. Bicchi: "Analysis and Control of Power Grasping," Proc. of IEEE/RSJ Int. Workshop on Intelligent Robots and Systems IROS'91, pp. 691–697, 1991.

[8] A. Bicchi: "Force Distribution in Multiple Whole-Limb Manipulation," Proc. of the IEEE Int. Conf. on Robotics and Automation, pp. 196–201, 1993.

[9] Y. Zhang, F. Gao and W.A. Gruver: "Determination of Contact Forces in Grasping," Proc. of the IEEE Int. Conf. on Intelligent Robots and Systems, pp. 1038–1043, 1996.

[10] T. Omata and K. Nagata: "Rigid Body Analysis of the Indeterminate Grasp Force in Power Grasps," Proc. of the IEEE Int. Conf. on Robotics and Automation, pp. 1787–1794, 1996.

[11] M. Kaneko and K. Tanie : "Contact Point Detection for Grasping an Unknown Object Using Self-Posture Changeability," IEEE Trans. on Robotics and Automation, Vol.10, No.3, pp.355–367, 1994.

[12] M. Kaneko and K. Honkawa: "Contact Point and Force Sensing for Inner Link Based Grasps," Proc. of the IEEE Int. Conf. on Robotics and Automation, pp. 2809–2814, 1994.

Cross-Country Capabilities of a Walking Robot: Geometrical, Kinematic and Dynamic Investigation

V.E.Pavlovsky and A.K.Platonov

Keldysh Institute of Applied Mathematics of RAS, Moscow, Russia

Abstract. The paper deals with the investigation of abilities of a walking robot to overcome the hard relief. The investigation has an aim to find the way for determining the optimal parameters of a walker chassis depending on a mission the robot has to realize.

1 Introduction

Experiments with laboratory prototype of walking robot (Okhotsimsky et al., 1998) evidently show that it is reasonable to set up the problem of optimal design the walking chassis, but this design problem depends on the mission which the robot has to realize. The paper deals with such design problems for walking vehicle which either is able to realize comfortable motion over the complicated terrain, or uses its cross-country abilities in their full-scale to overcome hard obstacles. Main focus of a paper is investigation the cross-country capabilities of a walking robot, then the problem of defining the optimal parameters of a walking chassis may be solved as corresponding inverse problem.

2 Geometric and Kinematic Conditions of Moving Over an Obstacle

To find optimal parameters of walking chassis it is necessary to compare first of all different geometric and kinematic conditions of a walking motion taking also into account the dynamic realization of that motion. We will analyze six-legged robots with the so-called insectomorphic legs, so each leg of a robot has 3 DOF. That scheme is shown on a Figure 1.

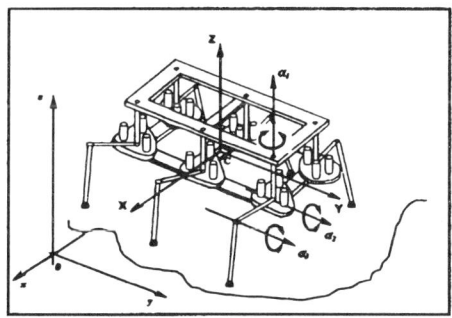

Figure 1. The scheme of a walker with rigid body.

The Figure 1 shows the absolute $Oxyz$ and moving $CXYZ$ coordinate systems which we will use, and shows degrees of freedom in each legs as well. We will assume that robot moves along the straight line which we will use as one of coordinate axes.

To determine the cross-country capabilities of a walking robot we will investigate the special class of obstacles. This class of obstacles is shown on a Figure 2a below. It is possible to describe that class with a help of a set of two parameters: R - the width of an obstacle, and H - the height (elevation) of an obstacle, $R \geq 0$, H may be of arbitrary value. The class of those objects allows to describe sufficiently wide set of real objects, such as rocks, pits, and so on. The Figure 2b shows the projection of an obstacle onto the plain π where the robot is moving. It is necessary to say that the R value on the Figure 2b is the width of an obstacle (the same as on Figure 2a), but the R_z is an efficient width of an obstacle, and $R_z \geq R$. Parameter R_z is most important when $H \neq 0$ and allows to simplify the analysis of correct mutual displacement of working zone of a leg (which is rather complicated) and an obstacle (this means the problem of avoidance of collisions between leg parts and an obstacle itself). Evidently, the robot has to overcome the obstacle (R_z, H), not the (R, H). Further we will assume that zones of different obstacles do not cross, which is equivalent to the situation where obstacles are not 'interfere' and we will treat the situation as presence of an isolated obstacles on the plain π.

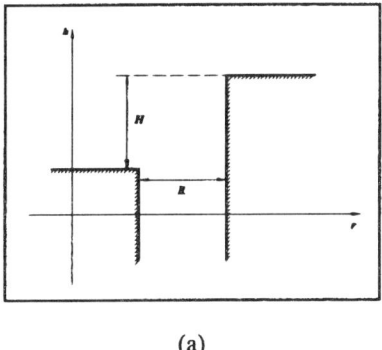
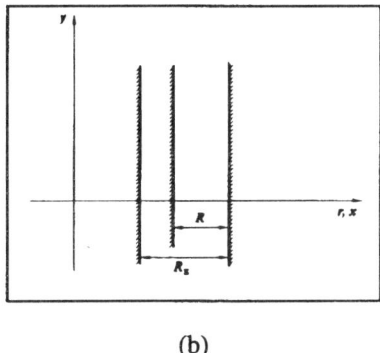

(a)   (b)

Figure 2. Parameters of an obstacle.

The main conditions which allow to determine the optimal values of chassis parameters are those conditions which coordinate the geometric and kinematic parameters of leg working zones while the legs step onto the plane π, with parameters of body trajectories and obstacle parameters. Those conditions are the set of inequalities which include parameters mentioned above by the following way.

Let B, G be the parts of a leg as it is shown on a Figure 3. Let $N_l, \Delta h, \Delta y$ be parameters which determine the coordinates of a stepping point of a leg in a robot coordinate system $CXYZ$, let a be a one half of a body width, as it is defined on Figure 1 and Figure 3. It is assumed that axe Ox is a vertical projection of CX-axe. The N_l function defines abilities of a robot to realize given steps while moving, and so behaviour of this function allows to define maximal parameters of obstacles.

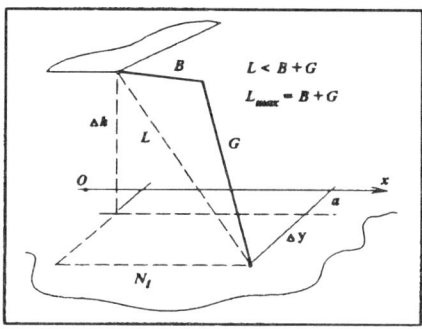

Figure 3. The geometry of a leg.

Evidently, there is a simple geometry equation for calculation the maximal value of N_l parameter which is the following:

$$N_l = \sqrt{(B+G)^2 - (\Delta y - a)^2 - \Delta h^2} \qquad (1)$$

and we will assume that $\Delta y > a$. Then it is possible with using the equations like (1) find conditions which combine parameters of a walker and of an obstacle.

For doing this it is necessary to analyze the geometric and kinematic properties of a motion of a walker inside and near the obstacle zone R_z, as it is shown on Figure 4 where a part (subset) of sequences of stepping points, i.e. a part of stepping track (traces track) of a robot, located near the obstacle, is drawn. Let S_k ($k = 2i, 2i+1, 2i+2, 2i+3$) be the stepping points themselves, ε be a stability margin (parameter which guarantees the stability of a motion and allows to use more simple kinetostatic equations, not dynamic), and l_e is a corresponding linear stability margin (the distance along the Ox-axe from the projection of a center of gravity of a walker body to the corresponding boundary of supporting polygon defined by points S_k), as it is also defined on Figure 4.

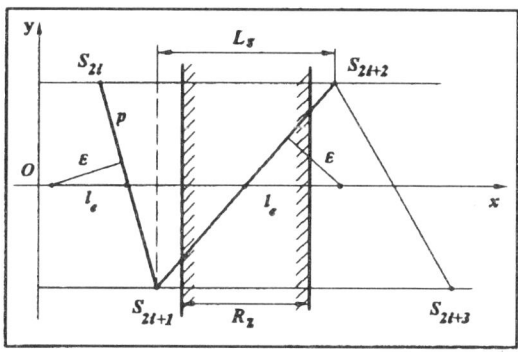

Figure 4. The parameters of stepping points around the obstacle.

With using (1) and taking into account the scheme given on a Figure 4 it is possible to write down those conditions in a following form of a system of inequalities:

$$x_{2i+2} \leq x_{2i} + \left[\frac{x_{2i+1} - x_{2i}}{2} + X_{2M} - le(S_{2i}, S_{2i+1}) + N_{2i+2}\right]$$

$$x_{2i+2} \leq x_{2i} + 2\left[-\frac{x_{2i+1} - x_{2i}}{2} - X_2 - le(S_{2i+1}, S_{2i+2}) + N_{2i}\right] \quad (2)$$

$$x_{2i+3} \leq x_{2i+1} + \left[\frac{x_{2i+2} - x_{2i+1}}{2} + X_{2M-1} - le(S_{2i+2}, S_{2i+1}) + N_{2i+3}\right]$$

$$x_{2i+2} \leq x_{2i} + 2\left[-\frac{x_{2i+2} - x_{2i+1}}{2} - X_1 - le(S_{2i+2}, S_{2i+3}) + N_{2i+1}\right]$$

Here x_k are the coordinates of S_k, X_1, X_2, X_{2M-1}, X_{2M}, are constant coordinates of points where legs are mounting to the body, of the rear part and of the front part of a body respectively, taken in the moving coordinate system and equal to $\pm b$ (b is one half of a robot body length), functions l_e and N_k, ($k = 2i, 2i+1, 2i+2, 2i+3$), are defined above. It is first necessary to note, that system (2) is an geometrical estimation of corresponding kinematics conditions (obtained in similar form of inequalities), which in addition depend on the values of ratio of body and legs linear speeds. And at last it is most important to note that in the system (2) the values like $(x_{2i+2} - x_{2i+1})$ directly define the size R_z of an obstacle which the robot can overcome, and if R_z be larger then more complicated (R_z, H) - obstacle the robot can overcome.

The system (2) defines conditions which allow to formulate the following conclusions. First, it shows, that on depending of an obstacle parameters (R_z, H) to maximize R_z the motion of a walker near an obstacle have to be varied as its shown on Figure 5. As R_z will be larger, the more significant variation has to be done: the stepping track have to approach to symmetrical one in respect to Ox-axe, and body height above the upper zone of obstacle have to be smaller and has to be constructed as shown on Figure 5. Second, the system (2) are those conditions sought for determining optimal values of parameters of walker kinematic scheme in coordination with the mission of walker motion. It is possible to formulate that conclusion in a following form. Let the motion shown on Figure 5 be the comfortable motion since the walker body motion is simple - it is translational motion. If in this case (R_z, H) are given, then it is possible to say that the system (2) defines such values as $<B, G, a, b, \Delta y, \mathcal{E}>$ which will allow to realize comfortable motion of a walker while moving over the given obstacle.

3 Comfortable Motions

The system (2) may be efficiently interpreted with using the graphic drawing of that system. Those scheme is built on plane of (R, H) parameters and is shown on Figure 6a. Note, this chart shows values of parameters of those obstacle which may be efficiently overcame by the walker when it realizes comfortable motions.

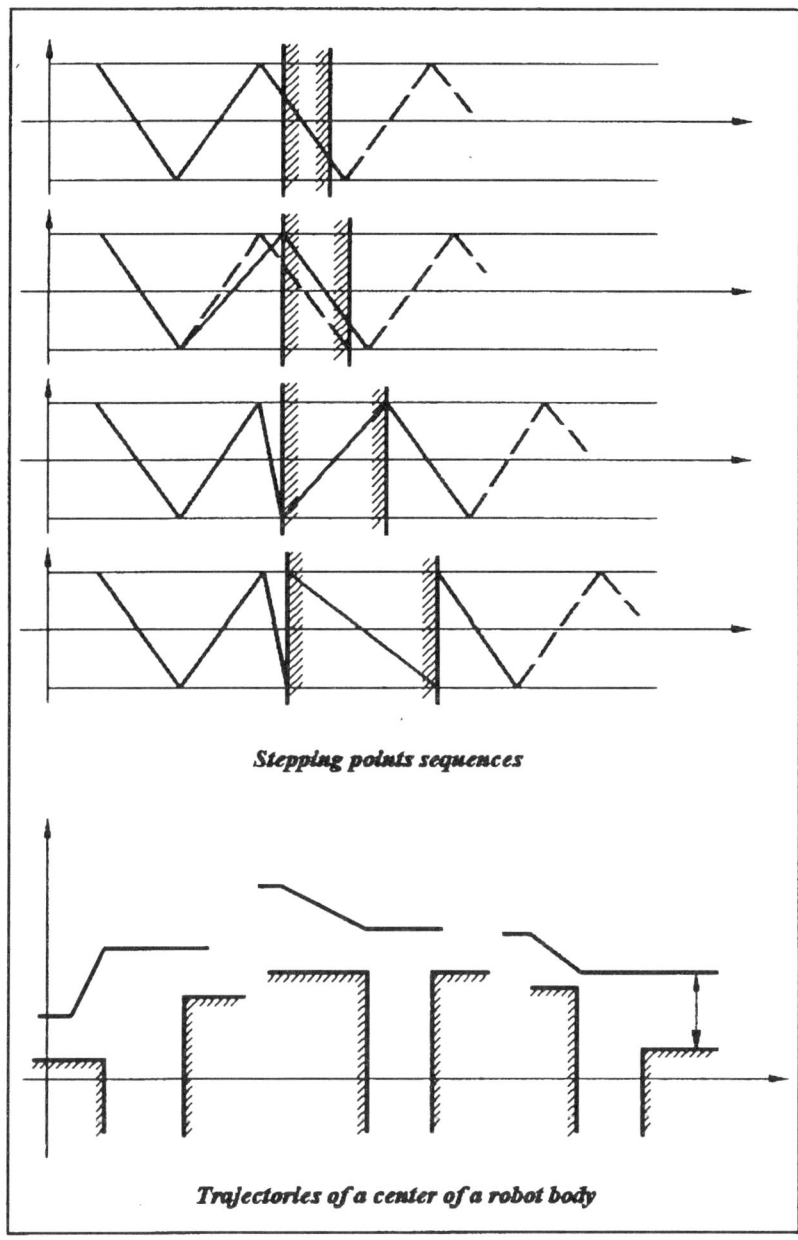

Figure 5. Modes of motion: stepping points arrangements and trajectories of a walker body in a vertical plane.

Figures 6a,b are drawn for a walker with the following parameters: $B+G=160$, $b=120$ (values are given in some conventional units), boundaries on charts are parameterized by the B/G ratio as well as by ε and h_p, which is a height margin over the obstacle, the region with maximal volume corresponds to $B=G$, $\varepsilon=10$, $h_p = 0$.

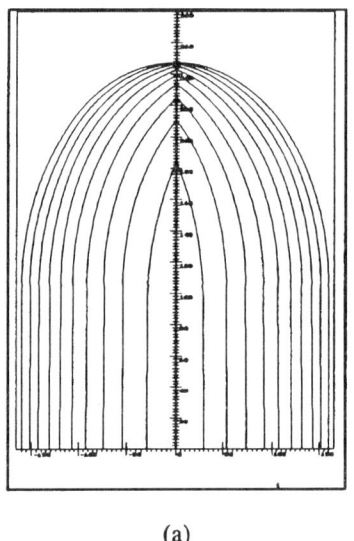

(a)

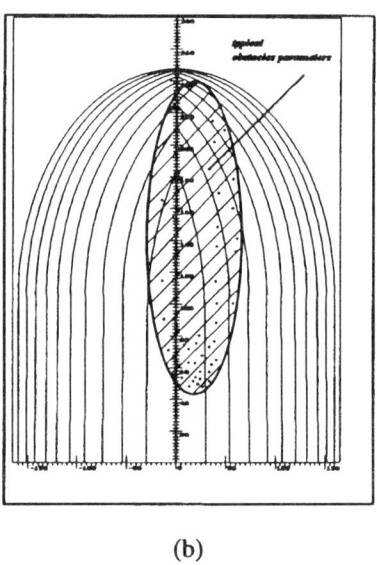

(b)

Figure 6. Regions of a comfortable walking.

This scheme may be also used in the following way. In a case, for example, when the robot is designed to move on terrain with known sizes of typical obstacles, it is possible to estimate the corresponding sizes (parameters) of a robot chassis which will allow to realize comfortable motions over that terrain.

Such drawing is shown on Figure 6b where the region of typical parameters of obstacles is shown as well as previously mentioned regions of those motions of robot which can be realized comfortably. On that figure points located inside the obstacle region show the estimation of (R,H) and (R_z,H) parameters of Martians objects obtained from photos of planet surface (units are conventional). Such estimation may be used in different ways, e.g. for preliminary designing of a walking chassis intended for working on rough terrains, for example, this can be done for such missions as planet exploration if typical charcteristics of roughness of terrain are known. In this way such estimation were used for defining parameters of a walker for simulation the walking chassis of a Martian rover as it is shown on Figure 6b.

Note also that it is possible to compare diagrams like those from Figure 6a,b with ones drawn for wheeled chassis. Diagrams for wheeled chassis are similar ones but have rectangular boundaries and show that sizes of walking chassis will be significantly less then for wheeled chassis for the same sets of obstacles. As a result it is possible to say that walking chassis is strictly preferable for moving on roufgh terrain. Diagrams like those on Figure 6a,b allow to confirm this numerically.

4 Simulation of a Complicated Case

Next it is possible to extend this investigation to take into account the more wide set of modes of robot motion. Here it is assumed that robot has to use its motion abilities in their full scale. First, it is possible to take into account the inclination of a robot body while the robot is moving over the obstacle. Second it is interesting to compare different schemes of walking chassis design, namely to compare walkers with rigid bodies and with articulated bodies (Okhotsimsky et al., 1998, Blazevic et al., 1999). To investigate those variants the special simulation system was developed. This system was developed under the MATLAB environment (Blazevic et al., 1999), and allows to simulate general cases of climbing motion over the obstacles, some examples of that simulation are given as partial screenshots on Figure 7.

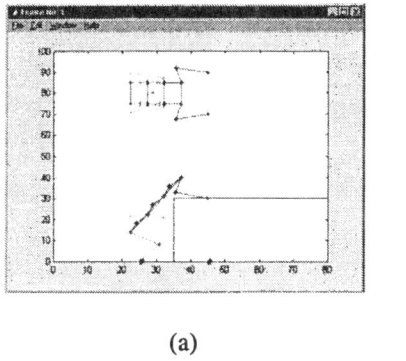
(a)

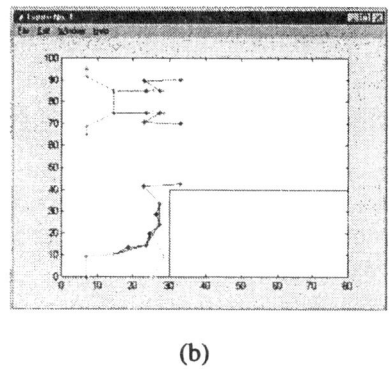
(b)

Figure 7. Simulation system screenshots.

Simulation system also allows to find the regions on (R,H)-plane of those obstacles which the robot can overcome. New regions are similar to those shown on Figure 6 but are larger in sizes and include (cover) previous ones. The Table 1 given below shows the comparison of two walking chassis intended for climbing over the high obstacle, similar results were given in (Blazevic et al., 1999). The main parameters of robots legs and bodies were the same and were equal to: $B+G=160$ ($B=80$, $G=80$), $b=120$ (lenght of a body is equal to 240), but the first machine has the rigid body, and the second one has the articulated body. Those machines have parameters very close to parameters of machine given above while calculating the Figure 6a,b, the difference is only in dividing the leg into it parts (B and G).

The most important result is that Table 1 shows that the second machine can overcome the obstacle height of which is 30% larger then in the first case. It is also interesting to compare data from the Table 1 with the maximal height of obstacles obtained from Figure 6a, where that maximal value is equal to 160 conventional units. This difference is the result of using the comfortable motion while calculating the Figure 6a, and using the inclination and articulating of a body while calculating data in the Table 1.

Robot type	Length of a leg	Length of a body	Height of an obstacle
with rigid body	160	240	240
with articulated body	160	240	320

Table 1. Comparison of climbing abilities of different types of robots.

The simulation system allows also to take into investigation the dynamic parameters of a motion and to find dynamic characteristics of a motion or take dynamically possible motions of a robot while some restrictions are set.

5 Conclusion

As a result of investigation it is possible now to compare numerically different shemes of a walkers chassis and to choose an optimal one to extend the cross-country capabilities of a walking robot. The results of such comparison of robots with rigid and articulated bodies were shown above.

In particular, those results show that sizes (heights) of obstacles which could be overcame by typical walking chassis under modes of comfortable motion, uncomfortable motion with using inclination of a body, and with using body articulation, are under ratio $1 : 1.5 : 2$.

It is also possible to compare walkers with wheeled vehicles to find a relations between those schemes for vehicles intended for realization the same missions of moving over the hard relief.

Finally, it is possible to use the system developed as some kind of CAD system to choose the appropriate scheme of a walker which will be able to fulfill the given mission on a given environment (terrain). For doing this it is possible to put the parameters of typical obstacles of a terrain onto the (R,H)-regions like shown above. So, if typical parameters of typical obstacles are given or known on some way, it is possible to choose such parameters of a robot, as type of body (rigid or articulated) and main geometrical parameters of a body and legs like sizes of a body and lengths of leg parts, to build the robot which will be successfully used in work in that environment.

References

D.E.Okhotsimsky, A.K.Platonov, V.E.Pavlovsky, A.V.Lensky, A.A.Kiril'chenko and V.S.Yaroshevsky. (1998). Concept, Design and Control of Six-Legged Walking Robot. In *Proceedings of the 1-st Int. Symposium on Climbing and Walking Robots CLAWAR'98*. Brussels, Belgium, 1998, 361-366.

P.Blazevic, A.Iles, D.E.Okhotsimsky, A.K.Platonov, V.E.Pavlovsky and A.V.Lensky. (1999). Development of Multi-Legged Walking Robot with Articulated Body. In *Proceedings of the 2-nd Int. Symposium on Cimbing and Walking Robots CLAWAR'99*. Portsmouth, UK, 1999, 205-212.

Optimization of Robot Gripper Parameters Using Genetic Algorithms

Stanislaw Krenich and Andrzej Osyczka

Department of Mechanical Engineering, Cracow University of Technology, Poland

Abstract. In this paper the multicriteria optimization model of the robot gripper is built. In this model the decision variables are geometrical dimensions of the gripper, which are under side constraints and those constraints which are yielded by the structure of the gripper. The objective functions are: (i) the difference between the maximum and minimum griping forces for the assumed range of the gripper ends displacement, (ii) the force transmission ratio between the gripper actuator and the gipper ends and (iii) the length of all the elements of the gripper. All functions are computationally expensive functions, thus a special Genetic Algorithm based multicriteria optimization method has been developed. Using this method a two-stage optimization process is proposed in which at each stage one bicriterion optimization model is solved. The presented example shows the effectiveness of the proposed approach.

1 Introduction

Genetic Algorithms are widely used to solve different design optimization problems (see review paper by Parmee, 1999) including multicriteria design optimization problems (see Osyczka and Kundu, 1995 and Osyczka et al., 1997) and mechanism synthesis (see Boudreau and Turkkan, 1996, Roston, 1997, Boudreau and Gosselin, 1999 and Osyczka et al., 1999). The main advantage of the use of GAs in multicriterion optimization problems is that while running the GA program the full set of Pareto optimal solutions (nondominated solutions) can be obtained and the designer has a full picture of the possible compromise solutions. This advantage becomes very important while solving optimization models of a robot gripper, in which several criteria occur. Moreover, the objective functions which represent these criteria are computationally expensive functions. Thus a special Genetic Algorithm based method has been developed and used to optimize robot grippers. The method uses the tournament selection mechanism which does not require evaluation of fitness values in order to create a new population of chromosomes for the next generation. The tournament is arranged in this way that objective functions are evaluated only for feasible solutions. The presented results show the effectiveness of the proposed method.

2 Problem Formulation

The multicriteria optimization problem of robot grippers is formulated as follows:
Find $\mathbf{x}^* = [x_1^*, x_2^*, ..., x_N^*]$ which will satisfy the K inequality constraints

$$g_k(\mathbf{x}) \geq 0 \qquad k = 1, 2, ..., K \tag{1}$$

and minimize objective functions:

$$f(x^*) = \min [f_1(x), f_2(x),...,f_N(x)] \qquad (2)$$

where: $x = [x_1, x_2,...,x_I]$ is the vector of decision variables, the elements of which represent dimensions of the gripper elements.

$f_1(x), f_2(x),..., f_N(x)$, are the objective functions.

Generally, several objective functions can be used to evaluate the robot gripper design. The most important are: (i) the difference between maximum and minimum griping forces for the assumed range of the gripper ends displacement (ii) the force transmission ratio between the gripper actuator and the gipper ends and (iii) the length of all the elements of the gripper. The decision making process with more than two criteria is usually a very difficult task especially when the set of Pareto optimal solutions is large. Thus in the paper a two-stage optimization process is proposed. At both stages one of the two bicriterion optimization models is solved and two Pareto sets are presented to the designer. Both sets can be illustrated graphically in the space of the objectives. At the first stage from the first set of Pareto solutions the designer decides how to change one of the objective functions into a constraint and, with this constraint, at the second stage new Pareto optimal solutions are generated and illustrated graphically. From this set the designer may easily choose the most preferable solution.

3 Constraint Tournament Selection Method

Several Genetic Algorithm (GA) based methods for solving multicriteria optimization problems have been developed recently (see review papers by Fonseca and Fleming, 1995 and Coello, 1999). These methods, which are quite effective in solving typical nonlinear models, are less effective while solving highly constrained models with computationally expensive objective functions. Robot gripper models belong to the latter models. In this paper a new method for solving multicriteria constrained optimization problems is used.

The general idea of the constraint tournament selection method is as follows:

(i) If both chromosomes are not in the feasible region the one which is closer to the feasible region is taken to the next generation and the values of the objective functions are not calculated in this case.

(ii) If one chromosome is in the feasible region and the other one is out of the feasible region the one which is in the feasible region is taken to the next generation and the values of the objective functions are calculated only for the feasible solution.

(iii) If both chromosomes are in the feasible region, the values of the objective functions are calculated for both chromosomes. If one chromosome dominates another one the dominant one is taken to the next generation. If both chromosomes are non-dominated one of them is chosen at random and taken to the next generation.

Graphical illustration of this method is presented in Figure 1. The detailed description of the method is presented in the paper (Osyczka and Krenich, 2000).

Note that using this method the objective functions are calculated only for feasible solutions. This makes the process of calculations more effective especially for problems for which the objective functions are computationally expensive functions.

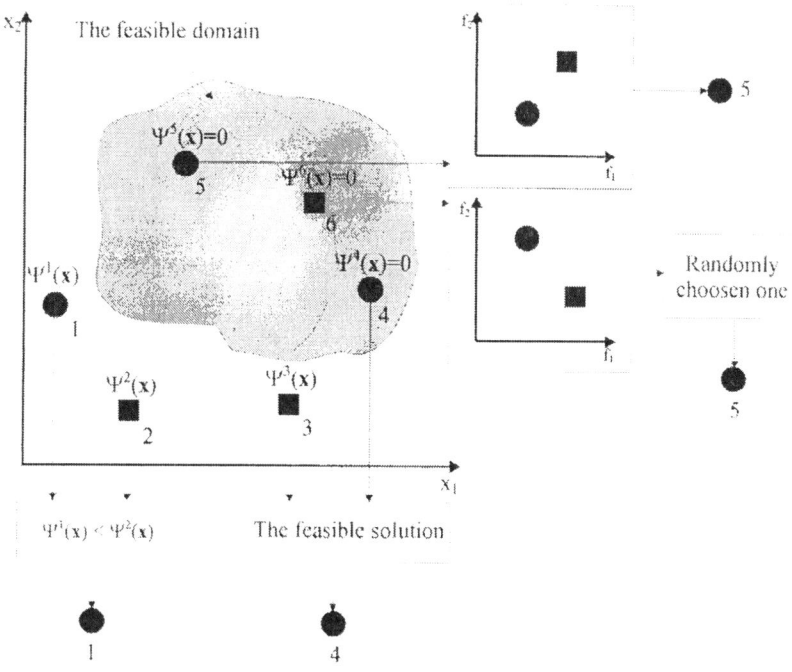

Figure 1. Graphical illustration of the constraint tournament method.

4 Geometrical Dependencies in the Robot Gripper.

Let us consider an example of a robot gripper the scheme of which is presented in Figure 2.

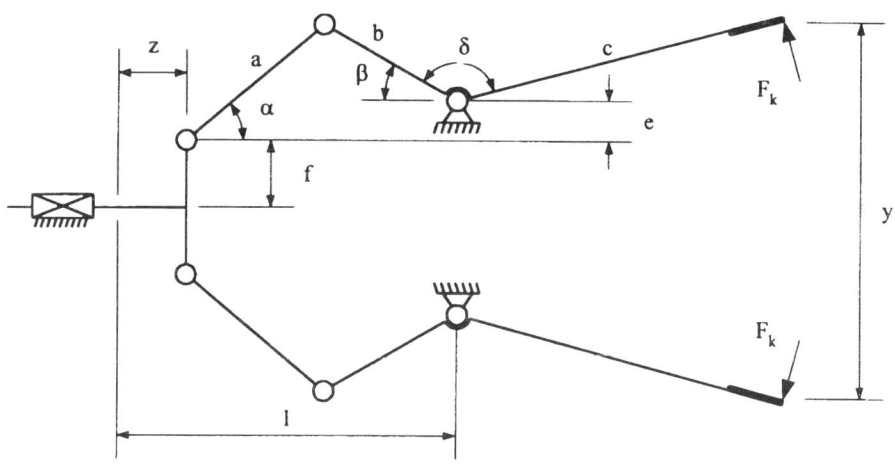

Figure 2. Scheme of a robot gripper mechanism.

The geometrical dependencies of the gripper mechanism are:

$$g^2 = (1-z)^2 + e^2, \quad g = \sqrt{(1-z)^2 + e^2}, \quad b^2 = a^2 + g^2 - 2 \cdot a \cdot g \cdot \cos(\alpha - \phi),$$

$$\alpha = \arccos\left(\frac{a^2 + g^2 - b^2}{2 \cdot a \cdot g}\right) + \phi, \quad a^2 = b^2 + g^2 - 2 \cdot b \cdot g \cdot \cos(\beta + \phi),$$

$$\beta = \arccos\left(\frac{b^2 + g^2 - a^2}{2 \cdot b \cdot g}\right) - \phi, \quad \phi = \operatorname{atan}\left(\frac{e}{1-z}\right)$$

The dependencies between the forces are:

$$R \cdot \sin(\alpha+\beta) \cdot b = F_k \cdot c, \quad R = \frac{P}{2 \cdot \cos(\alpha)}, \quad F_k = \frac{P \cdot b \cdot \sin(\alpha + \beta)}{2 \cdot c \cdot \cos(\alpha)}$$

On the basis of these dependencies two optimization models are built and presented below. The optimization problem was treated as a continuous programming problem, but using the genetic algorithm based method problems of any type, for example, integer, discrete and mixed programming problems can be solved. In both models the vector of decision variables is:
$\mathbf{x} = [\,a, b, c, e, f, l, \delta\,]^T$, where a, b, c, e, f, l, are dimensions of the gripper and δ is the angle between the elements b and c (see Figure 2).

5 Optimization Process.

Stage 1
For the robot gripper presented in Figure 2 the following objective functions are considered in Stage 1:
- $f_1(\mathbf{x})$ is the function which describes the difference between the maximum and the minimum griping forces for the assumed range of the gripper ends displacement, and can be evaluated as follows:

$$f_1(\mathbf{x}) = \max_z F_k(\mathbf{x}, z) - \min_z F_k(\mathbf{x}, z)$$

- $f_2(\mathbf{x})$ is the function which describes the force transmission ratio between the gripper actuator and the gipper ends:

$$f_2(\mathbf{x}) = \frac{P}{\min_z F_k(\mathbf{x}, z)}$$

Both objective functions depend on the vector of decision variables and the displacement z. For the given vector $\mathbf{x}$ the values of $f_1(\mathbf{x})$ and $f_2(\mathbf{x})$ are evaluated using a procedure which makes these functions computationally expensive.

From the geometry of the gripper the following constraints can be derived:
1. $g_1(\mathbf{x}) = Y_{min} - y(\mathbf{x}, Z_{max}) \geq 0$,
2. $g_2(\mathbf{x}) = y(\mathbf{x}, Z_{max}) \geq 0$,
3. $g_3(\mathbf{x}) = y(\mathbf{x}, 0) - Y_{max} \geq 0$,

4. $g_4(\mathbf{x}) = Y_G - y(\mathbf{x},0) \geq 0$,
5. $g_5(\mathbf{x}) = (a+b)^2 - l^2 - e^2 \geq 0$,
6. $g_6(\mathbf{x}) = (l-Z_{max})^2 + (a-e)^2 - b^2 \geq 0$,
7. $g_7(\mathbf{x}) = l-Z_{max} \geq 0$,

where: $y(\mathbf{x},z)=2\cdot[e + f + c\cdot\sin(\beta+\delta)]$ displacement of the gripper ends,
Y_{min} – minimal dimension of the griping object,
Y_{max} – maximal dimension of the griping object,
Y_G – maximal range of the gripper ends displacement,
Z_{max} – maximal displacement of the gripper actuator.

The optimization process was carried out using the following data:
a) The geometric constraints:
$10 \leq a \leq 250$, $10 \leq b \leq 250$, $100 \leq c \leq 300$, $0 \leq e \leq 50$,
$10 \leq f \leq 250$, $100 \leq l \leq 300$, $1.0 \leq \delta \leq 3.14$
$Y_{min} = 50$, $Y_{max} = 100$, $Y_G = 150$, $Z_{max} = 50$, $P = 100$.
b) The parameters for GA:
 – Length of the string for each decision variable 18 bits.
 – Crossover rate $R_c = 0.4$, mutation rate $R_m = 0.02$.
 – Population size = 400, number of generations = 400.

The experiments were carried out for several different initial populations. The set of Pareto optimal solutions presented in Figure 3 is the result of these experiments.

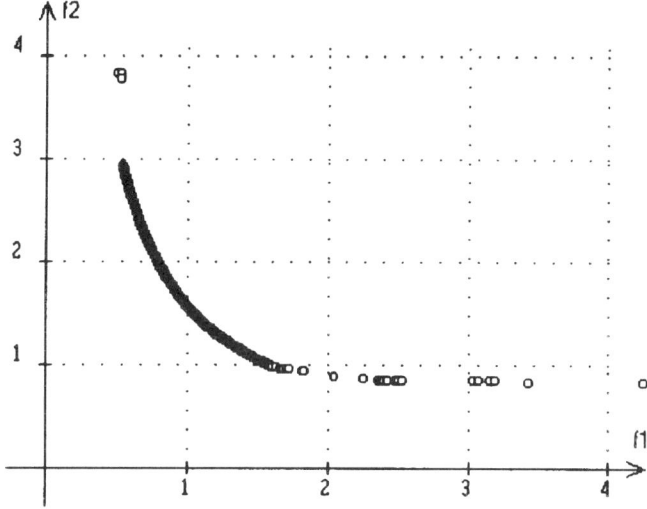

Figure 3. The set of Pareto optimal solutions obtained at Stage 1.

On the basis of these results the designer realises what the range of the assumed objective functions is and may proceed to Stage 2.

Stage 2.

At this stage the designer decides which of the objective functions will be treated as the constraint and on which level. Assuming that the first objective will be treated as the constraint we have:

$$g_8(x) = FR - (\max_z F_k(x,z) - \min_z F_k(x,z)) \geq 0,$$

where: FR is the assumed upper limit for the difference between the maximum and minimum griping forces.

The remaining constraints are the same as at Stage 1.

Now instead of the first objective function from Stage 1, a new function representing the length of all the elements of the gripper is introduced. Thus we have the following objective functions:

- $f_1(x)$ is the function which describes the length of all the elements of the gripper, and can be evaluated as follows:

$$f_1(x) = a + b + c + e + f + l,$$

- $f_2(x)$ is the function which describes the force transmission ratio between the gripper actuator and the gipper ends:

$$f_2(x) = \frac{P}{\min_z F_k(x,z)}$$

For the above model the optimization process was carried out using the same data as at Stage 1. The results of the experiments are shown in Figure 4 and Figure 5, where Figure 4 presents the set of Pareto optimal solutions for FR = 1 whereas Figure 5 presents the set of Pareto solutions for FR = 2.5. Four representative solutions from the Pareto set for FR = 2.5 are presented in Table 1.

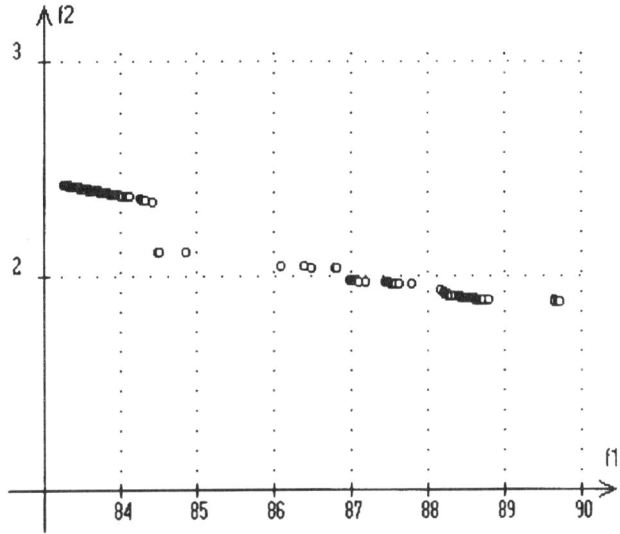

Figure 4. Set of Pareto optimal solutions for the optimization process at Stage 2 and for FR = 1.

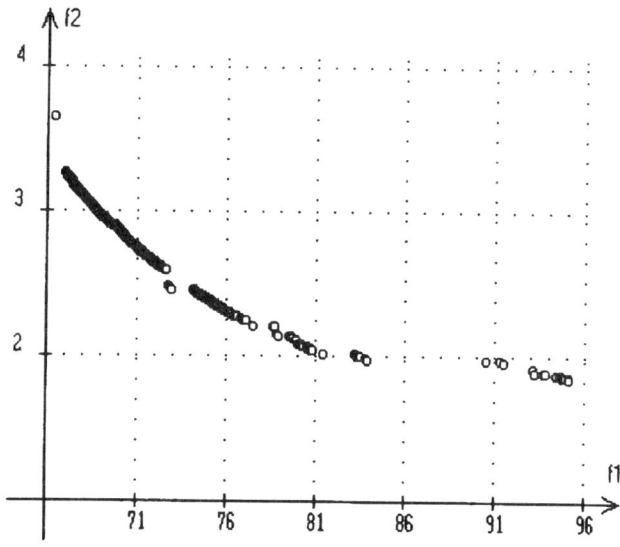

Figure 5. Set of Pareto optimal solutions for the optimization process at Stage 2 and for FR = 2.5.

Table 1. Four solutions from the set of Pareto optimal solutions at Stage 2 with FR = 2.5.

Item	$f_1(x)$	$f_2(x)$	a	b	c	e	f	l	δ
1	66.44	3.65	131.90	116.10	200.80	0.06	13.75	201.80	2.18
2	74.32	2.44	130.00	174.00	200.00	0.009	10.03	229.20	2.34
3	83.82	1.90	147.20	208.70	200.00	0.45	11.00	270.90	2.38
4	95.13	1.85	163.80	220.00	200.00	0.015	70.09	297.50	2.66

From the results presented above it is clear that the choice of FR has a great influence on the results at this stage. The designer may run the genetic algorithm for other values of FR and in each case the set of Pareto optimal solutions is generated automatically.

6 Conclusions

In the paper the constraint tournament selection method is used to solve multicriteria optimization models of robot grippers. From the results obtained it is clear that this method provides the designer with a new and very effective tool for solving fairly complicated tasks considering both the complexity of the optimization model and the decision making problem. Some experiments carried out on grippers of other structures indicate that the method can also be used for selecting the best structure of the gripper. This will be the subject of further investigations.

Acknowledgements

This study was sponsored by Polish Research Committee (KBN) under the Grant No. 7 T07A 013 14.

References

Boudreau, R. and Gosselin, C. M. (1999): The Synthesis of Planar Parallel Manipulators with a Genetic Algorithm. *Journal of Mechanical Design. ASME*, vol. 121, pp. 533-537.

Boudreau, R. and Turkkan, N. (1996): Solving the Forward Kinematics of Parallel Manipulators with Genetic Algorithm. *Journal of Robotic Systems*, Vol. 13, No. 3, pp. 111-125.

Coello, C.A.C., (1999). A comprehensive survey of evolutionary-based multiobjective optimization techniques. *Knowledge and Information Systems*, Vol. 1, No. 3, pp. 269-308.

Fonseca C. M. and Fleming P. J., (1995): An overviev evolutionary algorithms in multiobjective optimization. *Evolutionary Computation* 3(1) pp. 1-16.

Osyczka, A. and Krenich S. (2000). A New Constraint Tournament Selection Method for Multicriteria Optimizatin Using Genetic Algorithm. In: *Proceedings of Congress of Evolutionary Computing*, San Diego, pp. 501-509.

Osyczka, A. and Kundu, S. (1995). A new method to solve generalized multicriterion optimization problem using the genetic algorithm. *Structural Optimization*, 10(2), pp.94-99.

Osyczka, A., Krenich S., and Karaś, J. (1999). Optimum design of robot grippers using Genetic Algorithms. In: *The 3rd World Congress on Structural and Multidiscipilinary Optimization*, Buffalo 17-22 May,USA.

Osyczka, A., Tamura, H. and Saito, Y. (1997). Pareto set distribution method for multicriteria design optimization using genetic algorithm. *Engineering Design and Automation Conference*, Bangkok, Thailand, March 18-21.

Parmee, C. (1999). A review of evolutionary/adaptive search in engineering design. *Evolutionary Optimization*, 1(1) pp.13 – 39.

Roston, G., P. (1997). Hazards in genetic design methodology. *In: Evolutionary Optimization in Engineering Design*, Dasgupta & Michalewicz, Eds. Springer–Verlag, pp. 135-132.

Design of Spatial Fixed-Sequence Manipulator for Precise and Approximate Reproduction of Gripper Predetermined Positions

Vigen Arakelian and Marc Dahan

LMARC (CNRS) – Institut de Productique, Besançon, France

Abstract. The synthesis of manipulation systems by the predetermined positions of the gripper is one problem of the most actual in modern robotics. One of the methods used to solve this problem is the creating one degree-of-freedom actuators with open kinematic chain allowing for either a precise reproduction of a limited number of gripper predetermined positions or an approximate reproduction of an unlimited number of gripper predetermined positions.

In the present work, such an approach has been developed and a spatial fixed-sequence manipulator designing method is examined based upon the well-known kinematic synthesis methods. We did not trait in details the methods of kinematic synthesis, as our objective is to show how can be used the results of these methods for designing one degree-of-freedom spatial manipulation systems reproducing the gripper predetermined positions.

1 Introduction

The problem dealing with the synthesis of manipulation systems by the predetermined positions of the gripper is one of the most actual in modern robotics. Two methods may be used to solve this problem: 1) by creating multi-degree-of-freedom mechanical systems with open kinematic chain allowing for a precise reproduction of an unlimited number of gripper predetermined positions[1], 2) by synthesis one degree-of-freedom actuators with open kinematic chain allowing for either a precise reproduction of a limited number of gripper predetermined positions or an approximate reproduction of an unlimited number of gripper predetermined positions.

The second method is considered to be the most optimum from the point of view of simplified control and minimum energy expenditure. It widens the field of application of spatial fixed-sequence manipulators and permits to use efficiently the theory of geometrical kinematics applied to mechanism synthesis.

The manipulation system with the actuator of Bennett mechanism constructed at the *Institut de productique de Besançon* (Fig.1) is one of the examples of design of such fixed-sequence manipulators [1,2]. It performs previously set manipulating functions: provides the predetermined trajectory of the manipulated object through three predetermined positions of the gripper. The choice of three positions of the manipulated object is motivated by the fact that

---

[1] Under "unlimited number of predetermined positions" should be understood predetermined N positions whose number is not limited by the conditions described by the relative motion of actuator's links.

they correspond to three processing steps: loading of blank, machine working and removal of finished part.

Figure 1. Fixed-sequence manipulator with actuator of Bennett mechanism.

The merit of this solution resides in that such a manipulating system with cyclic control is equipped with one drive only. A simplified design of this type and the control system improves the operational reliability of the mechanical system and reduces steeply the cost of this latter.

It should be noted also that when designing such manipulating system, links with variable parameters are often introduced therein [3], thus permitting to reproduce not a trajectory but a set of curves.

In this work, such an approach has been developed and a spatial fixed-sequence manipulator designing method is examined based upon the kinematic synthesis methods [4-12]. We shall not deal with the details of kinematic synthesis, as our objective is to show how can be used these methods for designing one degree-of-freedom spatial manipulation systems reproducing the gripper predetermined positions.

2 Statement of the Problem

In Fig.2 is illustrated a spatial fixed-sequence manipulator [3] which incorporates a one degree-of-freedom spatial mechanism *RRRSR* (1-5) and a two-link group (6-7), bears a gripping device (8) and forms with the links (3-4) of the actuator *ABCDEF* a parallelogram pantograph.

The system functions as follows: with rotation of the input rocker (2), the center «P» of the gripping device (8) describes a trajectory which copies the coupler curve of the point «C» of the spatial actuator *ABCDEF*. The ratio of similitude of these two curves is therewith expressed by the relationship: $\chi = EG/ED$.

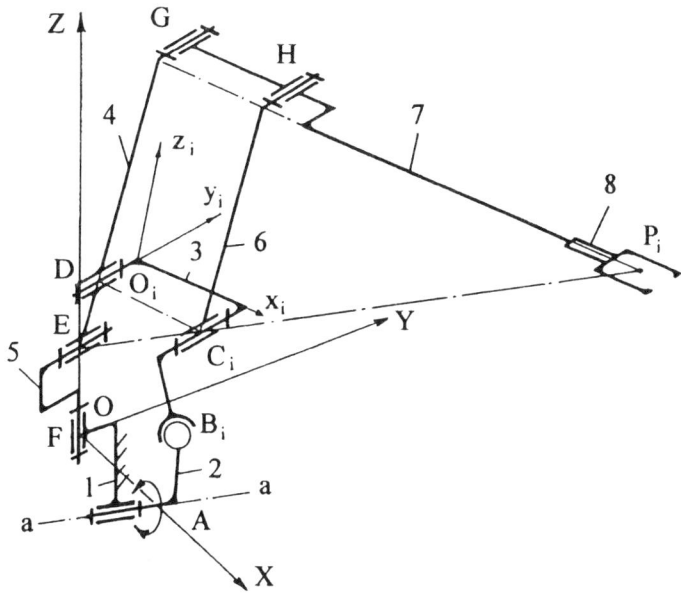

Figure 2. Spatial fixed-sequence manipulator.

In other words, the design of the manipulator consists of a one degree-of-freedom mechanism the trajectory of the coupler-point of which is reproduced by the gripping device due to the pantograph translator.

The objective of the synthesis resides in the following: the positions of the center "P" of the gripping device (8) are given, it is required to determine also the parameters of the actuator *ABCDEF* which provide the displacement of the gripping device through the given positions.

Let us consider the problem dealing with the kinematic synthesis of the manipulator for precise and approximate reproduction of the displacements of the gripping device. For precise reproduction, we shall consider two cases: reproduction through three and four positions.

3 Kinematic Synthesis of Fixed-Sequence Manipulator Actuator

3.1 Through Three Gripper Positions

Lets are given three positions P_1, P_2 and P_3 of the gripper center "P" relatively to fixed system of coordinates *OXYZ*. Taking the ratio of similitude χ for design considerations, we obtain the parallelogram pantograph *EDGHC* and determine the three positions C_1, C_2 and C_3 of the coupler point C. At the three predetermined positions of the gripper, the position of the third coupler point can be set arbitrarily. Thus, setting the three positions of the coupler plane *BCD* we obtain the positions B_1, B_2 and B_3 of the center of the spherical joint B. It is evident that the determination of the length l_{AB} of the link 1 and of the center of its rotation does not present any difficulty as any three points lay on the circle (if they are not on a straight line).

We shall not treat in details the well-known methods of kinematic synthesis [5-9] but shall mention only that the coordinates of the center A of the revolute joint lays on the intersection of the axis a-a defined by the equation[2]

$$(X_{B_1} - X_{B_j})X_A + (Y_{B_1} - Y_{B_j})Y_A + (Z_{B_1} - Z_{B_j})Z_A = 0.5(R_{B_1}^2 - R_{B_j}^2), \tag{1}$$

where

$$R_{B_1}^2 = X_{B_1}^2 + Y_{B_1}^2 + Z_{B_1}^2 \;;\; R_{B_j}^2 = X_{B_j}^2 + Y_{B_j}^2 + Z_{B_j}^2, \; (j = 2, 3),$$

and the planes of the circle that may be defined as a plane drawn through the points B_1, B_2 and B_3:

$$D_1 X_A + D_2 Y_A + D_3 Z_A - D_0 = 0, \tag{2}$$

where

$$D_1 = \begin{vmatrix} Y_{B_1} & Z_{B_1} & 1 \\ Y_{B_2} & Z_{B_2} & 1 \\ Y_{B_3} & Z_{B_3} & 1 \end{vmatrix}, \; D_2 = \begin{vmatrix} Z_{B_1} & X_{B_1} & 1 \\ Z_{B_2} & X_{B_2} & 1 \\ Z_{B_3} & X_{B_3} & 1 \end{vmatrix}, \; D_3 = \begin{vmatrix} X_{B_1} & Y_{B_1} & 1 \\ X_{B_2} & Y_{B_2} & 1 \\ X_{B_3} & Y_{B_3} & 1 \end{vmatrix},$$

$$D_0 = \begin{vmatrix} X_{B_1} & Y_{B_1} & Z_{B_1} \\ X_{B_2} & Y_{B_2} & Z_{B_2} \\ X_{B_3} & Y_{B_3} & Y_{B_3} \end{vmatrix}.$$

The sense of the axis a-a is determined by the expressions:

$$\cos\alpha_x = \frac{D_1}{\pm(D_1^2 + D_2^2 + D_3^2)^{1/2}}; \; \cos\alpha_y = \frac{D_2}{\pm(D_1^2 + D_2^2 + D_3^2)^{1/2}}; \; \cos\alpha_z = \frac{D_3}{\pm(D_1^2 + D_2^2 + D_3^2)^{1/2}}.$$

---

[2] It is defined by the intersection of two planes that are perpendicular to the segments $B_1 B_2$, $B_1 B_3$ and passes through their midpoints.

3.2 Through Four Gripper Positions.

Prior to considering this problem, let us determine the maximum possible number of precisely realizable positions of the gripper of the manipulator under study. It should be noted that the maximum possible number of precisely realizable positions depends upon the displacements of the link 3 of the actuator $ABCDEF$. This number is equal to the ratio of the number of sought parameters k of the kinematic chain to be synthesized to the number of conditions of the constraints imposed on the motion of the link over this chain [9]. For the manipulator under consideration, after having prescribed the ratio of similitude χ, the objective of synthesis is confined to the determination of the parameters of the dyad (2-3), i.e. to the synthesis of the spatial dwell linkage of the type "SR" [6,8,9].

The dyad of the type "SR" is determined by $k=9$ constant parameters involving the coordinates x_B, y_B, z_B of the spherical joint center B, the coordinates X_A, Y_A, Z_A of the center A of the circle described by the point B, the angles between any two fixed axes of coordinates and the axis a-a of the revolute joint, as well as by the distance $r = l_{AB}$. As far as for this case $s = 2$, the maximum number of predetermined positions $N_{max} = k/s = 4$.

As distinct from the previous case, the position of the point B herewith is not taken arbitrarily, as otherwise it will not be possible to provide the disposition of four points on one circle.

We shall not treat here the details of kinematic synthesis, as the aim of this work is to examine the application of synthesis methods in designing fixed-sequence manipulators. Therefore we shall present only the final conditions of synthesis [5-9]. So that four positions B_i ($i = 1,2,3,4$) are on one circle, it is necessary that the following conditions will be satisfied [9]:

$$\begin{vmatrix} X_{B_1} & Y_{B_1} & Z_{B_1} & 1 \\ X_{B_2} & Y_{B_2} & Z_{B_2} & 1 \\ X_{B_3} & Y_{B_3} & Z_{B_3} & 1 \\ X_{B_4} & Y_{B_4} & Z_{B_4} & 1 \end{vmatrix} = 0 \quad \text{and} \quad \begin{vmatrix} R_{B_1}^2 & Y_{B_1} & Z_{B_1} & 1 \\ R_{B_2}^2 & Y_{B_2} & Z_{B_2} & 1 \\ R_{B_3}^2 & Y_{B_3} & Z_{B_3} & 1 \\ R_{B_4}^2 & Y_{B_4} & Z_{B_4} & 1 \end{vmatrix} = 0 . \quad (3)$$

The coordinates x_B, y_B, z_B of the point B in the moving system of coordinates (see Fig.2) are connected with the coordinates X_B, Y_B, Z_B in the fixed system of coordinates by the well-known transformation formulas describing the motion of the link from the zero position to the ith position:

$$\begin{bmatrix} X_{B_i} \\ Y_{B_i} \\ Z_{B_i} \end{bmatrix} = \begin{bmatrix} X_{O_i} \\ Y_{O_i} \\ Z_{O_i} \end{bmatrix} + T_i \times \begin{bmatrix} x_B \\ y_B \\ z_B \end{bmatrix}, \quad (i = 1,2,3,4) \quad (4)$$

where T_i is the 3x3 orthogonal rotation matrix [5,9].

The conditions (3), with the formula (4) taken into account, are transformed into a cubic equation system relatively to x_B, y_B, z_B. The cubic equation system with three unknowns defines the locus of points of the link 3, which have four positions on one circle. A spherical joint center can be arranged on any one of these points. Knowing the positions of the point B, it is possible to determine the parameters of the link 2 (in the same way as for the previous case).

3.3 Through N Positions of Gripper.

If the number of gripper positions is higher than the maximum number of given positions of the actuator $ABCDEF$ (i.e. $N > 4$), their precise realization is not possible. However, in such cases it is possible to reproduce these positions approximately. The methods for solving such problems are well known and elaborated in the approximation theory for mechanism synthesis [9-12].

We will examine an example of approximation theory application. Let N positions of the gripper are given consequently N positions of the link 3. Such points B_i should be searched which, at the indicated positions, are departing a little as far as possible from the circle. The approximated circle is defined as a line of intersection of the sphere with the plane. Such a sought point B in the positions considered must verge both towards the sphere and the plane. Therefore, the function which it is necessary to minimize is adopted the sum [9,10]:

$$S = \sum_{i=1}^{N} \Delta_{qi(sph)}^2 + \sum_{i=1}^{N} \Delta_{qi(pl)}^2 , \qquad (5)$$

where

$$\sum_{i=1}^{N} \Delta_{qi(sph)}^2 = \sum_{i=1}^{N} \left[(R_{B_i} - R_A)^2 - r^2 \right] = \sum_{i=1}^{N} (X_{B_i} X_A + Y_{B_i} Y_A + Z_{B_i} Z_A + H - 0{,}5 R_{B_i}^2)^2 ;$$

$$\sum_{i=1}^{N} \Delta_{qi(pl)}^2 = \sum_{i=1}^{N} (a X_{B_i} + b Y_{B_i} + c Z_{B_i} - 1)^2 .$$

The necessary conditions of the minimum sum (5) are:

$$\frac{\partial S}{\partial X_A} = 0; \quad \frac{\partial S}{\partial Y_A} = 0; \quad \frac{\partial S}{\partial Z_A} = 0; \quad \frac{\partial S}{\partial x_B} = 0; \quad \frac{\partial S}{\partial y_B} = 0;$$

$$\frac{\partial S}{\partial z_B} = 0; \quad \frac{\partial S}{\partial r} = 0; \quad \frac{\partial S}{\partial a} = 0; \quad \frac{\partial S}{\partial b} = 0; \quad \frac{\partial S}{\partial c} = 0; \tag{6}$$

These conditions bring about two equation systems:

$$\begin{bmatrix} \sum X_{B_i}^2 & \sum X_{B_i} Y_{B_i} & \sum X_{B_i} Z_{B_i} & \sum X_{B_i} \\ \sum X_{B_i} Y_{B_i} & \sum Y_{B_i}^2 & \sum Y_{B_i} Z_{B_i} & \sum Y_{B_i} \\ \sum X_{B_i} Y_{B_i} & \sum Y_{B_i} Z_{B_i} & \sum Z_{B_i}^2 & \sum Z_{B_i} \\ \sum X_{B_i} & \sum Y_{B_i} & \sum Z_{B_i} & N \end{bmatrix} \begin{bmatrix} X_A \\ Y_A \\ Z_A \\ H \end{bmatrix} = 0{,}5 \begin{bmatrix} \sum R_{B_i}^2 X_{B_i} \\ \sum R_{B_i}^2 Y_{B_i} \\ \sum R_{B_i}^2 Z_{B_i} \\ \sum R_{B_i}^2 \end{bmatrix} \tag{7}$$

$$\begin{bmatrix} \sum X_{B_i}^2 & \sum X_{B_i} Y_{B_i} & \sum X_{B_i} Z_{B_i} \\ \sum X_{B_i} Y_{B_i} & \sum Y_{B_i}^2 & \sum Y_{B_i} Z_{B_i} \\ \sum X_{B_i} Y_{B_i} & \sum Y_{B_i} Z_{B_i} & \sum Z_{B_i}^2 \end{bmatrix} \begin{bmatrix} a \\ b \\ c \end{bmatrix} = \begin{bmatrix} \sum X_{B_i} \\ \sum Y_{B_i} \\ \sum Z_{B_i} \end{bmatrix}, \quad (i = 1, \ldots, N) \tag{8}$$

These equations are interconnected by the variables x_B, y_B, z_B only. By assigning values to them, i.e. by fixing the position of the point B on the link 3, are obtained the parameters of the sphere and plane by the intersection of which is determined the approximating circle [9-12].

It should be noted that approximation synthesis of mechanisms are sufficiently elaborated and by using the known algorithms of computation iteration procedures [10] it is possible to design such manipulation systems for generation of spatial trajectory.

4 Conclusions

This work deals with the problem relating to the design of spatial fixed-sequence manipulators for precise and approximate reproduction of gripper predetermined positions. It is shown how to design such manipulation systems with the use of one degree-of-freedom actuator. The problem relating to precise reproduction of three and four positions of the gripper is treated on the example of *RRRSR* spatial actuator. It is suggested to use approximation synthesis methods for the design of spatial fixed-sequence manipulators reproducing approximately the spatial trajectories according to previously given positions of the gripper.

References

1. Hervé, J.M. and Dahan M. (1983) The two kinds of Bennett's mechanism. In *Proceedings of the Sixth World Cong. Theory of Machines and Mechanisms*, New Delhi, v.1, 116-119.
2. Dahan, M., Dalha, C. and Lexcellent, C. (1985) Propriétés et utilisation du mécanisme de Bennett. *Mech. and Mach. Theory*, 20, 189-197.

3. Arakelian, V.H. (1988) and all. Manipulator. Patent SU n. 1364467, B. I. n. 1, January 7.
4. Bottema, O. and Roth, B. (1979) *Theoretical kinematics*. New York: North-Holland Pub. 558p.
5. Roth, B. (1967) The kinematics of motion through finitely separated positions. *Journal of Applied Mechanics*, Series E, v. 34, n. 3, 591-598.
6. Roth, B. (1967) Finite-position theory applied to mechanism synthesis. *Journal of Applied Mechanics*, Series E, v. 34, n. 3, 599-605.
7. Dimentberg, F.M. (1982) *Theory of spatial mechanisms*. Naouka: Moscow, 336p.
8. Tsai, L.W. and Roth, B. (1972) Design of dyads with helical, cylindrical, spherical, revolute and prismatic joints. *Mech. and Mach. Theory*, v.7, n.1.
9. *Kinematics, dynamics and precision of mechanisms*. Ed. by Kreynin, Moscow, 1984, 224p.
10. Sarkissyan, Y.L. (1982) *Approximation synthesis of mechanisms*. Naouka: Moscow, 304p.
11. Gupta, K.C. (1974) *On a class of approximation problems in kinematics*. PhD dissertation, Stanford University, Stanford.
12. Gupta, K.C., and Roth B. (1975) A general approximation theory for mechanisms synthesis. *Journal of Applied Mechanics*, Series E, v. 42, n. 2, 451-457.

Chapter II

CONTROL OF MOTION

A Powered-Caster Holonomic Robotic Vehicle for Mobile Manipulation Tasks

Robert Holmberg and Oussama Khatib

The Robotics Laboratory, Department of Computer Science, Stanford University

Abstract. Mobile manipulator systems hold promise in many industrial and service applications including assembly, inspection, and work in hazardous environments. The integration of a manipulator and a mobile robot base places special demands on the vehicle's drive system. For smooth accurate motion and coordination with an on-board manipulator, a holonomic vibration-free wheel system that can be dynamically controlled is needed. In this paper, we present the design and development of a Powered Caster Vehicle (PCV) which is shown to possess the desired mechanical properties. To dynamically control the PCV, an new approach for modeling and controlling the dynamics of this parallel redundant system is proposed. The experimental results presented in the paper illustrate the performance of this platform and demonstrate the significance of dynamic control and its effectiveness in mobile manipulation tasks.

1 Introduction

Our experimental work in mobile manipulation (Khatib et al., 1996) has started with the development of the Stanford Robotics Platforms. In collaboration with Oak Ridge National Laboratories and Nomadic Technologies, we designed and built (Khatib et al., 1999b) two holonomic mobile manipulator platforms. Each platform was equipped with a PUMA 560 arm, and a base which consists of three "lateral" orthogonal universal-wheel assemblies (Pin and Killough, 1994), allowing the base to translate and rotate holonomically in relatively flat office-like environments. The Stanford Robotics Platforms provided a unique testbed for the development, implementation, and demonstration of various mobile manipulation control strategies, collision avoidance, and cooperative manipulation (Khatib et al., 1999a). The experiments conducted with these platforms have also illustrated the limitations of the holonomic base, and highlighted the need to advance its capabilities. The work presented in this paper is part of the commercial efforts of Nomadic Technologies in mobile robots and our continuing research in mobile manipulation.

A holonomic system is one in which the number of degrees of freedom are equal to the number of coordinates needed to specify the configuration of the system. In the field of mobile robots, the term holonomic mobile robot is applied to the abstraction called the robot, or base, without regard to the rigid bodies which make up the actual mechanism. Thus, any mobile robot with three degrees of freedom of motion in the plane has become known as a holonomic mobile robot.

Many different mechanisms have been created to achieve holonomic motion. These include various arrangements of universal or omni wheels (La, 1979) and (Carlisle, 1983), double universal wheels (Bradbury, 1977), Swedish or Mecanum wheels (Ilon, 1971), chains of spherical (West and Asada, 1992) or cylindrical wheels (Hirose and Amano, 1993), orthogonal wheels (Killough and Pin, 1992), and ball wheels (West and Asada, 1994). All of these mechanisms, except for some types with ball wheels, have discontinuous wheel contact points which are a great source of vibration; primarily because of the changing support provided; and often additionally because of the discontinuous changes in wheel velocity needed to maintain smooth vehicle motion.

These mechanisms tend to have poor ground clearance due to the use of small peripheral rollers and/or the arrangement of the mechanism leaves some of the support structure very close to the ground. The design and actuation of these mechanisms has been driven by kinematic concerns for minimum actuation and minimal sensing to make to the implementations of odometry and control mathematically exact. Yet, many of these designs have multiple rollers with the contact points of the wheel on the ground moving from one row to the other. These contact points are often assumed to remain stationary in the middle of each wheel. This emphasis on minimal design has led to many three wheeled designs which are more likely to tip over, or at least lift a wheel, as performance and payload is increased. Also, the minimal use of actuators often led to complex mechanical transmissions to distribute the power to the driving elements. The designs discussed are mechanically complex; often with many moving parts, some active, some passive.

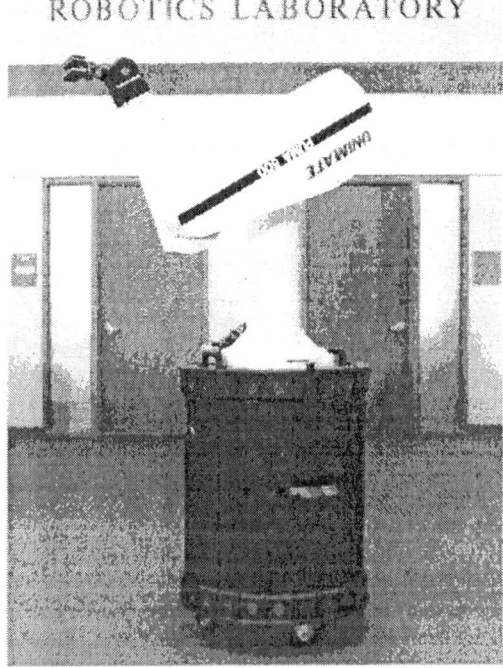

Figure 1. Nomadic XR4000 and PUMA 560

Just as a kinematic approach was used in the design of these holonomic mechanisms, the control of these mechanisms was looked at from a purely kinematic perspective. Many of the designs incorporate passive rollers without sensing of their motions, so that the dynamics of these elements cannot be accounted for. Without dynamic control, it is difficult to perform coordinated motion of a mobile base and dynamically controlled manipulator.

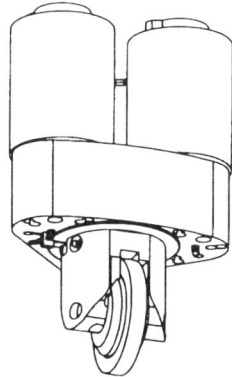

Figure 2. Powered Caster Module

We present here another type of holonomic vehicle mechanism which we will refer to as a *powered caster vehicle* or PCV. The desired motion of the PCV, especially at higher speeds, is often disturbed by the closely coupled dynamics of its wheel modules. A dynamically-controlled, holonomic mobile robot is particularly desirable in a mobile manipulation system for many reasons. A holonomic robot makes for easier gross motion planning and navigation. It allows for full use the null space motions of the system to improve the workspace and overall dynamic endpoint properties. A dynamically controlled mobile robot is especially important when used as the base "joints" of a mobile manipulation system so that the dynamic forces developed by the manipulator can be decoupled with *forces* generated in the base "joints".

We will present the design fundamentals of a working PCV mechanism, the Nomadic Technologies XR4000, shown in Figure 1. We will also present the new framework for efficient dynamic control of a PCV. The experimental results presented in the paper will show the benefit of this control framework and its impact on the integration of the PCV in a full mobile manipulation system.

2 Design

The PCV concept provides an effective approach for the development of holonomic mobility for a number of reasons. The contact points between the wheels and the ground move in a continuous manner and thus do not induce vibrations from shifting support points or discontinuous wheel velocities. The location of each contact point is well known so that control is more exact. Each wheel mechanism contains a single nonholonomic wheel which is large enough for good ground clearance. One final point which has not been adequately addressed previously, is that the PCV is the only holonomic mechanism which can be designed to effectively use currently available pneumatic tires — and consequently benefit from the suspension, traction, and wear properties of this well developed technology. Because there are no passive and more importantly no unmeasured bodies in a powered caster design the dynamics of the system can be accurately modeled.

A PCV is composed of $n \geq 2$ powered caster modules as illustrated in Figure 2. The modules could vary in size and power from module to module, but without loss of generality, we will assume that all the modules are identical. The PCV design is defined by the strictly positive geometric parameters: wheel radius(r), caster offset(b), and wheel module placement(h, β) (see Figure 3). Along with the mass and inertia of each component in the design, parameters which affect the system dynamics include the gear ratios and motor sizes. Values for the geometric parameters must be selected so that the area swept out by each wheel does not intersect any other. The wheels should have a large enough radius to surmount anticipated obstacles. The dynamic tradeoffs involve the geometry as well as the motors and gearing. Careful selection must be made to result in a mechanism which has good acceleration while maintaining the ability to reach the desired top speed. At the same time, by choosing components so that motor and gearbox speeds are kept low, mechanical noise due to high component speeds can be minimized.

The PCV mechanism shown in Figure 1, a Nomadic Technologies XR4000 mobile robot, was designed to be a high performance holonomic vehicle for mobile robotics and mobile manipulation. It has four 11 cm diameter wheels with 2 cm caster offset. It can accelerate at 2 m/s^2 on most surfaces and has a top speed of 1.25 m/s. The controller of the XR4000 used herein was modified at Stanford University by replacing the standard PWM motor amplifiers with current controlled motor amplifiers.

3 Dynamic Modeling

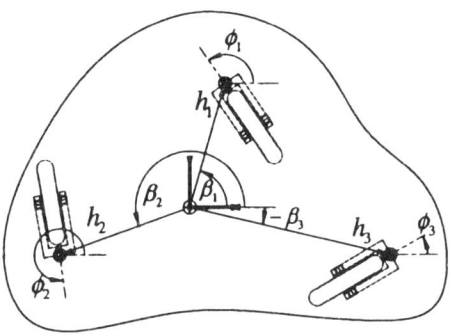

Figure 3. PCV Geometry

Typically, the dynamic equations of motion for a parallel system with nonholonomic constraints such as a PCV are formed in one of two ways: the unconstrained dynamics of the whole

A Powered-Caster Holonomic Robotic Vehicle for Mobile Manipulation Tasks

system can be derived and the the constraints are applied to reduce the number of degrees of freedom (Campion et al., 1993); or the system is cut up into pieces, the dynamics of these subsystems are found, and the loop closure equations are used to eliminate the extra degrees of freedom. For our four-wheeled XR4000 robot, using the first method, we will obtain 11 equations for the unconstrained system and 8 constraint equations for a total of 19 equations. The second method will yield 12 equations for the unconstrained subsystems and 9 constraint equations for a total of 21 equations. These systems of equations must then be reduced to 3 equations. Ideally, both these methods would yield the same minimal set of dynamic equations, but in practice it is difficult to reduce the proliferation of terms that are introduced in a large number of equations.

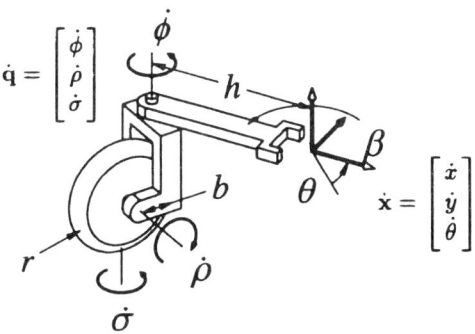

Figure 4. Powered Caster "Manipulator"

To get a more efficient form of the dynamic equations of motion we will use a method which uses compatible 3 DOF systems. We can model the PCV as a collection of cooperating manipulators such as shown in Figure 4.

Because of the parallel nature of the final mechanism we choose to write the relationship between joint speeds and local Cartesian speeds, $\dot{x}$, as

$$\dot{q} = J^{-1} \dot{x} \tag{1}$$

$$J^{-1} = \begin{bmatrix} -s\phi/b & c\phi/b & h[c\beta c\phi + s\beta s\phi]/b - 1 \\ c\phi/r & s\phi/r & h[c\beta s\phi - s\beta c\phi]/r \\ -s\phi/b & c\phi/b & h[c\beta c\phi + s\beta s\phi]/b \end{bmatrix}$$

As shown in Figure 4, $\dot{\phi}$ is the steering rate, $\dot{\rho}$ is the angular speed of rolling, and $\dot{\sigma}$ is the angular twist rate at the wheel contact. For compactness we use s· and c· as shorthand for $\sin(\cdot)$ and $\cos(\cdot)$. It is interesting to note that the first two rows of J^{-1} express the nonholonomic constraints due to ideal rolling while the third row is a holonomic constraint: $\theta = \sigma - \phi$.

Using the standard joint space dynamics and the Jacobian in eqn. 1, we can express the operational space dynamics (Khatib, 1987) of the ith manipulator as

$$\Lambda_i(q_i)\ddot{x} + \mu_i(q_i, \dot{q}_i, \dot{x}) = F_i \tag{2}$$

with
$$\Lambda_i = J_i^{-T} A_i J_i^{-1}$$
$$\mu_i = J_i^{-T} \left(A_i \dot{J}_i^{-1} \dot{x} + b_i \right)$$

where Λ is the operational space mass matrix, μ is the operational space vector of centrifugal and Coriolis terms, and $\mathbf{F}$. is the force/torque vector at the origin of the end effector coordinate system. Since our manipulator is simple and not redundant we compute J^{-1} directly, thus avoiding an inversion operation which is traditionally required.

If we choose the end effector frames of the various manipulators such that they are coincident while the wheel modules are correctly positioned with respect to one another, then, using the augmented object model of Khatib (1988), we can write the overall operational space dynamics of the mobile base as the simple sum of the dynamics of the individual modules.

$$\Lambda \ddot{x} + \mu = \mathbf{F} \tag{3}$$

with
$$\Lambda = \sum_i^n \Lambda_i \quad ; \quad \mu = \sum_i^n \mu_i \quad ; \quad \mathbf{F} = \sum_i^n \mathbf{F}_i$$

Here, Λ, μ, and $\mathbf{F}$ have the same meanings as before but now represent the properties of the entire robot.

With this algorithm we have determined the operational space dynamic equations of motion directly. For our four-wheeled XR4000 robot we generate 12 equations, 3 for each i in eqn. 2, which are then added in groups of four to give the required 3 operational space equations. Using the symbolic dynamic equation generator AUTOLEV to create Λ and μ, the number of multiplies and additions are reduced from 8180 and 2244, to 2174 and 567.

4 Dynamically Decoupled Control

The control and dynamic decoupling of the PCV is achieved by selecting the operational space control structure (Khatib, 1987)

$$\mathbf{F} = \Lambda \mathbf{F}^* + \mu \tag{4}$$

where $\mathbf{F}$ is the operational space force which is to be applied to the PCV and $\mathbf{F}^*$ is the control force for our linearized unit mass system. As an example we can choose to implement a simple P-D controller

$$\mathbf{F}^* = -K_p(\mathbf{x} - \mathbf{x}_d) - K_v(\dot{\mathbf{x}} - \dot{\mathbf{x}}_d) + \ddot{\mathbf{x}}_d \tag{5}$$

with K_p, K_v the position and velocity gains and $\mathbf{x}_d$ and its derivatives the desired position, velocity and acceleration.

This approach requires that we know the operational space velocities, $\dot{\mathbf{x}}$, of the PCV and the actuated joint torques, Γ', necessary to produce the commanded operational space force, $\mathbf{F}$. The XR4000 powered casters (see Figure 2) have an encoder on each motor. The encoders together with knowledge of the gearbox kinematics allow us to calculate the positions and velocities of the steering and rolling joints of each module. We can write the relationships between the observed joint speeds and the operational speeds of the ith wheel as the *wheel constraint matrix*,

$\tilde{J}_i$, which contains the two nonholonomic constraints from "manipulator" model in eqn. 1. We use the notation, $\tilde{J}$, to indicate that the constraint matrix is a type of Jacobian inverse. It describes the mapping from Cartesian speeds to joint speeds; a mapping opposite that of the traditional Jacobian. The important distinction is that this "inverse" is directly derived from the kinematics of the robot and is not an algebraic inversion of an available Jacobian matrix. We will use $\dot{q}'_i = [\dot{\phi}_i \; \dot{\rho}_i]^T$ to designate the observed joint speeds of the ith wheel.

$$\dot{q}'_i = \tilde{J}_i \dot{x} \tag{6}$$

$$\tilde{J}_i = \begin{bmatrix} -s\phi_i/b & c\phi_i/b & h_i[c\beta_i c\phi_i + s\beta_i s\phi_i]/b - 1 \\ c\phi_i/r & s\phi/r & h_i[c\beta_i s\phi_i - s\beta_i c\phi_i]/r \end{bmatrix}$$

The overall motion of the joints in the robot can be described by gathering the wheel constraint matrices into the *constraint matrix*, $\tilde{J}$.

$$\dot{q}' = \tilde{J} \dot{x} \tag{7}$$

$$\dot{q}' = \begin{bmatrix} \dot{q}'_1 \\ \vdots \\ \dot{q}'_n \end{bmatrix} \quad ; \quad \tilde{J} = \begin{bmatrix} \tilde{J}_1 \\ \vdots \\ \tilde{J}_n \end{bmatrix}$$

The dual of this relationship describes the operational space force produced by the torques at the actuated joints.

$$\mathbf{F} = \tilde{J}^T \boldsymbol{\Gamma}' \tag{8}$$

To find the operational space velocities and actuated joint torques we need to find the inverse relationships to eqns. 7,8. One common approach is to use a generalized inverse (Muir and Neuman, 1986) of the the full constraint matrix C. Our approach instead involves two steps: resolving the wheel velocities at the contact points and then resolving the joint velocities. This provides a more physically intuitive solution to the inverse problem.

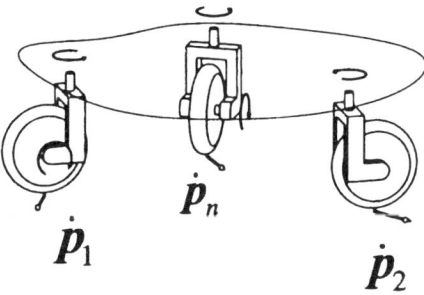

Figure 5. Contact point velocities

It may be easiest to visualize the contact point velocities as speeds, $\dot{p}$, the contact points would have in the world if the robot body were held fixed and the wheels were not in contact with the ground. This is illustrated in Figure 5.

The sensed contact points velocities can be calculated from the measured joint speeds with the one-to-one mapping below where C_q is square, full rank, block diagonal, and invertible.

$$\dot{\mathbf{q}} = J_q \, \dot{\mathbf{p}} \tag{9}$$

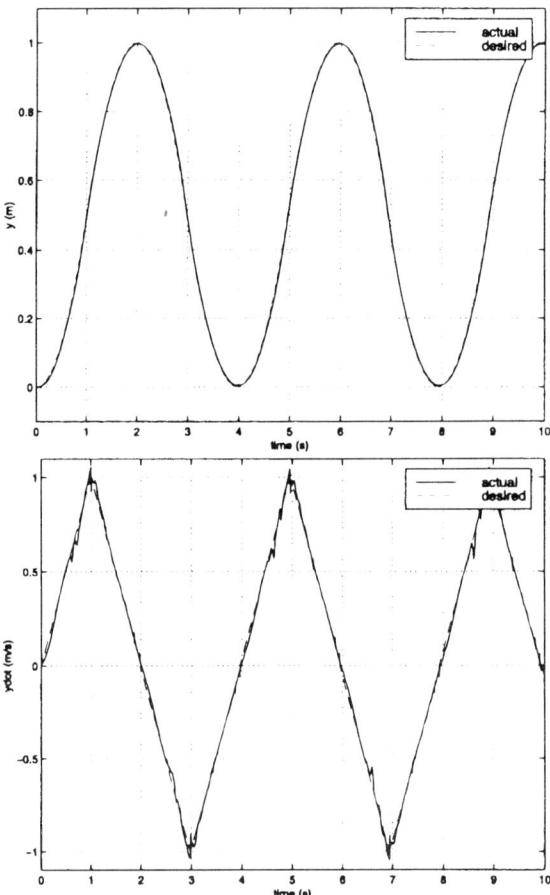

Figure 6. Position vs. time and velocity vs. time with dynamic compensation

When the robot obeys the ideal rolling assumptions there exists a robot velocity where the sensed contact speeds are identical to the consistent set of contact speeds, $\hat{\mathbf{p}}$, found with the kinematic relationship

$$\dot{\hat{\mathbf{p}}} = \tilde{J}_p \, \dot{\mathbf{x}} \tag{10}$$

However, as is to be expected, when there is some slippage and measurement noise, $\dot{\mathbf{p}} \neq \dot{\hat{\mathbf{p}}}$. By using the Moore-Penrose pseudo-inverse (indicated with a $^+$) of the non-square matrix $\tilde{J}_p$

we will minimize the total perceived slip by minimizing the differences between $\dot{\mathbf{p}}$ and $\hat{\dot{\mathbf{p}}}$. Our estimate of the robot velocity assuming that slip is minimized uses a generalized inverse of the constraint matrix and is

$$\dot{\mathbf{x}} = \tilde{J}^{\#} \dot{\mathbf{q}}'$$ (11)

with

$$\tilde{J}^{\#} = \tilde{J}_p^+ J_q^{-1}$$ (12)

We have tested the odometry of our XR4000 moving randomly for one minute in a 1.5m x 2.5m area and then returning to its starting position. When using the generalized inverse from eqn. 12 the dead-reckoning error was less than half as large as when the pseudo-inverse of the constraint matrix was used.

The dual of this result is just as ideal. There are many ways to distribute the effort among the joints to achieve a desired operational space force. By distributing the joint torques using the transpose of the generalized inverse in eqn. 12

$$\boldsymbol{\Gamma}' = \tilde{J}^{\#\mathsf{T}} \mathbf{F}$$ (13)

we minimize, in a least squares way, the contact forces developed by the wheels. The consequence is that the tractive effort is spread as evenly as possible among the wheels and the tendency for any one wheel to loose traction is minimized.

5 Experimental Results

To demonstrate the effectiveness of the proposed dynamic compensation, the XR4000 robot was commanded to move from zero to one meter in the y direction with 0.1 m/s^2 acceleration. The robot is to follow a straight line ($x = 0$) and to not rotate ($\theta = 0$). The gains used in these experiments were reduced by a factor of 10 from the gains used during normal motions so that dynamic disturbances would be more apparent.

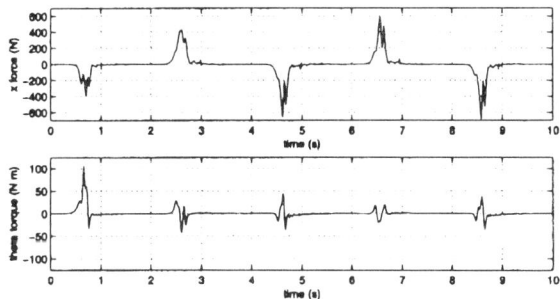

Figure 7. Coupling force compensation X and Theta

Each time the XR4000 moves in this task, all four wheels must flip directions, and cause large dynamic coupling forces as seen in Figure 7. Notice that the dynamic side forces reaches 400–600 Newtons and the dynamic coupling torque reaches 50–100 Newton-meters.

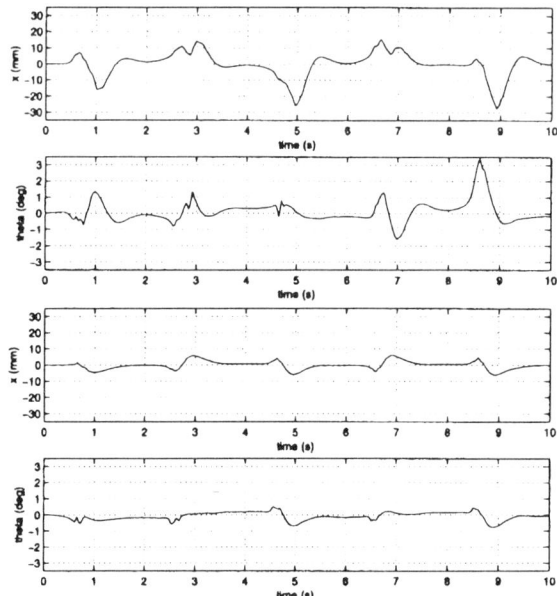

Figure 8. Position vs. time, without and with dynamic compensation

In Figure 8 the disturbance effects of the dynamic forces are shown. On the left, the robot is run without using dynamic compensation and has position errors on the order of 30 mm and 3°. On the right, the robot is run while implementing the proposed dynamic compensation and the errors are reduced to about 5 mm and 0.5°.

6 Conclusions

We have presented the design of a new wheeled holonomic mobile robot, the *powered caster vehicle*, or PCV, which is being produced as the XR4000 mobile robot by Nomadic Technologies. The design of the powered caster vehicle provides smooth accurate motion with the ability to traverse the hazards of typical indoor environments. The design can be used with two or more wheels, and as implemented with four wheels provides a stable platform for mobile manipulation.

We have also described a new approach for a modular, efficient dynamic modeling of wheeled vehicles. This approach is based on the augmented object model originally developed for the study of cooperative manipulators. The actuation redundancy is resolved to effectively distribute the actuator torques to minimize internal or antagonistic forces between wheels. This results in reduced wheel slip and improved odometry.

Using the vehicle dynamic model and the actuation and measurement redundancy resolution, we have developed a control structure that allows vehicle dynamic decoupling and slip minimization. The effectiveness of this approach was experimentally demonstrated for motions involving large dynamic effects.

The PCV dynamic model and control structure have been integrated into a new mobile manipulation platform integrating the XR4000 and a PUMA arm. The experimental results on the new platform have shown full dynamic decoupling and improved performance.

Acknowledgements

We gratefully acknowledge Nomadic Technologies Inc., where the development of the powered caster mechanism took place; for the resources devoted to this project, and to the work of all the individuals there, especially Anthony del Balso, Rich Legrand, Jim Slater and John Slater who were instrumental in the creation of the XR4000 mobile robot.

References

Bradbury, H. M. (1977). Omni-directional transport device. U.S. Patent #4223753.

Campion, G., Bastin, G., and d'Andréa-Novel, B. (1993). Structural properties and classification of kinematic and dynamic models of wheeled mobile robots. In *Proc. IEEE International Conference on Robotics and Automation*, 462–469.

Carlisle, B. (1983). An omni-directional mobile robot. In Rooks, B., ed., *Developments in Robotics*. Kempston, England: IFS Publications. 79–87.

Hirose, S., and Amano, S. (1993). The VUTON: High payload high efficiency holonomic omni-directional vehicle. In 6th *International Symposium on Robotics Research*.

Ilon, B. E. (1971). Directionally stable self propelled vehicle. U.S. Patent #3746112.

Khatib, O., Yokoi, K., Chang, K., Ruspini, D., Holmberg, R., and Casal, A. (1996). Proceedings of the ieee/rsj international conference on intelligent robots and systems. In *Vehicle/arm coordination and multiple mobile manipulator decentralized cooperation*, volume 2, 546–553.

Khatib, O., Brock, O., Yokoi, K., and Holmberg, R. (1999a). Dancing with juliet 1999. IEEE Robotics and Automation Conference Video Proceedings.

Khatib, O., Yokoi, K., Brock, O., Chang, K., and Casal, A. (1999b). Robots in human environments: Basic autonomous capabilities. *International Journal of Robotics Research* 18:684–696.

Khatib, O. (1987). A unified approach for motion and force control of robotic manipulators: The operational space formulation. *IEEE Journal of Robotics and Automation* RA-3(1):43–53.

Khatib, O. (1988). Object manipulation in a multi-effector robot system. In *Robotics Research 4 Proc. 4th Int. Symposium*, 137–144.

Killough, S. M., and Pin, F. (1992). Design of an omnidirectional and holonomic wheeled platform prototype. In *Proc. IEEE International Conference on Robotics and Automation*, volume 1, 84–90.

La, H. T. (1979). Omnidirectional vehicle. U.S. Patent #4237990.

Muir, P. F., and Neuman, C. P. (1986). Kinematic modeling of wheeled mobile robots. Technical Report CMU-RI-TR-86-12, The Robotics Institute, Carnegie-Mellon University, Pittsburgh, PA.

Pin, F. G., and Killough, S. M. (1994). A new family of omnidirectional and holonomic wheeled platforms for mobile robots. *IEEE Transactions on Robotics and Automation* 10(4):480–489.

West, M., and Asada, H. (1992). Design of a holonomic omnidirectional vehicle. In *Proc. IEEE International Conference on Robotics and Automation*, 97–103.

West, M., and Asada, H. (1994). Design of ball wheel vehicles with full mobility, invariant kinematics and dynamics and anti-slip control. In *Proceedings of the ASME Design Technical Conferences, 23rd Biennial Mechanisms Conference ASME*, 377–384.

Motion Coordination and Hybrid Position/Force Control of a Mobile Micromanipulator Actuated by Direct-Drive Vibromechanisms

Antoine Ferreira [1] and Patrice Minotti [2]

[1] Laboratoire de Vision et Robotique, Ecole Normale Supérieure d'Ingénieurs de Bourges, France
[2] Laboratoire de Mécanique Appliquée, University of Franche-Comté, France

Abstract. This article is concerned with motion control problem of a new miniature one-armed mobile micromanipulator actuated by multidegree of freedom ultrasonic actuators using active driving friction mechanism. The main control disturbances are due to the nonlinear and nonholonomic constraints governing the motion of the mobile platform in a small workplace and, also, to the fact that the dynamics of the micromanipulator and the mobile platform are highly coupled. It is proposed in this article to tackle the robustness issue by designing and implementing new schemes for motion coordination and hybrid position/force control.

1. Introduction

Modern flexible manipulation stations, which allow components of small dimensions to be automatically assembled, require high-precision micromanipulation robots integrating both transportation and manipulation skills. These mobile micromanipulators must be automated with the help of dedicated tools, in order to free humans from the tedious task of having to manipulate minuscule objects directly. In the field of small object handling systems, miniature mobile robots with micro-gripping, micromanipulation and transportation capabilities are one of the most important systems that can process small objects of different size. Very little work has been published at present regarding the question of control of mobile micromanipulation systems (Fukuda,1991) and (Munassypov,1996). This article is concerned with motion control problem of a new miniature one-armed mobile micromanipulator actuated by multidegree of freedom ultrasonic actuators reported in a previous work (Ferreira, 1998). The main control disturbances are due to the nonlinear and nonholonomic constraints governing the motion of the mobile platform in a small workplace and, also, to the fact that the dynamics of the micromanipulator and the mobile platform are highly coupled. Since the ultimate goal is to control kinematically and dynamically the motion of the micromanipulator's end-effector, it is proposed in this article to tackle the robustness issue by designing and implementing new schemes for motion coordination and hybrid force/position control.

2. Mechanics of the Micromanipulator

2.1. Description and Operating Principle of the Micromanipulator

Description. The prototype shown in Fig.1 has been realized in cooperation with the Applied Mechanical Laboratory R. Chaléat, France (Minotti and Ferreira, 1998). It consists of a mobile platform (1) driven by a linear ultrasonic motor (three planar reversible D.O.F: X-Y-θ_Z) and a manipulation tool driven by a spherical ultrasonic motor (2) (two rotation reversible D.O.F: θ_X, θ_Y). A mechanical hand (4), carrying an active gripper (6), is attached to a spherical rotor (2) in order to rotate the gripper in two orthogonal directions. The multidirectional motion of such actuators are performed by the control of separate elementary ultrasonic actuators (3). The end-effector designed for this manipulator is a two-fingered gripper actuated by shape memory alloy (SMA) wire and employs strain gage sensors to perform gripping force control.

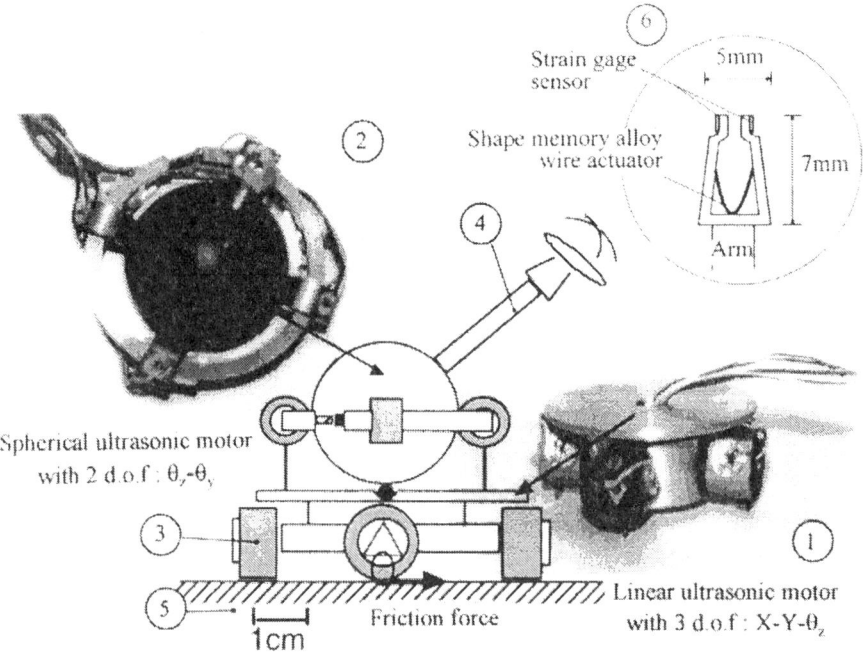

Figure 1. Mechanical configuration of a miniature mobile robot with micromanipulation capabilities having 5 d.o.f : (1) linear ultrasonic motor with 3 d.o.f. (X, Y, θ_Z), (2) spherical ultrasonic motor with 2 d.o.f. (θ_X, θ_Y), (3) elementary ultrasonic actuator, (4) manipulator's arm, (5) frame and (6) two-fingered gripper actuated by shape memory alloy (SMA) wire.

Multidegree of freedom ultrasonic actuators. Ultrasonic motor technology, called also vibromotor, features a lot of merits well adapted for micromechatronic systems. Their direct drive actuation enables precise motion control to be achieved since it provides an accurate force transmission method that does not include nonlinear behaviour such as backlash, hysteresis and dead zones. Furthermore, the high holding force obtained when the motor is deactivated ensures the decoupling of the different d.o.f. of the moving element and the mechanical stability of the end-effector. The physical dimensions of conventional mechanical solutions are not suited to the present objectives in the miniaturization of mechanisms. Ultrasonic motors make possible to form a robot equal in maximum load and lighter in weight, and possessing an integrated structure well adapted to narrow workspaces. Finally, the low voltages required by such vibromechanisms make it possible to built autonomous mobile robots. The elementary stator concept presented in Figure 2 consists of a vibrating metallic stator of simple geometry and piezoelectric ceramics. The alternating voltages $u1(t)=U_1\sin\omega t$ and $u2(t)=U_2\sin(\omega t+\Phi)$, where Φ is the phase-difference between applied voltages, are applied to the ceramics E_1 and E_2 respectively. An ideal resonant flexural traveling wave ($f_r=30kHz$) is generated along the radial direction when the phase-difference is set to $\Phi=\pi/2$. When the support is pressed to the stator by a given axial force F_N the traveling wave in the radial direction is transmitted, through friction force, into rigid body movement of the moving element. By changing the direction of the traveling wave ($\Phi=-\pi/2$), the direction of the linear motion can be reversed. Using this basic concept of motion, actuators with different shape, size and degrees of freedom can be built for providing the different motions in a plane or in space of the manipulator's end-effector.

Shape memory alloy (SMA) wire actuator. SMA-based actuators (Figure 3) feature strong force to weight ratios, controllability, direct drive capabilities, etc. These materials can memorize a certain shape and can retrieve that particular shape even when deformed. During the process of retrieving their memory shape, SMA produce large forces well adapted to gripping operations. To open the gripper, the SMA wire attempts to return to its remembered straight shape, thus forcing the gripper fingers apart.

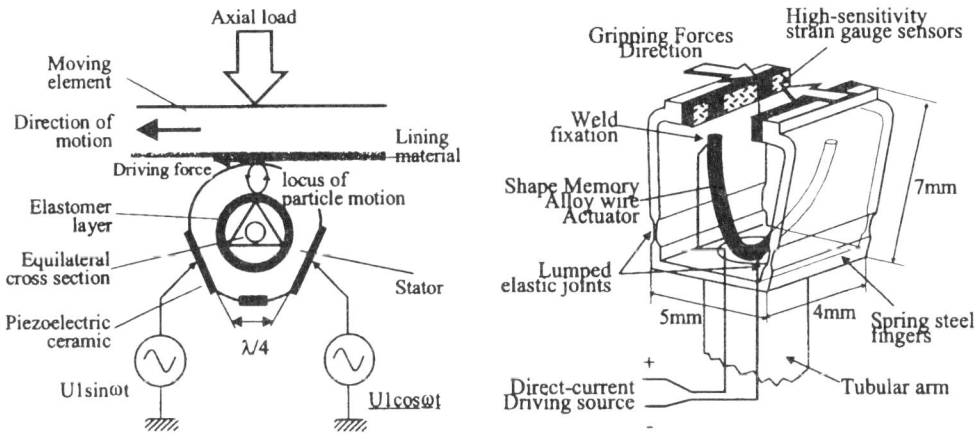

Figure 2. Elementary ultrasonic actuator concept

Figure 3. Two-fingered gripper actuated by SMA wire.

3. Configuration of the System

A view of the current configuration of the micromanipulator is shown in Figure 4. A dedicated user interface allows to pilot the robot with high flexibility. The first task to be tested is the task of transportation of the manipulation tool to some working areas (several squared micrometers size). Such movement can be regarded as a combined large and precise motion in an unlimited travel range. In our approach, we have chosen to detect accurately the initial location of both object and micro-robot by means of a vision system looking at the work surface through a high-definition CCD camera. For task oriented teleoperation, a vision-based feedback is realized on the workstation. The position and trajectory control are performed through position optical encoders. The signals are acquired on workstation by a Matrox video frame grabber. The workstation is connected with the micromanipulator controller by a RS232 line. Low-level control of the eight elementary ultrasonic actuators is performed by a MC68332 processor via specific HF power supply circuits. Two-axis position measurement of the mobile platform is performed by a laser vibrometer system (OFV-040, Polytec) and its servo-control boards execute the positioning of laser beams along X and Y axes. The measurement mechanism of the robot's arm orientation consists of two circular sliding guides and two high-optical encoders. These two position sensors are placed on each point of intersection of sliding guides to measure the orientation of the output shaft around X and Y axes. Finally, gripping forces and SMA driving current provides essential information about the grasped object.

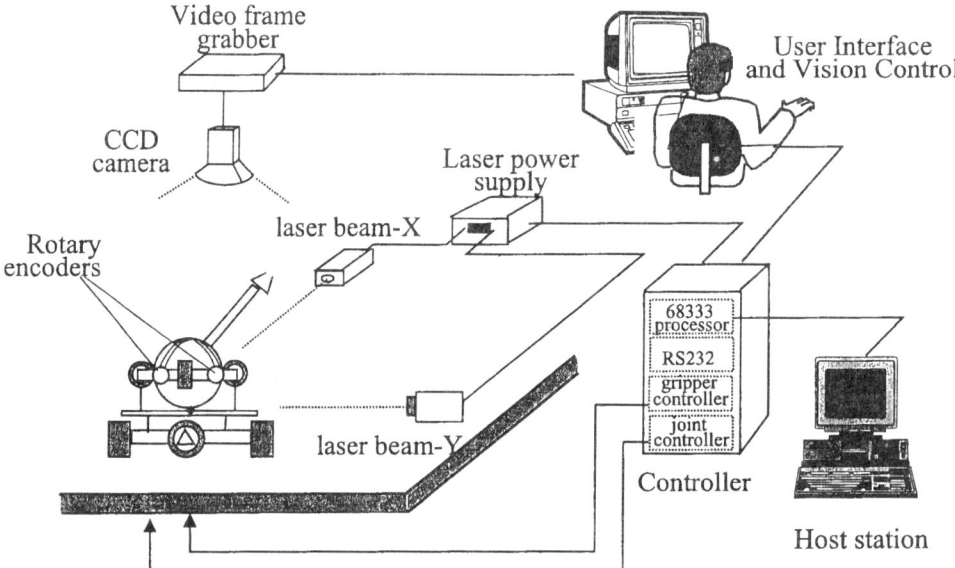

Figure 4. Setup of the mobile micromanipulator system.

4. Hybrid Force/Position Control and Motion Strategy

The proposed scheme of motion control makes reference to the one made by Nassal (1996) on macro mobile manipulators. It is based on the distribution of two different subtasks : fine end-effector motion control and configuration optimization. There are two control instances: the micromanipulator controller and the mobile platform controller. The manipulator controller has to ensure that the end-effector performs the desired motion, based on the current vehicle configuration, by using the kinematic decoupling scheme. The vehicle controller has to move the mobile platform to keep the manipulator in a decent configuration. The control architecture is shown in Figure 5. There are two instances that perform the different subtasks mentioned before. The manipulator controller uses the desired end-effector position and the current platform position to compute the frame of the end-effector with respect to the robot base. This frame is mapped to a vector of joint angles. These joint angles are evaluated by the mobile platform, which uses a cost function for that purpose. The cost function is mapped to a gradient that serves as a vector error signal to the *coordinated motion controller*. The controller computes a desired platform position that is executed by the platform controller. The current platform position is fed back to the manipulator controller using the kinematic decoupling scheme. These four building blocks establish a control loop for the configuration of the micromanipulator.

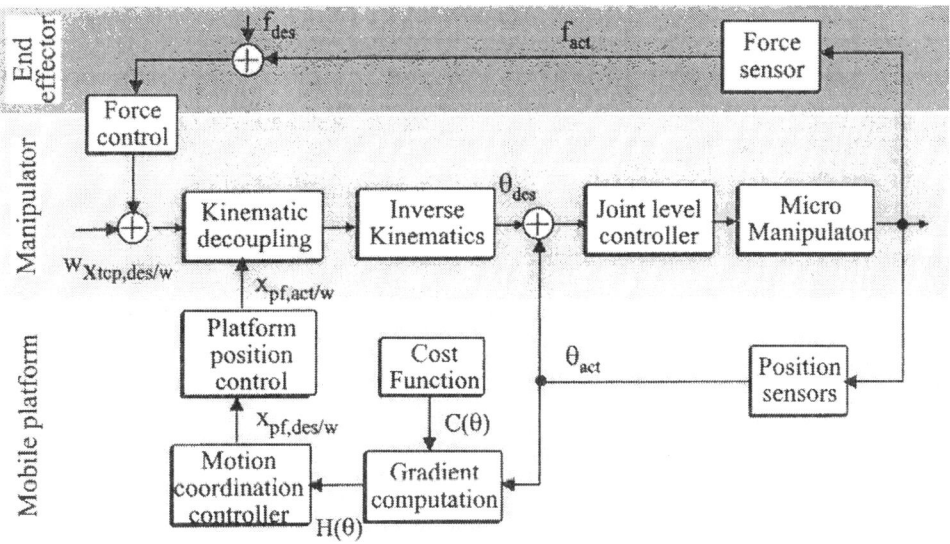

Figure 5. Low-level control structure for motion coordination.

4.1. Kinematic Decoupling Scheme

The *kinematic decoupling scheme* requires the mapping of the end-effector position in the world coordinates to the end-effector portion in manipulator coordinates. In the velocity domain, the velocity of the end-effector is given by:

$$\dot{x}_{e/w} = \underbrace{\left[T_1(x_{pf/w}) J_m \quad T_2 \right]}_{J} \cdot \begin{bmatrix} \dot{\theta}_m \\ \dot{x}_{pf/w} \end{bmatrix}. \quad (1)$$

It is a vector sum of the velocity induced by the joint motion ($\dot{\theta}_m$) and the velocity induced by the platform motion ($\dot{x}_{pf/w}$) where $T_1(x_{pf/w})$ is a transformation matrix to transform the end-effector motion from a platform-fixed coordinate system to world coordinates, which depends on the current configuration of the platform $x_{pf/w}$, T_2 denotes the transformation matrix to transform the platform motion into end-effector motion, J_m is the manipulator Jacobian and J is the Jacobian of the whole mobile manipulator. The joint motion is computed using the inverse problem :

$$\dot{\theta}_m = J_m^{-1} T_1^{-1} \left[\dot{x}_{e/w} - T_2 \dot{x}_{pf/w} \right], \quad (2)$$

which results in a non-redundant system where the current platform velocity is subtracted from the desired end-effector velocity.

4.2. Cost Function

The *cost function* ensures that the configuration of the micromanipulator is within the workspace. It considers the avoidance of the joint limits of the manipulator :

$$C_{\text{limit}}(\theta) = \sum_{i=1}^{n} \frac{1}{\theta_i - \theta_i^{\min}} + \frac{1}{\theta_i^{\max} - \theta_i}, \quad (3)$$

where the cost value goes to infinity when the joint limit is approached. Another important constraint of the manipulator's workspace are singularities. Even in proximity to singularities control of a manipulator is getting more difficult. However, the design of the micromanipulator's structure does not induces such additional constraints.

4.3. Gradient Computation

The information that is required to control the mobile platform is the direction in which the vehicle has to move to get the lowest cost for the manipulator configurations, assuming that the kinematic decoupling scheme is active, i.e., the mobile platform does not move the micro end-effector. The desired quantity is the gradient.

$$H = \left.\frac{\partial C(\theta)}{\partial x_{pf/w}}\right|_{x_{tcp/w}=cst} = \left.\frac{\partial C(\theta)}{\partial \theta} \cdot \frac{\partial \theta}{\partial x_{tcp/tcp}} \cdot \frac{\partial x_{tcp/tcp}}{\partial x_{pf/w}}\right|_{x_{tcp/w}=cst} \quad (4)$$

The computation of this gradient requires the computation of the differentiation of : (i) the cost function with respect to the joint angles, (ii) the joint angles with respect to the end-effector and, (iii) the end-effector motion with respect to the platform motion. The *gradient H* represents the current direction of the platform motion that maximally increases the cost. Since this is only static information, a controller has been developed in order to control the platform motion to result in an effective coordinated motion. In this case study, the vehicle motion is unconstrained (no obstacles in the workspace).

4.4. Coordinated Motion Controller

The task of the *coordinated motion controller* is to keep the gradient small by moving the vehicle appropriately. This is achieved by using the gradient as an error signal and by controlling the components of the gradient independently. The design of the coordinated motion controller includes different motion states, see Figure 6. As proposed by Nassal (1996), these states are separated by calculating the velocity platform V_{pf} and the velocity end-effector V_{tcp} and comparing them to threshold values. For each of these cases a specific controller has been designed for the different motion states as described in the following.

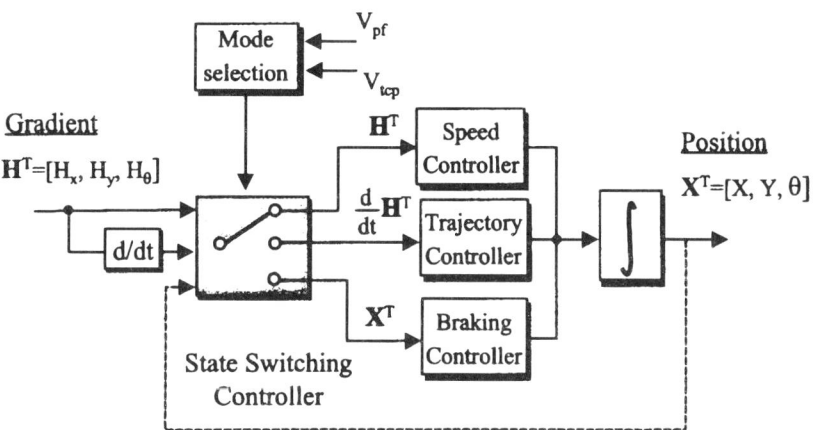

Figure 6. Structure design of the coordinated motion controller.

Speed limited controller. The *speed limited controller* is designed to realize a threshold behavior for the platform motion. The mobile platform should start moving if only the gradient is beyond a definite value. As soon as the vehicle speed goes beyond a certain value (constant speed), the controller switches to the *Trajectory Controller* in order to match with the desired position ($x_{ref,w}, y_{ref,w}, \theta_{Z,ref}$).

Trajectory controller. The trajectory controller is designed to get a good tracking behavior that keeps the mobile platform near the optimal mechanical configuration. The main difficulty is to calculate, precisely, the set of forces parameters allowing to control the platform's trajectory even if some particles are disturbing the motion of the vehicle.

Braking controller. Finally, in order to stop the manipulator in an optimal configuration, the mobile platform should stop precisely (micrometric precision) without oscillations and slippage. For that, a third controller named *braking controller* is designed in order to solve the inverse problem, i.e. how to calculate the minimum number of pulses **T** applied to the actuators to minimize the position error ε of the mobile platform. It can be solved through the condition :

$$\text{minimize} \sum_{i=1}^{4} t_i \text{ subject to: } \begin{bmatrix} \varepsilon_X \\ \varepsilon_Y \\ \varepsilon_\theta \end{bmatrix} = \begin{bmatrix} P_{1X} & P_{2X} & P_{3X} & P_{4X} \\ P_{1Y} & P_{2Y} & P_{3Y} & P_{4Y} \\ P_{1\theta} & P_{2\theta} & P_{3\theta} & P_{4\theta} \end{bmatrix} \times \begin{bmatrix} t_1 \\ t_2 \\ t_3 \\ t_4 \end{bmatrix}. \quad (5)$$

The speed limited and trajectory controllers use the gradient as the first input, and the temporal derivation of the gradient as the second input. However, for the braking controller the mobile platform position is fed back instead of the derivation of the gradient. The state switching concerns the controller for each component of the gradient $\mathbf{H}^T=[H_x,H_y,H_\theta]$ and of the position $\mathbf{X}^T=[x,y,\theta]$.

5. Experiments

The following experiments present the results that were achieved by applying the coordinated motion controller. Figures 7 a-c show the performance of the motion coordination strategy. The parameter ξ, which is settled to 2.5, represents the threshold parameter used in the speed limited controller. Figure 7-a shows the cost values for the manipulator that executes the trajectory tracking rises very rapidly and keeps a low level. Obviously, due to the thresholding, the cost values do not return to their initial state. Figure 7-b show the course of the gradient, which is similar to the course of the cost value. The Figure 7-c shows the end-effector position with respect to the robot manipulator base. These plots show that only small-range motion are necessary to get a rapidly cost value. Figure 8-a and -b shows that the mobile platform and the manipulator's arm, respectively, can make small incremental movements in a defined orientation when the *braking controller* is controlling the final positioning control.

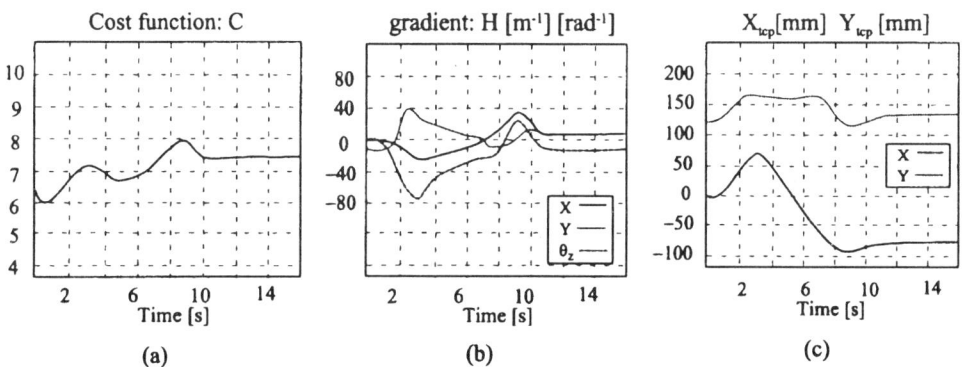

Figure 7. Time evolution of the : a) cost function, b) gradient and c) position.

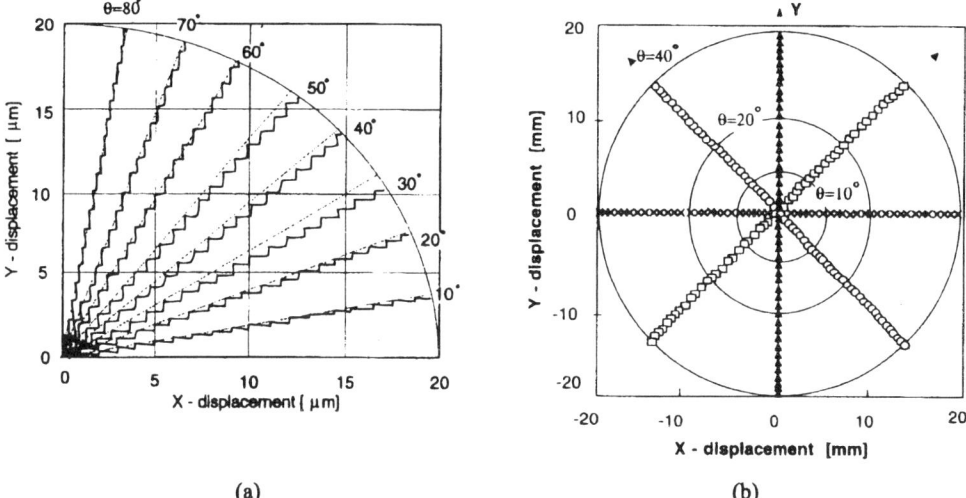

Figure 8. a) Experimental mobile platform's positioning results in a working area and b) experimental manipulator arm's positioning results.

6. Conclusion

A mobile robot micromanipulator system has been developed to investigate manipulation operations applied to objects with millimeter size dimensions. A new approach for coordinated motion control of the micro-robot movements has been applied taking into account the kinematic decoupling of the different motion tasks. The mobile platform is controlled in a such a way to optimize the manipulator configuration before to initiate the micro-gripping task. Further work is being made in the design and control of redundant trunk-like micromanipulator with flexible joint structure composed of a set of binary actuators (Baptiste et al., 1994). For the movement from an initial point to a final one, while the system is strongly redundant, the main concern consists in searching the best final configuration of the manipulator motion leading to an optimized mechanical configuration of the end-effector. The motion control strategy developed in this project allows the easy integration of such a kind of new trunk-like manipulator.

7. References

Baptiste, P. and Taillard, J-P. (1994). Command method of a trunk-like microrobot. In: *Proccedings of the 2nd Japan-France Cong. On Mechatronics*, Takamatsu, Japan, 723-726.

Ferreira, A. (1998). Design and control of a mobile micro manipulator driven by ultrasonic motors with multi-degrees of freedom. *Journal of Advanced Robotics*, Vol.12, N°2, 115-133.

Fukuda, T. *et al.* (1991). Design and dextrous control of micromanipulator with 6 d.o.f. In *Proceedings of the IEEE Int. Conf. Robotics and Autom.*, Sacramento, CA, 1628-1633.

Minotti, P. and Ferreira A.(1998). Les Micromachines. Hermes Editor, Paris, France.

Munassypov, R. et al. (1996). Development and control of piezoelectric actuators for a mobile micromanipulation system. In: *Proceedings of the 5th Int. Conf. On New Actuators*, Bremen, Germany, 213-216.

Nassal, U. M. (1996). Motion coordination and reactive control of autonomous multi-manipulator system, *Journal of Robotic Systems* 13(11), 737-754.

Coordination Control of
a Human/Manipulator System

J. H. Chung and S. A. Velinsky

Department of Mechanical & Aeronautical Engineering
University of California-Davis, Davis, California 95616

Abstract. In this paper, a new type of coordination called operator manipulator coordination control (OMCC) is developed. A primary goal is the development of technology to constitute a robotic human assist system to allow an individual worker to accomplish tasks with which the size of the workpiece is too large or heavy for an unassisted worker to handle. Implicit in the human assisted scheme is force feedback and control. In OMCC, an H-infinity optimal controller integrates the system and the human operator with explicit force control. Since the human operator is modeled as a time-varying environment, the control method has the effect of robustfying the closed loop system performance to the disturbances applied by the human operator. The experimental results illustrate the effectiveness of the proposed control algorithm.

1 Introduction

Many manipulation tasks cannot be performed using a single robot manipulator or the performance of the manipulation process can be improved by multiple robot manipulators. However, due to the complexity of existing coordination schemes and current technology, multiple coordinated robot manipulators can typically only perform simple tasks in a well designed environment (Su and Stepaneko, 1995). Therefore, a human/manipulator coordination is a new and attractive use of robot manipulators where two coordinated manipulators are necessary but performance of the task requires high level system intelligence.

Much research has been done on mechanical systems with human-robot interaction, such as Hardyman, a master-slave manipulator system, a robot for man-robot cooperation, an Extender, etc. Fukuda et al. (1990) have proposed a manipulator, which is designed for handling heavy objects in cooperation with a human. Kazerooni (1990) has proposed the extender or the manipulator system to extend the strength of the human arm. However, there exists a limited amount of literature that considers human factors. Ikera and Inooka (1990) investigated a variable impedance control method of a robot that was proposed for cooperation with a human. Xi and Tarn (1998) developed a heterogeneous function-based coordination or a human intervention control based on position feedback.

In this paper, a new type of coordination control called operator manipulator coordination control (OMCC) will be developed, which is based on robust explicit force regulation. Therefore, implicit in the human assisted scheme is force feedback and control. The primary goal is the development of technology to constitute a robotic human assist system to allow an individual worker to accomplish tasks with which the size of the workpiece is too large or heavy for an unassisted worker to handle. The implementation of force control allows the manipulator to operate effectively while the end effector, or an object that the end effector is holding, is in contact with the environment. Such feedback permits decisions to be made about lifting location and force, as might be used in human assisted operations

In OMCC, the human operator takes the lead of a task execution. In other words, the operator arm will exert a force on the manipulator to initiate a task. The task trajectory can be applied arbitrarily by the operator. Also, the impedance of the human arm is considered variable and the weight of an object may be unknown to the manipulator, which can create problems in compliant motion control. In OMCC, an H-infinity optimal controller integrates the system and the human operator with explicit force control. Since the human operator is modeled as a time-varying environment, the control method has the effect of robustfying the closed loop system performance to the disturbances applied by the human operator. The experimental results illustrate the effectiveness of the proposed control algorithm.

2 Explicit Force Based OMCC

Explicit force control describes a strategy that compares the reference and measured force signals, processes them, and provides an actuation signal directly to the plant. The reference force may also be fedforward and added to the signal going to the plant. In general, an explicit force control system consists of a plant, a controller, a feedforward transfer function, and a force feedback filter. Active damping, if present, is included in the plant. The controller is usually some subset of PID control.

In OMCC, it is assumed that the human operator moves at a reasonably slow speed such that the contact point with the object is always located within the workspace of the manipulator. It is obvious that position control is not a suitable choice for the current objective since any position error may result in a separation or cause a large contact force. A variation of the hybrid control scheme (Raibert and Craig, 1981) is adopted. In this approach that utilizes explicit forces, the exact geometry of the surface of the moving object is not required.

The nominal explicit force control law is applied as follows:

$$m\ddot{x} = F_r - K_i \int F_e dt - K_d \dot{x}. \qquad (1)$$

The integral control was chosen due to its zero steady state error and low-pass filter characteristics. The active damping term is very effective in maintaining stable contact and avoiding bounces. The desired contact force is set to zero.

3 H_∞ Controller Design

In the human/manipulator coordination system, the human operator is modeled as a trajectory planner that actually is a source of uncertainty. The H_∞ controller integrates the system and the human operator with explicit force control. Since the human operator is modeled as a time-varying environment, the control method has the effect of robustfying the closed loop system performance to the disturbances applied by the human operator. As shown in the Figure 1, the H_∞ controller modifies the reference force trajectory by operating on the force error as follows:

$$m\ddot{x} = \Im_{H_\infty}[F_r - K_i \int F_e dt] - K_d \dot{x}. \qquad (2)$$

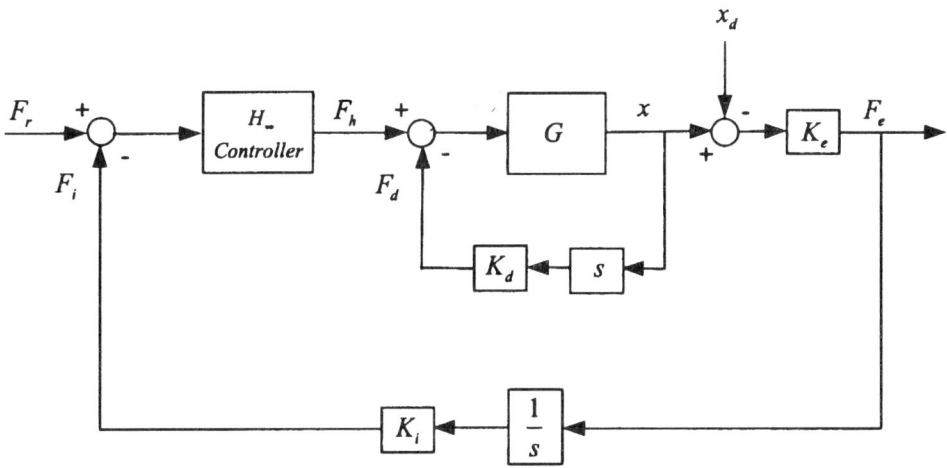

Figure 1 Robust explicit force-based OMCC

Note that the trajectory applied by the human operator is injected to the loop to generate the contact force. The control design is carried out using the Matlab Robust Control Toolbox v.4.2c which solves a weighted mixed sensitivity problem with γ - iteration

$$\max_\gamma \left\| \begin{bmatrix} \gamma W_1 S \\ W_2 T \end{bmatrix} \right\|_\infty < 1 \qquad (3)$$

where S is the sensitivity function and T is the complementary sensitivity function. The weighting functions W_1 and W_2 are chosen to reflect the design objectives. The original plant G is augmented with the weighting functions. The weighting functions are not completely arbitrary, as they are subject to some constraints. The following weighting functions are selected in OMCC:

$$W_1 = \frac{150(1+0.05s)^2}{(1+0.2s)(1+0.008s)} \quad (4)$$

$$W_2 = \frac{s^2}{40000}. \quad (5)$$

4 Experiment

Experiments were conducted using the linear positioning table shown in Figure 2. The apparatus consists of a Hansen HLE 100 linear track and a roller carriage. The load plate (or slide) is joined to a brushless dc servo motor by a nonslip drive cable system and JR3 force sensor is integrated onto the plate to provide the operator interface. This permits fast and accurate movement of the cable. The position of the load plate is measured by an optical encoder attached to the brushless dc motor which results in a linear resolution of 19,000 q-counts per meter.

In the experiment, a human operator leads the manipulator to track a reference trajectory in task space. Figures 4a and 4b depict the trajectory applied by the operator and the control performance of OMCC using the nominal control. Figures 5a and 5b show the trajectory and the control performance of the robust force-based OMCC using H_∞ control.

It is well understood in explicit force-based OMCC that a good controller should be able to regulate the reaction force close enough to zero between the human operator and the robotic manipulator. It is not difficult to see the superior tracking performance of the robust OMCC over the nominal control by comparing Figures 4b and 5b. It is observed in Figure 5b that the reaction force is bounded to a very small value so that smooth coordination between the human operator and the manipulator can be achieved. Note that a faster trajectory with higher amplitude was applied for the robust controller and Figure 4b represents the tracking performance of the nominal control with the best tuned feedback gains. The feedback gains for the nominal control are $K_i = 50$ and $K_d = 100$.

5 Conclusions

A new robust coordination control for a human/manipulator system was developed based on integral derivative control utilizing H_∞ control theory and it was demonstrated experimentally. The experimental results showed excellent force regulation and tracking performance of the developed controller. Future work will focus on extending the technique to general multi-input multi-output systems and implementing the new control law.

References

Su, C. and Stepaneko, Y. (1995). Adaptive sliding mode coordinated control of multiple robot arms attached to a constrained object. *IEEE Transactions on System, Man, and Cybernetics* V25(N5):871-878.

Fukuda, T., Fujisawa, Y. and Arai, F. (1990). Man-robot cooperation work type of manipulator. *In Proc. 8th Ann. Conf. of Robotics Society of Japan* 655-656.

Kazerooni, H. (1990). Human robot interaction via the transfer of power and information signals. *IEEE Transactions on System, Man, and Cybernetics* V20(N2).

Kashihara, A., Hirashima, T., and Toyoda, J. (1995). A cognitive load application in tutoring. *User Modeling and User-Adapted Interaction* 4:279–303.

Ikera, R. and Inooka, H. (1990). Variable impedance control of a robot for cooperation with a human. *IEEE ICRA* 3097-3102.

Xi, N. and Tarn, T. J. (1998). Heterogeneous function-based human/robot cooperations. *IEEE ICRA* 1296-1301.

Raibert, M. H. and Craig, J. J. (1981). Hybrid position/force control of manipulators. *ASME Journal of Dynamic systems, Measurement, and Control* V103(N2):126-133.

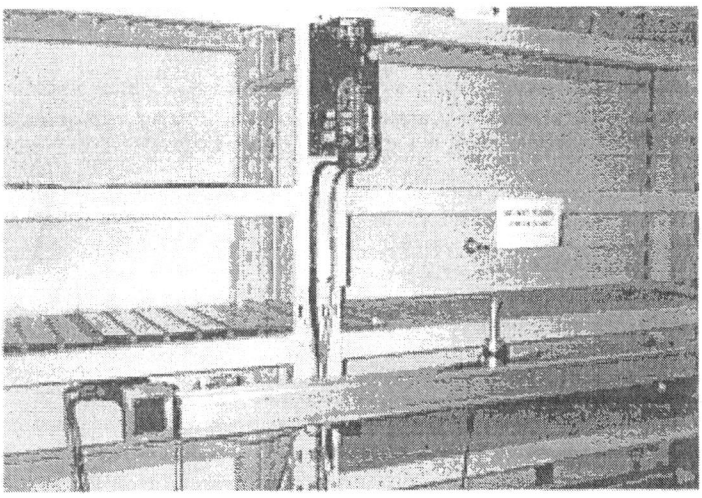

Figure 2 Simple human/machine coordination system

Figure 3 Operator interface

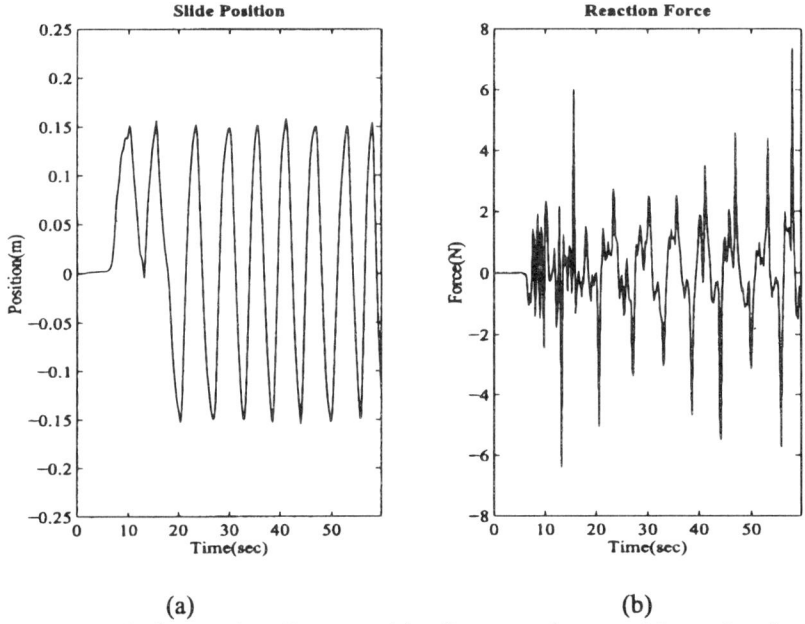

(a) (b)

Figure 4 Nominal control performance (a) reference trajectory (b) reaction forces

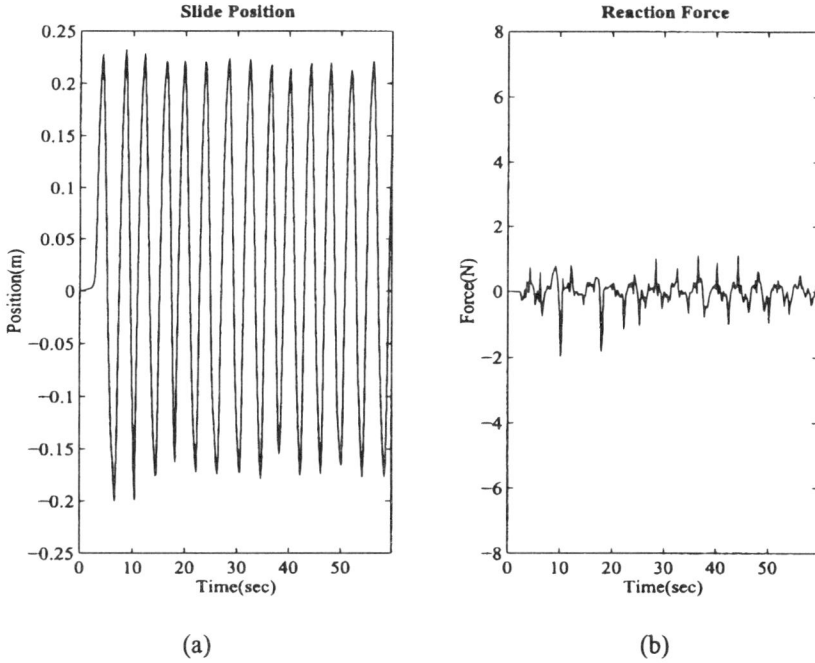

Figure 5 Robust control performance (a) reference trajectory (b) reaction forces

Stability of Cooperating Manipulators with Symmetric Position/Force Control and Time Delay

Anatoli Schneider [1], Igor Zeidis [2], Klaus Zimmermann [2]

[1] Department of Automation, Fraunhofer Institute for Factory Operation and Automation, Magdeburg, Germany
[2] Department of Technical Mechanics, Ilmenau Technical University, Germany

Abstract. The mathematical model and stability of motion of two cooperating manipulators is considered. The high order equations of this model make the mathematical analysis of the stability more difficult. By using a symmetrical control scheme with time delay in a feedback loop, we have obtained the domains of stability and non-stability for such parameters of the system as coefficients of gains, stiffness, of force sensors, time delay in the control loop, and mass of load.

1 Introduction

The problems of coordinated control of a few manipulators, which are coupled through an external object, acquire actual meaning in accordance with the appearance of new tasks in the area of automation of technological processes.

The technological operation present a special interest, which realises the manipulation of non-rigid objects (such as film materials, thin metal sheet and so on), their expansion, bending, and attachment to another object with defined tension. In all of these cases for coordinated movement of the coupled manipulators, it is necessary to control and correct their movements in accordance with information about reaction forces acting on the system. It is possible to measure these forces with the force sensors mounted between manipulators and the objects they hold, and to use their signals in the control loop.

By synthesis of control laws it is possible to use various control schemes based on measurements of the reaction forces. Correction of manipulator movement by position/force control is done with the help of feedback on position and force.

The most widespread case of consideration of dynamic movement is the movement of two manipulators connected through an elastic or lumped-elastic element, whose mass is negligible. In such cases, as a rule, the time delay in the control system is not taken into account, reducing the order of the characteristic equation dawn to four.

In the article, Kim et al. (1989), a system of two master-slave manipulators is considered, coupled with each other through a lumped-elastic force sensor by two types of control, positioning control and positioning/force control, and without time delay in the control system is. Applying the direct Lyapunov method it is shown that the system is stable in both cases. In few articles (see, e.g., Kazerooni et al., 1988, Kokkins, 1989, Kopf, 1989, Wen et al., 1992) more general formulations of the problem are shown, but the influence of the time delay on the system's stability is not considered.

The present paper considers not only the cooperating manipulators, but also the load by linear positioning/force laws of control. The model takes into account the time delay in the control loop. The symmetrical scheme of control is investigated, in which control laws of both manipulators are similar. In this case it is possible to receive areas of stability and non-stability of the dynamic system depending on its parameters, such as coefficients of gains, stiffness of force sensors, time delay in the control loop, and mass of load.

2 Model of Connected Manipulators

We will study the one-dimensional, translational motion of two manipulators which hold a load (Figure 1). Each manipulator consists of an arm (1) with mass, $m_1(m_2)$, a gear train (DC motor (2) and reductor (3)). The output power gear (4) is connected to the rack attached to the arm. The manipulator arm is moved along the horizontal axes OX.

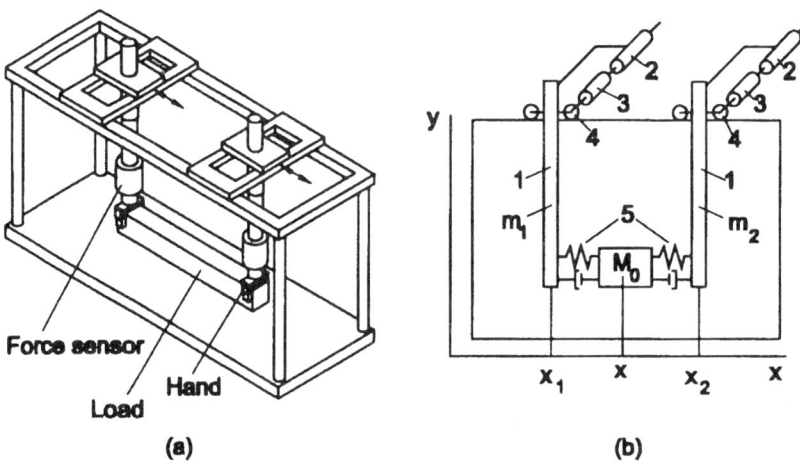

Figure 1. General View of the Research Manipulators (a) and Schematics for the One-dimensional Motion of the Manipulators (b).

A single-component force sensor (5) is mounted to the lower end of each manipulator arm to measure the horizontal force component acting on it.

The force sensor ends have plates, which hold an undeformable object with mass M_0. We will neglect the moments produced by the forces, which act from side of the manipulator relative to the centre of gravity of the load. We will model a single-component force sensor by a massless lumped elastic element.

2.1 Mathematical Model

By x_1, x_2, we denote the coordinates of the point where the sensor is mounted on the appropriate manipulator by x_1, x_2, and by x the coordinate of the centre of gravity of the load. We use the Lagrange second order equations to derive the equations of motion for the system:

$$\frac{d}{dt}\left(\frac{\partial T}{\partial \dot{q}_v}\right) - \frac{\partial T}{\partial q_v} = Q_v \quad v = 1,...,n \tag{1}$$

where T – kinetic energy of the system, Q_v – generalised forces, q_v – Lagrange coordinates, n – number of degrees of freedom.

We take Cartesian coordinates x, x_1, x_2 for the Lagrange coordinates. Thus the kinetic energy of the system can be expressed as:

$$T = \tfrac{1}{2}(m_1 \dot{x}_1^2 + m_2 \dot{x}_2^2 + M_0 \dot{x}^2 + J_{G1} \dot{\varphi}_1^2 + J_{G2} \dot{\varphi}_2^2) \tag{2}$$

where $\varphi_i (i=1,2)$ - is a rotational angle of the motor rotor, J_{G_i} – is the moment of inertia of the rotor of each motor.

The linear displacement $x_1 (x_2)$ of the manipulator arm is related to the rotational angle φ of the motor as:

$$x_i = (r_i / j_i)\varphi_i = \rho_i \cdot \varphi_i$$

where r_i is a radius of the outer pinion in the gear train, j_i is the gear train reduction ratio. The generalised forces Q_v may be presented as:

$$Q_v = \frac{\partial U}{\partial x_v} - \frac{\partial R}{\partial \dot{x}_v} + X_v \quad v = 1, 2, 3. \tag{3}$$

The force function reads

$$U = -\tfrac{1}{2}k_1\left[x - a - (x_1 + l_{01})\right]^2 - \tfrac{1}{2}k_2\left[x_2 - l_{02} - (x+a)\right]^2 \tag{4}$$

where k_i is the stiffness of each force sensor, $2a$ is the size of the load, l_{0i} is the length of the elastic element of the force sensor in unloaded state.

The energy dissipation in the system may be expressed by the Rayleigh dissipation function

$$R = \tfrac{1}{2} \left[d_{21}\dot{x}_1^2 + d_{22}\dot{x}_2^2 + b_1(\dot{x} - \dot{x}_1)^2 + b_2(\dot{x}_2 - \dot{x})^2 \right] \qquad (5)$$

where the positive constants $d_{2i} = \dfrac{C_{2i}}{\rho_i^2}$, $C_{2i} = \dfrac{M_{pi} - M_{ni}}{\dot{\varphi}_i}$ for a specific motor may be calculated by using the values for the starting torque M_{pi}, nominal voltage U_i, nominal torque M_{ni}, nominal angular velocity $\dot{\varphi}_i$, and b_i is a coefficient of the force sensor viscosity. The generalised forces $X_i = d_{1i}U_i$ are proportional to the voltage U_i applied to each of the motors. They perform work along the possible displacement δx_i ($i = 1,2$). Here $d_{1i} = \dfrac{C_{1i}}{\rho_i}$, $C_{1i} = \dfrac{M_{ni}}{U_{ni}}$.

By substituting equations (3), (4) and (5) into equation (1) we obtain the equation system:

$$\begin{aligned} M_0 \ddot{x} + (b_1 + b_2)\dot{x} + (k_1 + k_2)x - b_1\dot{x}_1 - k_1 x_1 - b_2 \dot{x}_2 - k_2 x_2 - k_1 l_1 + k_2 l_2 &= 0 \\ M_1 \ddot{x}_1 + (d_{21} + b_1)\dot{x}_1 + k_1 x_1 - b_1 \dot{x} - k_1 x + k_1 l_1 - d_{11} U_1 &= 0 \\ M_2 \ddot{x}_2 + (d_{22} + b_2)\dot{x}_2 + k_2 x_2 - b_2 \dot{x} - k_2 x - k_2 l_2 - d_{12} U_2 &= 0 \end{aligned} \qquad (6)$$

where $M_i = m_i + \dfrac{J_{Gi}}{\rho_i^2} = \dfrac{J_i}{\rho_i^2}$, $l_i = a + l_{0i}$, $i = 1,2$.

To study the dynamics of such a system we need the control laws for the voltage, U_1 and U_2 applied to the motors.

2.2 Control Laws

The general relation, including the signals of feedback of position, velocity, and force, and the delay in feedback loop may be expressed in the form:

$$U_i(t + T_i) = -k_{pi}(x_i(t) - x_{pi}) - k_{vi}\dot{x}_i(t) - k_{Fi}(F_i(t) - F_{pi}) \qquad i = 1,2 \qquad (7)$$

where $k_{pi}, k_{vi}, k_{Fi} \geq 0$ are coefficients of feedback about position, velocity, and force for each manipulator, x_{pi} is the programmed position of the manipulator, F_{pi} are programmed values of forces acting from side of the manipulator on the load, whereas $F_{p1} = -F_{p2} = F_p \geq 0$, T_i is the time delay in each feedback loop.

From now on we consider the position of the manipulator and the load related to the coordinate system as shown in Figure 1.
To linearize the equations (7) we express $U(t+T)$ in the form of a series in T and consider only the linear part

$$U(t+T) = U(t) + T\dot{U}(t) \tag{8}$$

3 Stability of the System of Motion

To study the stability we rewrite the equations of motion (6) and the control law (7) considering the expression (8) in deviations from the stationary regime holding for the deviations from the old symbols x, x_1, x_2, U_1, U_2. We obtain the following equation system:

$$\begin{aligned}
M_0\ddot{x} + (b_1 + b_2)\dot{x} + (k_1 + k_2)x - b_1\dot{x}_1 - k_1 x_1 - b_2\dot{x}_2 - k_2 x_2 &= 0 \\
M_1\ddot{x}_1 + (d_{21} + b_1)\dot{x}_1 + k_1 x_1 - b_1\dot{x} - k_1 x - d_{11}U_1 &= 0 \\
M_2\ddot{x}_2 + (d_{22} + b_2)\ddot{x}_2 + k_2 x_2 - b_2\dot{x} - k_2 x - d_{12}U_2 &= 0 \\
U_1 + T_1\dot{U}_1 &= -(k_{F1}k_1 + k_{P1})x_1 - (k_{V1} + k_{F1}b_1)\dot{x}_1 + k_{F1}k_1 x + k_{F1}b_1\dot{x} \\
U_2 + T_2\dot{U}_2 &= -(k_{F2}k_2 + k_{P2})x_2 - (k_{V2} + k_{F2}b_2)\dot{x}_2 + k_{F2}k_2 x + k_{F2}b_2\dot{x}
\end{aligned} \tag{9}$$

The stability of the system (9) is given by the position of the roots of the corresponding characteristic polynomial in the complex plane. This polynomial is of 8-th order, which is a consequence of the fact that: 1) There are three second order differential equations describing the motion of the load and of both manipulators (with three degrees of freedom); 2) There are another two, first-order polynomials which make the mathematical analysis of the stability even more difficult. In spite of this, the analysis is possible in some special cases.

3.1 The Stability of a System with Position Control and without Delay in Feedback Loop

We set in control laws of the system (9) the coefficients of the force $k_{F1} = k_{F2} = 0$ and delay time $T_1 = T_2 = 0$. Then by substituting the equations for the control voltages U_1, U_2 in the system (9) we rewrite it in a matrix form:

$$M\ddot{\bar{y}} + K_V\dot{\bar{y}} + K_P\bar{y} = 0 \tag{10}$$

where M, K_V, K_P are the symmetric matrices and $\bar{y}, \dot{\bar{y}}, \ddot{\bar{y}}$ are vectors.
We use the direct method of Lyapunov and assume the Lyapunov function in the form:

$$V = \frac{1}{2}(\dot{\bar{y}}^T \cdot M \cdot \dot{\bar{y}} + \bar{y}^T \cdot K_P \cdot \bar{y})$$

where the sign "T" means the operation of transposition. The positive definiteness of the matrix M is obvious, and the positive definiteness of the matrix K_p follows from the fact that its principal diagonal minors are positive (Gantmacher, 1960). By the same reasoning the matrix K_V is also non-negative definite. Therefore, V is a positive definite quadratic form. The final form of derivation of Lyapunov function V, taking into consideration the system (10), has the form:

$$\frac{dV}{dt} = \dot{\bar{y}}^T (M\ddot{\bar{y}} + K_p \bar{y}) = -\dot{\bar{y}}^T K_V \dot{\bar{y}} \leq 0$$

So, we have shown, that with position control and no time delay, the motion of the system is stable for any values of the gain coefficients in the control systems of both manipulators, for any load mass, and for any characteristics of the force sensors. This result is a generalisation of the result obtained Kim et al. (1989).

3.2 The Stability of the System with Symmetrical Control Law and in Presence of Time Delay in Feedback Loop

Now we consider the system of two equal manipulators, which have symmetrical control laws and the equal feedback gains of positions, velocities and forces corresponding:

$$M_1 = M_2 = M \quad k_1 = k_2 = k \quad b_1 = b_2 = b \quad d_{11} = d_{12} = d_1$$
$$d_{21} = d_{22} = d_2 \quad k_{V1} = k_{V2} = k_V \quad k_{F1} = k_{F2} = k_F \quad T_1 = T_2 = T$$

For the sake of the stability analysis let us introduce non-dimensional variables by the formulas:

$$\bar{t} = \rho \sqrt{\frac{k}{J}} \cdot t, \quad \bar{x} = \frac{k\rho}{C_1 U_0} x, \quad \bar{x}_i = \frac{k\rho}{C_1 U_0} x_i, \quad \bar{U}_i = \frac{U_i}{U_0}, \quad (i = 1,2) \quad (11)$$

Here the bar points to non-dimensional variables.
By express dimensional variables with non-dimensional variables and substituting them into the system of equations (9) we obtain the following system in non-dimensional variables (bar will be left out):

$$\mu\ddot{x} + 2\beta\dot{x} + 2x - \beta\dot{x}_1 - x_1 - \beta\dot{x}_2 - x_2 = 0$$
$$\ddot{x}_1 + (\alpha + \beta)\dot{x}_1 + x_1 - \beta\dot{x} - x - U_1 = 0$$
$$\ddot{x}_2 + (\alpha + \beta)\dot{x}_2 + x_2 - \beta\dot{x} - x - U_2 = 0 \quad (12)$$
$$\tau\dot{U}_1 + U_1 + (f+s)x_1 + (v+\beta f)\dot{x}_1 - fx - \beta f\dot{x} = 0$$
$$\tau\dot{U}_2 + U_2 + (f+s)x_2 + (v+\beta f)\dot{x}_2 - fx - \beta f\dot{x} = 0$$

Here $\alpha = \dfrac{d_2}{\sqrt{Mk}} = \dfrac{C_2}{\rho\sqrt{Jk}}$, $\beta = \dfrac{b\rho}{\sqrt{Jk}}$, $\mu = \dfrac{M_0\rho^2}{J}$, $f = k_F d_1 = \dfrac{k_F C_1}{\rho}$,

$v = \dfrac{k_v d_1 \rho}{\sqrt{Jk}} = \dfrac{k_v C_1}{\sqrt{Jk}}$, $\tau = T\rho\sqrt{\dfrac{k}{J}}$, $s = \dfrac{k_p d_1}{k} = \dfrac{k_p C_1}{\rho k}$ are non-dimensional expressions

determining the behaviour of the system.

This mathematical model is based on the assumption of the same characteristics for both manipulators and identical positions, velocities and force feedback coefficients in both control laws. This fact makes the analysis of the 8th order characteristic polynomial simpler, because it is factored into two polynomials of 3rd and 5th order, respectively.

Further, for simplicity, we will consider the force sensors as only elastic but not viscous-elastic (availability of viscosity only increases the zone of stability). According to the Hurwitz criterion, the asymptotic stability condition of the system may be written in the form:

$$0 \leq S \leq S^* \tag{13}$$

where the limit value S^* is defined from the corresponding squared equation and has the form:

$$S^* = \dfrac{1}{2}\left[\dfrac{1}{\tau}(\alpha+v)(f+2) + \alpha(\alpha+v) - (\dfrac{2}{\mu}+1)(f-\alpha\tau) - \sqrt{D}\right] \tag{14}$$

where

$$D = (f-\alpha\tau)^2\left[\dfrac{1}{\tau^2}(\alpha+v)^2 + (\dfrac{2}{\mu}+1)^2\right] + 2(\alpha+v)(\dfrac{2}{\mu}-1)\left[\alpha(2f-\alpha\tau) - \dfrac{1}{\tau}f^2\right]$$

Figure 2(a) shows the relationship between limit gains k_p^* and time delay T in the control loop (curve 1) obtained in correspondence with equations (13) and (14).

The numerical parameters of the manipulator used for the calculation of this curve are as follows (see Gorinevsky et al., 1997):

$$C_1 = 2{,}64 \cdot 10^{-3} \text{ N·m/V} \quad C_2 = 6 \cdot 10^{-5} \text{ N·m·s} \quad \rho = 2{,}94 \cdot 10^{-4} \text{ m}$$

$$J = 6{,}96 \cdot 10^6 \text{ kg·m}^2$$

sensor stiffness is $k = 10^5$ N/m; load $M_0 = 10$ kg; velocity gain $k_v = 5 \cdot 10^4$ V·s/m; force gain $k_F = 1{,}0$ V/N.

The range of the time delay T is selected from $0{,}01$ s to $0{,}1$ s which corresponds to real values of delay for various robot systems.

With $T \ll 1$, the equation of the neutral curve (14) separating the field of stability from the field of instability becomes simpler, and the corresponding condition for asymptotic stability (13) may be written in the form:

$$0 \le k_p < \frac{1}{T}(\frac{1}{\rho} \cdot \frac{C_2}{C_1} + k_v) - k \cdot k_F \qquad (15)$$

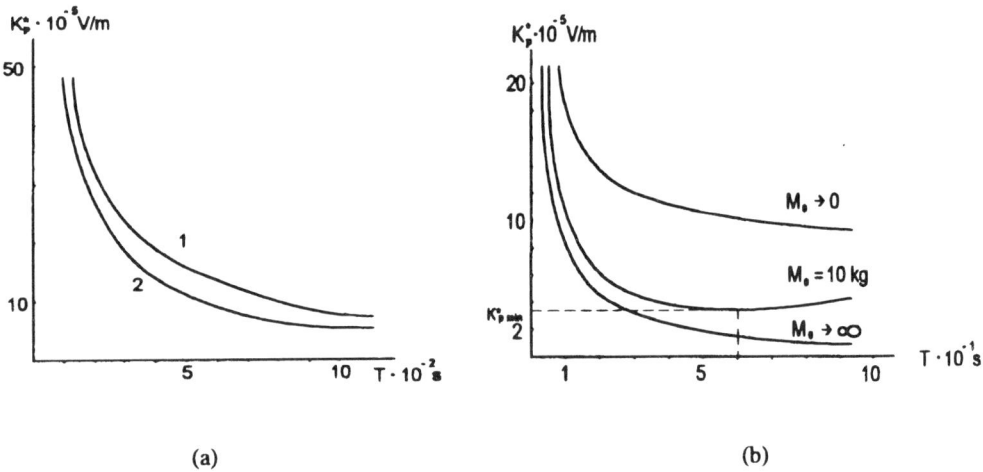

Figure 2. The Relationship between Limit Gains k_p^* and Time Delay T (a) and the Neutral Curve $k_p^* = k_p(T)$ Separating the Stable and the Unstable Domain (b).

Since the values k_v are much bigger than the values of $\frac{1}{\rho} \cdot \frac{C_2}{C_1}$, the inequality (15) may be written as

$$0 \le k_p < \frac{k_v}{T} - k \cdot k_F \qquad (16)$$

Figure 2(a) displays also the corresponding curve (2) specified by the above variables (k_v, k, k_F). Since the curve (2) is located lower than the curve (1), it is clear that by fulfilling the condition (16), the condition (13) is also satisfied.

As follows from condition (16) the domain of stability is diminished with the increase of gain k_F and the stiffness of force sensor k. That is to say, the system with a "soft" sensor is more stable than one with a "rigid" one. The stability domain increases with increase of velocity feedback gain k_V. The system is always stable in absence of a time delay ($T \to 0$).

Figure 2(b) shows the neutral curve $k_p^* = k_p(T)$ separating the stable and the unstable domain. The values of other parameters are fixed and they are the same as in Figure 2(a). The curve has a flat minimum given by parameters $T \approx 0.6$ s, $k_{p\min}^* = 3.94 \cdot 10^5$ V/m. Thus, the system is stable with any time delay if the position gain k_p is smaller than $k_{p\min}^*$. Certainly, the values of time delay are limited to such values for which the relationship (8) is fulfilled. The curve has also a horizontal asymptote.
With increase of the value of the load, M_0, the stable domain narrows monotonically (Figure 2(b)).

4 Conclusions

The analysis of a mathematical model of two, one-degree-of-freedom manipulators holding a load allows us to make the following conclusions:
1. Using position control and in the absence of time delay in the feedback loop, the system is stable with any gains, mass of load, arms, and force sensor stiffness.
2. The linear symmetrical control system with the position and force feedback loop is also stable in absence of time delay.
3. With time delay in feedback loop and linear symmetrical control laws, the range of position and force feedback gains in which the motion of manipulators are asymptotically stable has an upper bound. The stability domain diminishes with the growth of the sensor stiffness (k) and mass (M_0) of load. Thus, the system stability may be violated by an increase of position (k_p), force (k_F) feedback gains, stiffness k of force sensor, mass of load M_0, and by an increase of time delay T. An increase of velocity feedback gain k_V stabilises the control system.

References

Gantmacher F.R. (1960). The Theory of Matrices. Chelsea Publishing Company; New York.
Gorinevsky D.M., Formalsky A.M., Schneider A.Yu. (1997). Force control of robotics systems, CRS Inc.
Kazerooni H., and Tsay T.I. (1988). Control and stability analysis of cooperating robots, *American Control Conference, Atlanta*.
Kim K.I., and Zheng Y.F. (1989). Two strategies of position and force control for two industrial robots handling a single object, *Robotics and Autonomous Systems* 5: 395-403.
Kokkinis T. (1989). Dynamic hybrid control of cooperating robots by nonlinear inversion. *Robotics and Autonomous Systems* 5: 359-368.
Kopf C.D. (1989). Dynamic two arm hybrid position/force control. *Robotics and Autonomous Systems* 5: 369-376.
Wen J.T.,and Kreutz-Delgado K. (1992). Motion and force control of multiple robotic manipulators. *Automatica* 28/4: 729-743.

Remote Control of Periodic Robot Motion

Tamás Insperger and Gábor Stépán

Department of Applied Mechanics, Budapest University of Technology and Economics, Hungary

Abstract. Remote control of robots often leads to the presence of time delay in the information transmission of the signals in the control loop. Analytical methods are available for the calculation of the maximum critical time delays and control gains when stationary end positions of robots, or constant contact forces between actuators and environment are still stable. When the desired trajectory is periodic, or the desired contact force varies periodically, the non-linearities of the robotic structure take an important role even in the local stability behavior about the desired motion. The non-linear characteristics and the periodic path together lead to parametric excitation, i.e., the stiffness, damping and gain parameters may vary periodically in time. The stability behavior of these systems become intricate in the presence of great time delays. Stability charts are constructed which explain the stability properties of remote periodic force control.

1 Introduction

The design of robot structures and their control usually neglects the presence of time delays. This approach is often acceptable, in spite of the fact, that time delays always appear in real structures. There are three important sources of these delays: (1) the sampling time of the digital controller; (2) the delay of the signals in the information transmission system; (3) the mechanical structure itself. In case (1), the sampling delay is usually in the range of $10^{-3} - 10^{-2}$ seconds and it is combined with a so-called zero order holder. In case (2), the time delay is constant and may vary from 10^{-6} to 1 second depending on the distance between the controlled robot and the controller, e.g. in master/slave systems. Consequently, this information delay is often negligible, but it may be crucial, for example, in space applications (see Vertut et al., 1976). In case (3), time delay may arise when the actuator is in elastic contact with the environment along a contact surface (see Stépán, 1997), and the delay is inversely proportional to the relative velocity of the contact surfaces. In this study, case (2) is examined only, but the results can partly be extended for the digital control case (1) due to the physical similarities between the two cases.

There are existing methods to estimate the critical values of the delays in case of robot position/force control when the desired position or contact force is fixed in time (see Stépán and Steven, 1990). The linear variational system of the equations of motion at a trivial solution leads to a system of linear autonomous differential-difference equations. In these equations, the time delay is responsible for the presence of the difference in time, while the differential part comes from the time derivatives in Newton's laws. In spite of the finite degrees of freedom, this system has an infinite dimensional nature, its phase space is infinite dimensional in mathematical sense. Still, the linear system allows us to get conclusion on the local stability of the robot control strategy. These investigations are supported by methods developed for autonomous, i.e. time independent systems.

When closed trajectories, periodic motion, or periodic time-varying contact forces are desired, the non-linearities of the robot manipulator structure and those of its elastic and viscous elements result time-varying parameters in the linear system for the small perturbations of the desired motion. If the desired motion in time is 'slow', there is not much difference with respect to the stability limits regarding time delays. In other words, the greater the time delay in the information transmission is, the slower the motion should be for stable operation. This is observed when elderly people having slow reflexes (consequently great time delays) are advised to move slower to avoid fall-over, i.e. to maintain stability (see Cooper and Kojeca, 1994).

The stability properties of those systems where relatively fast periodic motion has to be controlled remotely is still quite unexplored. The analytical study of these systems may partly explain unexpected losses of stability, while it may also lead to new stable control parameter domains since the time-periodic parameters sometimes stabilize otherwise unstable equilibria (see Insperger and Horváth, 2000). However, this physical observation has not been approved yet in case of delayed or retarded dynamical systems.

As a general introductory example, consider the remote control of a robot having the well-known general nonlinear system of equations of motion in the form

$$\mathbf{M}(\mathbf{y})\ddot{\mathbf{y}} + \mathbf{f}(\mathbf{y},\dot{\mathbf{y}}) = \mathbf{Q}(\mathbf{y},\dot{\mathbf{y}}),$$

with the general coordinate vector $\mathbf{y}$, the general mass matrix $\mathbf{M}$, coriolis, centrifugal, gravitational, elastic, viscous and other forces all included in $\mathbf{f}$, and with the control force denoted by $\mathbf{Q}$ (see Spong and Vidyasagar, 1989). If the desired motion is periodic with period T, that is $\mathbf{y}_d(t) = \mathbf{y}_d(t+T)$, and the control force is constructed from the computed torque $\mathbf{Q}_d(t)$ and the simplest PD compensator, the linear system of equations of motion with respect to the small perturbation $\mathbf{x} = \mathbf{y} - \mathbf{y}_d$ assumes the form

$$\ddot{\mathbf{x}}(t) + \mathbf{B}(t)\dot{\mathbf{x}}(t) + \mathbf{C}(t)\mathbf{x}(t) = \mathbf{P}(t)\mathbf{x}(t-\tau) + \mathbf{D}(t)\dot{\mathbf{x}}(t-\tau).$$

The coefficient matrices may all be periodic with period T originated in the desired motion period. Due to the information delay τ occurring in the control, the position and velocity errors contain this delay in the compensator. Thus, the above system is a parametrically excited delay-differential equation. In the subsequent sections, preliminary experimental work, a new stability criterion, and stability charts for remote periodic force control are summarized.

2 Preliminary experiments

The Newcastle robot was an excellent tool to study hybrid position/force control strategies (see Stépán and Steven, 1990). In case of a one degree of freedom force control implementation, different digital sampling delays were introduced to check the critical proportional gain P_{cr} applied for the force error at the limit of stability. The mechanical model of the structure is presented in Figure 1. The precisely identified parameters of the robot are the mass $m = 2500$ [kg], the viscous damping factor $c=32$ [Ns/mm], and the linear stiffness of the force sensor $k = 44.5$ [N/mm]. The experimental stability chart presented in the plane (τ,P) of the delay and the gain parameters in see Stépán and Steven (1990) confirmed the analytical estimation for the critical proportional gain P_{cr}:

$$0 < P < P_{cr} = (2c)/(3k\tau).$$

For sampling times in the range 7.5–30 [ms], the critical gains are between 19 and 62. After the loss of stability, the resulting vibration frequencies were in the surprisingly low range of 3.5–6 [Hz] in accordance with the estimated value

$$f = \sqrt{c/(6m\tau)}/\pi.$$

These results cannot be extended directly for great time delays and time periodic contact forces. This gives the motivation for the following analysis.

3 Stability analysis of remote force control

Let us consider the non-linear spring characteristic $k\,y + k_3\,y^3$ in the force control model of the Newcastle robot as shown in Figure 1. The softening spring has a negative coefficient $k_3 = -6.6$ [N/mm^3].

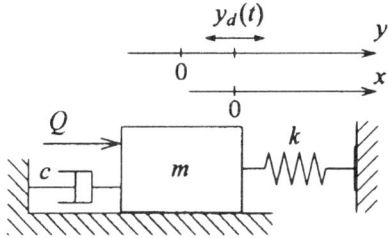

Figure 1. Mechanical model

The equation of motion of force control reads (see Craig, 1986)

$$m\,\ddot{y} + c\,\dot{y} + k\,y + k_3\,y^3 = -PF_e + (Q_d + F_e), \tag{1}$$

where the force error Fe is the difference of the sensed and the desired forces

$$F_e(t) = F_s(t) - F_d(t) = k\,y(t-\tau) + k_3\,y^3(t-\tau) - (k\,y_d(t-\tau) + k_3\,y_d^3(t-\tau)),$$

and the computed torque assumes the form

$$Q_d(t) = m\,\ddot{y}_d(t) + c\,\dot{y}_d(t) + k\,y_d(t) + k_3\,y_d^3(t).$$

For the small perturbation x defined by $y = y_d - x$, the substitutions and linearization leads to the linear time-periodic delay-differential equation

$$m\,\ddot{x}(t) + c\,\dot{x}(t) + (k + 3k_3\,y_d^2(t))\,x(t) = -(P-1)(k + 3k_3\,y_d^2(t-\tau))x(t-\tau).$$

Introduce dimensionless time t by $t = t/\tau$, and by abuse of notation, drop the tilde immediately. We get the following equation

$$\ddot{x}(t)+4\pi\zeta(\tau f_n)\dot{x}(t)+4\pi^2(\tau f_n)^2(1+3r_k y_d^2(t))x(t)$$
$$=-(P-1)4\pi^2(\tau f_n)^2(1+3r_k y_d^2(t-1))x(t-1), \qquad (2)$$

where

$f_n = (\sqrt{k/m})/2\pi$ is the natural frequency of the undamped, uncontrolled oscillating system,

$\zeta = c/(2\sqrt{km})$ is the relative damping factor,

$r_k = k_3/k$ is a ratio characterising the spring non-linearity.

Clearly, for linear spring characteristic $r_k = 0$, the system is time independent, and the stability chart can be given in closed form as explained in Section 2 for the preliminary experiments (see Figure 2).

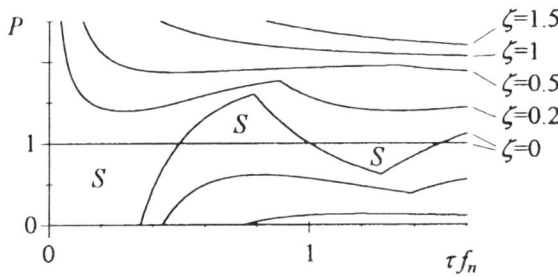

Figure 2. Stability charts for linear spring and various damping values

The stability of fixed point $x(t) \equiv 0$ of equation (2) for a non-linear spring with $r_k \neq 0$ cannot be determined in closed form. The delayed term $x(t-1)$ is approximated as follows

$$x(t-1) \approx \int_{-\infty}^{0} x(t+\vartheta) \cdot w_n(\vartheta) d\vartheta, \qquad (3)$$

where $w_n(\vartheta)$ is a special weight function series coming from the product of a polynomial and an exponential expression

$$w_n(\vartheta) = (-1)^n \frac{n^{n+1}}{n!} \vartheta^n e^{n\vartheta}.$$

The function $w_n(\vartheta)$ satisfies the following properties

$$\int_{-\infty}^{0} w_n(\vartheta) d\vartheta = 1, \qquad \lim_{n \to \infty} w_n(\vartheta) = \delta(\vartheta+1),$$

where δ is the Dirac distribution. It was proved by Fargue (1973) that (3) converges to $x(t-1)$ as n tends to infinity. Figure 3 shows the weight function with parameters $n = 2, 10, 50, 100$. It can be seen, that the greater the approximation parameter n is, the more correct the approximation is.

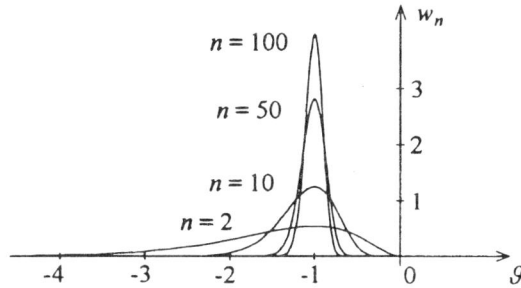

Figure 3. Weight functions

Approximation (3) can be applied in (2). A long calculation (derivations and partial integration) yields a finite dimensional system of differential equations with a time periodic coefficient matrix

$$\frac{d}{dt}\mathbf{z}(t) = \mathbf{A}(t)\mathbf{z}(t), \tag{4}$$

where $\mathbf{z}(t) = \mathrm{col}(z1\ z_2 \ldots z_{n+3})$ and $c(t) = 4\pi^2 (1+3\ r_k\ y_d^2(t))$ in the coefficient matrix

$$\mathbf{A}(t) = \begin{bmatrix} 0 & 1 & 0 & 0 & \cdots & 0 \\ -(\tau f_n)^2 c(t) & -4\pi\zeta(\tau f_n) & -(P-1)(\tau f_n)^2 c(t-1) & 0 & \cdots & 0 \\ 0 & 0 & -n & -1 & \cdots & 0 \\ \vdots & \vdots & \vdots & \ddots & \ddots & \vdots \\ 0 & 0 & 0 & \cdots & -n & -1 \\ (-1)^n (n)^{n+1} & 0 & 0 & \cdots & 0 & -n \end{bmatrix}.$$

System (4) is asymptotically stable if and only if all the characteristic multipliers of the principle matrix $\mathbf{C}$ are in modulus less than one, that is they are inside the unit circle of the complex plane. In general, the principal matrix can not be calculated in closed form. However, if the desired contact force is piecewise constant, the function $c(t)$ is also piecewise constant

$$F_d(t) = \begin{cases} F_1 & \text{if } 0 \le t < T/2 \Rightarrow c(t) \equiv c_1 \\ F_2 & \text{if } T/2 \le t < T \Rightarrow c(t) \equiv c_2 \end{cases}. \tag{5}$$

In this case, the principal matrix takes the form

$$\mathbf{C} = \exp(\mathbf{A}_4 t_4)\exp(\mathbf{A}_3 t_3)\exp(\mathbf{A}_2 t_2)\exp(\mathbf{A}_1 t_1)$$

where the functions $c(t)$ and $c(t-1)$ are constant in each time interval of length t_i belonging to the coefficient matrix $\mathbf{A}_i$ ($i=1,2,3,4$ and $t_1 + t_2 + t_3 + t_4 = T$).

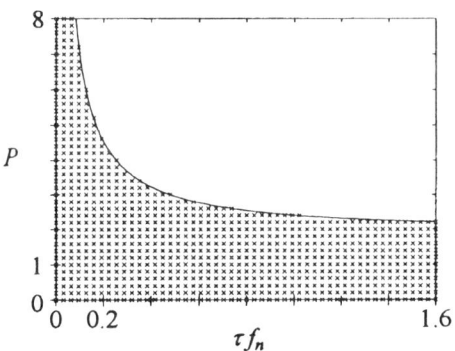

Figure 4. Stability chart for constant desired contact force

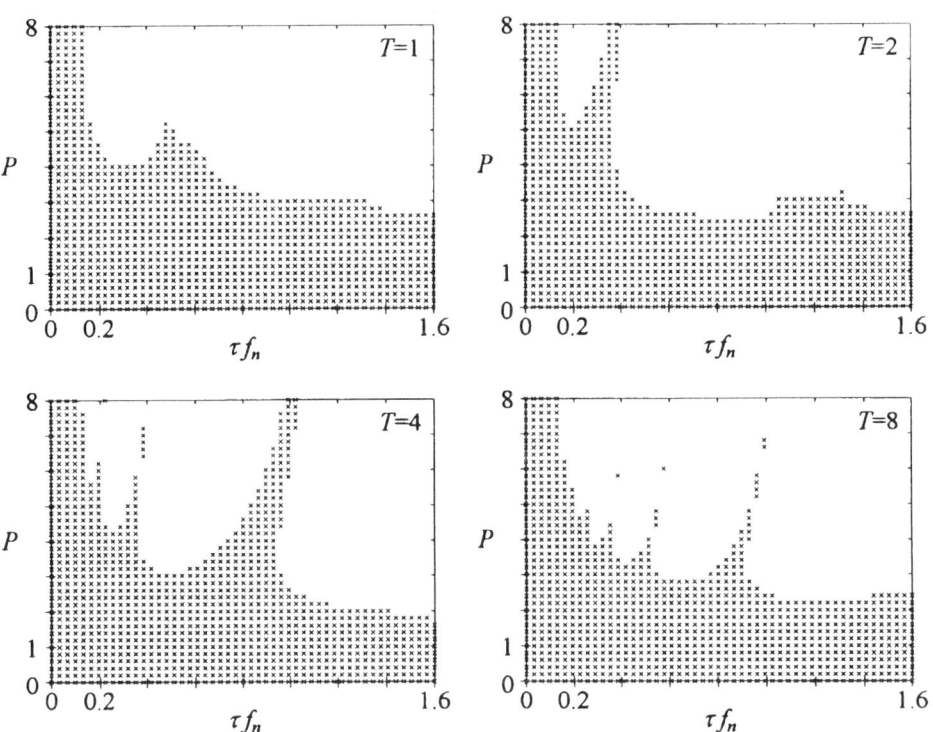

Figure 5. Stability charts for time periodic desired contact forces with various periods T

4 Conclusions

The system is analysed with the parameters of the Newcastle robot (see in Section 2), the modal parameters are $f_n = 0.67$ [Hz], $\zeta = 1.52$. The desired periodic contact force switches between $F_1=21.4$ [N], $F_2=44.5$ [N] in (5). The stability charts are shown in the $(P,(\tau f_n))$ plane of dimensionless parameters, for various time periods ($T = 1, 2, 4, 8$ [s]). The stable points are marked with crosses in the stability charts of Figures 4 and 5 In case of constant desired force, the $n = 100$ approximation gives an acceptable result in the presented parameter domain with errors under 1% relative to the exactly known stability limit (solid line in Figure 4).

As Figures 5 clearly show, the maximal gains decrease for increasing time delays. For long time periods T, the stability properties approach the stationary case shown in Figure 4. For small periods T, i.e. for fast changes in the desired contact force, certain ranges of the delay (and natural frequency) show great improvement in the stability properties. These effects, together with the periodic changes in the contact force may greatly help to improve accuracy limited by the Coulomb friction in the structure.

Acknowledgments

This research was supported by the Hungarian National Science Foundation under grant no. OTKA T030762/99, and the Ministry of Education and Culture grant no. MKM FKFP 0380/97.

References

Cooper, J. M., Kojeca, D. M. (1994). Relationship of Leg Strength, H-Reflexes and Balance in Young and Elderly Adults – Preliminary Study, In *Proceeidngs of Twelve's International Sypmosium on Biomechanics in Sport*, Budapest, 157-158.

Craig, J. J. (1986). *Introduction to Robot Mechanics and Control*, Reading, Addison-Willey.

Fargue, D. (1973). Réducibilité des Systémes héréditaires á des systémes dinamiques, *C. R. Acad. Sci. Paris*, 277B: 471-473.

Insperger, T., Horváth, R. (2000). Pendulum With Harmonic Variation of the Suspension Point, 2000, *Periodica Polytechnica – Mechanical Engineering*, **44**, Budapest, to appear.

Spong, M. W., Vidyasagar, M. (1989). *Robot Dynamics and Controll*, Singapure, Wiley & Sons.

Stépán G. (1997). Nonlinear Oscillations and Shimmying wheels, In *Proceedings of Symposium on New Applications of Nonlinear and Chaotic Dynamics in Mechanics*, Ithaca, 373-387.

Stépán, G., Steven A. (1990). Theoretical and Experimental Stability Analysis of a Hybrid Position-Force controlled robot, In *Proceedings of Eight Symposium on Theory and Practice of Robots and Manipulators*, Cracow, 53-60.

Vertut, J., Charles, J., Coiffet, P. and Petit. M. (1976). Advance of the New MA23 Force Reflecting Manipulator System, In *Proceedings of the Second Symposium on Theory and Practice of Robots and Manipulators*, Udine, 50-57.

Dynamic Control of Multiple Joint Manipulators Interacting with Dynamic Environment

Atanasko Tuneski[1] and Miomir Vukobratovic[2]

[1] The Faculty of Mechanical Engineering, Skopje, Macedonia
[2] Robotics Laboratory, Mihailo Pupin Institute, Belgrade, Yugoslavia

Abstract. The dynamic adaptive control of multiple robot manipulators handling a dynamic object motion of which is constrained by the dynamic environment, when object and/or environment parameters are not known in advance, is proposed. It may be implemented when: (i) there is no good understanding of all physical effects incorporated in the multiple robots/object/environment system; (ii) the parameters of the system are not precisely known; (iii) the system parameters do vary in a known regions about their nominal values. The proof that controller is asymptotically stable is based on the Lyapunov stability theory.

1 Introduction

The problem addressed in this article is the synthesis of an dynamic adaptive control of multiple compliant manipulation on dynamical environment. Only a few papers consider the problem of multiple robot compliant manipulation on dynamic object (see Luo et al., 1993, Yukawa et al., 1993), and several adaptive control methods for multiple manipulators in the contact tasks were proposed (see Hu and Goldenberg, 1993, Su and Stepanenko, 1993, Yao et al., 1992, 1993).

The dynamic control law synthesized here has the following characteristics: (i) it may be implemented when there is uncertainty in the system parameters and they do vary in a known regions about their nominal values; (ii) it is composed of an identification part (parameter update law), and a control law part; (iii) in deriving the dynamic control law, the inverse dynamics controller structure is adopted. The proof that the proposed dynamic controller is asymptotically stable is based on the Lyapunov stability theory.

2 System Model and Task Setting

The whole system consists of k rigid joint robots, a dynamical tool or object, and a constraint dynamical environment. The coordinated robots have to simultaneously: (i) move the object along a predetermined known trajectory on the constraint environment; (ii) exert a prespecified contact force on the environment; (iii) keep the value of the internal force between a certain minimum and maximum. In order to facilitate the formulation of the system dynamics, the following reasonable assumptions are made: (i) the dynamic manipulated object has any size and/or geometrical shape, and equivalent dynamic characteristics in all contact points. (ii) the constraint environment is fixed; (iii) each robotic mechanism is non-redundant and all coordinated robots have the same number of joints. The object has k connections with the multiple manipulators and one connection with the constraint environment. The object together with $(k+1)$ connections can be represented by a dynamic system of $(k+1)$ rigid bodies (Fig.1), that is, each of the k connections between the object and the k manipulators can be represented as a local rigid body, and in the mass center of each body contact, gravitational, damping and elasticity forces act as external forces.

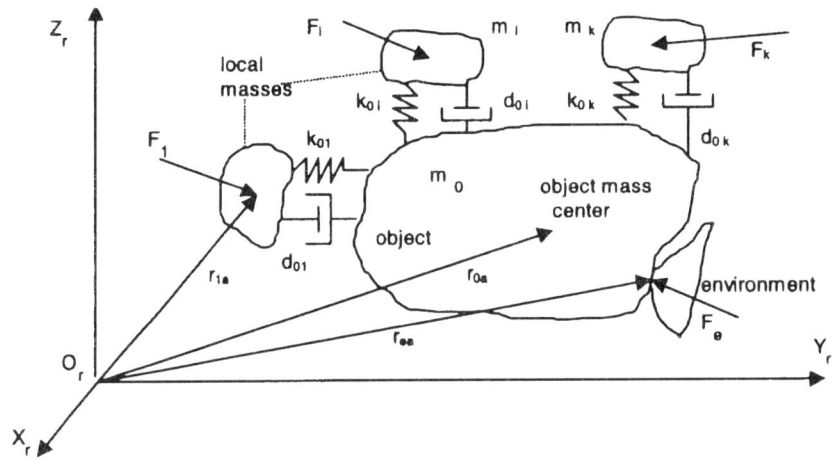

Figure 1. Elastic system of $(k+1)$ rigid bodies

If m denotes a mass of the free object, then after the contacts are established, the manipulated object has the following mass:

$$m_0 = m - \sum_{i=1}^{k} m_i \tag{1}$$

where $m_i, (i = 1, 2, ..., k)$, denotes the mass of the i-th local rigid body at the contact between the i-th robot end-effector and the object (Fig.1). Two states of dynamic system of $(k+1)$ bodies may be defined: the unloaded state, that is, the state when no force system acts on the system of $(k+1)$ elastically connected rigid bodies, and the loaded state, i.e. the state when a system of forces acts on the dynamic system of $(k+1)$ bodies, and they are moving from the unloaded state. Using the Lagrange equations of motion, the mathematical model of multiple non-redundant rigid robot manipulators performing cooperative work on a single dynamical object of which the motion is constrained by a dynamical environment has been derived (see

Vukobratovic and Tuneski, 1998). Here, the equations describing the multiple compliant manipulation model are briefly reported:

$$N(q)\ddot{q}+n(q,\dot{q},Y_{0a},\dot{Y}_{0a})=\tau \in R^{6k\times 1} \quad (2)$$

$$\ddot{Y}_{ca}=\dot{J}(q)\dot{q}+J(q)\ddot{q} \quad (3)$$

$$W_{0a}(Y_{0a})\ddot{Y}_{0a}+w_{0a}(q,\dot{q},Y_{0a},\dot{Y}_{0a})=F_e \quad (4)$$

$$W_{ca}(Y_{ca})\ddot{Y}_{ca}+w_{ca}(q,\dot{q},Y_{ca},\dot{Y}_{ca})=F_c \quad (5)$$

$$F_e=G_e(\mu^*,Y_{0a},\dot{Y}_{0a},\ddot{Y}_{0a})=(S^T(Y_{0a}))^{-1}[M(Y_{0a})\ddot{Y}_{0a}+L(Y_{0a},\dot{Y}_{0a})] \quad (6)$$

where

$$N(q)=(H(q)+J^T(q)W_{ca}(g(q))J(q))\in R^{n_s\times n_s}, \qquad q=[q_1^T,....,q_k^T]^T\in R^{n_s\times 1}$$

$$n(q,\dot{q},Y_{0a},\dot{Y}_{0a})=C(q,\dot{q})\dot{q}+G(q)+J^T(q)W_{ca}(g(q))\dot{J}(q)\dot{q}+$$
$$+J^T(q)w_{ca}(g(q),\dot{g}(q),J(q)\dot{q},Y_{0a},\dot{Y}_{0a})\in R^{n_s\times 1}$$

$$H(q)=diag\{H_1(q_1),...,H_k(q_k)\}\in R^{n_s\times n_s}$$

$$n_s=\sum_{i=1}^{k}n=6k \qquad \tau=[\tau_1^T,......,\tau_k^T]^T\in R^{n_s\times 1}$$

$$C(q,\dot{q})\dot{q}=[(C_1(q_1,\dot{q}_1)\dot{q}_1)^T,...,(C_k(q_k,\dot{q}_k)\dot{q}_k)^T]^T\in R^{n_s\times 1};$$

$$G(q)=[G_1^T(q_1),....,G_k^T(q_k)]^T\in R^{n_s\times 1}; \qquad J(q)=diag\{J_1(q_1),...,J_k(q_k)\}\in R^{n_s\times n_s};$$

$$f=[f_1^T,.....,f_k^T]^T=[-F_1^T,.....,-F_k^T]^T=-F_c\in R^{n_s\times 1}$$

$Y_{0a}\in R^{6\times 1}$ are the absolute coordinates of the mass center of the manipulated object for the loaded dynamic system state; $Y_{ca}=(Y_{1a}^T,....,Y_{ka}^T)^T\in R^{6k\times 1}$ are the absolute coordinates of the mass centers of the local rigid bodies at the contacts between the robot end effectors and the object for the loaded dynamic system state; $F_e\in R^{6\times 1}$ are the generalized forces at the contact between the environment and the manipulated object; $W_{0a}(Y_{0a})\in R^{6\times 6}$ is a matrix composed of the first six rows and the first six columns of the matrix $W_a(Y_a)$ representing the kinetic energy matrix; $w_{0a}(Y_{0a},\dot{Y}_{0a})\in R^{6\times 1}$ is a column matrix representing the potential and dissipative energy matrix for the object; $W_{ca}(Y_{ca})=W_{ca}^T(Y_{ca})=(W_{1a}(Y_{1a}),........,W_{ka}(Y_{ka}))\in R^{n_s\times n_s}$ represents the kinetic energy matrix of the local rigid bodies at the contacts between the robots and the object; $n=6$ is the number of the degrees of freedom of the i-th robot mechanism; k is the number of the multiple robot manipulators; $q_i\in R^n$ is the vector of the i-th robot joint coordinates; $w_{ca}(Y_{ca},\dot{Y}_{ca})=(w_{1a}^T(Y_{1a},\dot{Y}_{1a}),........,w_{ka}^T(Y_{ka},\dot{Y}_{ka}))^T\in R^{n_s\times 1}$, where $w_{ia}(Y_{ia},\dot{Y}_{ia})\in R^{6\times 1}$, $i=1,2,...,k$, represents the potential and dissipative energy matrix for the local rigid bodies at the contacts between the the robots and the object; $F_{ca} = (F_{1a}^T, \ldots, F_{ka}^T)^T \in R^{6k \times 1}$ are the generalized forces at the contacts between the manipulators and the manipulated object; $H_i(q_i) \in R^{n \times n}$ is the symmetric, bounded, positive inertia matrix of the i-th robot manipulator; $C_i(q_i, \dot{q}_i)\dot{q}_i \in R^{n \times 1}$ is the vector of centrifugal and Coriolis terms; $G_i(q_i)$ is the vector of gravitational terms; $\tau_i \in R^{n \times 1}$ are the applied joint driving forces; $J_i(q_i) \in R^{n \times n}$ is the generalized Jacobian matrix; $f_i = -F_i \in R^{n \times 1}$ is the vector of forces and moments exerted by the i-th robot end-effector on the object; $G_e \in R^{n \times 1}$ is nonlinear vector function, μ^* represents viscous friction, inertial, centrifugal, gravitational and elastic contributions; $M(Y_{0a}) \in R^{n \times n}$ is the non-singular inertia matrix representing contribution of the environment; $L(Y_{0a}, \dot{Y}_{0a}) \in R^{n \times 1}$ collects terms which represent viscous friction, Coriolis, centrifugal, gravitational and elastic contributions; $(S^T(Y_{0a})) \in R^{n \times n}$ is continuous matrix.

The goal of the multiple robot dynamic adaptive control can be formulated in the following way: assuming that the constrained, closed set $\overline{V}_\theta \in R^r$ of possible values of the unknown multiple robots and/or object parameters has been given by

$$\theta(t) \in \overline{V}_\theta, \quad \overline{V}_\theta \in R^r, \quad \forall t \geq t_0$$

then, define the control law $\tau(t)$ for $t \geq t_0$ to satisfy the control goals:

$$\lim_{t \to \infty}(q(t) - q_p(t)) = 0 \qquad \lim_{t \to \infty}(Y_{0a}(t) - Y_{0a}^p(t)) = 0 \qquad (7)$$

where the index "p" denotes nominal values of the corresponding vectors. The satisfaction of the control goals (7) also implies a satisfaction of the following control goals [see Ekalo and Vukobratovic, 1998]:

$$\lim_{t \to \infty}(F_c(t) - F_c^p(t)) = 0; \qquad \lim_{t \to \infty}(F_e(t) - F_e^p(t)) = 0 \qquad (8)$$

In addition to the control goals (7), (8) it is also necessary to ensure satisfaction of the internal force constraint. In the cooperative manipulation of a dynamical object, the internal force is defined as a part of the contact forces which is canceled within the object and therefore does not influence the object motion. The desired value of the internal force P, the maximum internal force, and the minimum internal force are given by the task planner.

3 Dynamic Control Law Synthesis

The deviation character of the real multiple robot motion and object motion from the programmed one may be described by:

$$\eta(t) = \begin{bmatrix} \eta_1(t) \\ \eta_2(t) \end{bmatrix} = \begin{bmatrix} q(t) - q_p(t) \\ Y_{0a}(t) - Y_{0a}^p(t) \end{bmatrix} \in R^{6(k+1) \times 1} \qquad (9)$$

$$\eta_1(t) = (q(t) - q_p(t)) \in R^{6k \times 1}, \qquad \eta_2(t) = (Y_{0a} - Y_{0a}^p(t)) \in R^{6 \times 1}$$

The family of functions representing the transient processes may be given by the vector differential equation (see Vukobratovic and Tuneski, 1998, 1999):

$$\ddot{\eta}(t) = P(\eta(t), \dot{\eta}(t)) \qquad (10)$$

where $P(\eta(t), \dot{\eta}(t)) \in R^{6(k+1) \times 1}$ is a vector function, continuous over all the set of its arguments, such that (10) has a trivial solution $\eta(t) \equiv 0$. The function P may be adopted in the form:

$$P(\eta(t), \dot{\eta}(t)) = -\Gamma_1 \dot{\eta}(t) - \Gamma_2 \eta(t) \qquad (11)$$

where the matrices

$$\Gamma_1 = diag[\gamma_{v1}, \gamma_{v2}, \ldots, \gamma_{v(6k+6)}]$$
$$\Gamma_2 = diag[\gamma_{p1}, \gamma_{p2}, \ldots, \gamma_{p(6k+6)}]$$

have to be chosen such that the eigenvalues $\lambda_1, \lambda_2, \ldots, \lambda_{12k+12}$ of the matrix

$$\Gamma = \begin{bmatrix} 0_{6(k+1)} & I_{6(k+1)} \\ -\Gamma_2 & -\Gamma_1 \end{bmatrix} \qquad (12)$$

have negative real parts. In this way an asymptotic stability of the solution of the equation (11) is achieved.

We propose the following dynamic control law:

$$\tau(t) = \hat{N}(q) \ddot{q}_p + P^*(\eta, \dot{\eta}) + \hat{n}(q, \dot{q}, Y_{0a}, \dot{Y}_{0a}) = \hat{N}(q) \ddot{q}_p + \Gamma_2^* \eta_1 + \Gamma_1^* \dot{\eta}_1 + \hat{n}(q, \dot{q}, Y_{0a}, \dot{Y}_{0a}) \qquad (13)$$

and the parameter adaptation law:

$$\dot{\hat{\theta}} = K \Phi^T \hat{N}^{-1} \eta^* \qquad (14)$$

where: $P^*(\eta, \dot{\eta}) \in R^{(6k) \times 1}$ is a vector function composed of the first $6k$ rows of the vector function $P(\eta, \dot{\eta}) \in R^{6(k+1) \times 1}$; $\theta = (\theta_1, \ldots, \theta_r)$ is the r-dimensional vector of the unknown robots and object parameters; $K = diag[k_1, \ldots, k_r]$, $k_i > 0$, $i = 1, 2, \ldots, r$ is $(r \times r)$ positive definite adaptation gain matrix; $\Gamma_1^* = diag[-\gamma_{v1}, -\gamma_{v2}, \ldots, -\gamma_{v(6k)}]$, $\Gamma_2^* = diag[-\gamma_{p1}, -\gamma_{p2}, \ldots, -\gamma_{p(6k)}]$ are constant diagonal matrices, (see (11)); $\Phi(q, \dot{q}, \ddot{q}, Y_{0a}, \dot{Y}_{0a}) \in R^{(6k) \times r}$ is a known matrix of functions of $(q, \dot{q}, \ddot{q}, Y_{0a}, \dot{Y}_{0a})$; $\tilde{\theta} = (\theta - \hat{\theta}) \in R^r$ is the parameter error vector, $\hat{\theta} = (\hat{\theta}_1, \ldots, \hat{\theta}_r)^T \in R^r$ are the estimates of the parameters, such that a key feature of rigid robot dynamics, which is linearity of manipulator parameters, may be used to write the error equation in the form

$$\ddot{\eta}_1 + \Gamma_2^* \eta_1 + \Gamma_1^* \dot{\eta}_1 = \hat{N}^{-1}(q) [\tilde{N}(q) \ddot{q} + \tilde{n}(q, \dot{q}, Y_{0a}, \dot{Y}_{0a})] = \hat{N}^{-1}(q) \Phi(q, \dot{q}, \ddot{q}, Y_{0a}, \dot{Y}_{0a}) \tilde{\theta}$$

where $(\hat{\cdot})$ is the estimate of $(\cdot)$; $\tilde{n}(q,\dot{q},Y_{0a},\dot{Y}_{0a}) = n(q,\dot{q},Y_{0a},\dot{Y}_{0a}) - \hat{n}(q,\dot{q},Y_{0a},\dot{Y}_{0a})$ and $\tilde{N}(q) = N(q) - \hat{N}(q)$ represent errors in the multiple robots dynamic model (2) used in the synthesis of the control law (13) arising from the errors in the multiple robots and/or object parameters, and

$$\ddot{\eta}_1 + \Gamma_2^* \dot{\eta}_1 + \Gamma_1^* \eta_1 = \hat{N}^{-1}(q)[\tilde{N}(q)\ddot{q} + \tilde{n}(q,\dot{q},Y_{0a},\dot{Y}_{0a})]$$

is the error equation obtained by equating (2) and (13); $\eta^* = \xi \dot{\eta}_1 + \psi \eta_1$, where $\xi = diag(\xi_1, \xi_2, ..., \xi_{6k}), \xi_i > 0$, and $\psi = diag(\psi_1, \psi_2, ..., \psi_{6k}), \psi_i > 0$, $i = 1, 2, ..., 6k$, are positive constants chosen such that the transfer function

$$\frac{\eta^*}{\hat{N}^{-1} \Phi \tilde{\theta}} = \frac{\xi \dot{\eta}_1 + \psi \eta_1}{\ddot{\eta}_1 + \Gamma_2^* \dot{\eta}_1 + \Gamma_1^* \eta_1} \tag{15}$$

is strictly positive real (SPR) function. The condition that the transfer function (15) is strictly positive real ensures the possibility to implement the Popov-Kalman-Yakubovitch lemma.

The following theorem may be established:

THEOREM. Let us suppose that: **(i)** we have a perfect structural model of the cooperative manipulation represented by the equations (2)-(6); **(ii)** the constrained, closed set $\overline{V}_\theta \in R^r$ of possible values of the multiple robots and/or object parameters has been given by $\theta(t) \in \overline{V}_\theta, \forall t \geq t_0$; **(iii)** the desired multiple robot trajectories q_p satisfy the persistent excitation condition

$$\alpha I_r \leq \int_{t_0}^{t_0+\rho} U(q_p, \dot{q}_p, \ddot{q}_p, Y_{0a}^p, \dot{Y}_{0a}^p) U^T(q_p, \dot{q}_p, \ddot{q}_p, Y_{0a}^p, \dot{Y}_{0a}^p) dt \leq \beta I_r$$

for all t_0, where $U = (\hat{N}^{-1} \Phi)^T$, r is the number of the unknown system parameters, $\alpha, \beta,$ and ρ are all positive, I_r is the *(rxr)* unit matrix; **(iv)** the estimates of the multiple robots and/or object parameters lie within sufficiently small region such that the matrix $\hat{N}(q)$ remains positive definite and invertible (i.e. $\theta_{i(min)} \leq \hat{\theta}_i \leq \theta_{i(max)}$), where the lower bound $\theta_{i(min)}$ and the upper bound $\theta_{i(max)}$ are a priori known. Then, if the control law (13) and the parameter adaptation law (14) are introduced into the multiple robots dynamic model (2),

(i) the control goals (7) and (8) are satisfied;
(ii) the errors of the parameter estimation will converge to zero;
(iii) it is possible to satisfy the force constraints.

Proof. By equating (2) and (13), it follows

$$\ddot{\eta}_1 + \Gamma_2^* \dot{\eta}_1 + \Gamma_1^* \eta_1 = \hat{N}^{-1}(q)[\tilde{N}(q)\ddot{q} + \tilde{n}(q,\dot{q},Y_{0a},\dot{Y}_{0a})] = \hat{N}^{-1}(q)\Phi(q,\dot{q},\ddot{q},Y_{0a},\dot{Y}_{0a})\tilde{\theta}$$

where $(\hat{\circ})$ denotes the estimate of $(\circ)$;

$$\tilde{N}(q) = N(q) - \hat{N}(q), \quad \Phi(q,\dot{q},\ddot{q},Y_{0a},\dot{Y}_{0a}) \in R^{n_i \times r}, \quad \tilde{\theta} = (\theta - \hat{\theta}) \in R^r,$$
$$\tilde{n}(q,\dot{q},Y_{0a},\dot{Y}_{0a}) = n(q,\dot{q},Y_{0a},\dot{Y}_{0a}) - \hat{n}(q,\dot{q},Y_{0a},\dot{Y}_{0a})$$

For the i-th element of the deviation vector $\eta_1 \in R^{6k \times 1}$, (9), an error equation may be written as:

$$\ddot{\eta}_{1i} - \gamma_{vi}\dot{\eta}_{1i} - \gamma_{pi}\eta_{1i} = (\hat{N}^{-1}\Phi\tilde{\theta})_i \quad (16)$$

where $\Gamma_1^* = diag[-\gamma_{v1}, -\gamma_{v2}, ..., -\gamma_{v(6k)}]$ and $\Gamma_2^* = diag[-\gamma_{p1}, -\gamma_{p2}, ..., -\gamma_{p(6k)}]$, (see (13)), $(\hat{N}^{-1}\Phi\tilde{\theta})_i$ means the i-th element of the $(6k \times 1)$ vector $(\hat{N}^{-1}\Phi\tilde{\theta})$. The state-space representation of the error equation (16) is given by:

$$\dot{x}_i = A_i x_i + B_i(\hat{N}^{-1}\Phi\tilde{\theta})_i \quad (17)$$

$$\eta_i^* = C_i x_i \qquad A_i = \begin{bmatrix} 0 & 1 \\ -\gamma_{pi} & -\gamma_{vi} \end{bmatrix} \in R^{2 \times 2} \qquad B_i = \begin{bmatrix} 0 \\ 1 \end{bmatrix} \qquad C_i = [\psi_i \ \xi_i]$$

η_i^* is the i-th element of the vector $\eta^* = \xi\dot{\eta}_1 + \psi\eta_1$ (see (15)), and $x_i = [\eta_1 \ \dot{\eta}_1]^T$ is the state vector. If the positive constants $\xi = diag(\xi_1, \xi_2, ..., \xi_{6k}), \xi_i > 0$, and $\psi = diag(\psi_1, \psi_2, ..., \psi_{6k}), \psi_i > 0$, $i = 1, 2, ..., 6k$, are chosen such that

$$\frac{\eta^*}{\hat{N}^{-1}\Phi\tilde{\theta}} = \frac{\xi\dot{\eta}_1 + \psi\eta_1}{\ddot{\eta}_1 + \Gamma_2^*\eta_1 + \Gamma_1^*\dot{\eta}_1}$$

is strictly positive real (SPR) function, then by the Kalman-Yakubovitch-Popov lemma the symmetric, positive definite matrices $P_i \in R^{2 \times 2}$ and $Q_i \in R^{2 \times 2}$ do exist such that

$$A_i^T P_i + P_i A_i = -Q_i \quad (18)$$
$$P_i B_i = C_i^T$$

where the matrices A_i, B_i, and C_i are the matrices of the state-space realization (17) of the error equation (16).

The error equation of the whole system of k multiple manipulators in the state-space form is given by:

$$\dot{X} = AX + B\hat{N}^{-1}\Phi\tilde{\theta} \quad (19)$$
$$\eta^* = Cx$$

$A = blockdiag(A_1, A_2, ..., A_{nk}) \in R^{(2nk) \times (2nk)}$ \qquad $B = blockdiag(B_1, B_2, ..., B_{nk}) \in R^{(2nk) \times (nk)}$
$C = blockdiag(C_1, C_2, ..., C_{nk}) \in R^{(nk) \times (2nk)}$ \qquad $X = (x_1 \ x_2 \ .... x_{nk})^T \in R^{2nk \times 1}$

If a symmetric positive definite matrix Q

$$Q = diag(Q_1, Q_2, ..., Q_{nk}) \in R^{(2nk) \times (2nk)} \quad (20)$$

is chosen and the Lyapunov equation

$$A^T P + PA = -Q \tag{21}$$

is solved, then the unique symmetric positive definite matrix P

$$P = diag(P_1, P_2, \ldots, P_{nk}) \in R^{(2nk) \times (2nk)} \tag{22}$$

may be determined. The following Lyapunov function candidate is chosen

$$V(X, \tilde{\theta}) = X^T P X + \tilde{\theta}^T K^{-1} \tilde{\theta} \tag{23}$$

where $K = diag[k_1, k_2, \ldots, k_r]$, $k_i > 0$, $i = 1, 2, \ldots, r$, is $(r \times r)$ positive definite adaptation gain matrix.

The time derivative of $V(X, \tilde{\theta})$ along the trajectories (19) is

$$\dot{V}(X, \tilde{\theta}) = -X^T Q X + 2\tilde{\theta}^T \left[\Phi^T \hat{N}^{-1} \eta^* + K^{-1} \dot{\tilde{\theta}} \right] \tag{24}$$

The equation (24) suggests that, if the parameter update law is chosen to be

$$\dot{\hat{\theta}} = K \Phi^T \hat{N}^{-1} \eta^*$$

then, since the parameter vector θ is constant, $\dot{\tilde{\theta}} = \dot{\hat{\theta}}$, the second term in (24) equals zero, and

$$\dot{V}(X, \tilde{\theta}) = -X^T Q X \tag{25}$$

which is nonpositive because Q is chosen to be a positive definite matrix (20). Since $X, \theta, \hat{N}^{-1}$, and W are bounded, from (19) it may be stated that $\dot{X}$ is bounded as well. Thus X is uniformly continuous, and so is $\dot{V}(X, \theta)$. From (23) and (25) it follows that

$$\lim_{t \to \infty} V(X, \theta) = V_c \tag{26}$$

does exist, and

$$V_c = V(X_0, \theta_0) - \int_0^\infty X^T Q X \, dt \tag{27}$$

Since V_c and $V(X_0, \theta_0)$ are finite terms, it may be stated that

$$\int_0^\infty X^T Q X \, dt$$

has a finite value. Since $X^T Q X$ is positive, uniformly continuous, and has a finite integral, using the well-known mathematical analysis result (see Rudin, 1976), it may be stated that

$$\lim_{t \to \infty} X^T Q X = 0 \tag{28}$$

The equation (28) leads to the conclusion that the state vector

$$X = (x_1 \, x_2 \ldots x_{nk})^T \in R^{(2nk) \times (1)}$$

where $x_i = [\eta_1 \, \dot{\eta}_1]^T$, will asymptotically converge to zero vector as $t \to \infty$:

$$\lim_{t\to\infty}(\eta_1) = \lim_{t\to\infty}(q(t) - q_p(t)) = 0$$
$$\lim_{t\to\infty}(\dot\eta_1) = \lim_{t\to\infty}(\dot q(t) - \dot q_p(t)) = 0$$

i.e. the control goals (7) are satisfied. The satisfaction of the control goals (7) also implies a satisfaction of the control goals (8), (see Vukobratovic and Tuneski, 1999). The maximum and minimum value of the internal force are given by the task planner. So, if the control goals (8) are satisfied, we may state that the internal force constraints are also satisfied.

Now the parameter error convergence will be proved. According to (14) and (19) the state-space representation of the entire system may be written as

$$\begin{bmatrix} \dot X \\ \dot{\tilde\theta} \end{bmatrix} = \begin{bmatrix} A & BU^T \\ KUC & 0 \end{bmatrix} \begin{bmatrix} X \\ \tilde\theta \end{bmatrix} \tag{29}$$

where $U = (\hat N^{-1}\Phi)^T$. The asymptotic stability of (29) has been already studied in the literature. Anderson, (1977), has shown that (29) is uniformly asymptotically stable if: (i) the linear system (A, B, C) satisfies the strictly positive real condition (SPR); (ii) U satisfies the persistent excitation condition

$$\alpha I_r \le \int_{t_0}^{t_0+\rho} U(q_p,\dot q_p,\ddot q_p,Y^p_{0a},\dot Y^p_{0a}) U^T(q_p,\dot q_p,\ddot q_p,Y^p_{0a},\dot Y^p_{0a}) dt \le \beta I_r$$

for all t_0, where $U = (\hat N^{-1}\Phi)^T$, r is the number of the unknown system parameters, $\alpha, \beta,$ and ρ are all positive, I_r is the $(r \times r)$ unit matrix.

The proof of the theorem is completed.

4 Conclusion

Dynamic adaptive control of multiple autonomous non-redundant rigid joint robot manipulators handling one dynamic object motion of which is constrained by a dynamic environment is synthesized. The manipulator dynamics, the object dynamics and the environment dynamics are taken into consideration, together with the material constraints imposed on the system and program constraints. The object which is held by the k robot end-effectors makes a frictional point contact on the constraint environment. The robot and/or object parameters are assumed to be unknown and they are identified in the control synthesis process. The proposed dynamic adaptive control law ensures movement of the object along a predetermined known trajectory on the constraint environment, while simultaneously exerting a prespecified contact force onto the environment. The main contribution of this paper is that the dynamic adaptive coordinated control of multiple robot manipulators handling a dynamic object motion of which is constrained by the dynamic environment, when object and/or environment parameters are not known in advance, is proposed. A proof that the proposed controller is asymptotically stable is based on the Lyapunov stability theory.

References

Anderson B.D.O., (1977). Exponential Stability of Linear Equations Arising in Adaptive Identification, *IEEE Transactions on Automatic Control*, AC-22, 83-88.

Ekalo Y., and Vukobratovic M. (1993). Robust and Adaptive Position/Force Stabilization of Robotic Manipulators in Contact Tasks, *Robotica*, Vol. 11, 373-386.

Hu Y., and Goldenberg A. (1993). An Adaptive Approach to Motion and Force Control of Multiple Coordinated Robot Arms", *ASME Journal of Dynamic Systems, Measurement and Control* 115, 60-69.

Luo, Z., Ito, K., and Ito, M., (1993). Multiple Robot Manipulators: Cooperative Compliant Manipulation on Dynamical Environments, *Proceedings of IEEE/RSJ Conference on Intelligent Robots and Systems*, 1927-1934.

Luo Z., and Ito M. (1993). Control Design of Robot for Compliant Manipulation on Dynamical Environments, *IEEE Transactions on Robotics and Automation*, Vol.9, No.3, 286-296.

Rudin, W. (1976). *Principles of Mathematical Analysis*, McGraw-Hill, New York.

Su C.Y., and Stepanenko Y. (1993). Adaptive Sliding Mode Coordinated Control Of Multiple Robot Arms Handling One Constrained Object, *Proceedings of American Control Conference*, 1406-1413.

Vukobratovic M., and Tuneski A. (1998). Mathematical Model of Multiple Manipulators: Cooperative Compliant Manipulation on Dynamical Environment, *Journal of Mechanism and Machine Theory*, Vol.33, 1211-1239.

Vukobratovic M., and Tuneski A., (1999). Contribution to the Adaptive Control of Multiple Compliant Manipulation on Dynamic Environments, *Robotica*, Vol.17, 97-109.

Yao B., Gao W.B., Chan S.P., and Cheng M. (1992). VSC Coordinated Control of Two Manipulator Arms in the Presence of Environmental Constraints, *IEEE Transactions on Automatic Control*, Vol. 37, 1806-1812.

Yao B., and Tomizuka M. (1993). Adaptive Coordinated Control of Multiple Manipulators Handling a Constrained Object, *Proceedings of IEEE International Conference on Robotics and Automation*, 624-629.

Yukawa T., Uchiyama M., and Obinata G. (1993). Stability and Handling Characteristics of the Control System for Dual-Arm Manipulators to Handle Flexible Objects, *Proceedings of IEEE/RSJ Conference on Intelligent Robots and Systems*, 1161-1164.

A Comparison Between PD Controls in Terms of Normalized and Unnormalized Quasi-Velocities

Krzysztof Kozłowski and Przemysław Herman

Poznań University of Technology
Chair of Control, Robotics, and Computer Science, ul. Piotrowo 3A, 60-965 Poznań, Poland

Abstract. This paper presents a new controls for manipulators whose dynamics is expressed in terms of quasi-velocities Jain A., Rodriguez G. (1995). In contrary to previous algorithms Herman P. (1997) these controls consider also gravitational forces. Robot dynamic algorithms in terms of quasi-velocities are recursive in nature and consists of two recursions: one starts from a base of the manipulator towards its tip and the second in opposite direction. Both recursions are described by using vector-matrix notation. The considered controls allow to achieve end point of trajectory in Cartesian space. The controls were tested on a two degrees of freedom manipulator.
Key Words. Robots; PD control; spatial algebra; Cartesian space; quasi-velocities; gravitational forces.

1 Nomenclature

$\mathcal{N}$ number of joins and number of degrees of freedom,
$\mathcal{O}_k$ origin of the frame attached to the k-th link
$\theta(k)$ k-th generalized position,
$\dot{\theta}(k)$ k-th generalized velocity $\dot{\theta} = \ell^T(\theta)\nu = [I - H\psi K]^T D^{-\frac{1}{2}}\nu$,
$l(k, j) \in \mathbf{R}^3$ vector from $\mathcal{O}_k$ to $\mathcal{O}_j$,
$(.)^T$ transpose operation,
$m(k)$ mass of the k-th link,
$p(k) \in \mathbf{R}^3$ vector from $\mathcal{O}_k$ to the k-th link's center mass,
$b(k) \in \mathbf{R}^6$ spatial bias forces vector calculated according to expression in Kozłowski K. (1995) hidden in Coriolis terms,
$M(\theta) \in \mathbf{R}^{\mathcal{N} \times \mathcal{N}}$ mass matrix of the manipulator $M(\theta) = m(\theta)m^T(\theta)$,
$C(\theta, \dot{\theta}) \in \mathbf{R}^{\mathcal{N}}$ vector of Coriolis and centrifugal forces in standard equations of motion,
$\nu \in \mathbf{R}^{\mathcal{N}}$ vector of normalized quasi-velocities $\nu = m^T(\theta)\dot{\theta} = D^{\frac{1}{2}}[I + H\phi K]^T\dot{\theta}$,
$\xi \in \mathbf{R}^{\mathcal{N}}$ vector of unnormalized quasi-velocities,
$C(\theta, \nu) \in \mathbf{R}^{\mathcal{N}}$ vector of Coriolis and centrifugal forces in diagonalized normalized equations of motion and $\nu^T C(\theta, \nu) = 0$,
$C(\theta, \xi) \in \mathbf{R}^{\mathcal{N}}$ vector of Coriolis and centrifugal forces in diagonalized unnormalized equations of motion,
$m(\theta) \in \mathbf{R}^{\mathcal{N} \times \mathcal{N}}$ spatial operator defined $\nu = m^T(\theta)\dot{\theta}$ where $m(\theta) = [I + H\phi K]D^{\frac{1}{2}}$,

$\ell(\theta) \in \mathbf{R}^{\mathcal{N} \times \mathcal{N}}$ spatial operator defined $\ell(\theta) = m^{-1}(\theta)$ where $\ell(\theta) = m^{-1}(\theta) = D^{-\frac{1}{2}}[I - H\psi K]$,

$T \in \mathbf{R}^{\mathcal{N}}$ vector of generalized forces with elements $\tau(k)$ and $T = [I + H\phi K]D^{\frac{1}{2}}\epsilon$,

$\epsilon \in \mathbf{R}^{\mathcal{N}}$ vector of normalized quasi-moments $\epsilon = \ell(\theta)T = D^{-\frac{1}{2}}[I - H\psi K]T$,

$\kappa \in \mathbf{R}^{\mathcal{N}}$ vector of unnormalized quasi-moments,

$H \in \mathbf{R}^{\mathcal{N} \times 6\mathcal{N}}$ projection operator for all joint axes,

$\phi \in \mathbf{R}^{6\mathcal{N} \times 6\mathcal{N}}$ rigid manipulator force transformation matrix,

$\psi \in \mathbf{R}^{6\mathcal{N} \times 6\mathcal{N}}$ articulated manipulator force transformation matrix,

$K \in \mathbf{R}^{6\mathcal{N} \times \mathcal{N}}$ shifted Kalman gain matrix,

$P \in \mathbf{R}^{6\mathcal{N} \times 6\mathcal{N}}$ articulated inertia matrix,

$D = HPH^T \in \mathbf{R}^{\mathcal{N} \times \mathcal{N}}$ articulated inertia about joint axes matrix.

2 Introduction

Globally diagonalized dynamics rarely exists in practice for multibody systems like manipulators Jain A., Rodriguez G. (1995). The authors of Jain A., Rodriguez G. (1995) have proposed, instead of transformation in configuration space, a diagonalizing transformation in the velocity space. They have presented a new equations of motion, which are called diagonalized Lagrangian robot dynamics. In this paper we consider two new controls (in terms of normalized and unnormalized) quasi-velocities. Similar controls for manipulator we can find in Jain A., Rodriguez G.; Herman P. (1995; 1997) but without including gravitational forces. This paper is organized as follows. In the third section diagonalized equation of motion and new control in quasi-velocities space are described. Simulation results are presented in the fourth section and in the fifth section - experimental results. The last section contains concluding remarks.

3 Diagonalized equations of motion and new controls

Consider a serial manipulator with $\mathcal{N}$ rigid links. The links of manipulator are numbered in increasing order from its tip to the base. The most outer link is link 1 and the most inner link is link $\mathcal{N}$. The basic properties of spatial operators were given by Jain and Rodriguez Jain A., Rodriguez G. (1995). New equations of motion in (θ, ν) coordinates according to Jain A., Rodriguez G. (1995) are (see Nomenclature):

$$\dot{\nu} + C(\theta, \nu) = \epsilon. \quad (1)$$

In general, we consider system with gravity and the equations of motion are written as (in terms of normalized quasi-velocities):

$$\dot{\nu} + C(\theta, \nu) + G_\nu(\theta) = \epsilon \quad (2)$$

where $G_\nu(\theta) = m^{-1}(\theta)G(\theta)$ and $G(\theta)$ - gravitational forces in standard equations of motion.

New unnormalized equations of motion in (θ, ξ) coordinates according to Jain A., Rodriguez G. (1995) are (see Nomenclature):

$$D\dot{\xi} + C(\theta, \xi) = \kappa. \quad (3)$$

In general we can consider system with gravity and the equations of motion are written as (in terms of unnormalized quasi-velocities):

$$D\dot{\xi} + C(\theta, \xi) + G_\xi(\theta) = \kappa \qquad (4)$$

where $G_\xi(\theta) = D^{\frac{1}{2}}m^{-1}(\theta)G(\theta)$ and $G(\theta)$ - gravitational forces in standard equations of motion.

In order to propose PD controls in Cartesian space we have to consider a 12-dimensional vector $x_d = [n_d, o_d, a_d, p_d]^T$, where "$T$" denotes transposition of a matrix. Actual trajectory is described as $x = [n, o, a, p]^T$, where x is function of θ. First three vectors n_d, o_d, a_d are 3 unit vectors which together form an orthonormal system attached to task coordinate. These three vectors (named normal, orientation and approach Canudas de Wit C., Siciliano B., Bastin G.(Eds); Sciavicco L., Siciliano B. (1996; 1996)) indicate the desired orientation that the task coordinate system ought to reach as a result of the control action. Vector p_d denotes the desired linear position of the task coordinate system with respect the inertial frame.

In this case we propose the following control law:

$$\epsilon = -c_1\nu + J_{A\nu}^T(\theta)e - J_{A\nu}^T(\theta)c_2 J_{A\nu}(\theta)\nu + G_\nu(\theta) \qquad (5)$$

where c_1 and c_2 are positive diagonal control gain matrices and $J_{A\nu}^T(\theta)$ is an analytical Jacobian which depends on ν and θ (5) realises PD control in Cartesian space in terms of normalized quasi-velocities.

The second control law is described as:

$$\kappa = -c_1\xi + J_{A\xi}^T(\theta)e - J_{A\xi}^T(\theta)c_2 J_{A\xi}(\theta)\xi + G_\xi(\theta) \qquad (6)$$

where c_1 and c_2 are positive diagonal matrices and $J_{A\xi}^T(\theta)$ is an analytical Jacobian which depends on ξ and θ (6) realises PD control in Cartesian space in terms of unnormalized quasi-velocities. Error matrix e in (5) and (6) has the following form:

$$e = \begin{bmatrix} c_n(n \times n_d) + c_o(o \times o_d) + c_a(a \times a_d) \\ c_p(p_d - p) \end{bmatrix}. \qquad (7)$$

where c_n, c_o, c_a and c_p are 3×3 positive definite matrices.

Recall that Sciavicco L., Siciliano B. (1996) for classical description the equations of motion are:

$$M(\theta)\ddot{\theta} + C(\theta, \dot{\theta}) + G(\theta) = u \qquad (8)$$

PD control in operational space has the following form:

$$u = J_A^T(\theta)c_P\tilde{x} - J_A^T(\theta)c_D J_A(\theta)\dot{\theta} + G(\theta) \qquad (9)$$

where $\tilde{\theta} = \theta_d - \theta$ and $\tilde{x} = x_d - x$ are the joint error and the operational space error, respectively, and J_A is an analytical Jacobian.

Comparing the PD controls in quasi-variables with classical one we see that the first term is not present in classical case. It is concerned with joint quasi-velocities. The main difference between PD controls in normalized and unnormalized variables rely on other quasi-velocity using in it. The both quasi-velocities are related as $\xi = D^{-\frac{1}{2}}\nu$ Jain A., Rodriguez G. (1995). For that reason the dynamics of both systems is not the same. In controls in each quasi-moment we take into consideration in general all joint velocities.

These controls were written in MATLAB as SIMULINK programmes and here we present results for KARI-2 rigid robot (double pendulum in horizontal plane).

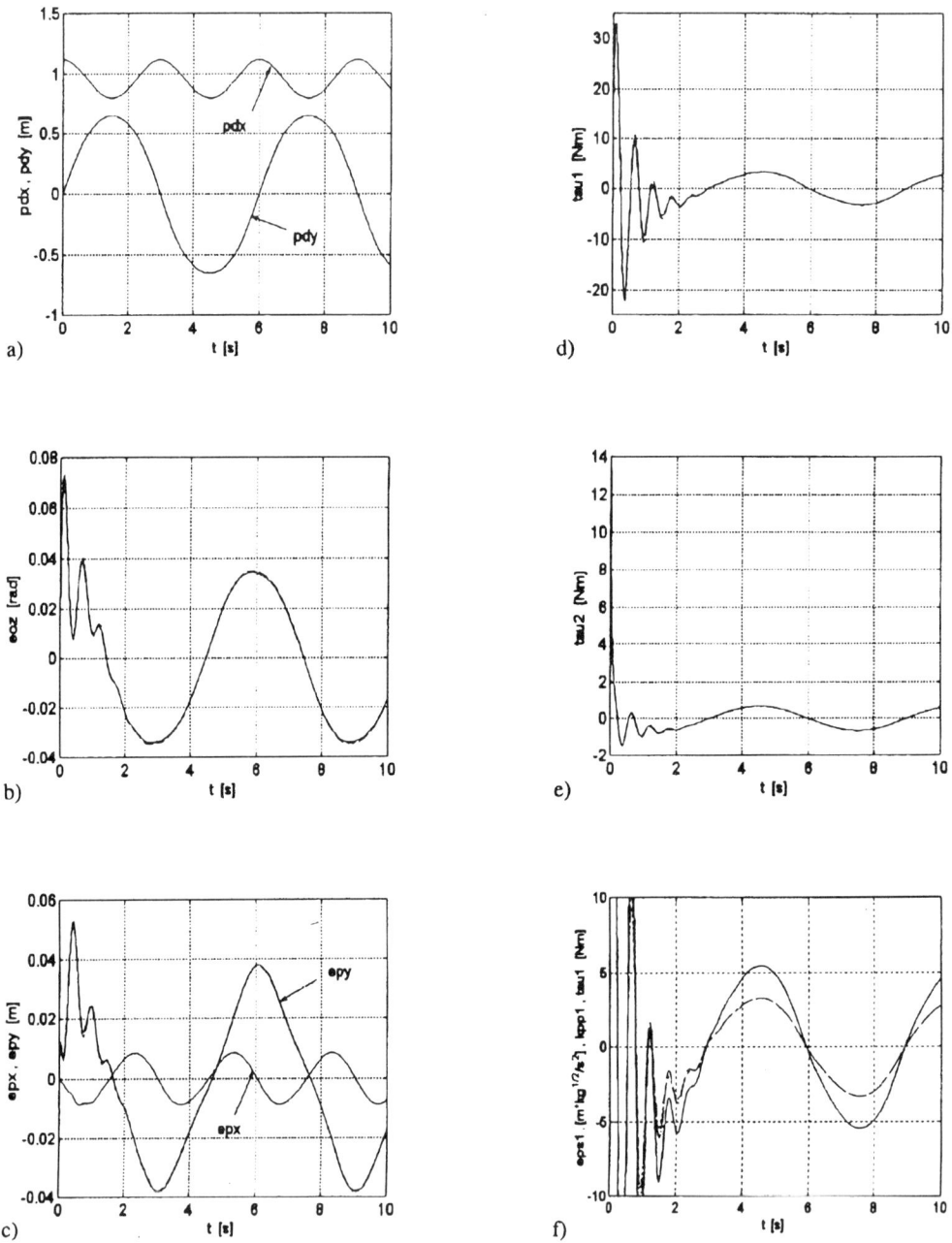

Figure 1. Results of simulation: a) desired trajectories in Cartesian space, b) orientation errors eoz, c) position errors epx, epy, d) moment in joint 1 near the tip $tau1$, e) moment in joint 2 near the base $tau2$, e) comparison between moment $tau1$ and quasi-moments $eps1$-normalized case, $kpp1$-unnormalized case. Lines: dotted line - $tau1$, solid line - normalized case, dashed line - unnormalized case.

4 Simulation results

In this section we present simulation results for manipulator KARI-2 consisting of two degrees of freedom (double pendulum) in horizontal plane. The KARI-2 robot is characterized by the following set of manipulator parameters:

- links masses: $m_1 = 0.75 kg, m_2 = 7.92 kg$;
- link inertias: $J_1 = 0.359 kgm^2, J_2 = 2.597 kgm^2$;
- distance axis of rotation - mass center: $p_1 = 0.28m, p_2 = 0.5235m$;
- length of links: $l_1 = 0.525m, l_2 = 0.595m$,
- static moment: $m_1 p_1 = 0.21 kgm, m_2 p_2 = 4.146 kgm$.

For simulation we have chosen a desired trajectory (in Cartesian space):

$$p_{dx} = l_1 \cos(\theta_1 + \theta_2) + l_2 \cos\theta_2 \tag{10}$$

$$p_{dy} = l_1 \sin(\theta_1 + \theta_2) + l_2 \sin\theta_2 \tag{11}$$

where $\theta_1 = \frac{\pi}{4}\sin(\frac{\pi}{3} \cdot t)$ and $\theta_2 = \frac{\pi}{10}\sin(\frac{\pi}{3} \cdot t)$. The following configuration values starting joint variables: $\theta_1 = \theta_2 = 0[rad]$. Point $\theta_2 = 0$ is not the same as kinematic singular point and arises from that the real manipulator starts from its own zero point. Simulation was performed in SIMULINK (fixed step size $0.005[s]$ and Dormand-Prince numerical integration method) with control coefficients (for normalized and unnormalized case):

$$c_1(1) = c_1(2) = 1, c_2(1) = c_2(2) = c_2(3) = 50, c_{oz} = c_n = c_o = c_a = 1500, c_{px} = c_{py} = 1500. \tag{12}$$

The differences between curves with step size $0.005[s]$ and $0.001[s]$ are hardly to distinguish. We also used two input coefficients equal 0.001 and two output coefficients equal 1000, respectively, because in experiment we had to use the same values (they are need for correct work of the manipulator KARI-2). For classical case $c_P \tilde{x}$ equals e and $c_D = c_2$.

The choice of these coefficients is very important because if c_1, c_2 have small values, then variables tend to steady values slowly. If they have large values, then variables quickly tend steady state. Sometimes the system may be not physically realized.

The simulation results obtained from SIMULINK are presented in Figure 1 from a to f. In Fig.1a the desired trajectories p_{dx}, p_{dy} in Cartesian space are shown. Figure 1b shows orientation errors. they are about $0.07-0.04[rad]$. Figure 1c shows position errors in directions x and y. They are about $1[cm]$ and $5 - 4[cm]$ respectively. We can get smaller values of errors if we increase control coefficients but it is more difficult in this case to compare simulation and experimental results. In Fig.1d and e we can see moments $tau1$ in the first joint (near the tip - joints are numbered in opposite direction as usual) and $tau2$ (near the base). After a short time these moments are eqaul about $4[Nm]$ and $1[Nm]$, respectively. In Fig.1f a comparison between real moment and quasi-moments (normalized and unnormalized) are shown. Errors and moments for new PD controllers are similar here.

5 Experimental results

The same kind of robot i.e. KARI-2 was investigated experimentally with using dSPACE-tools. Instead of dynamical model of manipulator a DSP card was inserted.

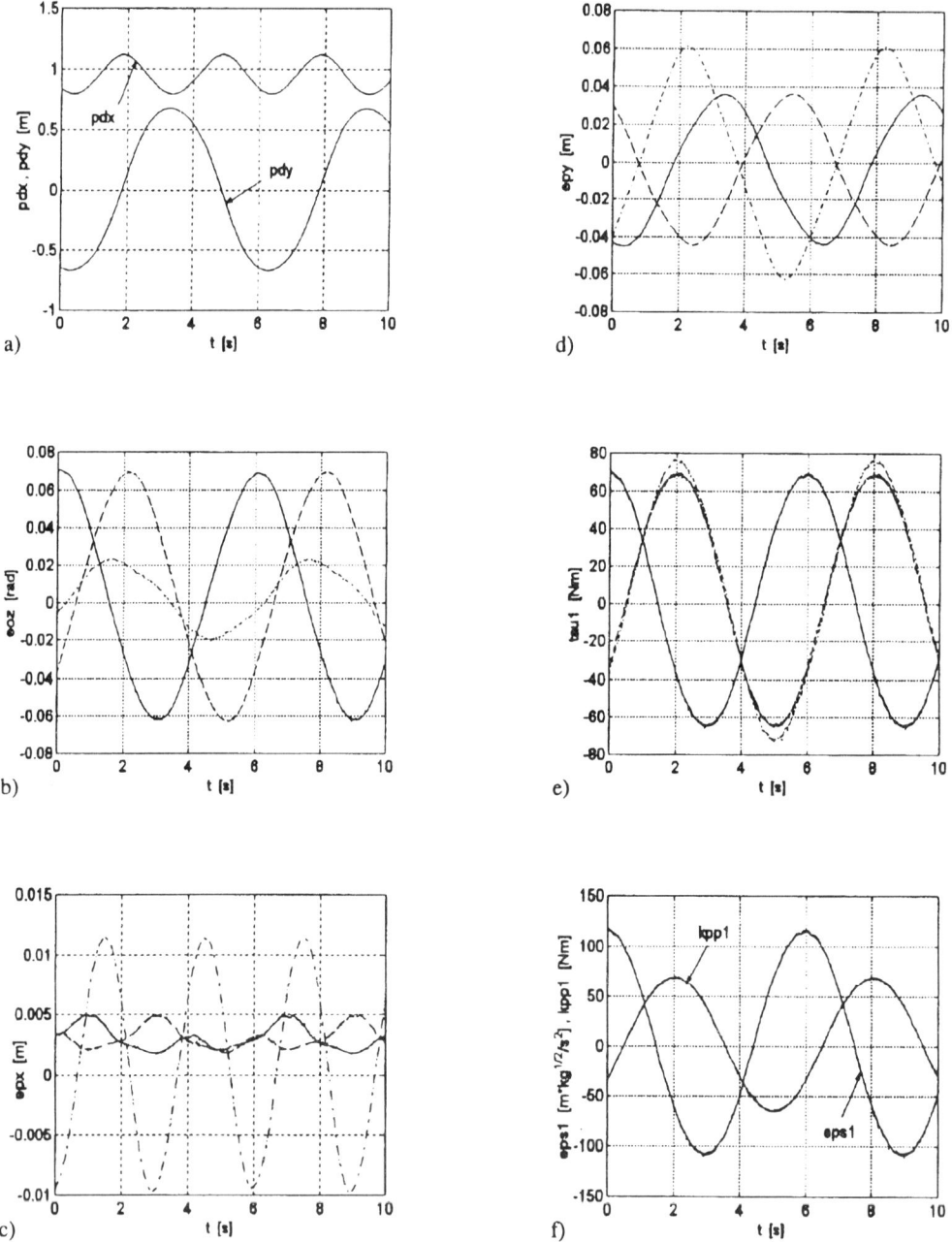

Figure 2. Results of experiment: a) desired trajectories in Cartesian space, b) orientation errors eoz, c) position errors epx, d) position errors epy, e) moment in joint 1 near the tip $tau1$, e) comparison between moment $tau1$ and quasi-moments $eps1$-normalized case, $kpp1$-unnormalized case. Lines: dashdot line - classical case, solid line - normalized case, dashed line - unnormalized case.

Experimental results was performed also with SIMULINK-programme with control coefficients (fixed step size $0.001[s]$ and Dormand-Prince numerical integration method):

$$c_1(1) = c_1(2) = 1, c_2(1) = c_2(2) = c_2(3) = 1, c_{oz} = c_n = c_o = c_a = 1500, c_{px} = c_{py} = 1500. \tag{13}$$

We also used two input coefficients equal 0.001 and two output coefficients equal -1000 respectively. Minus arises from internal connections in real manipulator. For classical case $c_P\bar{x}$ equals e and $c_D = c_2$.

In this case besides above given coefficients are also important input and output coefficients for real measurement system. They have influence for results obtained from experiment. In particularly if they have large values errors are smaller but if they are too large for measurement system the manipulator cannot work properly or work at all.

The experimental results obtained with dSPACE-tools are presented in Figure 2 from a to f. In Fig.2a the desired trajectories p_{dx}, p_{dy} in Cartesian space are shown. Figure 2b shows orientation error. It is about $0.04 - 0.06[rad]$. For classical case this error is smaller. Figures 2c and 2d show position errors in directions x and y. They are about $0.5 - 1[cm]$ and $4 - 6[cm]$ respectively. For quasi-variables errors are smaller than for classical case. In Figure 2e we can see moment $tau1$ in second joint. The moments are comparable in all cases but its value is higher than in simulation. It causes from that we cannot increase c_1 and c_2 to high value because KARI-2 does not work properly in such situation. In Fig.2f a comparison between real moment and quasi--moments (normalized and unnormalized) are shown. In normalized case $eps1$ has higher value than real moment and in unnormalized case $kpp1$ has similar value as $tau1$.

6 Concluding remarks

Results of research obtained in simulation and experimental way (with using SIMULINK and dSpace) indicate that we can realise effective (precise) control in quasi-velocity space. We examined that PD controllers in terms of quasi-velocities in Cartesian space give comparable results. These results are similar as for classical PD controller and similar as in literature Niemeyer G., Slotine J.-J. (1991). In literature PD controller for fast sinusoidal trajectory (faster as here and for other type of manipulator) has position errors in directions $x - y$ about $6 - 7[cm]$. In simulation moments have smaller values than in experiment. It results from higher values of coefficients in simulation. Real KARI-2 cannot work properly for these values equal about 10. We must also remember that the manipulator is real but control takes place an an abstract space. From that reason, we transform physical variables into quasi-velocity space and at the end we transform quasi-moments into physical moments, which inputs of manipulator. Other kinds of controls are also possible. We have in mind unnormalized quasi-velocities control, normalized and unnormalized quasi-velocities control with quasi-accelerations, tracking control (in the same workspace). Results of further investigations (simulation and experiment) will be presented in next papers.

References

Canudas de Wit C., Siciliano B., Bastin G.(Eds). (1996) *Theory of Robot Control*. Springer-Verlag Ltd., London, 1996.

Herman P. (1997) *New Robot Control Algorithms Using Articulated Body Inertia*. Ph.D. Dissertation, Poznan University of Technology (in Polish).

Jain A., Rodriguez G., Diagonalized Lagrangian Robot Dynamics. In *IEEE Transactions on Robotics and Automation*, Vol.11, No 4. pp.571-584.

Kozłowski K., Robot Dynamics Models in Terms of Generalized and Quasi-Coordinates: a Comparison. In *Appl. Math. and Comp. Sci.*, Vol.5, No 2. pp.305-328.

Sciavicco L., Siciliano B., *Modeling and Control of Robot Manipulators*. The McGraw-Hill Companies, Inc., New York.

Niemeyer G., Slotine J.-J., Performance in Adaptive Manipulator Control. In *The Int. Journal of Robotics Research*, Vol.10, No.2. 1991 pp.149-161.

Chapter III

SYNTHESIS AND DESIGN

The Modular Design of a Long-Reach, 11-Axis Manipulator

Jorge Angeles[1], Alexei Morozov[1], Leonid Slutski[1], Oscar Navarro[2], and Laurent Jabre[3]

[1] Department of Mechanical Engineering & Centre for Intelligent Machines, McGill University, Montreal, Quebec, Canada
[2] Integración de Procesos Industriales, León, Gto., Mexico
[3] CAE Electronics Ltd., St. Laurent, Quebec, Canada

Abstract. Introduced in this paper is a modular approach to the design of an 11-axis manipulator consisting of a four-axis submanipulator for the gross posing of the base of a second, seven-axis submanipulator. The base undergoes motions that keep the first axis of the second submanipulator vertical. The two modules making up the seven-axis submanipulator are intended, respectively, for the positioning of the center of its spherical wrist and for the orientation of the wrist. Positioning is accomplished with a four-axis redundant submanipulator of serial, isotropic architecture, while orientation by means of a parallel, isotropic architecture. The design philosophy and methodology adopted at the outset are outlined here.

1 Introduction

The need to service and maintain aircraft and other large structures like bridges and buildings has called for a new generation of manipulators that are characterized by a long reach and a highly redundant kinematic architecture (Hiller, 1996). The challenge to robot designers here is to produce a mechanical system capable of accurate tasks in the presence of a flexible structure. Current designs of such robots exist, but they are limited to tasks that are error-tolerant, e.g., cleaning (Schraft and Wanner, 1993). In this task, the end-effector is supplied with a highly compliant tool, namely, a cylindrical brush rotating about its axis, that does the cleaning, the inherent compliance being thus capable of compensating positioning errors. Other tasks required in the servicing and maintenance of aircraft are more demanding in terms of accuracy. For example, aircraft stripping requires more accuracy in the execution of the task, in that the tool is rigid and sharp, positioning errors thus becoming dangerous, for they can lead to damage of the fuselage.

We introduce here a modular approach to the design of the mechanical structure of an 11-axis robot to accomplish accurate positioning and velocity-controlled tasks in the presence of a flexible substructure. The manipulator is designed as a cascade of three modules, the proximal one being termed the gross manipulator, and comprising four revolute axes aimed at realizing four-dof positioning tasks of the type proper of what is called SCARA—Selective-Compliance Assembly Robot Arm—system. The gross manipulator is responsible for a long reach and a high flexibility. The two other modules, comprising a seven-axis smaller manipulator, are responsible for the accurate positioning of the end-effector tool. Of these, the intermediate module is a four-dof architecture responsible for the positioning of a point of its terminal link, which plays the role of the center of the three-dof spherical wrist. Both the intermediate module and the spherical wrist are designed with an isotropic architecture for highest positioning accuracy.

2 Design Conditions

The large, four-axis manipulator will be responsible for the gross posing—positioning and orientation—of the *base* of the small manipulator, and hence, need not be highly accurate, for which reason this manipulator is called *gross*. The main requirements of the gross manipulator pertain, then, to its motion capabilities, its actuation, and its structural behavior. We outline these requirements in the section below.

On the contrary, the small manipulator is responsible for the fine posing of the end-effector tool, and hence, should be as accurate as possible, i.e., this part of the manipulator should have a kinetostatically- and structurally-robust architecture. These two conditions can be met with a redundant architecture, which calls for both a minimum of seven axes for a general posing task and kinetostatic isotropy (Angeles, 1992). To meet the kinetostatic condition, a redundant architecture should allow one to choose a kinetostatically-optimum posture while performing a given task. The structural condition is met by the nature of the link lengths, which are substantially shorter than those of the gross manipulator. Besides, of course, special attention should be given to the mechanical design of the various links in order to enhance their stiffness. An outline of the design of the two modules of the small manipulator is given in Section 4.

3 The Gross Manipulator

This submanipulator should be capable of giving a plane, the base of the small submanipulator, three independent translations and one rotation about the normal to the plane; this normal is to be maintained in a vertical orientation. The type of motion to be generated is characteristic of those produced by SCARA systems (Furuya and Makino; Makino and Furuya; Makino and Furuya, 1980; 1981; 1982). As Hervé (1978) showed, this type of motion constitutes a *group* in the algebraic sense and can be produced with a combination of Π joints (Hervé and Sparacino, 1992). A Π joint is one that produces a relative translation of the two coupled links along a circular trajectory, and hence, is produced by a parallelogram mechanism, whereby the two coupled links are the base and the coupler link of the underlying four-bar linkage. We use a concatenation of two such joints to produce two independent translations in a vertical plane. The vertical plane is rotated about a vertical axis, in turn, by means of a revolute joint with vertical axis fixed to the base of the gross manipulator. The arbitrary orientation of the end-link of the gross manipulator in a horizontal plane is attained, finally, by means of a second vertical revolute joint. We thus obtain the architecture of Fig. 1, the rendering obtained by means of *Pro/ENGINEER*.

3.1 Kinematics of the Gross Manipulator

To control the motion of the gross manipulator at the joint-rate level, we need its Jacobian matrix relating the rates of the four joints to the twist of the end-link. The joint rates of the first and third joints are straightforward, for these joints are revolutes, their joint variables being labelled θ_1 and θ_3 (Fig. 2), respectively. If we let $c_i \equiv \cos\theta_i$ and $c_{ij} \equiv \cos(\theta_i + \theta_j)$, with similar definitions for s_i and s_{ij}, then the projections λ_1 and λ_2 in Fig. 2 are defined as:

$$\lambda_1 = l_1\sqrt{1 - s_1^2 c_2^2}, \quad \lambda_2 = l_2\sqrt{1 - s_{13}^2 c_4^2},$$

The Modular Design of a Long-Reach, 11-Axis Manipulator

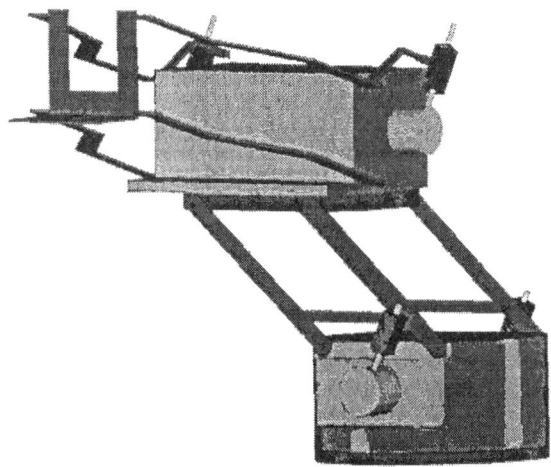

Figure 1. A rendering of the gross manipulator

where l_1 and l_2 are the lengths of the long links of the parallelogram mechanism. The intermediate joint rates, or joint variables, for that matter, i.e., those associated with the two Π joints, are less obvious, for they pertain to translations, but are produced by link rotations. We shall designate the rotation of two of the parallel links of the parallelogram mechanism of this joint as the joint variable, thereby ending up with the joint variables θ_2 and θ_4.

Moreover, the twist $\mathbf{t}_M$ of the end-link has four components, namely, the three components of the translational velocity vector $\dot{\mathbf{p}}$ of the *operation point* P of the end-link, and its scalar angular velocity $\dot{\phi}$:

$$\mathbf{t}_M \equiv \begin{bmatrix} \dot{\phi} \\ \dot{\mathbf{p}} \end{bmatrix}. \tag{1}$$

The 4×4 Jacobian matrix $\mathbf{J}_M$ of the gross manipulator thus takes the form

$$\mathbf{J}_M = \begin{bmatrix} 1 & 0 & 1 & 0 \\ \mathbf{e}_1 \times \mathbf{p}_1 & \mathbf{e}_2 \times \mathbf{r}_2 & \mathbf{e}_3 \times \mathbf{p}_3 & \mathbf{e}_4 \times \mathbf{r}_4 \end{bmatrix}, \tag{2}$$

with the definitions below: $\mathbf{e}_i$, for $i = 1, \ldots, 4$, denote the unit vectors parallel to the four axes. Notice that $\mathbf{e}_2$ and $\mathbf{e}_4$ are unit vectors lying in a horizontal plane and normal to the vertical planes of the first and the second parallelograms, respectively. We thus have

$$\mathbf{e}_1 = \mathbf{e}_3 = \begin{bmatrix} 0 \\ 0 \\ 1 \end{bmatrix}, \quad \mathbf{e}_2 = \begin{bmatrix} s_1 \\ -c_1 \\ 0 \end{bmatrix}, \quad \mathbf{e}_4 = \begin{bmatrix} s_{13} \\ -c_{13} \\ 0 \end{bmatrix}. \tag{3}$$

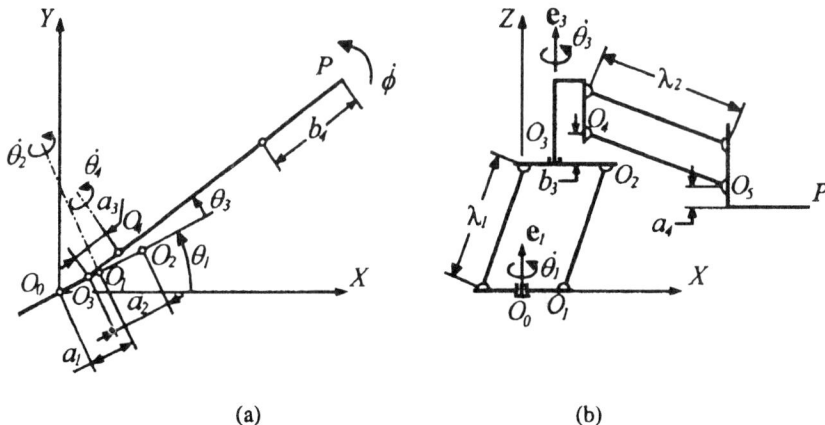

Figure 2. The kinematic chain of the gross manipulator: a) a view in the horizontal plane XOY; b) a view in the vertical plane XOZ

Further, $\mathbf{p}_i (i = 1, 3)$, denoting, respectively, the vectors directed from O_0 and O_3 to P, are given by

$$\mathbf{p}_1 = \begin{bmatrix} (a_1 + l_1 c_2 - a_2)c_1 + (a_3 + l_2 s_4 + b_4)c_{13} \\ (a_1 + l_1 c_2 - a_2)s_1 + (a_3 + l_2 s_4 + b_4)s_{13} \\ l_1 s_2 + b_3 - l_2 c_4 - a_4 \end{bmatrix}, \quad \mathbf{p}_3 = \begin{bmatrix} (a_3 + l_2 s_4 + b_4)c_{13} \\ (a_3 + l_2 s_4 + b_4)s_{13} \\ b_3 - l_2 c_4 - a_4 \end{bmatrix}. \quad (4)$$

Moreover, vectors $\mathbf{r}_i$ for $i = 2$ and 4, are directed from O_1 to O_2 and from O_4 to O_5, correspondingly:

$$\mathbf{r}_2 = \begin{bmatrix} -l_1 c_1 s_2 \\ l_1 s_1 s_2 \\ l_1 c_2 \end{bmatrix}, \quad \mathbf{r}_4 = \begin{bmatrix} l_2 c_{13} c_4 \\ l_2 s_{13} c_4 \\ l_2 s_4 \end{bmatrix}. \quad (5)$$

All four joints are to be actuated by means of electric motors. The actuation of the first and third joints does not pose any major challenge to the designer, for it can be implemented using a conventional speed reduction—a gear transmission or a harmonic drive. The second and fourth joints, i.e., the Π joints, are more challenging: due to gravity, these joints are subjected to much higher loads than the other two gross manipulator joints, which thus brings about an *unbalanced* gross manipulator actuation system. We decided to set up a criterion for the design of this system: *use, inasmuch as this is possible, identical motors to drive all four gross manipulator joints, while avoiding unbalanced wrenches—the concurrent action of force and moment—acting on the manipulator base that would impose large static loads on the base and hence, would affect the stability of the whole manipulator.*

The criterion led to a special design described in a subsection below. Briefly stated, the actuation of the gross manipulator is based on a cascade of two *tilt-pan* motion generators, with each providing one rotation about a horizontal axis, the tilt, and one about a vertical axis, the pan. Structural conditions call for an implementation of each Π joint by means of two identical parallelogram mechanisms, on vertical parallel planes.

3.2 The Tilt-Pan Mechanisms

We describe here the proximal tilt-pan mechanism, i.e., that closer to the base, its distal counterpart being a fairly faithful replica of the former. With the aim of generating two independent motions with substantially different load conditions—one motion takes place in a vertical plane; the other in a horizontal plane—by means of identical motors, we decided to use a differential gear train that would admit as inputs the motions of two identical DC motors, their outputs providing for a pan motion about a vertical axis and a tilt motion about a horizontal axis. Now, contrary to the pan motion, the tilt motion is generated indirectly, by means of a Whitworth mechanism that transforms the translational motion of a slider into the rotational motion of a revolute joint. Furthermore, since the slider will be driven from a turning shaft at one of the two outputs of the differential mechanism, a conversion has to take place from rotational to translational motion, which is most conveniently done by means of a ball-screw transmission. A sketch of the Whitworth mechanism is displayed in Fig. 3. Moreover, the transmission of motion from the horizontal axes of the output of the differential gear trains to the axes of the ball-screw mechanisms is done by means of two universal joints.

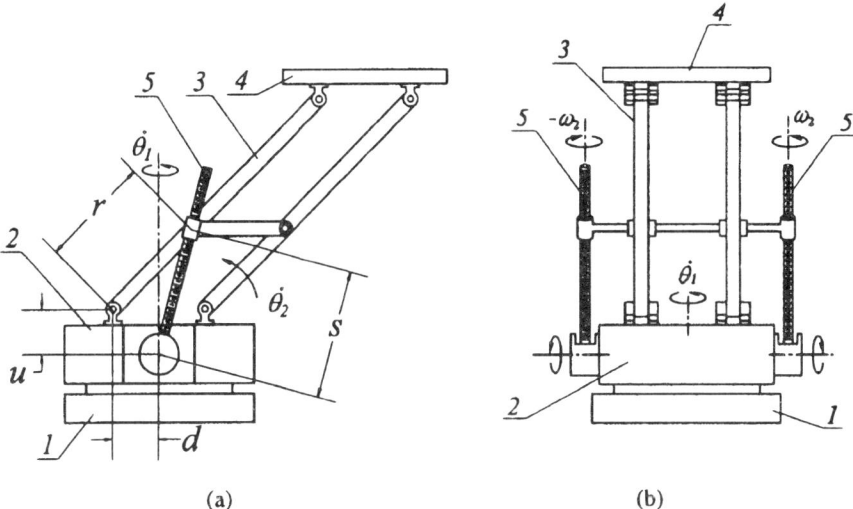

Figure 3. The Whitworth-differential mechanism: (a) a side view; (b) a front view, both showing the fixed base (1); the differential unit (2); the driven link (3); the upper platform (4); and the ball-screw shafts (5)

The conversion of the translational velocity $\dot{s}$ of the ball-screw nut along the ball-screw shaft into the angular velocity $\dot{\theta}_2$ of the driven link is derived from the relation $\theta_2 = \theta_2(s)$:

$$\frac{s^2 - u^2 - (d+r)^2}{2r} \tan^2 \frac{\theta_2}{2} - 2u \tan \frac{\theta_2}{2} + \frac{s^2 - u^2 - (d-r)^2}{2r} = 0. \tag{6}$$

Hence,

$$\dot{\theta}_2 = \theta_2'(s)\dot{s}, \quad \theta_2'(s) = \frac{1}{r}\sin\left[\arccos\left(\frac{r^2 + s^2 - u^2 - d^2}{2rs}\right)\right]. \tag{7}$$

Moreover, the velocity $\dot{s}$ is linearly related to the output velocity ω_2 of the differential mechanism via the lead λ of the ball screw:

$$\dot{s} = \lambda\omega_2, \tag{8}$$

thereby completing the instantaneous kinematic relations of the proximal tilt-pan mechanism. The corresponding relations for the distal mechanism follow exactly the same pattern, if with minor modifications.

4 The Small Manipulator

The small manipulator is a hybrid serial-parallel system, composed of a four-dof redundant arm, with a serial architecture, called the *Cuatro Arm*, and a three-dof spherical wrist, called the *Agile Wrist*. The Cuatro Arm is in charge of the positioning of a point of its end-link, where the center of the Agile Wrist will be placed. Each of these two subsystems, the Cuatro Arm and the Agile Wrist, constitutes one module of the small manipulator, as described below.

4.1 The Positioning Module

The Cuatro Arm was designed with a redundant, *isotropic* architecture (González-Palacios et al., 1993) for maximum positioning accuracy. What isotropy brings about is a robotic architecture that allows manipulator postures where the underlying Jacobian matrix $\mathbf{J}_m$ is isotropic. A matrix, in turn, is isotropic, when its singular values are nonzero and identical. Details on the Cuatro Arm are available in (Arenson et al., 1998).

The Cuatro Arm has the architecture displayed in Table 1 and illustrated in Fig. 4a, in which the Denavit-Hartenberg (DH) notation (Hartenberg and Denavit, 1964) has been followed: a_i denotes the distance between neighbouring axes; b_i–the offset between common normals to axes; α_i–the twist angles; and θ_i–the joint variables. Both Table 1 and Fig. 4a pertain to the isotropic posture.

i	a_i (mm)	b_i (mm)	α_i (deg)	θ_i (deg)
1	$444\sqrt{2}/3$	1486	-120.0	90.000
2	$444\sqrt{2}/3$	$-444\sqrt{3}/9$	60.0	109.471
3	$444\sqrt{6}/3$	$-888\sqrt{3}/9$	90.0	-125.264
4	222	$222\sqrt{3}/3$	0.0	-144.736

Table 1. DH parameters of the Cuatro Arm

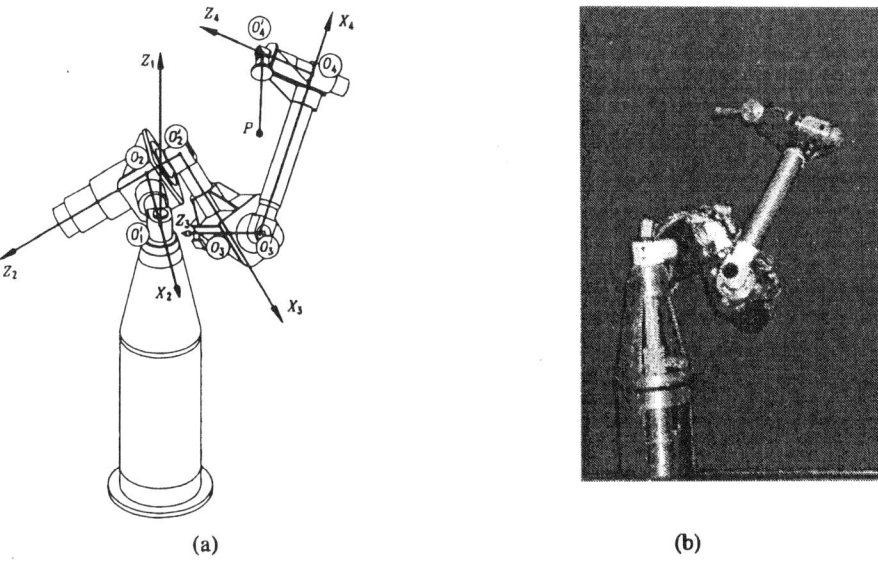

Figure 4. The Cuatro Arm: a) Denavit-Hartenberg frames, b) a photograph of the prototype

At the isotropic posture, the Cuatro Arm has the Jacobian

$$\mathbf{J}_m = \frac{\sqrt{3}}{3} \begin{bmatrix} -1 & 1 & 1 & 1 \\ 1 & -1 & 1 & 1 \\ 1 & 1 & -1 & 1 \end{bmatrix}. \qquad (9)$$

Details on how the foregoing architecture was derived from a prescribed Jacobian at the isotropic posture are to be found in (González-Palacios et al., 1993). A photograph of the Cuatro Arm is displayed in Fig. 4b.

4.2 The Orientation Module

This module was designed with the architecture of Laval University's *Agile Eye* (Gosselin and Lavoie; Gosselin and Hamel, 1993; 1994), who aimed at an isotropic, parallel manipulator. Under this architecture, a manipulator was designed for the fast orientation of a small, light camera. Our version of the Laval University manipulator, in consequence, bears the name of *Agile Wrist*.

The Agile Wrist has been undergoing design modifications in order to adapt it to shot-peening tasks. The dimensions of the shot-peening device call for in-depth design modifications to the Agile Eye design; however, the original kinematic structure, i.e., the manipulator architecture, remains unchanged. This architecture consists of two main elements, the base and the moving plates. The former is to be rigidly attached to the end-link of the Cuatro Arm, the latter is free to undergo arbitrary rotations with respect to the former about a point that remains fixed on the base plate. This point is the center of the wrist. Furthermore, the two plates are coupled by means of three identical *legs*, each constituted of two links coupled to each other by means of a revolute joint, and to each plate by one more revolute. Moreover, the axes of the three leg-revolute joints are concurrent at the center of the wrist, each axis making an angle of 90° with its neighbour. The wrist rendering is displayed in Fig. 5.

Figure 5. A rendering of the Agile Wrist

The Agile Wrist design was conducted using *Pro/ENGINEER*. The geometrical relations across mechanism parameters were established, which can be used for the dimensional synthesis of the wrist. An optimum synthesis of the driving unit was performed as well with the aim of making it as light as possible. A design task was conducted to determine the shapes of the mechanism moving links, so as to enhance the load-carrying capacity of the Agile Wrist, which, for the same volume, will carry a static load five times bigger than that of the Agile Eye.

5 Conclusions

A modular approach to the design of a long-reach manipulator for accurate positioning tasks was introduced in this paper. The approach was applied to the design of an 11-axis manipulator consisting of a "gross" and a "small" submanipulators. The gross submanipulator is capable of motions proper of SCARA architectures. The small manipulator is composed of two modules: the positioning module, the Cuatro Arm, is a serial submanipulator with four redundant axes for the positioning of a point of its terminal link; the orientation module, the Agile Wrist, is a parallel, spherical manipulator with three dof for the orientation of the end-effector tool. Both the Cuatro Arm and the Agile Wrist are endowed with an isotropic architecture for maximum posing accuracy. The manipulator outlined here is a downscaled replica of a hydraulically-actuated system to be developed for aircraft servicing and maintenance.

Acknowledgements

The work reported here is being supported by NSERC (Canada's Natural Sciences and Engineering Research Council) under Strategic Project 215729-98. The work of Messrs. Martin St.-Jean and Ary Pizarro-Chong, partly funded by the NSERC Undergraduate Summer Scholarship Programme, is dutifully acknowledged. The first author completed this work while on a Visiting Professorship at Nanyang Technological University (NTU), of Singapore. The support received from NTU, especially from the School of Mechanical and Production Engineering, is highly acknowledged.

References

Angeles, J. (1992). The design of isotropic manipulator architectures in the presence of redundancies. *The International Journal of Robotics Research* 11:196–200.

Arenson, N., Angeles, J., and Slutski, L. (1998). Redundancy-resolution algorithms for isotropic robots. In Lenarčič, J. and Husty, M.L., eds, *Advances in Robot Kinematics: Analysis and Control*. Dordrecht/Boston/London: Kluwer Academic Publishers. 425–434.

Furuya, N., and Makino, H. (1980). Research and development of selective compliance assembly robot arm. I. Characteristics of the system. *Journal of the Japan Society of Precision Engineering/Seimitsu Kogaku Kaishi* 46: 1525–1531.

González-Palacios, M. A., Angeles, J., and Ranjbaran, F. (1993). The kinematic synthesis of serial manipulators with a prescribed Jacobian. *Proceedings of the IEEE International Conference on Robotics and Automation*. Atlanta. 1:450–455.

Gosselin, C. M., and Hamel, J.-F. (1994). The agile eye: A high-performance three-degree-of-freedom camera-orienting device. *Proceedings of the IEEE International Conference on Robotics and Automation*. San Diego. 781–786.

Gosselin, C. M., and Lavoie, E. (1993). On the kinematic design of spherical three-degree-of-freedom parallel manipulators. The International Journal of Robotics Research 12:394-402.

Hartenberg, R., and Denavit, J. (1964). *Kinematic Synthesis of Linkages*. New York, Toronto: McGraw-Hill.

Hervé, J.M. (1978). Analyse structurelle des mécanismes par groupes de déplacements. *Mechanism and Machine Theory* 13:437–450.

Hervé, J. M., and Sparacino, F. (1992). STAR, a new concept in robotics. *Proceedings of the IMACS/SICE International Symposium on Robotics, Mechatronics and Manufacturing Systems'92*. Kobe. 1:176–183.

Hiller, M. (1996). Modelling, simulation and control design for large and heavy manipulators. *Robotics and Autonomous Systems* 19:167–177.

Makino, H., and Furuya, N. (1981). Motion control of a jointed arm robot utilizing a microcomputer. *Proceedings of the 11th International Symposium on Industrial Robots*. Tokyo. 405–412.

Makino, H., and Furuya, N. (1982). SCARA robot and its family. *Proceedings of the 3rd International Conference on Assembly Automation and 14th IPA Conference*. Stuttgart, West Germany. 433–444.

Schraft, R. D., and Wanner, M. C. (1993). The cleaning robot "Skywash". *Industrial Robot* 20: 21–24.

Structure Synthesis of Parallel Manipulators

Victor Glazunov, Alexander Kraynev, Gaguik Rashoyan, Anna Terekhova, and Marina Esina

Mechanical Engineering Research Institute, Moscow, Russia

Abstract. The structural synthesis of parallel manipulators according to their special areas is considered.

1 Introduction

Parallel manipulators (see, e. g., Stewart, 1965) having high positioning precision and load-carrying are classified at first by Hunt (1983) considering schemes with equal numbers of kinematic subchains and degrees of freedom of the final output link. Then this classification was extended by Glazunov et al. (1990). The classification of l-coordinate manipulators (of the type of the Stewart platform) including six subchains with two spherical and one sliding kinematic pairs was obtained by Koliskor (1982). Reciprocal screws are used by Mohamed and Duffy (1985) for kinematic analysis of parallel mechanisms with six degrees of freedom. Sugimoto (1990) considered over-constrained mechanisms using screw products of unit vectors of kinematic pairs. Then the similar approach using closed screw groups was extended for parallel manipulators (Glazunov et al., 1991). Hunt (1983) discussed special configurations of parallel manipulators. It is shown by means of screw groups that special configurations make form of continuous areas (Glazunov et al., 1990, 1991, and Kraynev and Glazunov, 1991), screw groups leading to neighboring special configurations are described (Glazunov and Rashoyan, 1990) and then this approach is extended for manipulators with parallel structure in each kinematic subchain (Glazunov et al., 1999). In this article we consider structural synthesis of parallel manipulators according to their special zones.

2 The Classification of Parallel Manipulators

When a rigid body (the final output link) can move absolutely free without any constraints then it has six degrees of freedom. If we connect this body with the base by n kinematic subchains then its degree of freedom is: $W = 6 - \Sigma D_i$ where W – the degree of freedom of the final output link, D_i – the number of constraints of i-th kinematic subchain (i = 1...n, n – the number of kinematic subchains), $D_i = -6m_i + 5p_{5i} + 4p_{4i}...$, m_i - the quantity of mobile links situated between the base and the output link, p_{5i}, p_{4i}...- the number of kinematic pairs with 1, 2... degrees of freedom of i-th subchain.

Thus we get the basic kinematic schemes of parallel manipulators when we vary the degree of freedom of the output link and the quantity of the connecting kinematic subchains. Furthermore we consider the number of mobility in each kinematic subchain and this approach we express on the Table 1 (for example the note 644(2) by W = 2 and m = 3 means that a parallel manipulator has 2 degrees of freedom, 3 connecting kinematic subchains containing correspondingly 6, 4 and 4 one-mobility kinematic pairs and father 2 different variants of disposition of actuators – either both are situated in the six-degrees-of-freedom subchain or one of them is disposed in six-degrees-of-freedom subchain and another – in any four-degrees-of-freedom kinematic subchain).

When considering only the quantity of kinematic subchains and the degree of freedom we have 57 base schemes of parallel manipulators. When considering also the numbers of actuators in each kinematic subchain we get 132 base schemes. Each scheme of parallel manipulator corresponds to different kinematic schemes (some of them can have parallel disposed links in connecting kinematic subchains). The Figure 1 presents a parallel manipulator corresponding to base scheme 666 (W = 6, n = 3). There 1 – the base, 2 – the final output link, 3 – the actuators, A_i, B_i, C_i, - (i = 1...3) - the points correspondingly of the base, of the actuators and of the final output link. One of three kinematic subchains contains 3 actuators, the second contains 2 actuators, and the third contains 1 actuator.

Table 1. The classification of parallel manipulators.

W (DOF)	n (the number of kinematic chains)				
	6	5	4	3	2
6	666666(1)	66666(1)	6666(2)	666(3)	66(3)
5	666665(1)	66665(2)	6665(4)	665(6)	65(5)
4	666655(1)	66655(2)	6655(5)	655(6)	64(4)
		66664(1)	6664(2)	664(4)	55(3)
3	666555(1)	66555(2)	6555(4)	555(3)	54(4)
		66654(1)	6654(3)	663(2)	63(3)
			6663(1)	654(6)	
2	665555(1)	65555(2)	6644(1)	662(1)	53(3)
		66554(1)	6651(1)	653(3)	62(2)
			6554(3)	644(2)	44(2)
			5555(2)	554(4)	
1	655555(1)	65554(1)	5554(2)	544(2)	43(2)
		55555(1)	6553(1)	553(2)	52(2)
			6544(1)	643(1)	61(1)
				652(1)	

3 How to Exclude Non-Controlled Finite Mobility

In order to exclude non-controlled finite mobility of the final output link we have to consider the disposition of non-actuated pairs of parallel manipulator. We use the theorem of A.P. Kotelnikov that only closed screw groups express movement groups. Correspondingly all structural formulas are connected with closed screw groups. There are eight closed screw groups: the one-member group (it could be expressed as a one-mobility kinematic pair), the two-member group (this is expressed by two sliding kinematic pair), three three-member group (they are expressed by spherical mechanisms, by spatial mechanisms with only sliding pairs, and by plane mechanisms or by mechanisms with screw kinematic pairs of the same pitch and direction), the four-member

Structure Synthesis of Parallel Manipulators

group (this is expressed by three non-parallel to one plane sliding pairs and one rotating pair perpendicular to one plane), and the six member group (this express all spatial movements). Closed groups include all screw products of the main members of these groups. The own structure formula corresponds to each such group.

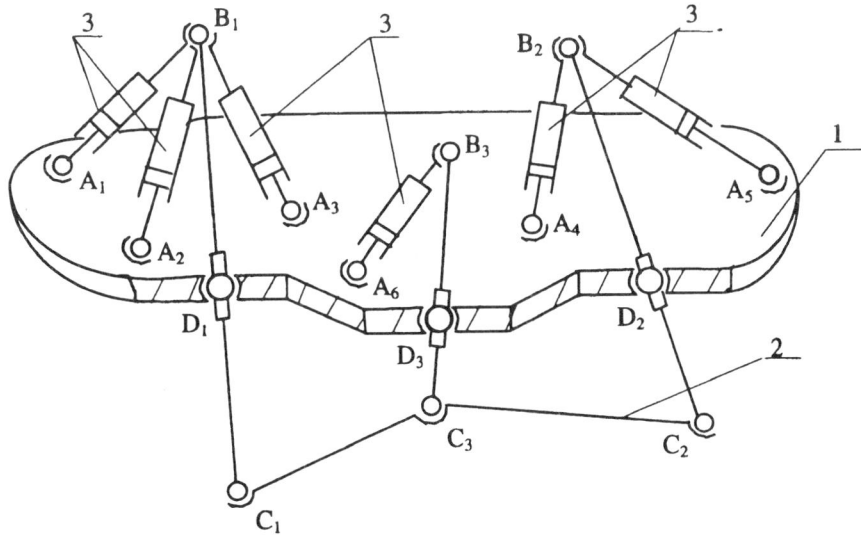

Figure 1. The parallel manipulator.

Let us consider the four-member group because it is used in mechanisms not so often as other

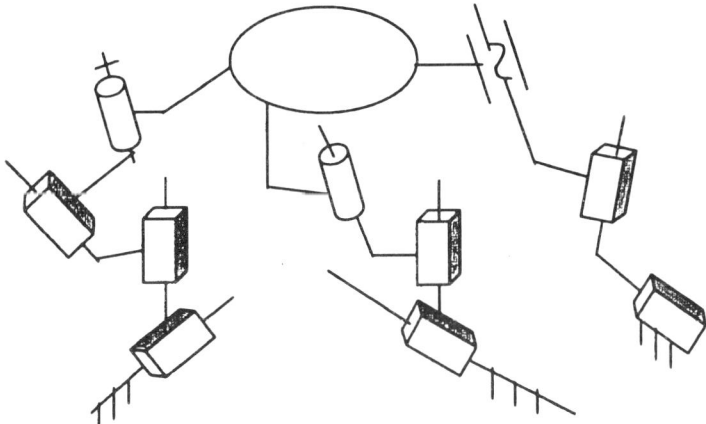

Figure 2. The parallel mechanism corresponding to four-member screw group.

groups. The Figure 2 presents a parallel mechanism with three degrees of freedom as it contains three kinematic chains connecting the base and the final output link, each chain contains one pair allowing to rotate around axe perpendicular to the same plane (rotation pair, cylinder pair and screw pair correspondingly) and three or two sliding pair. Similar mechanisms are described by structural formula: $W = 4 - D_i$ where W – the degree of freedom of the final output link, D_i – the number of constraints of i-th kinematic subchain (i = 1...n, n – the number of kinematic subchains), $D_i = -4m_i + 3p_{5i} + 2p_{4i}...$, m_i - the quantity of mobile links situated between the base and the output link, p_{5i}, p_{4i}...- the number of kinematic pairs with 1, 2... degrees of freedom of i-th chain. (These mechanisms don't correspond to the classification of the Table 1.)

Thus in order to exclude non-controlled mobility we are to dispose non-actuated kinematic pairs so that there are no twists determined by described above closed screw groups. For example the parallel manipulator presented on the Figure 3 contains two connecting kinematic subchains and

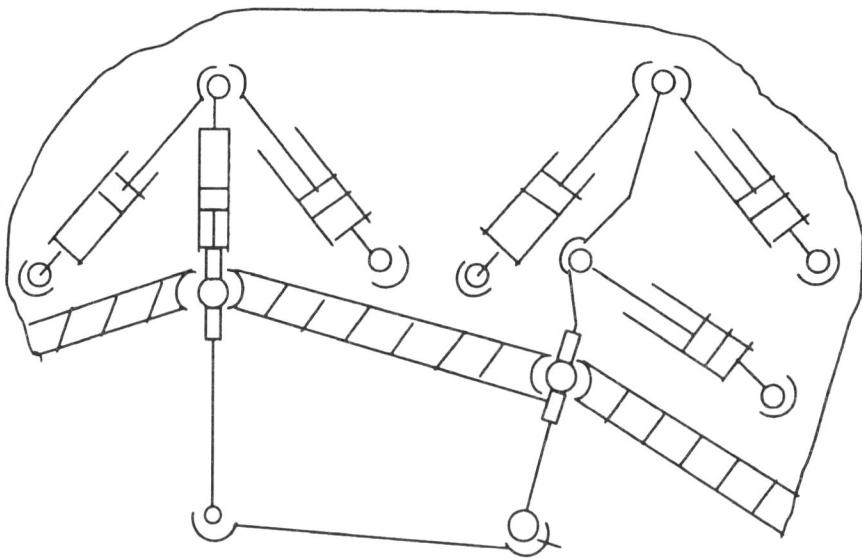

Figure 3. The parallel manipulator with two connecting kinematic subchains.

corresponds to base scheme 66 (W = 6, n = 2). Each subchain has three actuators. In order to exclude non-controlled finite mobility one of two spherical pairs situating between the output link and connecting subchains is of two degrees of freedom.

4 Special Areas of Parallel Manipulators

As parallel manipulators in special configurations have a infinitesimal non-controlled mobility we consider these configurations as a change of their structure. In general the six-member group of

Structure Synthesis of Parallel Manipulators

the unit wrenches $\mathbf{R}_i(\mathbf{r}_i, \mathbf{r}^o_i)$ (i=1,...6) acts to the output link of such manipulators. Determinant Δ composed of the screw coordinates of these wrenches is not equal zero. The wrench axis corresponding to i-th actuator is situated in the plane $(A_iB_iC_i)$, (Figure 1) passes through center joint C_j and is directed perpendicular to its virtual displacement.

At singular configurations $\Delta = 0$, we have to find the screw-gradient taking out the manipulator from special configuration most quickly and the five-member group of twists "orthogonal" to it and leading to neighboring infinitesimal close special configurations (see Kraynev and Glazunov, 1991, Glazunov and Rashoyan, 1990, Glazunov et al., 1999). For their determination it is necessary to consider an increment of the determinant Δ noted as $d\Delta$ due to six elementary displacements of the output link. The increments of the wrench screw coordinates can be represented as product of screw coordinate of the twist and some scalar factor depending on a mechanism position only. For example in case of the movement of the output link of the manipulator (Figure1) along the twist Ω^ξ $(d\xi,0,0,0,0,0)$ we have: $dr^\xi_{ix} = d\xi\, K^\xi_{ix}$, $dr^\xi_{iy} = d\xi\, K^\xi_{iy},\ldots, dr^{o\xi}_{iz} = d\xi\, K^{o\xi}_{iz}$, where (i = 1,...,6), $K^\xi_{ix},\ldots, K^{o\xi}_{iz}$ - coefficients, depending on the position of the manipulator; dr^ξ_{ix} ,..., $dr^{o\xi}_{iz}$ - increments of the wrench screw coordinates due to displacement along twist Ω^ξ. An increment of the determinant due to displacement along a twist Ω $(d\xi, d\eta, d\zeta, d\xi^o, d\eta^o, d\zeta^o)$ depends linearly on screw coordinates of this twist: $d\Delta = d\xi\, L^\xi + d\eta\, L^\eta + d\zeta\, L^\zeta + d\xi^o\, L^{\xi o} + d\eta^o\, L^{\eta o} + d\zeta^o\, L^{\zeta o}$, where the coefficients $L^\xi,\ldots, L^{\zeta o}$ are the partial derivatives $\partial\Delta/\partial\xi$, $\partial\Delta/\partial\eta$, $\partial\Delta/\partial\zeta$, $\partial\Delta/\partial\xi^o$, $\partial\Delta/\partial\eta^o$, $\partial\Delta/\partial\zeta^o$ and they express a gradient of a scalar function Δ of six arguments $\xi, \eta, \zeta, \xi^o, \eta^o, \zeta^o$.

Thus for the fastest leaving of a singular configuration it is necessary to give to the output link the final small displacement along the twist whose screw coordinates are correlated as mentioned coefficients. The five-member group of twists "orthogonal" to the twist-gradient leads to neighboring special configurations and thus we can find five-dimensional special area of parallel manipulator (see Glazunov and Rashoyan, 1990, Kraynev and Glazunov, 1991, Glazunov et al., 1999). In order to exclude non controlled mobility in configurations near to special we can find the screw-gradient and two plane perpendicular to its vector and moment correspondingly. Then we can find the components of the desired twist distorting minimally the trajectory. Also we can supply the manipulator of a supplementary actuator. For example we could dispose a rotation actuator at the point A_i or D_i.

5 Conclusion

Thus in this paper the approach to structure synthesis of parallel manipulators is represented which supposes to consider the degree of freedom of the final output link and the quantity of the kinematic subchains connecting the base and the output link. In order to exclude non-controlled mobility of the output link the theorem of A.P. Kotelnikov is used that only closed screw groups express movement groups. In order to avoid infinite small non-controlled motions in special configurations (which can make form of special zones) we have to supply the manipulator of a supplementary actuator or to plane the trajectory leading out of special configurations.

References

Stewart, D. (1965). A platform with six degrees of freedom. In *Proceedings of the Institution of Mechanical Engineers*, 371-386.

Hunt, K. (1983). Structural kinematics of in-parallel-actuated robot arms. In *Journal of Mechanisms, Transmissions, and Automation in Design*, 705-712.

Sugimoto, K. (1990). Existence criteria for overconstrained mechanisms design. In *Journal of Mechanisms, Transmissions, and Automation in Design*, 295-298.

Mohamed, M., and Duffy, J. (1985). A direct determination of the instantaneous kinematics of fully parallel robot manipulators. In *Journal of Mechanisms, Transmissions, and Automation in Design*, 226-229.

Glazunov, V., Koliskor, A., Kraynev, A., and Model, B. (1990). Classification principles and analysis methods for parallel-structure spatial mechanisms. In *Journal of Machinery Manufacture and Reliability*, 30-37.

Glazunov, V., Koliskor, A., and Kraynev, A. (1991). *Spatial Parallel Structure Mechanisms*. Moscow: Nauka.

Koliskor, A. (1982). Development and investigation of the industrial robots on the basis of l-coordinates. In *Journal Machines and Instruments*, 21-24.

Kraynev, A., and Glazunov, V. (1991). Design and analysis of spatial mechanisms with parallel structure. In *Proceedings of the Eighth World Congress on the Theory of Machines and Mechanisms*, 105-108.

Glazunov, V., and Rashoyan, G. (1990). The directions of motions of l-coordinate manipulators from special configurations. In *Journal of Machinery*, 9-12.

Glazunov, V., Kraynev, A., Rashoyan, G., and Trifonova, A. (1999). Singular zones of parallel structure mechanisms. In *Proceedings of the Tenth World Congress on the Theory of Machines and Mechanisms*, 2710-2715.

Design of Manipulators Under Dynamic and Kinematic Performances

S. Guerry and F.B Ouezdou

Laboratoire de Robotique de Paris, CNRS-UPMC-UVSQ.

Abstract. In this paper the improvement of a preliminary design process based on an iterative and interactive synthesis is proposed. The problem concerns the determination of dimentionnal parameters of a manipulator able to move a payload along a trajectory defined by a parametric curve under kinematic and dynamic performances (requested velocity and wrench at end-effector level with the take into account of bodies inertia properties). A numerical optimization process using a three steps objective function allows us to perform global and local criteria. An example with the synthesis of a 6R manipulator is given to illustrate the validity of the approach.

1 Introduction

Designing a manipulator is a process which can be splited into different stages. The first one concerns the kinematic synthesis which can be formulated as finding a manipulator able to manage a desired task while respecting a set of constraints. The kinematic synthesis, also called rigid body guidance problem of spatial six degree of freedom manipulators, does not have an exact solution for a trajectory following task, thus numerical methods are used. There are an infinite number of solutions and in order to decrease it, extraneous criteria are usually added to the problem. A first approach is to constrain the manipulator under other physical aspects, such as kinematic (Guerry et al., 1998) or static constraints (Raghavan and Roth, 1989), (Raghavan and Roth,). More general criteria such as isotropy (Angeles, 1992) or dexterity (Abdel-Malek, 2000) require minimal performances on a whole region of space instead of on a few points.

In this paper, the task consists of moving a payload over a curve at a given velocity. Thus, added criteria deal with achieving trajectory under kinematic, dynamic and static wrench performances. The next section introduces the task and details the geometric, kinematic and dynamic parts. The third section details the objective function used by the optimization function. The second step in the manipulator evaluation function uses the dynamic manipulator model to compute joint torque. This model and the assumption on the manipulator body are presented in the fourth section. The fifth section shows an example of a 6R manipulator synthesis.

2 Problem Description

The task requires the manipulator to be able to move a payload along a trajectory at a required velocity. The trajectory is defined by a parametric curve $r(p)$ of parameter p as given by equation 1:

$$r(p) = \begin{cases} x(p) = \sum_{i=0}^{deg(x)} x_i * p^i \\ y(p) = \sum_{i=0}^{deg(y)} y_i * p^i , \\ z(p) = \sum_{i=0}^{deg(z)} z_i * p^i \end{cases} \quad p_0 \leq p \leq p_1 \quad (1)$$

with $deg(P)$ the degree of polynom P and p_0 and p_1 two real numbers. The goal frame orientation is specified by a 3×3 rotation matrix $A(p) = \mathcal{F}(\vartheta(p), \varphi(p), \psi(p))$ with ϑ, φ and ψ Euler's parameters. An 4×4 homogeneous matrix, called $T_{ref}^{traj}(p)$, is build from this two components to describe the desired end-effector position and orientation at trajectory point with parameter p:

$$T_{ref}^{traj}(p) = \begin{bmatrix} A(p) & r(p) \\ 0 & 1 \end{bmatrix}, \quad p_0 \leq p \leq p_1 \quad (2)$$

The kinematic part imposes the end-effector velocity at each trajectory point. So, the linear velocity is constraint by $\mathbf{V}(p) = V(p).\mathbf{t}(p)$ where $\mathbf{t}(p)$ is a unit vector tangent to the curve at trajectory point of parameter p:

$$\mathbf{T}(p) = \begin{cases} \frac{dx(p)}{dp} \\ \frac{dy(p)}{dp} \\ \frac{dz(p)}{dp} \end{cases} \quad \mathbf{t}(p) = \frac{\mathbf{T}(p)}{\|\mathbf{T}(p)\|} \quad p_0 \leq p \leq p_1 \quad (3)$$

and $V(p)$ is the velocity magnitude required by the task. The angular velocity is build from ϑ, φ and ψ in the same way.

The last constraint required by the task is the wrench representing the payload. At each trajectory point of parameter p, the end-effector has to apply a wrench $\mathcal{W}(p)$ of general form $\mathcal{W}(p) = \{F_x(p), F_y(p), F_z(p), M_x(p), M_y(p), M_z(p)\}^T, p_0 \leq p \leq p_1$.

3 Objective Function

Manipulators are parameterized with the Khalil-Kleinfinger convention (Khalil and Kleinfiger, 1986) which is based on the usual Denavitt and Hartenberg one. Design vector x is formed by the manipulator's base position (3 parameters) and all structural parameters (3 parameters per joint):

$$x = (x_b, y_b, z_b, \alpha_0, d_0, r_0, \ldots, \alpha_i, d_i, r_i, \ldots, \alpha_{n-1}, d_{n-1}, r_{n-1}) \quad (4)$$

where n is the number of joints.

An objective function $F_O(x)$ is used to evaluate the manipulator defined by design vector x relatively to the task. An optimization process is then executed to find the design vector which minimize the objective function value.

Manipulator evaluation is splited into three consecutive steps. Evaluation can proceed to a step only if the previous ones have been already proceeded and have allowed the evaluation to go on. Each step produces a criterion value which is added to build the total objective function value.

3.1 Geometric Constraints

The first evaluation step deals with the task geometrical requirements and is splited into two part: checking *trajectory accessibility* and *configuration continuity*. Each part produce an error value which is added to create the geometrical criteria $C_g(x)$.

The first part checks if the manipulator end-effector is able to reach each trajectory point. This assumes that there is at least one solution to the inverse kinematic problem. In this case, the error value is zero and the evaluation can proceed with the second part. If the manipulator can't reach at least one trajectory point, the manipulator must be rejected with a high error value. To allow comparing manipulators, the error value given to a rejected manipulator is defined as the proportion of the trajectory with unreachable points.

The inverse kinematic problem have rarely a unique solution (for example a 6R manipulator has from 2 to 16 solutions). The second part checks if the manipulator follows the trajectory with the same configuration (i.e. if there is no discontinuity for each joint law along the trajectory). It may not be possible to find a joint law which is continuous. In this case, the manipulator must be rejected with a high error value. The error is again defined as the proportion of the trajectory which can't be reached with joint law continuity.

The geometrical criterion $C_g(x)$ is then defined by:

$$C_g(x) = \begin{cases} a_g + \frac{l_h}{l_t} & \text{if } l_h \neq 0 \\ b_g + \frac{l_t - l_c}{l_t} & \text{if } l_c \neq l_t \\ 0 & \text{else} \end{cases} \quad (5)$$

with l_h the length of the curve out of manipulator reach, l_c the length of longest curve portion which can be followed with configuration continuity and l_t the trajectory length. Coefficients a_g and b_g ensure that a manipulator which can't reach all the trajectory will evaluate with a higher value than a manipulator which reach all the trajectory but not with the configuration continuity.

If the manipulator fails one of these tests, the objective function value is set to the geometrical criterion value and the evaluation stops. In the other case ($C_g(x) = 0$), the joint law $q(p)$ is then build from the previous computation and the evaluation is allowed to proceed with the second step.

3.2 Joint Motion and Torque Laws

The first step makes sure that the manipulator fulfill geometric constraints and that a continuous joint position law $q(p)$ exists. Then, the second step can compute joint velocity $\dot{q}(p)$, acceleration $\ddot{q}(p)$ and torque $\Gamma(p)$ laws.

Computing the joint velocity law $\dot{q}(p)$ starts by computing the desired joint velocity $\dot{q}^{des}(p)$ for each trajectory point of parameter p:

$$\dot{q}^{des}(p) = J(x, q(p))^{-1} * \dot{X}_{task} \quad (6)$$

where J is the manipulator's Jacobian matrix and $\dot{X}_{task}$ is the twist specified by the task. Due to actuator limits, the velocity actually provided is $\dot{q}^{sat}$ given as the following:

$$\dot{q}^{sat}(p) = \begin{cases} \dot{q}_i^{min}(p) & \text{if } \dot{q}_i^{des}(p) < \dot{q}_i^{min} \\ \dot{q}_i^{des}(p) & \text{if } \dot{q}_i^{min} < \dot{q}_i^{des}(p) < \dot{q}_i^{max} \\ \dot{q}_i^{max}(p) & \text{if } \dot{q}_i^{max} < \dot{q}_i^{des}(p) \end{cases} \quad i = 1 \ldots n \quad (7)$$

where $\dot{q}_i^{max}$ is the maximum velocity for actuator i and $\dot{q}_i^{min}$ the maximum velocity in the other direction. The joint velocity law $\dot{q}(p)$ is build from $\dot{q}^{sat}$. Furthermore, a local kinematic error $e_k(x, p)$ is build to penalize manipulator with under-powered actuators:

$$e_k(x,p) = \frac{1}{n} \sum_{i=1}^{n} \frac{|\dot{q}_i^{des} - \dot{q}_i^{sat}|}{\dot{q}_i^M} \qquad (8)$$

with

$$\dot{q}_i^M = \max\left(\left|\dot{q}_i^{min}\right|, \left|\dot{q}_i^{max}\right|\right) \qquad (9)$$

Joint acceleration $\ddot{q}(p)$ are computed by the following equation (from Angeles, 1996):

$$\ddot{q}(p) = J(x, q(p))^{-1}\left(\ddot{X}_{task} - \dot{J}(x, q(p)).\dot{q}(p)\right) \qquad (10)$$

Joint torques $\Gamma(p)$ are computed from desired torques in the same way as joint velocity:

$$\Gamma^{des}(p) = M(q(p)).\ddot{q}(p) - S(q(p), \dot{q}(p)) - G(q(p)) - J^t(q(p)).\mathcal{W}(p) \qquad (11)$$

$$\Gamma_i^{sat}(p) = \begin{cases} \Gamma_i^{min}(p) & \text{si } \Gamma_i^{des}(p) < \Gamma_i^{min} \\ \Gamma_i^{des}(p) & \text{si } \Gamma_i^{min} < \Gamma_i^{des}(p) < \Gamma_i^{max} \\ \Gamma_i^{max} & \text{si } \Gamma_i^{des}(p) > \Gamma_i^{max} \end{cases} \quad i=1\ldots n \qquad (12)$$

$$\Gamma(p) = \Gamma^{sat}(p) \qquad (13)$$

with $M(q(p))$ the manipulator mass matrix, $S(q(p)), \dot{q}(p)$ representing Coriolis and centrifugals terms, $G(q(p))$ term issued from gravity and $J^t(q(p)).\mathcal{W}(p)$ the influence of the task wrench $\mathcal{W}(p)$ on joint torques.

A local torque error $e_t(x, p)$ is also build to penalize manipulator with under-powered actuators:

$$e_t(x,p) = \frac{1}{n} \sum_{i=1}^{n} \frac{|\Gamma_i^{des}(p) - \Gamma_i^{sat}(p)|}{\Gamma_i^M} \qquad (14)$$

$$\Gamma_i^M = \max\left(\left|\Gamma_i^{min}\right|, \left|\Gamma_i^{max}\right|\right) \qquad (15)$$

If the Jacobian matrix is singular at some trajectory point of parameter p, it is not possible to compute one of the previous quantities and the manipulator must be rejected with a high objective function value. In this case the "joint law" criterion $C_{jl}(x)$ is defined as the proportion of the curve where the Jacobian matrix is singular:

$$C_{jl}(x) = a_k + \frac{l_s}{l_t} \qquad (16)$$

where l_s is the length of the curve where the Jacobian matrix is singular. This criterion is added to the objective function value and the evaluation is stopped.

If all joint laws are computable in all trajectory point, global criteria $C_k(x)$ and $C_t(x)$ are build from local error $e_k(x, p)$ and $e_t(x, p)$ and are added to the objective function:

$$C_k(x) = \int_{p_0}^{p_1} e_k(x,p) dp \qquad C_t(x) = \int_{p_0}^{p_1} e_t(x,p) dp \qquad (17)$$

Then the evaluation is allowed to proceed with the last step.

3.3 Actuator Sizing

Previous steps have verified that the manipulator respects constraint which are necessary to manage the task. Constraints checked in this step are not imperative in the sense that if a manipulator breaks them it may still be a possible solution to the synthesis problem. The aim of these constraints is to make the manipulator which respects them better at managing the task. In this paper, joint velocity and torque laws are constrained to minimize their extremum. Manipulators will then manage the task with the smallest actuators.

As the optimization technique attempts to minimize the objective function value, extremum joint values are just added to the objective function. Kinematic $C_{ks}(x)$ and dynamic $C_{ds}(x)$ sizing criteria are then defined as follow:

$$C_{ks}(x) = \frac{1}{n}\sum_{i=1}^{n}\frac{\max_{p\in[p_0;p_1]}|\dot{q}_i(p)|}{\dot{q}_i^M} \tag{18}$$

$$C_{ds}(x) = \frac{1}{n}\sum_{i=1}^{n}\frac{\max_{p\in[p_0;p_1]}|\Gamma_i(p)|}{\Gamma_i^M} \tag{19}$$

$$\tag{20}$$

Any criteria can be defined in this way, for example a power sizing criteria $C_{ps}(x)$ may be expressed as:

$$C_{ps}(x) = \frac{1}{n}\sum_{i=1}^{n}\frac{\max_{p\in[p_0;p_1]}|\dot{q}_i(p).\Gamma_i(p)|}{\dot{q}_i^M.\Gamma_i^M} \tag{21}$$

3.4 Objective Function

To sum up, the objective function $F_O(x)$ has a different definition depending on the outcome of each of the previous steps:

$$F_O(x) = \begin{cases} C_g(x) & \text{if step 1 is not validated} \\ C_g(x) + C_{jl}(x) & \text{if step 2 is not validated} \\ C_g(x) + C_k(x) + C_t(x) + C_{ks}(x) + C_{ds}(x) & \text{else} \end{cases} \tag{22}$$

4 Dynamic Model

The dynamic model of equation 11 must be established for every manipulator evaluated. The manipulator dynamic model is parameterized by th manipulator structural parameters and is created with several simplifications. The first simplication deals with manipulator body shape. The ith manipulator body is modelised by a pipe like element with inner radius R_i^1 and outer radius R_i^2. The pipe axis go through the body origin $\mathcal{O}_i$ and the origin of the next body $\mathcal{O}_{i+1}$. The discs closing the pipe contains $\mathcal{O}_i$ and $\mathcal{O}_{i+1}$.

The second simplification deals with the actuation scheme. Indeed, due to their mass, actuators position in the kinematic chain have a great influence on the manipulator dynamic behavior. But taking actuator position as design parameters leads to a complex dynamic model. In order to limit this complexity, actuator position is fixed and the ith actuator is placed at the origin $\mathcal{O}_i$ of the ith body. Furthermore, actuator mass are fixed and provided as problem data.

5 Results

This approach is illustrated with the determination of a 6R manipulator able to move a payload of 6kg over the trajectory defined by equation 23 at the velocity defined by equation 24.

$$\mathbf{r}(p) = \begin{cases} x(p) = p + 1.5 \\ y(p) = -p^2 + p + 1.5 \\ z(p) = 0.5 * p \end{cases}, \quad A(p) = \begin{bmatrix} 1 & 0 & 0 \\ 0 & 1 & 0 \\ 0 & 0 & 1 \end{bmatrix}, \quad 0.0 \le p \le 1.0 \quad (23)$$

$$V(p) = \begin{cases} 3 * \frac{p-p_0}{p_1-p_0} & \text{if } p_0 \le p \le p_0 + \frac{p_1-p_0}{3} \\ 1 & \text{if } p_0 + \frac{p_1-p_0}{3} \le p \le p_0 + 2 * \frac{p_1-p_0}{3} \\ 3 * \frac{p_1-p}{p_1-p_0} & \text{if } p_0 + 2 * \frac{p_1-p_0}{3} \le p \le p_1 \end{cases} \quad (24)$$

Bodies and actuators parameters are given by table 1 and 2 respectively. The bodies density is $7800 kg/m^3$.

Table 1. Body inner and outer radius

i	1	2	3	4	5	6
R_i^1	0.100	0.075	0.065	0.025	0.025	0.010
R_2^i	0.125	0.100	0.075	0.035	0.030	0.015

Table 2. Actuators data

i	1	2	3	4	5	6
Γ_i^{min} (mN)	-135	-230	-115	-25	-25	-25
Γ_i^{max} (mN)	135	230	115	25	25	25
mass(kg)	1	0.8	0.7	0.6	0.5	0.4

A synthesis process was performed to determine the manipulator able to manage this task. The base position and structural parameters are given on table 3. Figure 1 presents the joint position, velocity and torque laws.

6 Conclusion

This paper has presented a synthesis process to determine a manipulator able to move a payload along a trajectory at a given velocity. This problem can be formulated as an optimization problem and a three-step objective function was build. The first and second step ensure that the

Design of Manipulators under Dynamic and Kinematic Performances

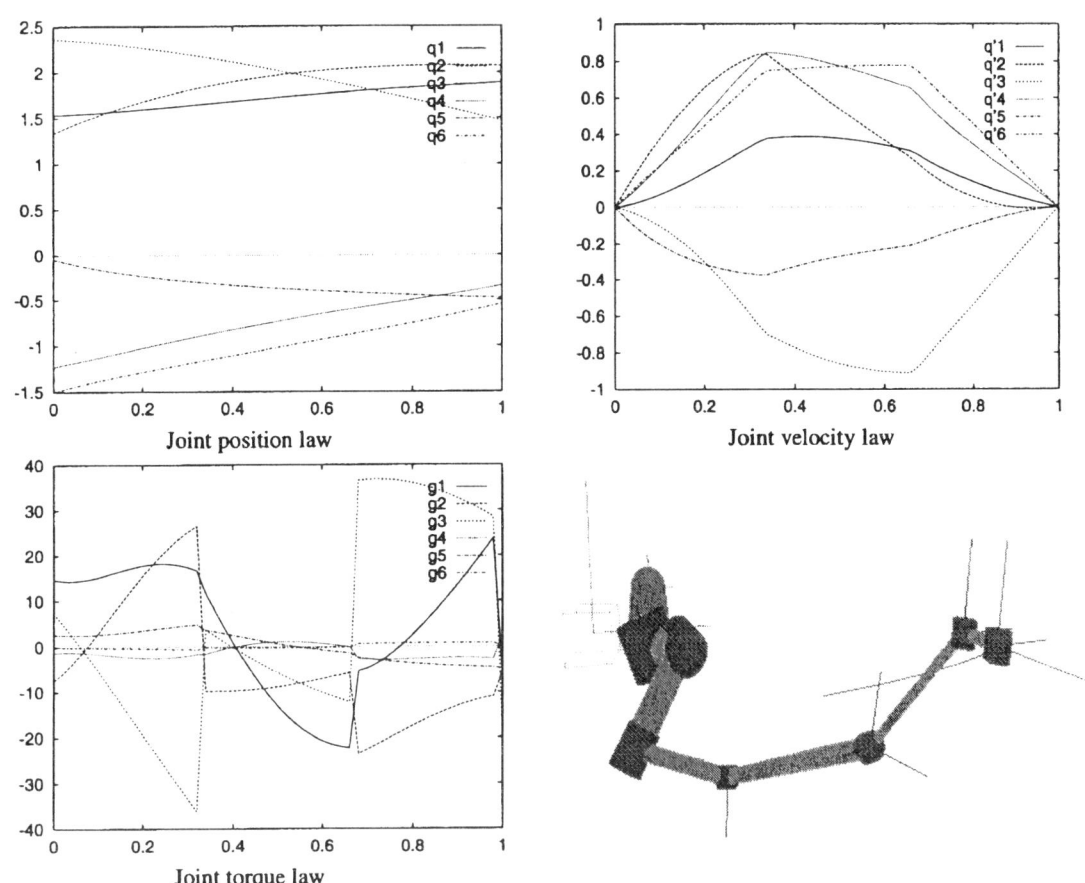

Figure 1. Joint law for 6R manipulator

Table 3. Manipulator structural parameters

Base	i	1	2	3	4	5	6
x_b 0.41	α_i	20.3	106.7	108.7	-73.8	-170.8	-73.9	0
y_b 0.48	d_i	0.62	0.14	0	0.41	0.81	0.26	0.1
z_b 0.35	r_i	-0.202	0.185	-0.465	0.296	0.067	0.355	0

manipulator is able to manage the task while the third adds constraints to find the best manipulator among all possible solutions. This allows to find the manipulator managing the task with the lowest actuators performances requirement and lead to actuator sizing.

Dynamic model used in the computation was keep simple to prevent complexity overwhelming. This mean that bodies are modelised by pipe like elements and actuators are fixed at a priori chosen position on the kinematic chain. A perspective to this work is to use an improved dynamic model to perform the dynamic simulation. It may then be possible to use bodies with more complex shape and to let actuation scheme be a design parameter. Furthermore, as these tools provide detailed data, more complex criteria, such as energy consumption or travel time, can be defined.

References

Guerry S., Régnier S., Ben Ouezdou F. (1998). Kinematic Design of Manipulator. In *Proceeedings of the twelfth CiSM-IFToMM Symposium Romansy 12*, 245–252. Paris.

Raghavan M. and Roth B. (1989). Kinematic analysis of the 6R manipulator of general geometry. In *International symposium on robotics research*, 263–269. August.

Raghavan M. and Roth B. On the Design of Manipulators for Applying Wrenches. In *Proceedings of IEEE Robotics and Automation Conference*, Vol 1, 438–443 Scottsdale, Arizona.

Angeles J. and Lopez-Cajun C. (1992) Kinematic isotropy and the conditioning index of serial robotic manipulators. In *The international journal of Robotics Research*, Vol 11, Nb 6.

Abdel-Malek K.A. and Yeh H.J. (2000). Local Dexterity Analysis for Open Kinematic Chains. *Mechanism and Machine Theory*. Vol 35. 131–154.

Khalil W. and Kleinfinger J.F. (1986). A new geometric notation for open and closed loop robots. In *IEEE Conference Robotics and Automation*, 75–79.

Angeles J. (1996). Fundamentals of Robotic Mechanical Systems. *Springer-Verlag*. September.

Influence of Leg Flexibilities on the Trajectory Planning of a 3-DOF Spherical Parallel Manipulator

Józef Knapczyk[1] and Grzegorz Tora[1]

[1] Cracow University of Technology, Poland

Abstract. The paper deals with a spherical 3-DOF parallel manipulator with three linear actuators considered as the linear springs. The spherical motion of the platform was decomposed into major displacement calculated by solving the direct position problem, and the minor displacement resulting from elastic deflections of the legs under a given load applied to the platform. The minor displacement was found by solving the force and moment equilibrium equations for the platform and for the legs with elastic deflections taken into account. The compliance matrix of the mechanism was derived. The simulation software has been implemented to investigate the effect of leg flexibilities on the platform trajectory planning and used to calculate the influence of the manipulator geometry and stiffness coefficients of the legs on the dynamic behavior. An orientation workspace was determined by using the ranges of the Euler angles. The actual workspace is restricted mainly by the limits on the leg lengths. The singular position was determined by $\det[J] = 0$. The inverse dynamics problem was solved for the actuator forces required to produce a given acceleration of the motion platform. The position, velocity and acceleration of the legs were computed by inverse kinematics. The Euler equation for the platform was used to determine the dynamic moment. Numerical example is given.

1 Introduction

The paper deals with a spherical 3-DOF parallel manipulator with three linear actuators, any of them is attached to the base with a universal joint and to the platform with a spherical joint. The fourth spherical joint between the platform and the base permits the spherical motion of the platform. The actuator drive systems contain compliant elements considered as the linear springs.

The spherical motion of the platform can be decomposed into major displacement resulting from mobility of the mechanism and the minor displacement resulting from elastic deflections of the legs under a given load applied to the platform. The major displacement was calculated by solving the direct position problem. The minor displacement was found by solving the force and moment equilibrium equations for the legs and for the platform with elastic deflections of the legs taken into account. The resultant displacement can be described as a finite spherical displacement. Additionally, the compliance of the mechanism may be described in the form of a linearized relationship involving the compliance matrix.

Compliant characteristics of parallel manipulator determine limits of its performance. The study of the compliant characteristics was subject of many papers, for example Di Gregorio (1999), Knapczyk (1998) and (1996). However, investigation into the matrix representation of the manipulator compliance is relatively limited.

2 Workspace of the motion platform

A mathematical description of the manipulator uses the following notation: A_i is i-th base point (the center of the joint connecting the leg with the base), i=1, 2, 3; B_i is i-th point of the platform, i=2, 2, 3; B_o is the center of spherical joint connected the platform with the base. The vectors listed below are described in the base coordinate system $Ox_b y_b z_b$ (Fig. 1):
$a_i = OA_i =$ const., $b_i = OB_i =$ var., $a_{ij} = a_j - a_i$, $b_{ij} = b_j - b_i$, $d_i = a_i - b_i$, $a_i° = a_i/a_i$, $b_i° = b_i/b_i$.
Let α, β, γ denote the three Euler angles of the platform, and the rotation matrix expressed by these angles is given below:

$$R = \begin{bmatrix} c\beta & s\beta s\gamma & s\beta c\gamma \\ s\alpha\beta & -s\alpha c\beta s\gamma + c\alpha c\gamma & -s\alpha c\beta c\gamma - c\alpha s\gamma \\ -c\alpha s\beta & c\alpha c\beta s\gamma + s\alpha c\gamma & c\alpha c\beta c\gamma - s\alpha s\gamma \end{bmatrix} \quad (1)$$

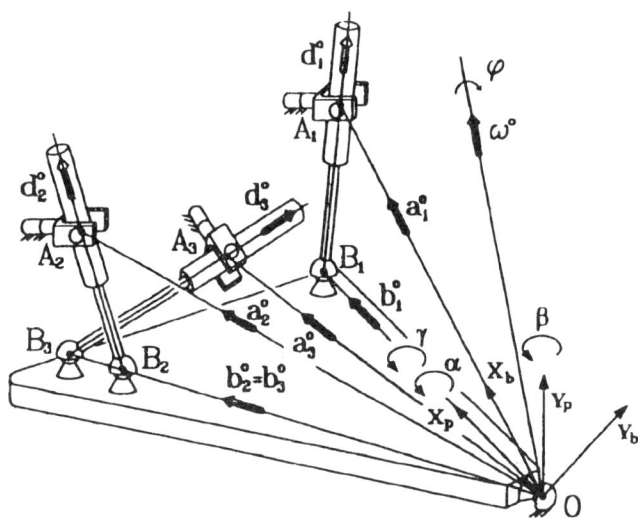

Figure 1. Kinematic scheme of a 3-DOF spherical parallel manipulator.

An orientation workspace is defined by the ranges of these angles. Note that if the three ranges are defined as $[0, 2\pi]$ then the workspace is called the maximal workspace. The actual workspace of this type manipulator is restricted mainly by the limits on the leg lengths. As the leg lengths are functions of the platform orientation, computing the workspace is a complex task. This problem has been addressed by fixing one of the orientation angles. The boundaries of the workspace can be generated geometrically. The solution technique used was as follows. Sets of values for α, β, γ were substituted into closure equation and the resulting d_i (i=1, 2, 3) were solved. The values of α, β, γ were changed incrementally in steps, and if the limits for d_i (i=1, 2, 3) are satisfied, the interior of the workspace is described by an array of these values.

In order to solve the direct velocity problem, the vector closure equation for i-th leg was differentiated with respect to time. Taking dot products of both sides of the resulting equation by d_i one can obtain vector equation which can be expressed in matrix form:

$$\mathbf{J}\,\omega = \mathbf{D}\,\dot{\mathbf{d}} \qquad (2)$$

where:

$$\mathbf{J} = [\mathbf{a}_1 \times \mathbf{b}_1 \quad \mathbf{a}_2 \times \mathbf{b}_2 \quad \mathbf{a}_3 \times \mathbf{b}_3]^T, \quad \dot{\mathbf{d}} = [\dot{d}_1 \quad \dot{d}_2 \quad \dot{d}_3]^T, \quad \mathbf{D} = \begin{bmatrix} d_1 & 0 & 0 \\ 0 & d_2 & 0 \\ 0 & 0 & d_3 \end{bmatrix}$$

Singular positions are determined by the condition $\det[J] = 0$, where either force or speed control of the actuators will become difficult. In our application, the inputs are α, β and γ. The result ca be plotted by varying α, β and keeping γ constant, as is seen by the curve in Fig. 5.

3 Elastokinematic analysis

Assuming linear elasticity of actuators, the compliance matrix is used to relate the small changes in the actuator reaction forces to the corresponding deflections by the formula:

$$\Delta \mathbf{d} = \mathbf{C}_d\, \Delta F \qquad (3)$$

where:
$$\mathbf{C}_d = \begin{bmatrix} c_1 & 0 & 0 \\ 0 & c_2 & 0 \\ 0 & 0 & c_3 \end{bmatrix},$$

c_i is compliance coefficient of i-th leg, $\Delta \mathbf{d}$ is the vector of leg deflections.

Assuming small angular displacement of the platform we can write:

$$\mathbf{J}\,\Delta\varphi = \mathbf{D}\,\Delta\mathbf{d} \qquad (4)$$

where: $\Delta\varphi$ is the vector of the platform displacement as a result of the elastic deformations of the legs represented by the vector $\Delta\mathbf{d}$.

By introducing compliance matrix of the manipulator:

$$\mathbf{C}_\varphi = \mathbf{J}^{-1} \mathbf{D}\, \mathbf{C}_d \left(\mathbf{J}^{-1} \mathbf{D}\right)^T \qquad (5)$$

we can describe the linearized relationship between the external moment acting on the platform and the platform angular displacement

$$\Delta\varphi = \mathbf{C}_\varphi\, \Delta M \qquad (6)$$

where: ΔM is the increment of the external moment.

In order to find the equilibrium position of the platform with the leg compliancies taken into account, a generalized energetic equilibrium condition is used. The orientation of the platform depends on its major displacement in the range of the workspace and minor compliant displacement caused by the load acting on the platform.

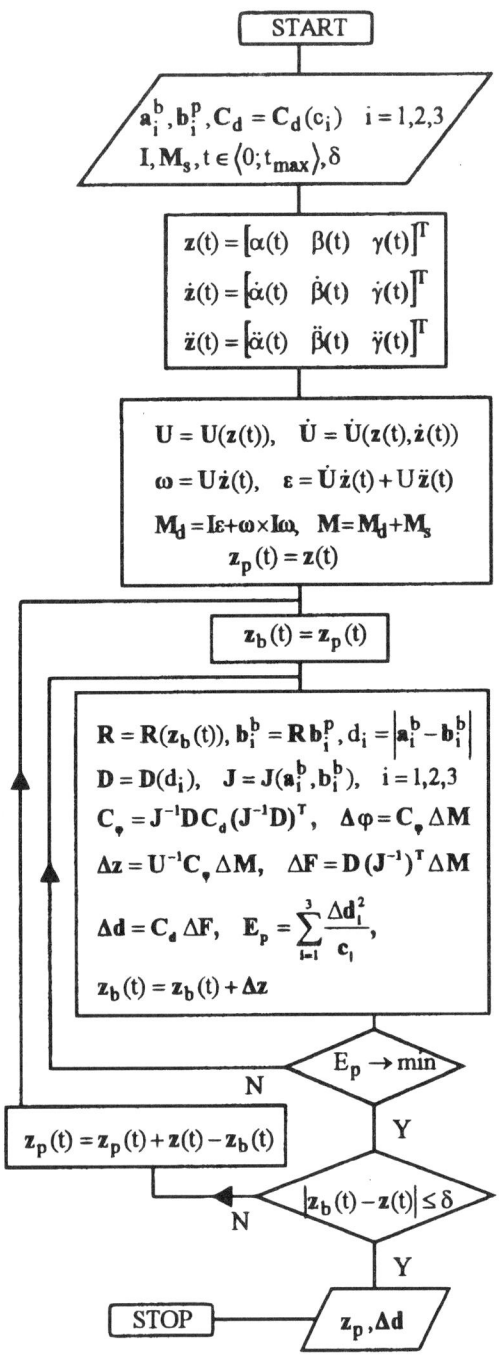

Figure 2. A general flow chart of the trajectory planning algorithm including the leg flexibilities and the platform dynamic moment

An equilibrium configuration of the manipulator corresponds to a minimum of the energy function E, representing the potential energy of the mechanism minus the work performed by external load.

One of the possible methods of solving this problem is an iterative approach, starting from a certain equilibrium position determined with the assumption of rigid legs, and then allowing for additional small deformations in the leg lengths resulting from the actual reaction forces. In the subsequent step the orientation of the platform is modified by adding the components of the computed minor displacement, are used to determine the actual loading of the legs. In each iteration step the value and derivatives of energy function are calculated, the condition of stable configuration is being verified. The equilibrium position corresponds to the local minimum of energy function computed with prescribed accuracy δ. A general flow chart of the elastokinematic analysis algorithm is presented in Fig. 2.

4 Inverse Dynamics Problem

The inverse dynamics problem is to solve for the actuator forces required to produce a given acceleration of the motion platform. The position, velocity and acceleration of the legs are computed by inverse kinematics. The Euler equation for the platform is used to determine the dynamic moment:

$$\mathbf{M}_d = \mathbf{I}\varepsilon + \omega \times \mathbf{I}\omega \tag{7}$$

where: $\omega = \mathbf{U}\dot{z}(t)$, $\varepsilon = \dot{\mathbf{U}}\dot{z}(t) + \mathbf{U}\ddot{z}(t)$

$$\mathbf{U} = \begin{bmatrix} 1 & 0 & c\beta \\ 0 & c\alpha & s\alpha s\beta \\ 0 & s\alpha & -c\alpha s\beta \end{bmatrix} \tag{8}$$

U is the transformation matrix for angular velocity of the platform from Euler coordinates into Cartesian coordinates:

$$z(t) = [\alpha(t)\ \beta(t)\ \gamma(t)]^T,\ \dot{z}(t) = [\dot{\alpha}(t)\ \dot{\beta}(t)\ \dot{\gamma}(t)]^T,\ \ddot{z}(t) = [\ddot{\alpha}(t)\ \ddot{\beta}(t)\ \ddot{\gamma}(t)]^T \tag{9}$$

where $z(t), \dot{z}(t), \ddot{z}(t)$ are the given trajectory of the platform motion.

The moment acting on the platform is the sum of static and dynamic moments

$$\mathbf{M} = \mathbf{M}_d + \mathbf{M}_s \tag{10}$$

where $\mathbf{M}_s = [M_s \sin\theta \sin\psi\ \ M_s \sin\theta\cos\psi\ \ M_s \cos\theta]^T$

$\mathbf{M}_s$ is the vector of the static external moment, θ and ψ are the orientation angles of this vector in the base system.

5 Numerical example

The input data for the considered manipulator were determined by measurement of a road-building machine. The coordinates of the joint centers are described in the base frame for the base joints, and in the platform frame for the platform joints.

$a_1^b = [2{,}47\ 0\ 0]^T$ [m], $a_2^b = [2{,}25\ 1{,}02\ 0]^T$ [m], $a_3^b = [2{,}72\ 0{,}13\ -0{,}46]^T$ [m],

$b_1^p = [2{,}40\ 0\ 0]^T$ [m], $b_2^p = [2{,}18\ 1{,}01\ 0]^T$ [m], $b_3^p = [2{,}41\ 1{,}07\ -0{,}03]^T$ [m],

$c_1 = c_2 = c_3 = 5*10^{-6}$ [m/N], $M_* = 1000$ [Nm],

$0{,}53$ [m] $< d_1 < 1{,}78$ [m], $0{,}53$ [m] $< d_2 < 1{,}78$ [m], $0{,}81$ [m] $< d_3 < 2{,}06$ [m],

$$I = \begin{bmatrix} 541 & -609 & 467 \\ -609 & 3133 & 96 \\ 467 & 96 & 3334 \end{bmatrix} \text{[kgm}^2\text{]}$$

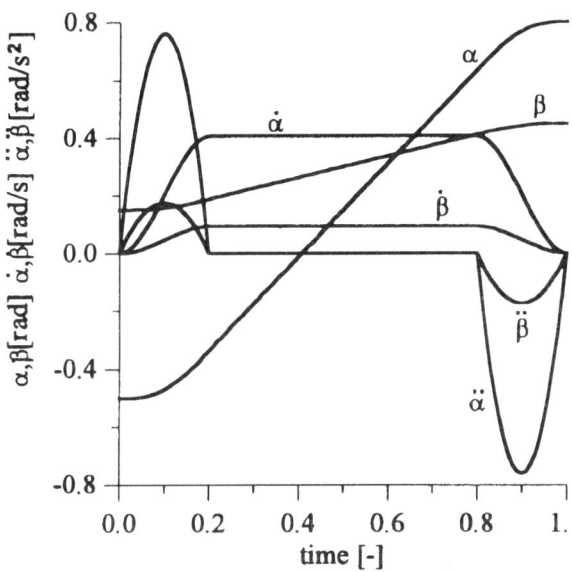

Figure 3. The generated trajectory of the platform orientation angles

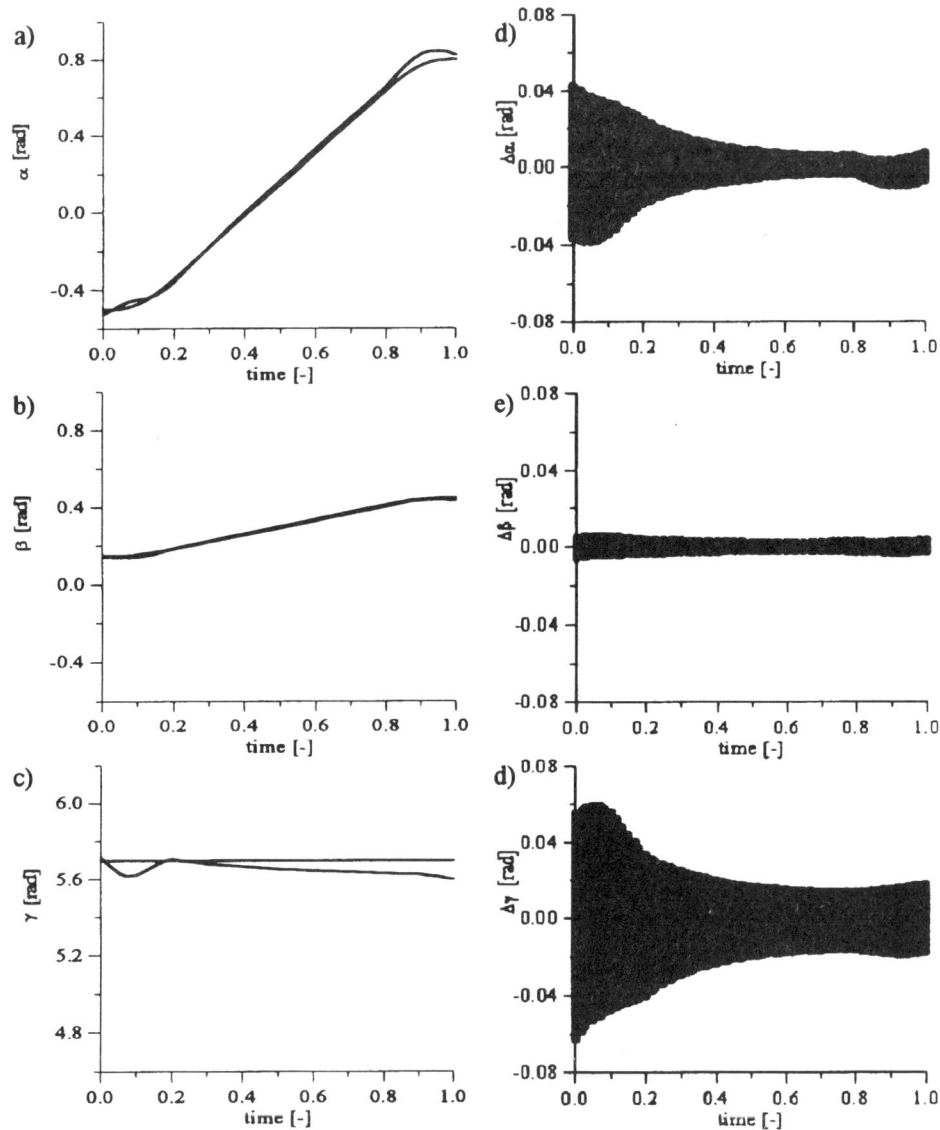

Figure 4a,b,c. The desired and actual trajectories including the elastic deflections of the leg; caused by the given moment (with specified orientation) acting on the platform;.

Figure 4d,e,f. The differences between the desired and actual trajectories calculated for the considered manipulator including the leg deflections under the moment with the orientation angles taken from the range $[0, 2\pi]$.

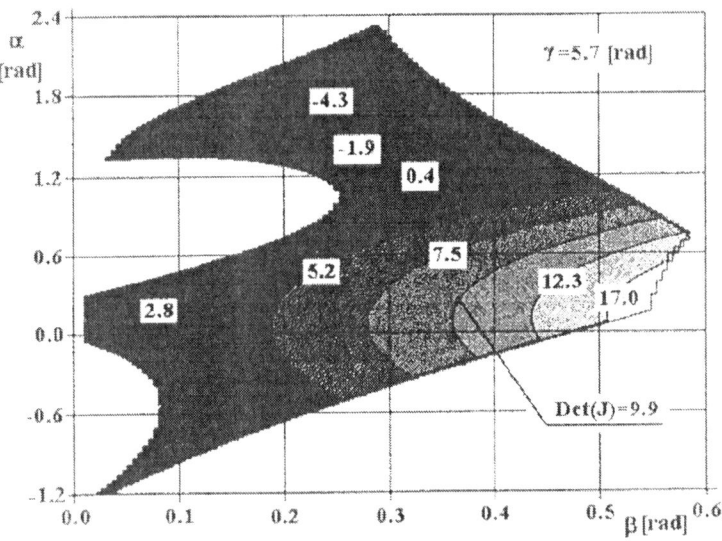

Figure 5. The crossection of the manipulator workspace with γ = 5,7 [rad] and the curves of determined values of Det (J).

Conclusion

The simulation scheme shown in Fig.2 has been implemented to investigate the effect of leg flexibilities on the platform trajectory planning. The respective computer algorithm may be also used to calculate the influence of the manipulator geometry and stiffness coefficients on the dynamic behavior. This model can be used to manipulator calibration.

References

Di Gregorio, R., and Parenti-Castelli, V. (1999). Influence of the Flexibility on the Kinetostatic Behavior of a 3-DOF Fully-Parallel Manipulator. *Proc. of the Tenth World Congress on the Theory of Machines and Mechanisms, Oulu.*

Knapczyk, J., and Dzierzek S., (1998). Elastokinematic Analysis of the 5-6 Parallel Mechanism with Translational Springs Supporting the Platform. *Proc. of RoManSy'98, Paris.*

Knapczyk, J., and Tora, G.(1996). An Inverse Force Analysis of the Spherical 3-DOF Parallel Manipulator with Three Linear Actuators Considered as Spring System. *Proc. of RoManSy'96, Udine.*

Yang, P., Waldron, K.J., Dutt, V., and Orin, D.E. (1997). Design of a Three Degree of Freedom Motion Platform for a Low-Cost Driving Simulator. *Fifth Applied Mechanisms and Robotics Conf. Cincinnati.*

STUDY ON THE SPECIFIC CHARACTERISTICS OF VARIOUS ACTUATORS

Teru Hayashi

Faculty of Engineering, Toin University of Yokohama, Japan

Abstract. The motive of the work for author was to search for the suitable actuators of the micromechanisms. Many kinds of actuators are known and the characteristics of each actuator had been discussed separately. However, their comparisons between deferent kind actuators have never been discussed. In this paper, for discussing the characteristics of any actuators on the same base the author offered three items. They are 'the equivalent pressure' for comparing amount of output forces, 'the acceleration factor' for comparing acceleration ability and 'the common diagrams of the two types of actuators' for comparing response characteristics.

1 Introduction

They are developing many kinds of actuators as not only the electromagnetic motor, ultrasonic motor, SMA actuator, etc. for various kinds of uses. Also they found out new principle actuators in living things. For use of the actuator, new domain as of the micromechanisms appeared. Then, there occurred the demand for establishing the estimate method of suitable actuator to choose for the use. For meeting the demand, the author offered three items. They are 'the equivalent pressure' for estimating the output force, 'the acceleration factor' for estimating the ability of acceleration and 'two types of actuator' for estimating the ability of control for any kinds of actuators.

2 Classification of actuators

The author surveyed presently known actuators and their media to generate their output forces or displacements. The result is shown in Table 1. There the bio-actuator is considered as a kind of electrostatic actuator.

3 The equivalent pressure of the actuator

The equivalent pressure p [Pa] is defined as the value that the output force F [N] divided by the area A [m^2] where the actuation force works. The value of p should be dependent to the kinds of the media of each actuator. Table 2 shows the ideal and actual values of equivalent pressures about media of some actuators. There, the contact parts of the gears and the friction drives are recognized as the kind of media because they are used the last output positions of actuators and take considerable spaces.

Table 1. Classification of actuators and their media.

Name of actuator	Media	Energy
1. Flux, Solenoid, Motor	Magne. flux	Electromagnetic-E.
2. Magnetostrictive actuator	Metal	"
3. Hydromagnetic actuator	Hydraulic	"
4. Electrostatic actuator	Electric flux	Electrostatic-E.
5. Electrostrictive-, Piezoelectric-actuator	Ceramic	"
6. Motor, Rocking actuator, Cylinder	Oil	Hydraulic pressure
7. Motor, Turbine, Roking act., Cylinder, Balloon	Air	Pneumatic pressure
8. Expansion actuator, SMA actuator	Metal	Thermal-E.
9. Mechanochemical actuator	Gel	Chemical-E.
10. Bio-actuator	Electric flux	"
11. Gear drive	Metal	Mechanical-E.
12. Friction drive	"	"

Table 2. Equivalent pressures p on some media.

	Medium	Equivalent Pres. [MPa]		References
		Ideal	Actual	
1.	Electromagnetic	0.4 - 1.0 [2]	0.005-0.02 [1]	B_r=1.6 [T]
2.	Electrostatic	-10 [2]	-0.001 [1]	300 [V], 0.25 [µm]
3.	Hydraulic pres.	5 - 50	0.7 - 5 [1]	
4.	Pneumatic pres.	-5	- 1 [1]	
5.	Ceramic	-10 (50)		
6.	Metal (SME)	-50 [2]		
7.	" (Gears)	-2 [2]	0.005-0.1 [2]	z =50
8.	" (Friction D.)	-20 [2]	-	σ_c =100 [MPa], µ =0.2
9.	Gel		- 0.4 [3]	
10.	Biomat.(Muscle)		-0.006 [4]	♦5 [nm], ^{l}12.5 [µm]

These values were obtained by estimating from the following materials as the followings.
(1) Catalogs of some companies, (2) Chronological Scientific Table (1993), (3) Hayashi T. et al (1993),
(4) Ishijima A. et al. (1990).

The fact that the ideal values are much larger than the actual values in the equivalent pressure of some actuators seems to give us interesting subjects. Comparing the output forces of various kinds of actuators, it is seen that oil pressure-, air pressure-, solid-actuators are advantageous and the friction drive is the more advantageous than gears.

4 The size effect of actuators

As each force has the size effect that is peculiar to generating phenomenon [1] of the force, the equivalent pressures of some medium have size effect. The output force of the electrostatic have the size effect of $[L^0]$, then the size effect of the equivalent pressure that is the value of force/area will become $[L^{-2}]$. Almost kinds of force have the size dimension of $[L^2]$, so the equivalent pressure of their media will have the dimension of $[L^0]$, which means the pressure is to be constants. Only about the electromagnetic medium in minute aria, permissible current must have the dimension of $[L^{1.5}]$ which is decided by the heat balance, and the electromagnetic force must have the dimension of $[L^3]$, then it's equivalent pressure must have the dimension of $[L^1]$. The size effects of equivalent pressure on some kinds of media of actuators are shown in Figure 1.

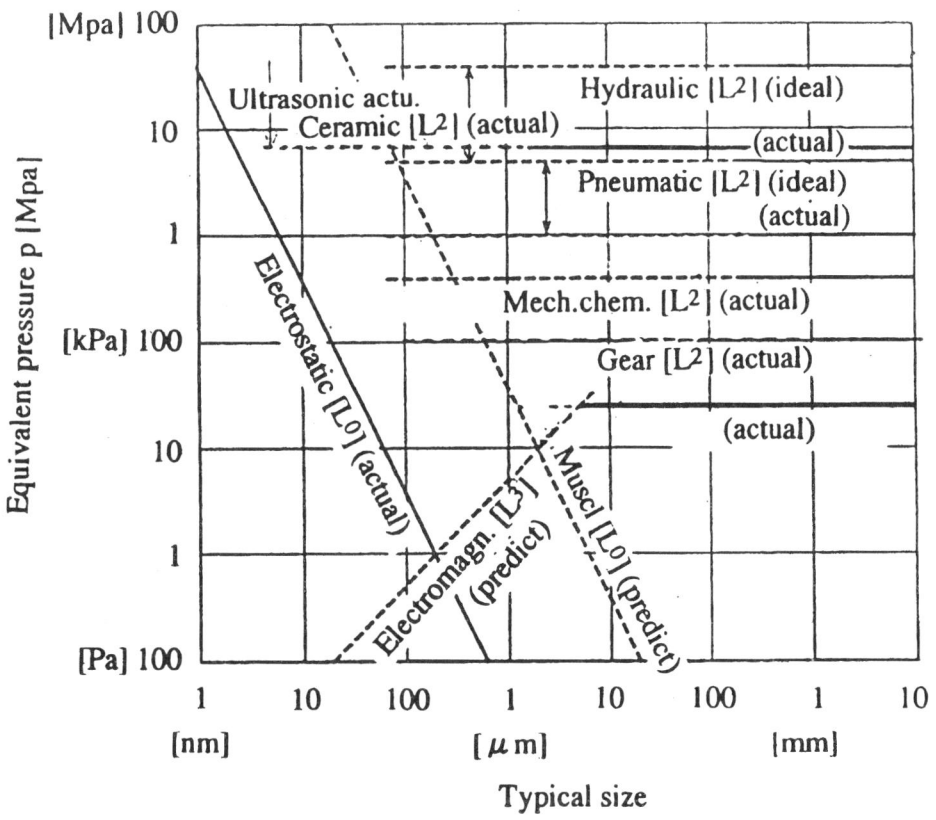

Figure 1. Size effects of equivalent pressure.

Judging from the figure, the domain smaller than 1[μm] the electrostatic force is seemed advantageous to electromagnetic force. Also, it is surmised that the oil pressure, air pressure, solid actuator and gears are effectively used to the domain of 1[μm].

Using the values of the equivalent pressure in Table 2 or in Figure 1 the output forces of actuators with any shapes and types will be estimated.

5 The acceleration factor

When the load with the mass of m_L is driven by the actuator with the mass of m_M and output force of F in impedance matched condition, the load is to be given the acceleration a as following amount.

$$a = f/2\sqrt{m_M \cdot m_L} = A/2\sqrt{m_L}, \text{ and } A = F/\sqrt{m_M}$$

There, A is defined as the accelerating factor of the actuator and it expresses the performance of the acceleration (A^2 is power rate). The result of the investigation about the relation between the equivalent pressures p and the accelerating factors A on some examples of actually used actuators are shown in Figure 2.

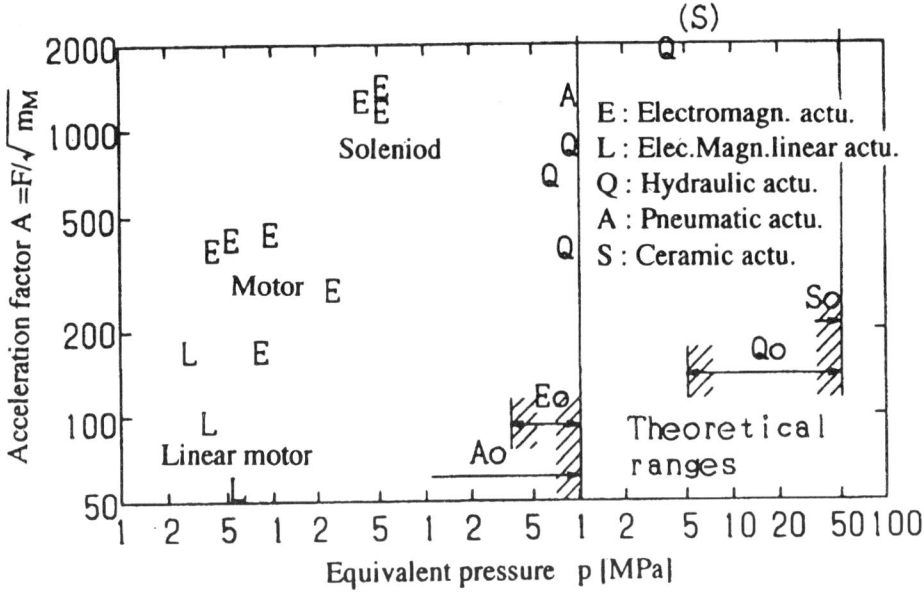

Figure 2. Acceleration factors on various actuators.

On the upper line of the figure the ideal values of equivalent pressures of the air (A)-, electromagnetic (E)-, oil (Q)-, ceramic S) - actuator are shown. It is seen that the liner motors(L) are left and under side in the figure (the position of the lowest performances) and the ultrasonic motors are right and upper side (the position of the highest performances).

6 Actuator for controlling

The characteristics of response must be discussed on the actuators for controlling. There, the actuators should be classified into two types. They are "the force type" in which the output force is proportional to the input signal and „the displacement type" in which the output displacement is proportional to the input signal.

6.1 In Case of Force Type Actuator

Figure 3 shows the control system with the Force Type actuator and the equations for the amplitude-frequency response on masses of actuator and load is connected through gears in impedance matched condition.

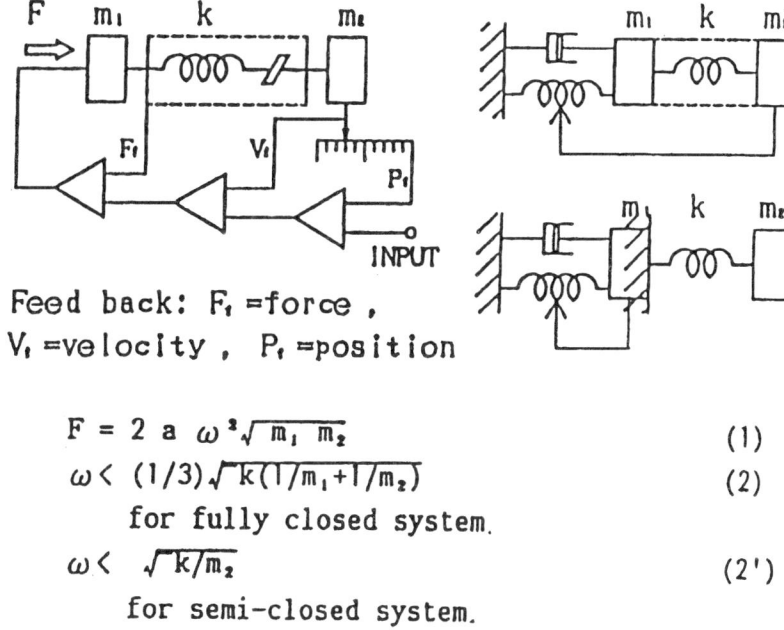

Feed back: F_t =force, V_t =velocity, P_t =position

$$F = 2a\omega^2\sqrt{m_1 m_2} \quad (1)$$

$$\omega < (1/3)\sqrt{k(1/m_1 + 1/m_2)} \quad (2)$$

for fully closed system.

$$\omega < \sqrt{k/m_2} \quad (2')$$

for semi-closed system.

Figure 3. Control system and it's equations on Force Type actuators.

Figure 4 shows the diagram of the amplitude-frequency response in general case. The characteristic line is changed along the inclined line in the diagram. The horizontal lines through the point O and O' show the limits lines in the case of perfect closed loop and semi closed loop respectively. The points O_1 and O_2 show the limit points when the output force of the actuator grows 10 times and when the geometrical size of the system is decreased 1/10 respectively.

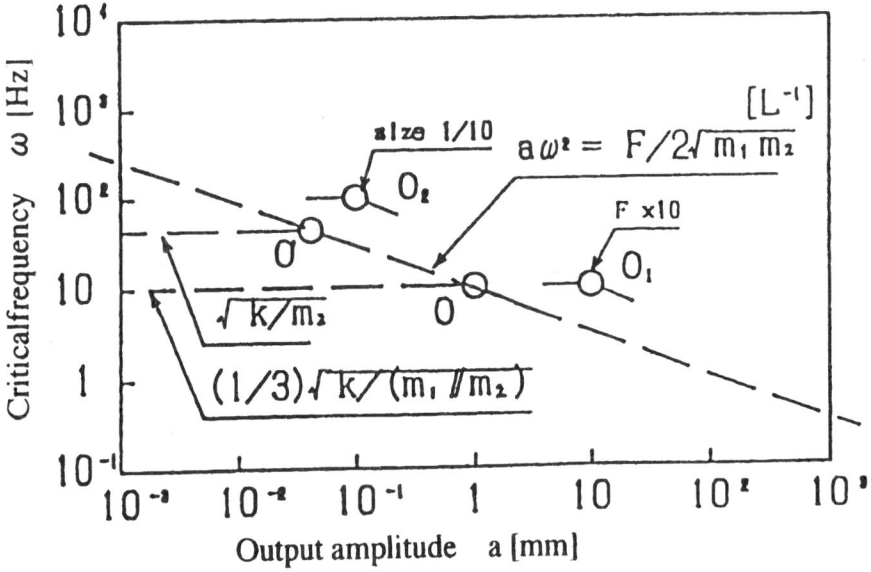

Figure 4. Characteristics of Force Type actuators.

6.2 In Case of Displacement Type Actuator

Figure 5 shows the control system with the Displacement Type actuator and the equations for the amplitude-frequency response on the system.

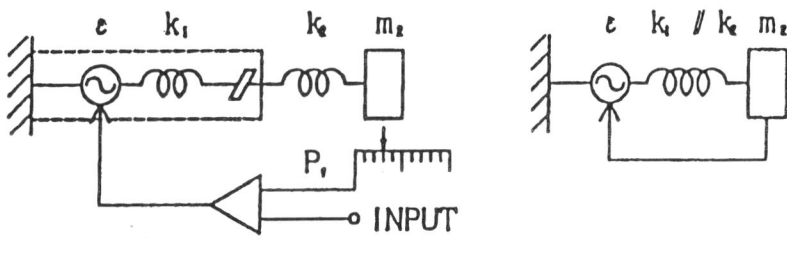

$$a = R \; l \; \varepsilon \tag{3}$$
$$\omega < \sqrt{(1/k_1 + 1/K_2)/m_2} \; (= \sqrt{E \; S/R \; l \; m_2}) \tag{4}$$

Figure 5. Control system and it's equations on Displacement Type actuators.

Figure 6 shows the diagram of the amplitude-frequency response in general case. The characteristic is changed along the inclined line in the diagram. The characteristic line is rectangular shape with the O point as an edge. The O point moves along the inclined line following to the multiplication factor of the output displacement. The points O_1, O_2 and O_3 show the limit position when the sectional area or the Yang's Modulus of the actuator increased to ten times, when the geometrical size of the system decreased to 1/10 and the amount of the output displacement increased to 10 times respectively. Using those diagrams the change of the characteristics of the actuator in any condition will be to estimate.

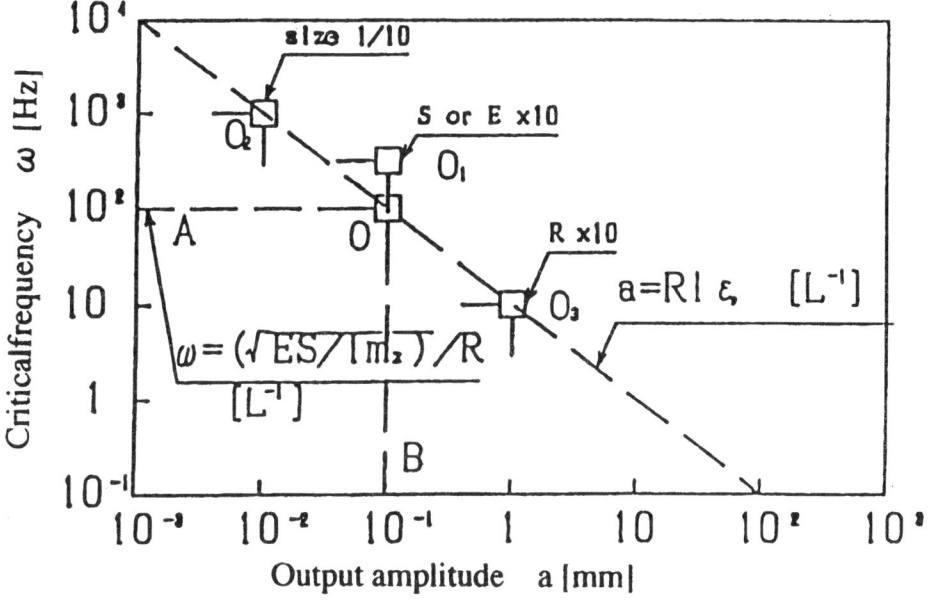

Figure 6. Characteristics of Displacement Type actuators.

6.3 The comparison of the performance on the actuator for controlling

For estimating the characteristics of the amplitude-frequency response, the diagrams on Force type- and on Displacement type-actuators can be used in pile.

Figure 7 shows the amplitude-frequency response diagram in which the points that express the characteristics on some examples. In the figure, the marks (□) show some examples of ceramic actuators and the inclined line XX shows their limit area, the marks (o) show that of electromagnetic actuators and the line YY shows their limit area and the marks (•) show that of oil pressure actuators.

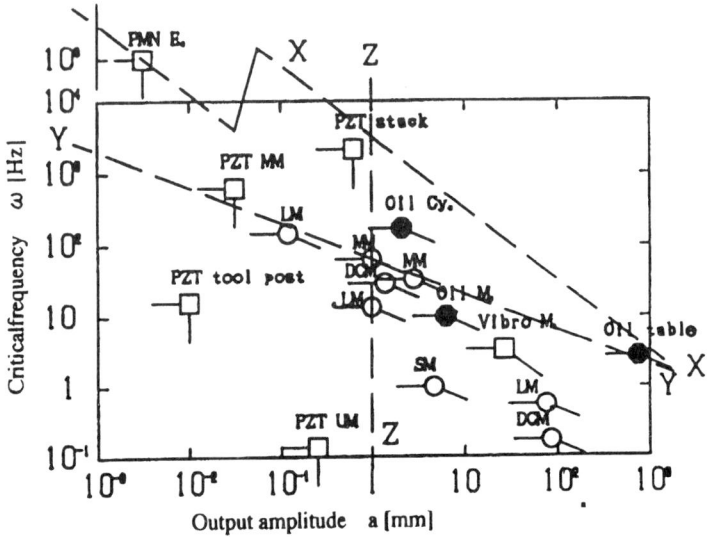

Figure 7. Characteristics on various kinds of actuators.

From the figure it is said that the oil pressure actuators are used in the domain of large force or large displacement and that the characteristics points of solid actuator are scattered that means that they are on the way of development and the high performance actuators are expected to be developed in future.

7 Conclusions

By the above discussion, defining the equivalent pressure, the acceleration factor and two types actuator, it became possible to compare on one table the characteristics of actuators with different principles, to estimate the performance and size effects of any kinds of actuators.

As the motive of the work for author was to search for the suitable actuators of the micromechanisms, the size effects of the actuators in minute area was the most important point in this work.

8 References

1. Hayashi,T. (1993). Special Feature of Micromechanism, Proceedings of 1st IFToMM International Micromechanism Symposium Tokyo, p.21.
2. Catalogs of some companies.
3. Chronological Scientific Table(1993), ed. National Astronomical Observatory, Maruzen. Co.
4. Hayashi,T., Yoshida,K., Terashima,Y., Ono,Y. (1993), High-Absorptive Electrolytic Polymer-Gel Actuator, Proceedings, IFToMM International Micromechanism Symposium.
5. Ishijima,A., Harada,Y., Yanagida,T. (1990) Measured the motion of biological molecular engine, Keisoku and Control, vol.29, No.12.

Micro-Manipulation and Adhesion Forces

D.S. Haliyo, Y. Rollot, S. Regnier and J.C. Guinot

Laboratoire de Robotique de Paris, Université Paris 6, Université Versailles Saint-Quentin-en-Yvelines,
CNRS URA 1778

Abstract. In the aim of achieving a task of manipulation of micro-devices with high resolution and accuracy, physical phenomena at the micro-scale were investigated. First studies on magnitude and appearance of forces at this scale let us believe that the use of those forces could be suitable for the fixed task. On the hypothesis that Newton's laws are applicable at the micro scale, a dynamical model of micro-manipulation, i.e. capture and release of micro-particles, was proposed, simulated and discussed 1; 2 (1; 2). Our goal is to take advantage of sticking effects instead of trying to minimize them. The manipulation will thus be achieved using adhesion forces to capture and release the object. The proposed model integrates all sticking effects taking place at this scale : Van der Waals forces, capillary forces and electrostatic forces as well as deformations at contact 3 (3). First results, obtained by simulation using silicon micro-spheres, show clearly that micro-manipulation using only adhesion forces is possible. We find a "probe initial acceleration window" allowing the manipulation depending on environmental conditions. The experimental set-up for simulation results validation is actually under construction. In this paper we present the last theoretical step before the realization of experiments as well as first experimental results. It is focussed on the determination of the dynamical behavior of a system composed of a piezoelectric actuator and a micro-probe. First studies allowed us to develop dynamical equations of the motion of the free surface of the ceramic and obtain its dimensions ensuring a minimum stress to be generated. The system proposed is then suitable for the task of capture and release of micro-objects. Some experimental adaptation will certainly be necessary to take into account oscillations of the probe for safe and accurate manipulations.

1 Introduction

The arise of new applications in micro-electronics or tele-surgery requires new know-how in accurate assembly of micro-systems. Although the manipulation in macro and atomic scales are well mastered, at the micro-scale, where both continuum and quantum physics are partially valid, a reliable manipulator is not yet realized. Such an objective requires the understanding of dominant phenomena at this scale. We focused on the study of contact (micro-tribology) and surface forces. In fact, this forces are less influenced than volumic forces by the reduction of dimensions. It is shown that sticking effects are stronger than weight for the objects smaller than $300 - 500 \, \mu m$4 (4) In some works, in order to use macro scale tools like grippers, the main concern is the elimination of those effects, for example by the reduction of contact area 5 (5), or by the use of forces of higher magnitude 6 (6). Instead of this approach, we propose a manipulation mode taking advantage of adhesion forces 1; 2 (1; 2). In this aim, a dynamical model of capture and release of micro objects is developed, taking into account Van der Waals (VdW), electrostatic and capillary forces. The chosen task consists of the capture of the upper sphere of a four-sphere pyramidal construction (fig 1). Static considerations lead to the choice of the materials and the surface geometries of the considered objects. In the aim of easing the task, the substrate is in polystyrene, spheres ("balls") in silicium, and the end-effector ("probe") in gold. Strong adhesion between the probe and the manipulated ball invokes a problem for the release. In order to weaken this force, the probe is rotated of θ along a horizontal axis. This reduces the adhesion force along the vertical axis by factor $\cos\theta$

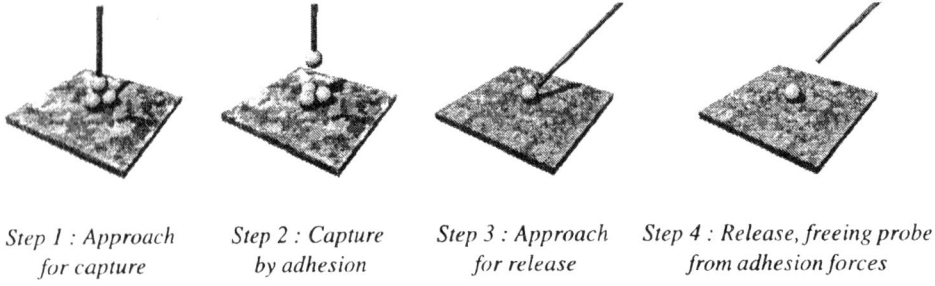

Step 1 : Approach for capture   Step 2 : Capture by adhesion   Step 3 : Approach for release   Step 4 : Release, freeing probe from adhesion forces

Figure 1. 4 steps of the proposed manipulation mode in non constraint environment

Under the hypothesis that Newton mechanics and Newton's law are applicable on the microscale 7 (7), the dynamical model is obtained by writing the dynamic equilibrium of each part of the system :

$$m_p \ddot{Y}_p = F_{ext} - F_{bp}^{VdW} - F_{bp}^{cap} - F_{bp}^{elec} - m_p g \tag{1}$$

$$m_b \ddot{Y}_{b1} = F_{bp}^{VdW} + F_{bp}^{cap} + F_{bp}^{elec} - \sqrt{3} F_{bb}^{VdW} - \sqrt{3} F_{bb}^{cap} - \sqrt{3} F_{bb}^{elec} - m_b g \tag{2}$$

$$m_b \ddot{D}_1 = \frac{\sqrt{3}}{2} F_{bb}^{VdW} + \frac{\sqrt{3}}{2} F_{bb}^{cap} + \frac{\sqrt{3}}{2} F_{bb}^{elec} - F_{bs}^{VdW} - F_{bs}^{cap} - F_{bs}^{elec} - m_b g \tag{3}$$

$$Y_p = D_1 + 2R_b + \frac{\sqrt{3}}{2}(D_2 + 2R_b) + D_3 \tag{4}$$

This model is simulated for given initial accelerations of the end effector. Results shows that the capture occurs for accelerations included in an interval, which depends on the choice of materials and environment. For example, in humid atmosphere, capillary forces and a loss of magnitude of the Hamaker constant are to be taken into account. Fig 2 shows initial acceleration windows ensuring the capture for different environments.

This graph clearly shows the necessity to be able to master important accelerations of the end-effector. We are aiming to construct a prototype manipulator. This will allow us to experiment the validity of the above described theory and lead to the elaboration of a reliable manipulator.

2 Technologic developments

The first point to take care of in the conception of a micro-manipulator is the precision in displacements and control. We use linear motors manufactured by *Newport, USA* with $50nm$ resolution on

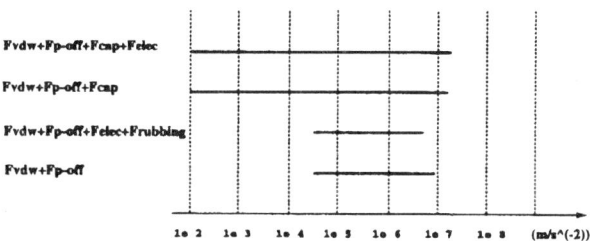

Figure 2. Acceleration windows for capture

3 axis and an additional $2nm$ resolution nano-translator (*PI, Germany*) on the vertical axis. The main concern of our work is the end-effector. According to the dimensions of the manipulated objects we will be working under optical microscope. The maximum distance between the lens of the microscope and the object is limited to $15mm$. The end effector must fit in this space. In addition, above mentioned actuators are of course not capable of creating a sufficient acceleration. A 4th actuator devoted to this usage will be added. The necessary output being the acceleration, the mass applied to it must be minimized in order to minimize the force it should develop. So it is preferable to mount the probe directly to the actuator. We will first discuss the nature of the probe, then the actuator itself.

2.1 The probe

It is the most delicate part of the manipulator. Except that it must have small dimensions, it must also have a force feedback function, to avoid damaging manipulated parts and to "feel" the contact, which is visually undetectable at this scale.

Atomic force microscopes, which are used to explore surface quality with nano-metric resolution seem to be a good solution to our problem. They consist of a silicon cantilever which's deflexion is measured by reflection of a laser beam or by piezoresistivity. The one we choose for our application (fig. 3) is a $600 \times 140 \times 10 \mu m$ monocrystalline silicon beam connected to a Wheatstone bridge. It's crystallagraphic orientation is $<100>$. The overall dimension of the device including connections is $5 \times 8 mm$ and it weights $0.1g$. The beam being manufactured by bulk micromachining process, it has a good surface quality, which allows to be in reasonably good agreement with perfect contact hypothesis between the probe and the manipulated object. Figure 3-(c) shows the tension/deflection relationship of the A.F.M. This relation is completely linear and allows very accurate measurements.

The dynamic behavior of the cantilever, given the magnitude of the accelerations that he is subject, is important to explore. In a uni-dimensional approximation its given by:

$$EI\frac{\partial^4 u_z(x,t)}{\partial x^4} + \rho S \frac{\partial^2 u_z(x,t)}{\partial t^2} - F_{ext} = 0 \qquad (5)$$

The probe is embedded to the actuator at its base section. Supposing that the actuators displacement, speed and acceleration is mastered at every instant, we obtain dynamic entries of the problem. In addition, this embedding implies the non-rotation of the base section :

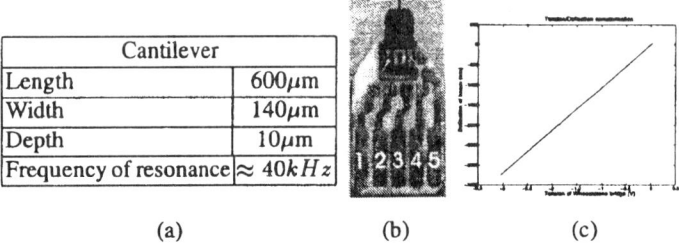

Cantilever	
Length	$600\mu m$
Width	$140\mu m$
Depth	$10\mu m$
Frequency of resonance	$\approx 40kHz$

(a)   (b)   (c)

Figure 3. Characteristics of the AMF used as end-effector (Nanosensors Gmbh)

- $u_z(0,t)$ $\dot{u}_z(0,t)$ et $\ddot{u}_z(0,t)$ are known for every instant
- $\left(\frac{\partial u_z(x,t)}{\partial x}\right)_{x=0} = 0$ $\forall t$

It is not possible to obtain an analytic solution to this equation in the light of above mentioned constraints. Also the uni-dimensional approximation is not really justified by the fact that the width is about a third of the length of the beam. We propose a finite elements approach. In order to validate the finite elements model, a frequency analysis has been conducted. The results obtained for the first two natural modes are in good agreement with experimental ones. Figure 4 shows the FE grid of the cantilever, with the results of the experimental and theoretical frequency analysis.

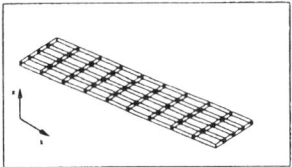

Resonance	Experimental	Finite elements
1st mode	$33,33\,kHz/45,45\,kHz$	$34,04\,kHz$
2nd mode	$196\,kHz/210\,kHz$	$215,39\,kHz$

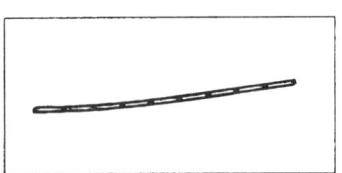
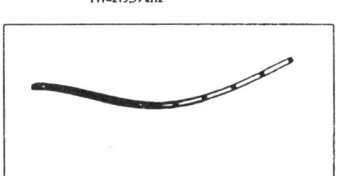

Figure 4. Results of frequency analysis - theoretical and experimental

2.2 The actuator

Let us summarize the characteristics that we except of the actuator:

- It must be able to develop an acceleration in the range $10 - 10^7 m.s^{-2}$.

- Giving the limited space, it must be of smallest possible dimensions.
- In addition to the acceleration, it must provide also a displacement of about $100nm$ in order to move away the probe out of the range of influence of the adhesion forces.

This constraints leave all classical actuator useless for our application. It is necessary to explore the possibilities of "active materials" such as memory shape alloys, or piezoelectric materials. In particular, piezoelectric materials have interesting properties:

- they have sufficient dynamical and kinematical capacities.
- an external force is not needed to return to their initial shape.
- they are used since many years and their behavior is well known.
- their dependence on environmental changes is negligible.

We will be using a piezoelectric ceramic of rectangular shape, which will be embedded to the structure of the manipulator at the upper face and to the probe at the lower face (fig. 5).

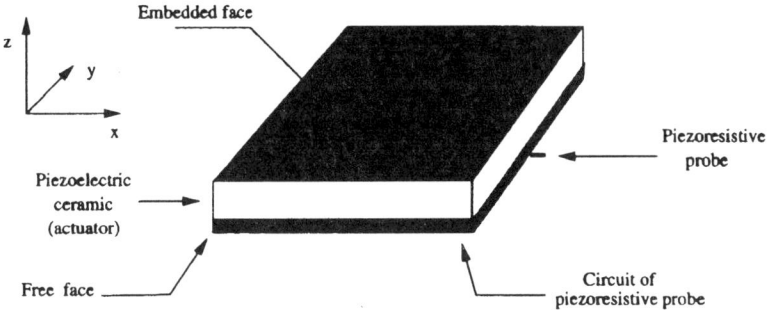

Figure 5. End-effector: Piezoelectric actuator with the probe

The direct piezoelectric equation is expressed by 8 (8):
$$T_\alpha = c_{\alpha\beta} S_\beta - e_{i\alpha} E_i \qquad (i = 1,2,3 \ \alpha, \beta = 1,2,,..6) \tag{6}$$

or
$$S_\alpha = s_{\alpha\beta} T_\beta - d_{i\alpha} E_i \qquad (i = 1,2,3 \ \alpha, \beta = 1,2,,..6) \tag{7}$$

with
$$T_\alpha = c_{\alpha\beta} S_\beta \text{ and } S_\alpha = s_{\alpha\beta} T_\beta \tag{8}$$

T,S,e and d being respectively the strain, stress, and two different expressions of piezoelectric tensors and E the electric field. c and s are the Hooke's law tensors. Neglecting horizontal deformations and under the hypothesis that the deformation of a point along the vertical axis is independent of its horizontal coordinates (x, y), static equilibrium of the lower face can be expressed by: $\quad A \cdot T_3(t)\mathbf{z} + \mathbf{F}_{ext}(t) = 0 \quad$ with A surface area of the ceramic

Writing the dynamic equilibrium of the probe, we obtain:

$$\mathbf{F}_{ext}(t) = \ddot{u}_z(t) \cdot m_p \qquad m_p \text{ and } u_z \text{ respectively mass and displacement of the probe}$$

h being the height of the ceramic, the piezoelectric equation becomes:

$$S_3(t) = s_{33}T_3(t) + d_{33}E_3(t) \quad \text{so} \quad h \cdot S_3(t) = s_{33}(h \cdot T_3(t)) + d_{33}(h \cdot E_3(t)))$$

Under the hypothesis of small deformations:

$$h \cdot S_3(t) = u_z(t) \quad \text{and} \quad (h + u_z(t)) \cdot E_3(t) = v(t)$$

where $v(t)$ is potential difference applied to the ceramic. We can thus express the acceleration of the lower face, where the probe is embedded, by:

$$\ddot{u}_z(t) = (d_{33}v(t) - u_z(t))\frac{A}{m_p.h.s_{33}} \tag{9}$$

which gives for the initial acceleration:

$$\ddot{u}_z(0) = \frac{Ad_{33}}{m_p.h.s_{33}}v(0) \tag{10}$$

Simulations of eq. 9 give the response of the ceramic for given profiles (fig 2.2). We used respectively a rectangular than a ramp signal of a period smaller, greater and equal to the natural period of the ceramic. Results are in good agreement with experimental results found in the literature 8 (8).

For accurate manipulation, it is important to avoid the oscillations of the probe. In the case of signals of periods different than the natural period of the ceramic, vibrations around the final position occurs. This must be definitely avoided, by using the signals of the same period of the ceramic's (fig 2.2-(3) and (6)). Additionally, the final position of the probe must be about $100nm$ higher of the initial position, as stated above. In this case, the appropriate signal is a ramp (fig 2.2-(6)). The final position of the actuator is given by $u(t_{final}) = v(t_{final}) \cdot d_{33}$.

By choosing the width of the signal as a multiple of the natural period of the ceramic and by regulating $v(t_{final})$ according to the desired displacement, we obtain the different profiles of signals ensuring the desired range of acceleration.

We notice that the magnitude of the acceleration is directly proportional to A, surface area of the ceramic and to the quotient $\frac{d_{33}}{s_{33}}$, and conversely proportional to the mass of the probe and the height of the ceramic. We chose A in order to cover all the surface of A.F.M cicuit to maximize the acceleration. Maximum tension that a ceramic can support depends on its height. The minimal value we can use allowing the desired amplitude is $1mm$. The quotient $\frac{d_{33}}{s_{33}}$ depends on the nature of the ceramic. In this order, different ceramics have been tested on simulations with signals of width of 1 to 10 periods and of amplitude $400V$. Values of initial accelerations for each case are presented at the table 1.

We shall now simulate the dynamic behavior whole end-effector. First, according to the desired acceleration, a command signal is chosen. The behavior of the ceramic is simulated and the displacement of its lower face is collected. Than this data is used to simulate the dynamic behavior of the probe, as the displacement data of the base section of the cantilever. The finite elements soft "CASTEM 2000" is used for this simulations. As excepted, a large range of accelerations, varying from a few ms^{-2} to more than $10^6 ms^{-2}$, are obtained. Figure 7 shows a low acceleration behavior. The magnitude of oscillations of the probe's tip are negligible beside the displacement

Micro-Manipulation and Adhesion Forces

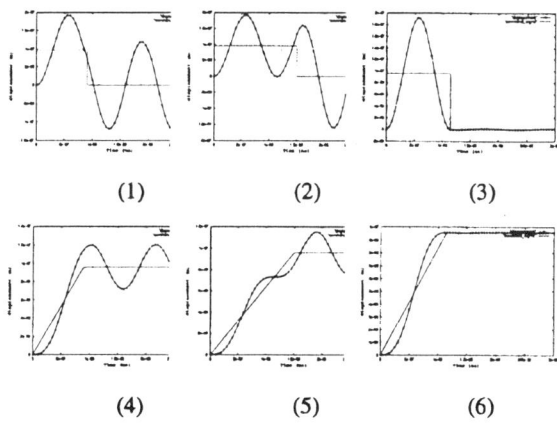

Figure 6. Simulations of the actuator - Natural period of the ceramic is $1143ns$

Type of ceramic	acceleration (1 period)	acceleration (10 periods)
P1-89 (8x5x1)	$5 \cdot 10^5$ $m.s^{-2}$	$4,7 \cdot 10^4$ $m.s^{-2}$
P1-91 (8x5x1)	$7,5 \cdot 10^5$ $m.s^{-2}$	$7 \cdot 10^4$ $m.s^{-2}$
P1-94 (8x5x1)	$9 \cdot 10^5$ $m.s^{-2}$	$9 \cdot 10^4$ $m.s^{-2}$

Table 1. Reachable accelerations for different ceramic and signals

of the system. Meanwhile, in a high-acceleration case (fig. 8), although an acceleration of $1,45 \cdot 10^6 ms^{-2}$ is reached, important vibrations occur. During this, the tip is subject to accelerations as high as $80 \cdot 10^4 ms^{-2}$. It can generate an important external force on the manipulated object, surpassing the pull-off force, and cause the loss of adhesion.

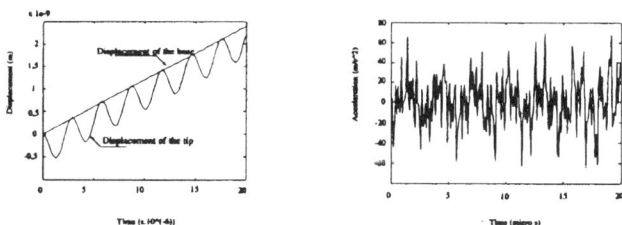

Figure 7. Low acceleration case

3 Conclusion

By simulations of the dynamical model of a micro-manipulation task using adhesion forces, we have obtained technological specifications of the manipulator. The necessity to create important

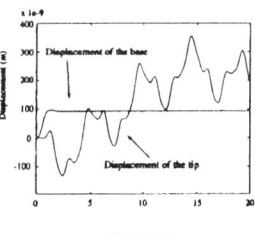 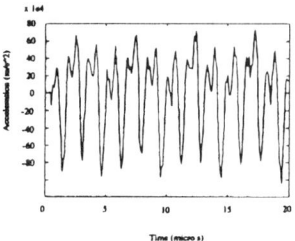

Figure 8. High acceleration case

accelerations of the end effector has been shown. In this aim, a system integrating a piezoelectric actuator and an atomic force microscope used as the probe is realized. Dynamic behavior of this system has been theoretically studied and command signals ensuring the desired accelerations has been established. Results shows important oscillations of the probe. Nevertheless, for low accelerations, oscillations are in a limited range and do not cause the loss of the manipulated object. They can also be compensated by the nano-translator. For high accelerations, oscillations and the acceleration variations of the probe is quite important. It is not sure that the cantilever can support without damage such a treatment. It will be necessary to improve the system in this aim. Let us recall that the command of the piezoelectric actuator is done on open-loop. The use of a closed-loop control algorithm can be efficient to eliminate the oscillations. As a second solution, different probes of more complex geometry and/or different materials can be used. Meanwhile, the piezoresistivity seems to be a precious characteristic in the development of such an end-effector. An electronic device capable of generating the desired command signals is under construction. Meanwhile some successful experiments have been conducted with a choice of material requiring small accelerations. It shows the feasibility of a adhesion based micro-manipulator.

Figure 9. Experimental site and the end-effector

References

Y. Rollot, S. Regnier, and J-C. Guinot, "Simulation of micro-manipulations : Adhesion forces and specific dynamic models," *International Journal of Adhesion & Adhesives*, vol. 19, pp. 35–48, 1999.

Y. Rollot, S. Regnier, and J-C. Guinot, "Micro-robotics : A dynamical model of micro-manipulation by adhesion," in *Proc. of the Twelfth CISM-IFToMM Symposium, Theory and Practice of robots and manipulators, Paris, France, july*, 1998, pp. 111–118.

B. V. Derjaguin, V.M. Muller, and YU. P. Toporov, "Effect of contact deformations on the adhesion of particles," *Journal of Colloid and interface science*, vol. 53, No. 2, pp. 314–326, 1975.

J. Israelachvili, *Intermolecular and Surface Forces*, Academic Press, 1991.

F. Arai, D. Ando, T. Fukuda, Y. Nonoda, and T. Oota, "Micro manipulation based on micro physics -strategy based on attractive force reduction and stress measurement-," in *proc. of the IEEE/RSJ Intelligent Robots System*, 1995, pp. 236–241.

G. Danuser, I. Pappas, B. Vögeli, W. Zesch, and J. Dual, "Manipulation of microscopic objects with nanometer precision : Potentials and limitations in nano-robot design," *International Journal of Robotics Research*, vol. submitted, 1997.

R.P. Feynman, "There's plenty of room at the bottom," *Talk at the annual meeting of the American Physical Society at the California Institute of Technology (Caltech), December 29th*, 1959.

K. Uchino, *Piezoelectric actuator and ultrasonic motors*, Kluwer Academic Publishers, Series Editor : Harry L. Tuller, 1997.

Universal Dental Robot
- 6-DOF Mouth Opening and Closing Training Robot WY-5 -

Hideaki Takanobu[1], Takeo Maruyama[1], Atsuo Takanishi[1,2]
Kayoko Ohtsuki[3] and Masatoshi Ohnishi[3]

[1] Department of Mechanical Engineering, School of Science and Engineering,
Waseda University, Japan
[2] Humanoid Robotics Institute, Waseda University, Japan
[3] Department of Oral and Maxillofacial Surgery, Yamanashi Medical University, Japan

Abstract. This paper describes the mechanism, control method, and training results of the 6 degrees-of-freedom (DOF) mouth-opening and -closing training robot WY-5 (Waseda Yamanashi No. 5) that is an application of the Universal Dental Robot (UDR). The mouth-opening training is indicated for the rehabilitation of the patients suffering from disturbance of the mouth-opening and -closing. The six linear actuators manipulate the u-shaped end-effector of UDR. This u-shaped end-effector is a moving platform of 6-DOF parallel manipulator. Each linear actuator has displacement, velocity, and force sensor to measure the position and orientation of the u-shaped end-effector. The WY-5 is a master-slave parallel robot that manipulates the patient's mandible that can't widely open. The doctor grasps the 2-DOF master manipulator, and the 6-DOF slave manipulator opens the patient's mandible according to the master manipulator's motion. As the result of therapy by using WY-5 for two female patients who cannot open their mouth widely, the mouth opening distance increased.

1 Introduction

In the dentistry field, the mastication robot (Takanobu and Takanishi, 1996), a patient robot (Miyairi, 1996), a CAD/CAM system (Uchiyama, 1996)(Sohmura and Takahashi, 1996), applications of computer simulations (Maki, 1995), and computed dental technology (Tsutsumi, 1996), have been researched. However in spite of the quantification of the robot, robots for dental use have not been developed. Therefore, authors have coined the term "Dental Robotics (Takanobu and Takanishi, 1998)."

The mouth-opening training is indicated for the rehabilitation of the patients suffering from disturbance of mouth-opening and -closing. In this paper, the authors present the mouth opening and closing training robot WY-5 (Waseda Yamanashi No. 5) that is an application of the UDR (Universal Dental Robot).

The conventional mechanical devices such as a jaw motion measurement system, a food texture measurement system, a mastication robot, or a wooden mouth opening and closing training device are designed for special or one purpose only. Therefore we must take into account of the coordination transformation, or the difference of precision between each mechanical system. Here if we have developed a Universal Dental Robot that can be generally used in the dental field, it will solve these problems. The UDR described in this paper is a multi purpose robot for dental use that can be used in the many situations such as the dental research of jaw movement, the training for the dental disorders, or the food texture's measurement. This paper especially focuses on the training for the dental disorders that is mouth opening and closing disorders.

The UDR is a kind of 6-DOF parallel mechanism. The six linear actuators manipulate the u-shaped end-effector of UDR. This u-shaped end-effector is a moving platform of the 6-DOF parallel manipulator. Each linear actuator has displacement, velocity, and force sensor to measure the position and orientation of the u-shaped end-effector. The authors have formulated the UDR's control method based on the parallel mechanism's computing method. The computing steps are as follows. First, setting of the target position and orientation of the end-effector. Second, computation of each linear actuator's connecting point to the end-effector in three-dimensional space. Third, computation of each actuator's length. Finally, the controller drives the six linear actuators based on third step's computation.

The WY-5 is a master-slave parallel robot that manipulates the patient's mandible that can't widely open. The doctor grasps the 2-DOF master manipulator, and the 6-DOF slave manipulator opens the patient's mandible according to the master manipulator's motion.

This paper consists of five chapters. Chapter 1 is an introduction. Chapter 2 shows the conventional training as the background of the robot mechanism. Chapter 3 describes the details of the mouth-opening and -closing training robot WY-5. Chapter 4 is robot training method and results, and chapter 5 is conclusions and future works.

2 Conventional mouth opening training method

This chapter provides findings on the mouth opening and closing training. The doctor qualitatively measures mouth-opening distance with their fingers as in Figure 1. The authors divided mouth opening and closing training into two elements, the patient and doctor models.

The patient model means the physical model of the patient, and includes the mechanical property of the mouth opening apparatus. It also includes the mechanical property of a rubber band that is wound around the load-side of the mouth opening apparatus if the apparatus is a conventional wooden device.

The doctor model means the maneuverability and mechanical property of the human operator. In conventional therapy for mouth opening and closing training, only two persons, the patient and operator, are aware of the real mechanical constraints involved in particular mouth opening and closing training. Therefore, other persons can understand the training situation only by means of a paper report, expressed feelings, or oral description.

3 Mouth opening and closing training robot WY-5

The mouth opening and closing training robot WY-5 is a parallel mechanism robot having 6-DOF. This chapter shows the mechanisms, sensors, controller, and safety systems of WY-5.

3.1 Mechanisms

Former mouth opening and closing training robots have one or 3-DOF because the authors designed them based on the ideal human mandible that is 3-DOF. However, the real patient's

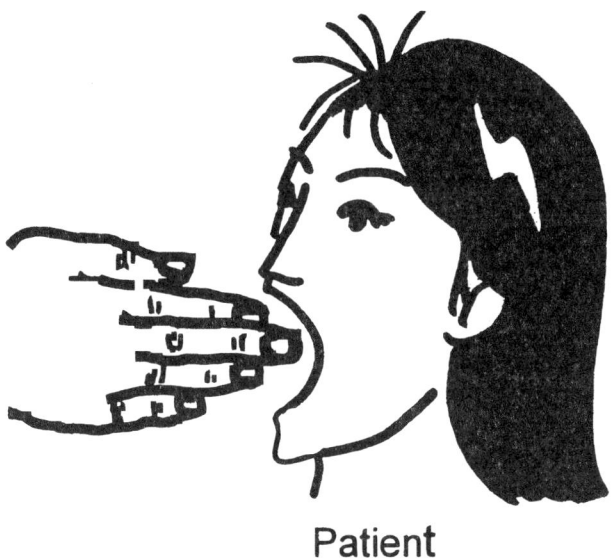

Figure 1. Measurement of mouth opening distance with fingers.

mandible has four or more DOF because their mandibles move to the sideways by the disorders on the muscles or the temporomandibular joint. Therefore, the authors developed a 6-DOF parallel robot WY-5 to adapt the real patient's natural mandible motion. The six linear actuators manipulate the u-shaped end effector that is inserted in the patient's mouth via mouthpiece. The instantaneous kinematics of WY-5 is a linear equation (Arai et al., 1992):

$$\dot{x}_0 = J^{-1} \dot{l}_0$$

where

$l_0 = (l_{01},...,l_{06})'$: link vector, l_{0i}: i-th link length

$x_0 = \begin{pmatrix} v_0 \\ \omega_0 \end{pmatrix}$: velocity vector of mouthpiece

J: Jacobian matrix

Figure 2 shows the comparison of the movable area of human mandible and WY-5's patient manipulator shown in Figure 3. The movable area of the human mandible is similar to the shape of a banana. The robot's movable area is 160 [mm] in grinding and 55 [mm] in opening directions. WY-5 is one of the parallel mechanisms therefore its movable area is similar to the typical parallel mechanism, that is Stewart platform. The robot's area includes a human area as shown in Figure 2.

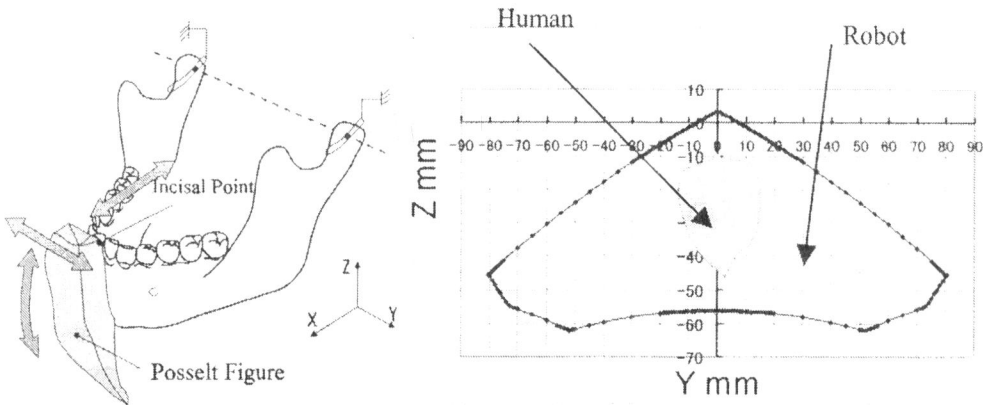

Figure 2. Comparison of WY-5 and human movable area.

Figure 4 shows the mechanism of the doctor manipulator. The doctor manipulator has 2-DOF to control the patient manipulator. Actual training needs 2-DOF, open/close and forward/backward, therefore 2-DOF is enough for this doctor manipulator.

The human opening and forward mandible motion occurs simultaneously. The robot can open the patient's mandible by utilizing both its mouth opening/closing DOF and forward/backward DOF.

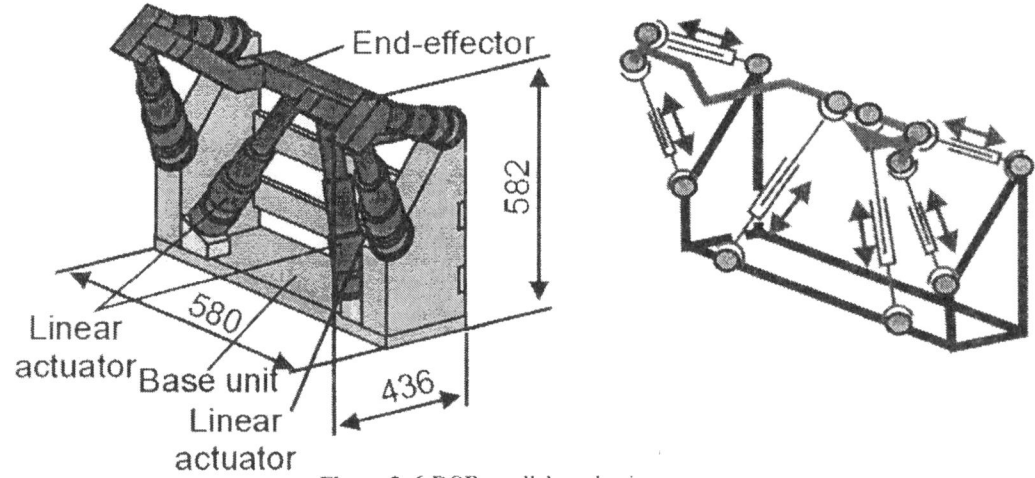

Figure 3. 6-DOF parallel mechanism.

3.2 Sensors

Most important factor during the therapy is the biting force acting on the patient via the mouth-opening gag. For example, the patient closes their mouth as if by a reflex action in case of an instantaneous large force acting on the mandible by tightly gripping the mouth opening gag. Therefore, the robot measures the biting force acting on it from the patient. This force sensor measures the translational force on XYZ axes. A 6 axes force-moment sensor measures the mouth opening force. Each DOF has displacement and velocity sensors.

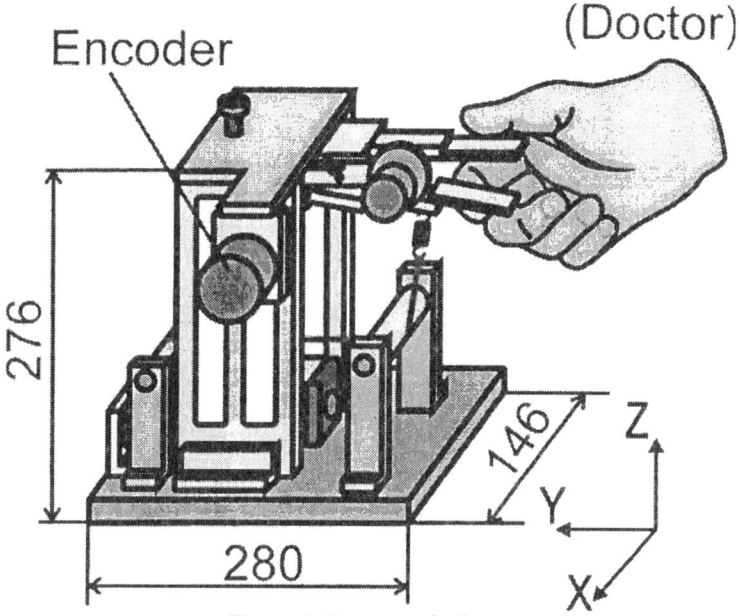

Figure 4. Doctor manipulator.

3.3 Controller

A personal computer controls the patient and doctor manipulator. The WY-5 is a force feedback-type master-slave system. The doctor manipulator's mouth opening angle and forward/backward displacements are sent to the patient manipulator. At the same time, patient manipulator's force information is sent to the doctor manipulator.

3.4 Safety systems

The authors have already denoted that the movable area of the WY-5 includes that of a human's. Therefore, the existence of safety systems and their effectiveness is important for the patient's safety. This section shows the configurations of the safety systems in this robot. There are four safety systems in this robot. Those are, an emergency motor stop switch, mechanical stoppers, electrical stoppers, and a software limit. The robot stops when at least one of these four safety systems runs, because these safety systems are connected in parallel. The concrete configurations of these safety systems, the action triggers, and special features are as follows.

An emergency motor stop switch is placed in the hand of the patient. Before the training, the authors rigidly enforce the patient's confirmation of the motor stopping function with this

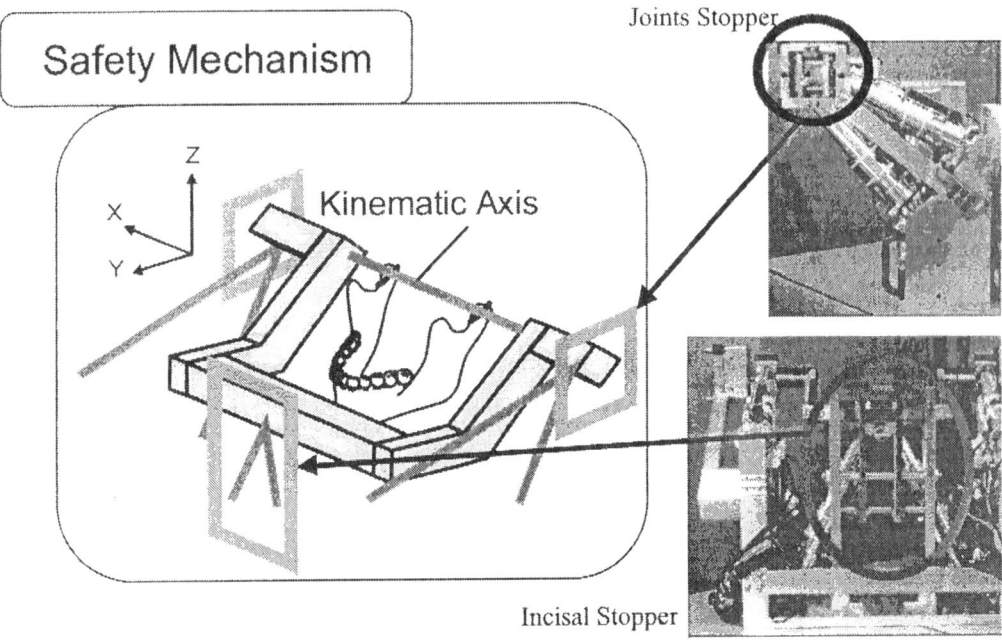

Figure 5. Mechanical safety system.

emergency switch. This confirmation frees the patient from fear. It is important to place this switch in the hand of the patient, because it is the only safety system under the patient's control.

The mechanical stoppers are discrete stoppers that protect the patient from excessive mouth opening angle and forward distance. As shown in Figure 5, there are two mechanical stoppers in WY-5: incisal stopper and joint stopper. The incical stopper constraints the motion of the front tooth to avoid the excessive mouth-opening angle. The joints stopper constraint the both side temporomandibular joints' motion to avoid the excessive forward motion of the joint.

The electrical stopper uses the fuses on each actuator to avoid excessive current.

The software limit means the threshold of the mouth opening force acting on the robot from the patient. The doctor inputs the force threshold before the training, and the motor stops in case the measured force passes the threshold. The doctor decide the concrete value of this threshold based on the training phase.

Total system configuration of the mouth-opening and -closing training robot WY-5 is shown in Figure 6.

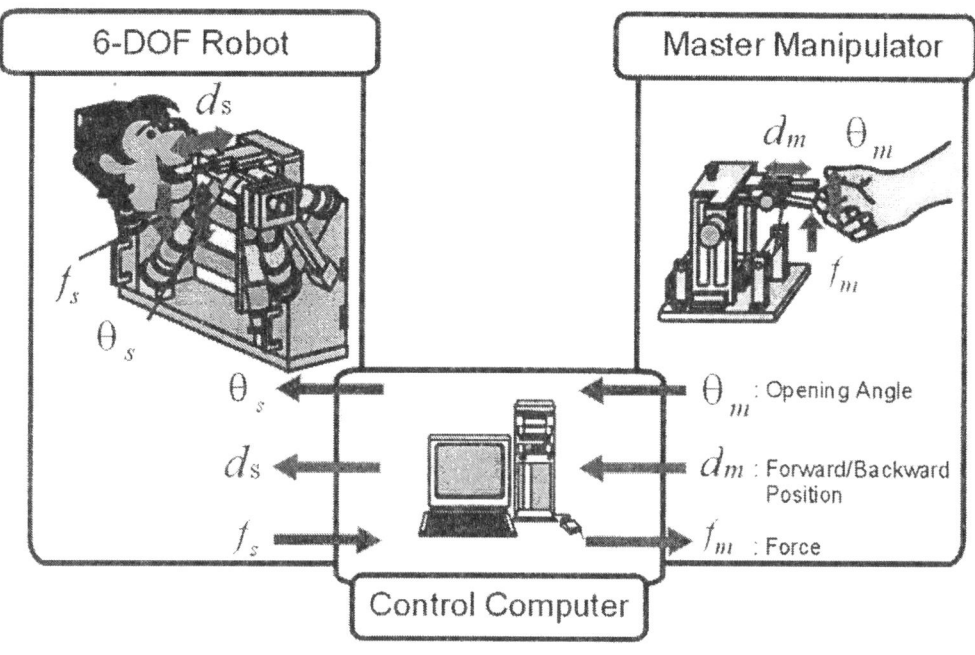

Figure 6. Total system of mouth-opening and -closing training robot WY-5.

4 Therapy method and results

The authors have done clinical trainings with WY-5. All patients agreed to use the training robot for mouth-opening and -closing training. The authors showed and explained the details of the robot for the patients. Two patients have done the clinical training with WY-5 as shown in Figure 7 and 8.

Patient A: -Female
    -58 years old
    -Fibrous adhesion of the temporomandibular joint
    -Training period is 5 [min], 5 times
    -Initial mouth opening distance is 21 [mm]
    -Final distance is 32 [mm]

Patient B: -Female
    -20 years old
    -Fibrous ankylosis
    -Training period is 15 [min], twice
    -Initial mouth opening distance is 15 [mm]
    -Final distance is 26 [mm]

The authors performed therapy on the above patients. The methods of the robotic clinical therapy are as follows.

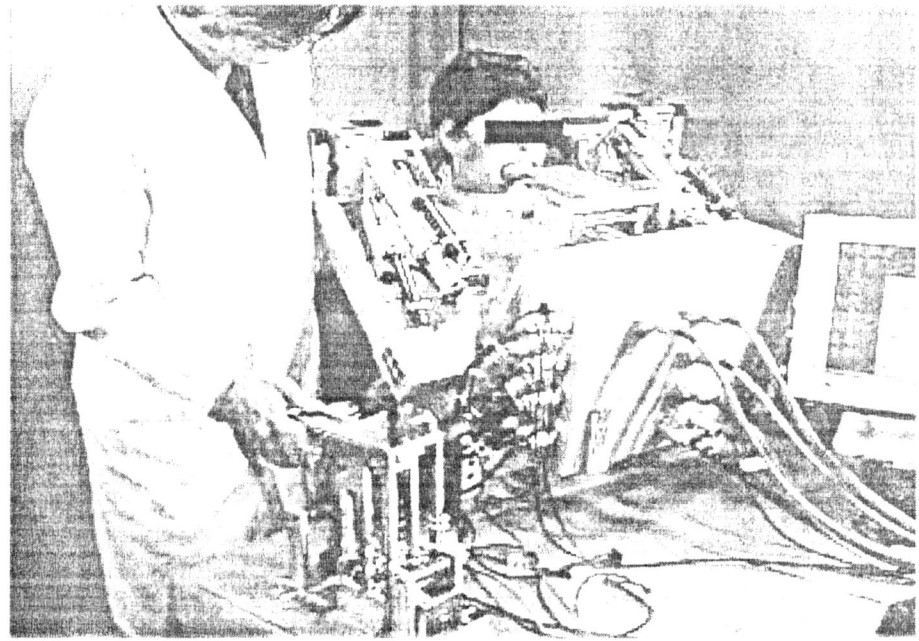

Figure 7. Training with WY-5 (Patient A).

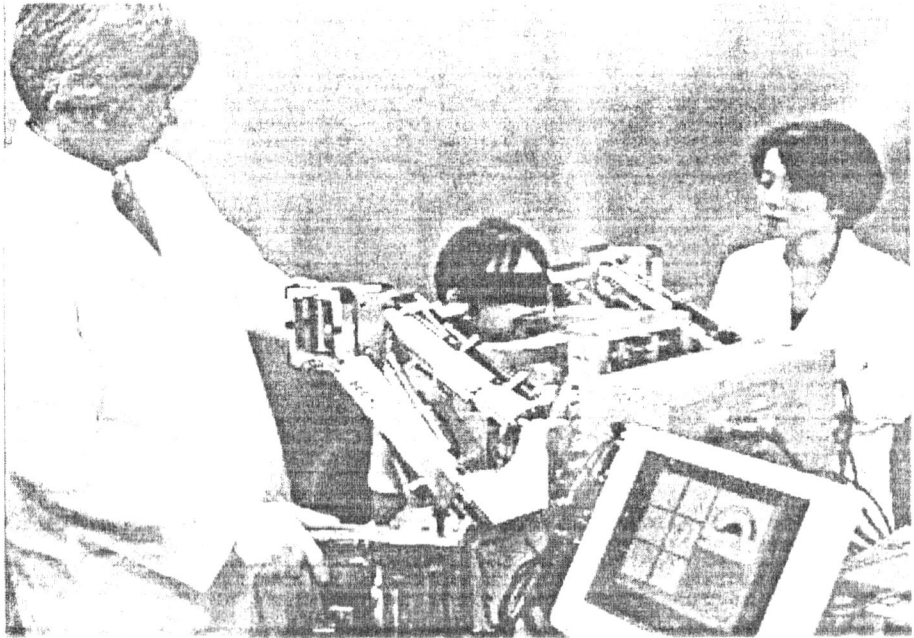

Figure 8. Training with WY-5 (Patient B).

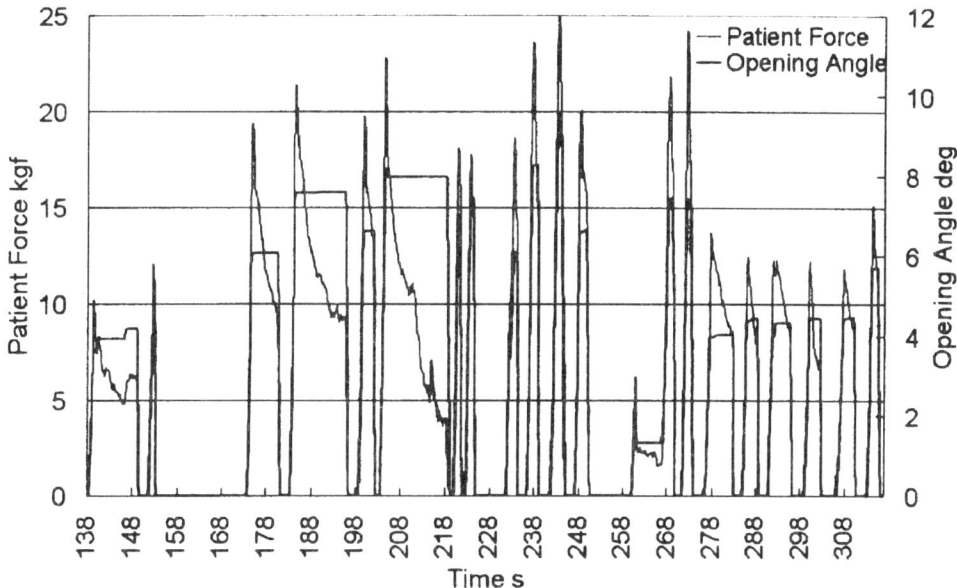

Figure 9. Treatment data (Patient B).

The tooth mark fixed on the patient manipulator was inserted into the patient's mouth, and the doctor manipulated the doctor manipulator of WY-5.

The doctor manipulated the doctor manipulator in forward/backward and open/close DOF, and other DOF of the patient manipulator are virtually compliance controlled. First patient's mouth moved asymmetrically, so WY-5 followed this patient's natural mouth motion.

Figure 9 shows the treatment data for patient B. Maximum force during this figure was about 25 [kgf] and the mouth opening angle pattern was similar to a trapezoid. These quantitative force and angle data will be the important index for more quantitative and standardized treatment with the robot.

5 Conclusions and future works

This paper is summarized as follows.
1) The authors have developed a 6-DOF mouth-opening and -closing training robot WY-5 (Waseda Yamanashi-5). WY-5 is a parallel mechanism robot.
2) The two patients' mouth opening distances increased from 21 to 32 [mm], and from 15 to 26 [mm] with WY-5.
3) The quantitative data of the training has been acquired by WY-5. Such quantitative data is useful and important for the standardization of mouth opening and closing training.
   In future, the authors are planning to develop a more human friendly training robot.

Acknowledgment

The authors would like to express great gratitude to the patients who agreed to use WY robots for mouth opening and closing training. This research was supported by a Grant-in-Aid for Scientific Research (No. 07557371, 11750217) and Electro-mechanic Technology Advancing Foundation.

References

Arai, T., Stoughton, R., Raju, G.J. (1992). Bilateral control for parallel-link manipulators. *IMACS/SICE international symposium on robotics,mechatronics and manufacturing systems'92*. 467-472.

Maki, K. (1995). Computer aides simulation technique present situation and future view, *The quintessence*. Vol. 14, No. 1, 197-207.

Miyairi, H. (1996). Evaluation of dental cutting technique using the patient robot. *Journal of the Robotics Society of Japan (RSJ)*. Vol. 14, No. 5, 28-32.

Sohmura, T., and Takahashi, J. (1996). Present condition of dental CAD/CAM system to fabricate prostheses. *Journal of the Robotics Society of Japan (RSJ)*. Vol. 14, No. 5, 33-37.

Takanobu, H., and Takanishi, A. (1996). Biomechanical aspect of a mastication robot. *CISM-IFToMM symposium on theory and practice of robots and manipulators (Ro.Man.Sy.11)*. 251-258.

Takanobu, H., and Takanishi, A. (1998). Dental robotics mouth opening and closing training robot. *CISM-IFToMM symposium on theory and practice of robots and manipulators (Ro.Man.Sy. 12)*. 427-434.

Tsutsumi, S. (1996). Computer introduction to dental technology from international viewpoints. *Shika gikou*, Vol. 24, No. 1, 60-70.

Uchiyama, Y. (1996). Oral device engineering system for the next generation as CAD/CAM. *Shika gikou*. Vol. 24, No. 1, 50-59.

Development of the Design of Polycrank Manipulator Without Joint Limits

Kazimierz Nazarczuk, Krzysztof Mianowski and Sławomir Łuszczak

Institute of Aeronautics and Applied Mechanics Warsaw University of Technology, Warsaw, Poland

Abstract. Some new results concerned with the design of POLYCRANK manipulators without joint limits are presented. Main features of such a solution are links in the form of light hollow cranks connected one to the other by cross-roller bearings. The first pre-prototype of POLYCRANK with six DOF's has been completed and tested. Three first DOF's, with parallel vertical axes are driven by electric Direct Drive units mounted coaxial in the base of the manipulator. Three last DOF's are driven by light motors with gears mounted on the last link of the horizontal chain with using parallelograms. High speed of cyclic gross motion in the convex workspace and good isotropic properties are the main advantages of this manipulator. Proper choosing of such mechanical properties like high natural frequency with good damping of vibrations, high stiffness, relatively low Lost Motion, is the basic requirement and it was the main purpose of investigations. Main disadvantages of the previous version are high cost of DD motors and of big diameter cross-roller bearings. A new one version with much more less cost is consider. One of the basic features of it's is a new serial-parallel modular arrangement of the arm and untypical spherical wrist.

1 Introduction

Some problems concerned with manipulators without joint limits have been investigated in the literature in the last decade by G. Johansson and S. Grahn (1995), and by W. Walishmiller at all. (1996). Some concepts of such a manipulators, have been presented by authors in papers (1995, 1996), while the first concept of the construction named POLYCRANK in papers (1998, 1999). Main links of this manipulator are in the form of skew hollow cranks connected by cross roller bearings with special skew parallelograms inside. Last time the authors has been investigated some new variants of such a designs. General concept of POLYCRANK manipulator with eight DOF's described by authors in (1999) is presented in Fig. 1. Three first DOF's in the form of planar kinematic chain with vertical axes consist links 1, 2 and 3 are actuated by driving system with special hollow shafts located coaxially in the base. Two first links are in the form of skew hollow cranks. First link is directly connected to first shaft. The second link is driven remotely by parallelogram located inside the link 1.

Third link is driven in similar way with using two serially connected parallelograms inside links 1and 2. Next three DOF's with parallel horizontal axes and two last DOF's (of the spherical wrist), are driven by motors with gears mounted on the link 3. Rotation of third DOF is limited by twisting the cables for these motors. Rotations of any other DOF's are unlimited. Kinematic scheme of

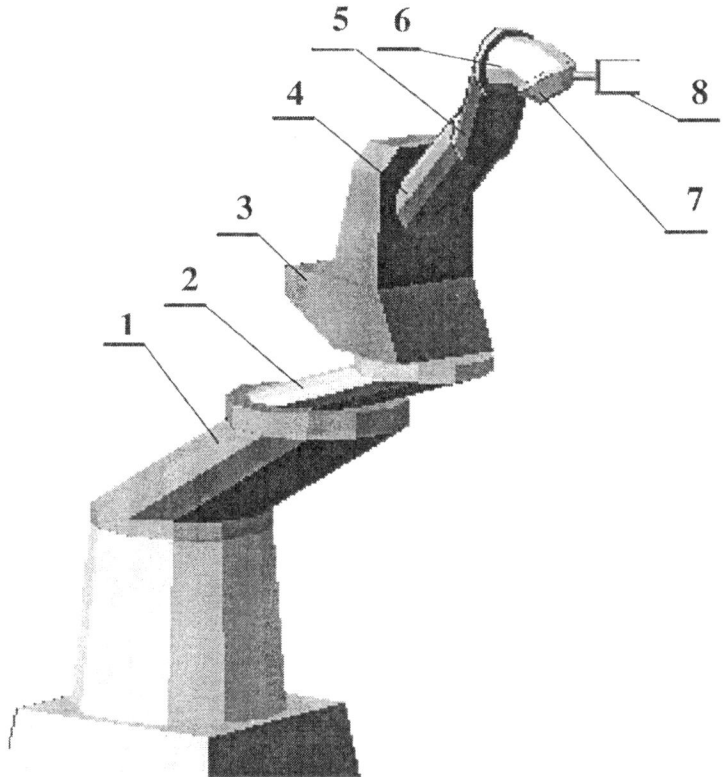

Fig. 1. Basic idea of POLYCRANK manipulator with eight DOF's.

this redundant manipulator is very simple and it has some specific properties. Applied parallelograms ensure kinematic decoupling between some DOF's i.e. orientations of three first DOF's are not dependent one to each other. Orientation of the wrist is dependent only on third and sixth to eight DOF's. Planar kinematic chain with vertical axes of three first DOF's, and planar kinematic chain with horizontal axes of three next DOF's with spherical wrist cause, that when eliminate two selected DOF's, usually it is obtain manipulator with six DOF's and analytic solution of second kinematic task. But it cannot be eliminate simultaneously fourth and fifth or seventh and eight without degeneration of manipulability. It has been investigated a number of various versions of manipulators with six or seven DOF's. One of the best solution of such a manipulator with seven DOF's can be obtain when fourth and fifth links with identical lengths are kinematically coupled by the special inverse gear, and are driven by fourth motor. Such a mechanism of the fourth DOF realizes straight-line vertical motion of the center point of the wrist. Because of this dynamic coupling between fourth and three first DOF's are very small. By proper mass distribution of the third link, it is easy to reduce dynamic coupling between third and two first DOF's. Non-redundant version of six DOF's POLYCRANK manipulator with simple two DOF's wrist has been realized and is tested just now.

2 Prototype of POLYCRANK manipulator

Design of the first prototype of POLYCRANK manipulator is presented in Fig. 2.

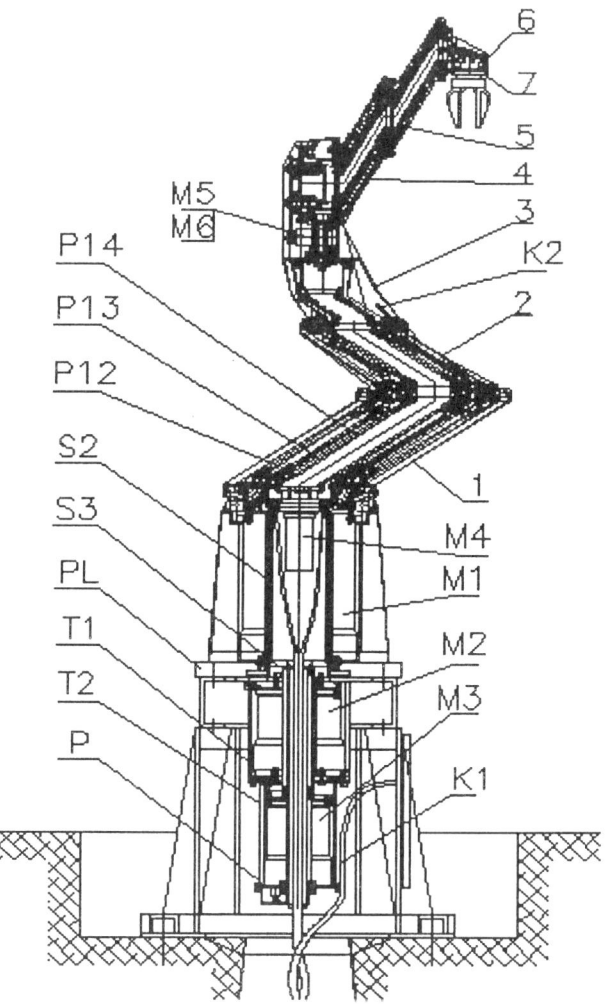

Fig. 2. Design of POLYCRANK manipulator

Three first DOF's in the form of planar kinematic chain consist links 1, 2 and 3 are actuated by DD motors M1, M2 and M3 located coaxially in the base. Two first links are in the form of skew hollow cranks. First link is directly connected to the rotor of motor M1. The second link is driven remotely by shaft S2 and parallelogram P12 located inside the link 1. Third link is driven by the motor M3 with the shaft S3 and two serially connected parallelograms P13 and P23, the last one not marked in the figure. Inside connecting rods of these parallelograms in special holes are located the cables K2 for motors M5 and M6 mounted on the link 3. Because of the cables, third DOF can rotate with ±2 rotations. Cables K1 are twisting in special hole in the ground. Rotation of third DOF measured by the potentiometer P is limited with the use of special buffers. Third link has the form of vertical column with gears for next DOF's. Fourth DOF is driven by the motor M4 with gear, which is located inside the shaft S3 and three serially connected parallelograms P14, P24 and P34. It is realized with the use two links 4 and 5 of the same lengths kinematically coupled by the inverse gear and skew parallelogram located outside the link 4. In such a way the end element of link 5 with wrist can realize straight-line vertical motion. Wrist consist of links 6 and 7 are driven by motors with gears M5 and M6 and two sets of parallelograms located inside links 4, 5, which have the form of skew cranks similar to links 1 and 2.

Fig. 3a Fig. 3b Fig. 3c

Fig. 3. General view POLYCRANK manipulator – some typical configurations.

Planar kinematic chain consist links 1, 2 and 3 can realize motion inside the full ring. Third link can be treated like a redundant DOF of the arm or for some tasks can be exploit for changing orientation of the end-effector.

Workspace of POLYCRANK manipulator arm with fourth DOF realized vertical straight-line motion has the form of full cylinder. Inside this cylinder, manipulator can realize various high-speed gross cyclic motions. It is worth noting, that orientation of the wrist is not dependent of any movements of first, second and fourth DOF's.

On the basis of this design, first prototype has been made described by the authors in the papers (1998, 1999). Some typical configurations of the prototype are shown in the figure 3. Initial experiments have been concerned of stiffness, mechanical hystherezis, and vibrations. It is worth noting, that minimal frequency of nature vibrations is higher then 30Hz. It was provided some long-time experiments of the cyclic motions with angular velocities in joints not exceed 1rot/s. There were shown, that basic elements and mechanisms has been designed properly to planned tasks. In particular elastic cables for fifth and sixth DOF's, located in very narrow special ducts inside connecting rods of parallelograms are working without any troubles

3 Some dynamic problems

Manipulator POLYCRANK with Direct Drives has been designed for high speeds. However mass of the third link with motors and gears for next DOF's is rather big, so in some trajectories, when this mass is moving on the straight line with high kinetic energy, going to the boundary of workspace, it can has very high acceleration in Cartesian space. This effect can occur even during free motion without driving forces.

Evident dynamic interactions in POLYCRANK manipulator are present only between two first DOF's. All other links can be treated as one mass located at the end of the second link as it has been reported in (1998). Inertia of this mass is much more then inertia of two first links, which are very light and are driven with the use DD motors located in the base. Because of it, this mass can realize gross free nearly straight-line motion with constant velocity, when the chain of two first links is far-away from singularities. Big curvature of considered trajectory is possible near singular configurations, so during high-speed motions, very high reaction force R23 between second and third links can occur. Some simulation results of such motion obtained with the use ADAMS program is shown in Fig. 4. In Fig. 4a time history of angular velocities of links 1+3 and value of reaction R23 is shown, while in Fig. 4b successive configurations of investigated planar chain with the vector of reaction R23 are presented. Motion is started with very high speeds in opposite directions w1=-w2 of two first links and velocity of third link w3=0. In singular configuration reaction R23 is very high equal up 4200N. Simulation

investigations have been shown, that effective reduction of exceed reaction force R23 can be obtained with the use proper balancing mass at opposite side of second link. Simulation result after adding such a balance is shown in Fig. 4c.

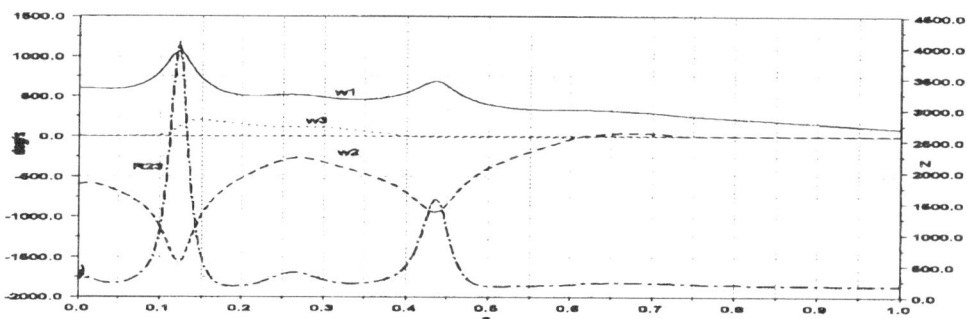

Fig. 4a

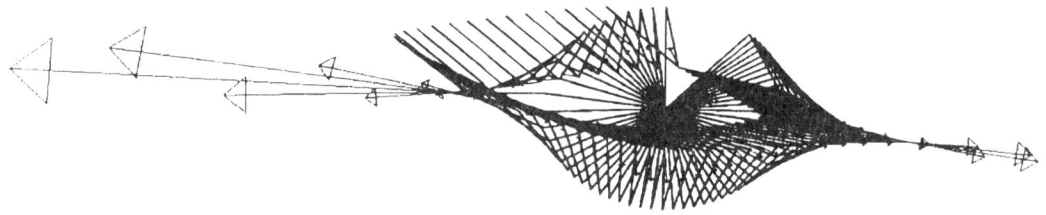

Fig. 4b

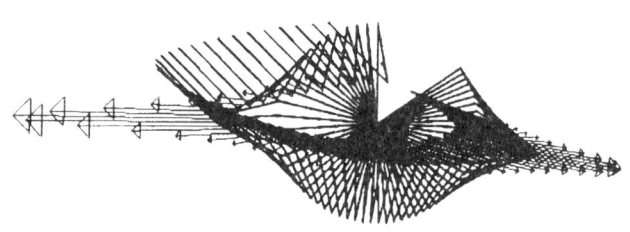

Fig. 4.c

Fig. 4. Simulation results of free motion of first planar kinematic chain (description in details in the text)

4 New concept of portal POLYCRANK manipulator

Investigations concerned with dynamic analysis of some versions connected with the basic idea of POLYCRANK manipulator shown, that it is possible to develop this concept in many variants with different properties to various applications and purposes. One example of the novel universal portal version of POLYCRANK manipulator is shown in Fig. 5. It can has seven or eight degrees of freedom. As a drives, electric motors with gears have been applied. Motors for second and third DOF's has been used as a counterbalance for second link. There are mounted in the platform moving in the space with constant orientation, so there are no any problems with twisting the cables for second and third motor. Motors for all next DOF's are mounted in the moving platform realised motion with constant orientation in the frame of the third link. These motors are used as a counterbalance for gravity compensation of all next links. Spherical wrist, it is novel simple construction with the use skew parallelogram inside. Every DOF's beside of the third are without joint limits. This version of the portal POLYCRANK manipulator has very big workspace with good isotropic kinematic and dynamic properties. It can be used for many various applications.

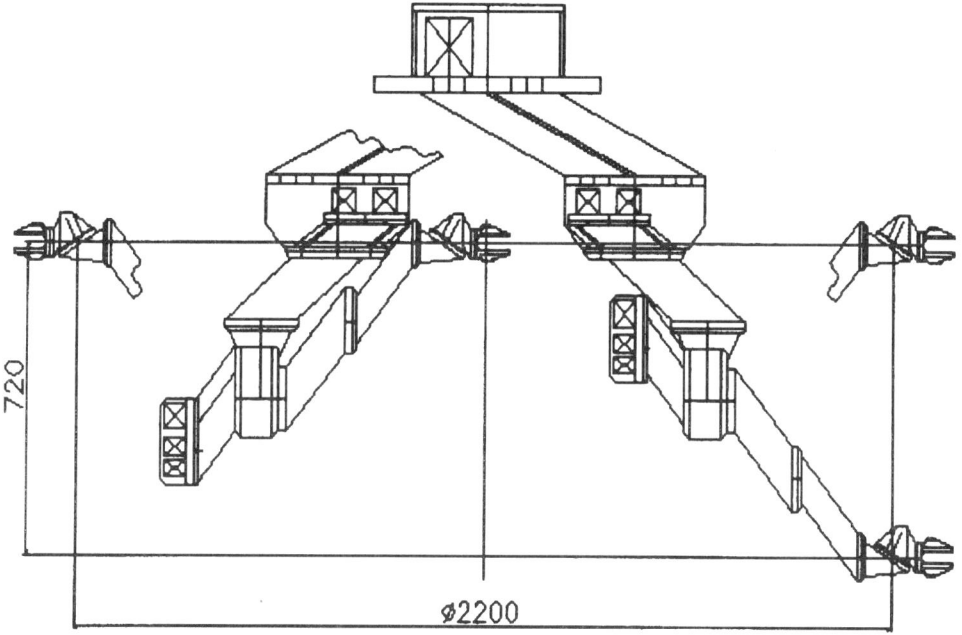

Fig. 5. New concept of portal version of POLYCRANK manipulator

5 Conclusions

Model-prototype of POLYCRANK manipulator is the invention and it can be used as a pattern to new construction of the fast robots without joint limits. Planar kinematic chain consists three first DOF's can be treated as a separated modul, which can be applied to the wide range of new technological processes like laser cutting and forming, scanning, etc. Portal version of POLYCRANK manipulator with with spherical wrist and very big workspace can be used for many new various applications.

6 Acknowledgements

This work was supported by the Polish Com. for Scien. Res (KBN)–grant No 8 T 11A 028 09 and CATID program of Warsaw University of Technology in 1999/2000.

References

Asada H., Youcef-Toumi K.: Direct Drive Robots, Theory and Practice, The MIT Press, 1987.
Chironis N.P.: Mechanisms, Linkages, and Mechanical Controls, McGraw-Hill Book Company 1965.
Holmberg R., Dickert S., Khatib O.: A New Actuation System for High-Performance Torque-Controlled Manipulators, Proc. Ro.Man.Sy 9, Springer-Verlag 1993.
Johansson G., Grahn S.: A New Robot Concept for High Speed Handling, Proc. 27th ISIR, Milano 1996.
Nazarczuk K.: Design of the Fast Manipulator with Eliminated joint Limits and Reduced Dynamic Interactions, Proc Ro.Man.Sy 10, Springer-Verlag 1995.
Nazarczuk K., Mianowski K.: Fast Manipulator with Invariant Inertia and Eliminated Joint Limits, Proc. 27th ISIR, Milan 1996.
Mianowski K.: A 3-D Lever Parallel Mechanism of a Robot With 6 Degrees Proc. of 6-th Int. Symp. Measurement and Control in Robotics, Brussels, 1996.
Mianowski K.: Dextrous Fully Parallel Manipulator With Six Degrees of Freedom, Proc Ro.Man.Sy 12, Springer 1998.
Nazarczuk K., Mianowski K.: POLYCRANK - Fast Manipulator Without Joint Limits, Proc Ro.Man.Sy 12, Springer 1998.
Nazarczuk K., Mianowski K., Wojtyra M.: Manipulator of POLYCRANK robot, Pomiary Automatyka Kontrola, No 8, Warsaw 1999 (in polish).
Walishmiller W., Lee H. Y., Bains N., Majarais B., Scott D. A.: Application of the TELBOT Robot in Hazardous Environment, Proc. 27th ISIR, Milano 1996.
Zieliński C., Szynkiewicz W.: Control System of POLYCRANK Robot, Pomiary Automatyka Kontrola, No 8, Warsaw 1999 (in polish).

Chapter IV

LEGGED LOCOMOTION

Emotion-based Walking of a Biped Humanoid Robot

Hun-ok Lim[1,3], Akinori Ishii[2] and Atsuo Takanishi[2,3]

[1] Department of System Design Engineering, Kanagawa Institute of Technology
1030 Shimoogino, Atugi, Kanagawa, 243-0292 Japan

[2] Department of Mechanical Engineering, Waseda University
3-4-1 Okubo, Shinjuku, Tokyo, 169-8555 Japan

[3] Humanoid Research Institute, Waseda University

Abstract. This paper describes the emotional walking of a biped humanoid robot based on its body motion. We consider three emotions such as happiness, sadness and anger. These emotions are defined by motion parameters of the biped robot. To express the emotions as walking motions, the patterns of the low-limbs, upper-limbs and head are preset using a polynomial. While the biped robot is emotionally walking according to the patterns, its stability is maintained by using the combined motion of the trunk and waist that is calculated by a body motion control method. To confirm the emotion expression, we have constructed a human-like robot WABIAB-RII (WAseda BIpedal humANoid robot-Revised II) which has forty-three mechanical degrees of freedom.

1 Introduction

Humans and animals use their legs to locomote with great mobility, but we do not yet have a full understanding of how they do so. One sign of our ignorance is the lack of man-made robots that use legs to obtain high mobility. Specially, two legged robots that are suitable for locomotion in our living environments such as offices and homes have not been studied so many up to date. We, therefore, describe both the scientific goal of understanding a human's walking in certain emotions and the engineering goal of building a biped humanoid robot useful for emotion expression.

The researches on human-like biped robots in Waseda University for the last three decades have had two main thrusts. One trust has been toward realizing the dynamic complete walking not only on an even or uneven terrain but also on a hard or soft terrain [1, 2]. The other thrust has been toward exploring robot-environment interaction. They achieved the dynamic walking under the unknown external force 100[N] acted on the waist of the biped robot [3, 4]. The human-following walking control was proposed which has the pattern switching technique based on the action criterion of human-robot physical interaction [5].

To date, almost of all researches on biped robots have covered the classical problems of walking control and energy consumption, but not included an issue of emotional walking behavior [6]. The balance of a biped robot was proposed by manipulating the projected center of gravity and the support area provided by the feet [7]. Honda's humanoid robot called P2 has been constructed and realized dynamic walking on even or uneven terrain [8].

In this paper, we focus on the emotional walking of a biped robot to understand human walking fully. Therefore we have constructed a biped robot called WABIAN-RII that has forty-three mechanical degrees of freedom and four passive degrees of freedom. A picture of WABIAN-RII is shown in Figure 1. A body motion control is described for the balance of the biped robot. Motion parameters having a great effect on three emotions are also discussed.

This paper is organized as follows. In Section 2, we describe a balance control. Section 3 describes motion parameters to express emotions. In Section 4, emotional experiments are discussed. Finally, conclusions are described in Section 5.

Fig. 1. The photo of WABIAN-RII.

2 Balance Control Method

In this section, a body motion control is described to balance a biped robot.

2.1 ZMP and Iteration

To define mathematical quantities, a inertia coordinate frame $\mathcal{F}_O$ is fixed on the floor. Also, a waist coordinate frame $\bar{\mathcal{F}}_O$ is attached on the center of the waist. $Q(x_q, y_q, z_q)$ is defined as the origin of the waist frame $\bar{\mathcal{F}}_O$ relative to $\mathcal{F}_O$.

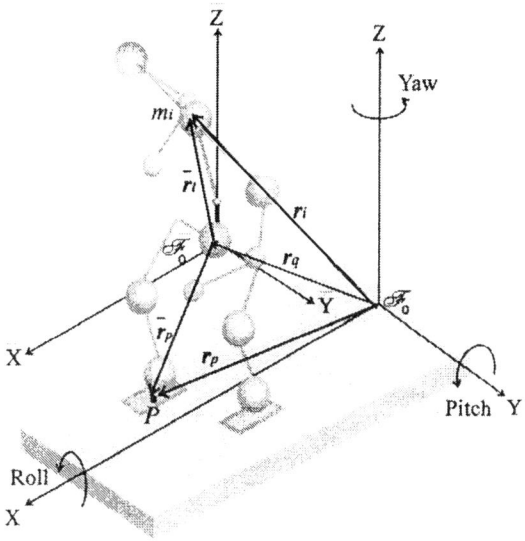

Fig. 2. Coordinate frames.

According to the mass distribution of the biped robot as shown in Figure 2, the moment balance around a contact point p on the floor surface can be obtained as

$$\sum_{i=1}^{n} m_i (r_i - r_p) \times (\ddot{r}_i + G) + T = 0, \qquad (1)$$

where m_i is the mass of the particle i. r_p is the position vector of the contact point p with respect to the world coordinate frame $\mathcal{F}_O$. r_i and $\ddot{r}_i$ denotes the position and acceleration vector of the particle i with respect to the world coordinate frame $\mathcal{F}_O$, respectively. G is the gravitational acceleration vector, T is the moment acting on the contact point p.

It is difficult to derive the solution from Equation (1) because the motion of the trunk and waist is interferential each other. Thus, the head and spine mass is regarded as a part of the neck mass, each arm mass as a part of the shoulder mass, and the thigh mass as a part of the hip mass. Also, we assume that the mass of the legs is concentrated on the ankle and knee joints. Considering the relative motion of the particles and the mass distributions as shown in Figure 3, the ZMP Equation (1) can be rewritten with respect to the $\bar{\mathcal{F}}_O$ as

$$\begin{aligned} m_s r_s \times \ddot{\bar{r}}_s + m_u (\bar{r}_{cs} - \bar{r}_{zmp}) \\ \times (\ddot{\bar{r}}_{cs} + \ddot{Q} + G) + m_h r_h \times \ddot{r}_h \\ + m_w (\bar{r}_{cw} - \bar{r}_{zmp}) \times (\ddot{\bar{r}}_{cw} + \ddot{Q} + G) \\ = -M. \end{aligned} \qquad (2)$$

where m_s and m_{cs} denote the mass of the shoulder including the arm and that of the neck including the head and spine, respectively. m_u is the total mass of the upper body

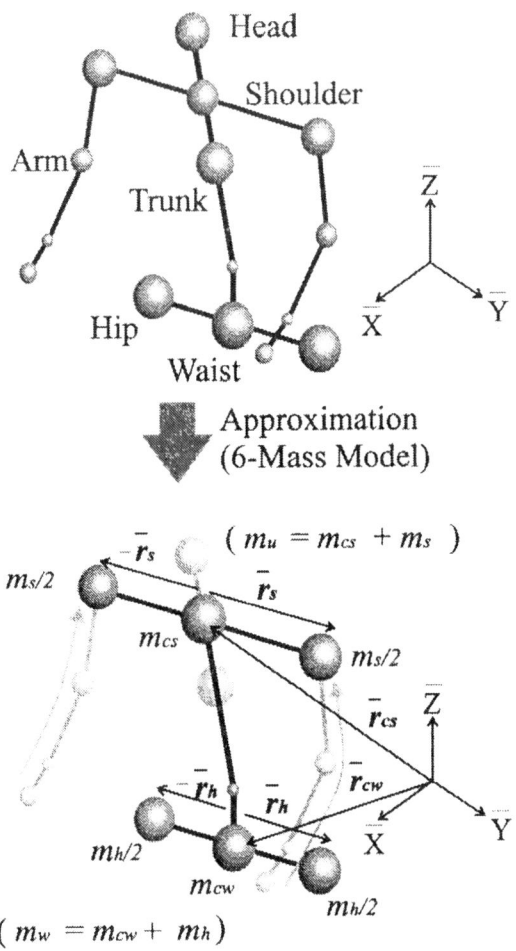

Fig. 3. Approximation model of the upper body

including the arms. m_h and m_{cw} are the mass of the hips and the waist, respectively. m_w is the total mass of the hips and waist. $\bar{r}_{cs}$ and $\bar{r}_{cw}$ are the position vectors of the neck and waist with respect to $\bar{\mathcal{F}}_O$, respectively. $\bar{r}_s$ is the shoulder position vector with respect to the neck, and $\bar{r}_h$ the hip position vector with respect to the waist. $\bar{r}_{zmp}$ is ZMP of the contract point p expressed in $\bar{\mathcal{F}}_O$. M denotes the moment generated by the lower-limb particles.

Equation (2) is interferential and non-linear because the waist is connected to the lower-limbs and the trunk by rotational joints. To obtain analytic solutions, we assume that neither the waist arm consisting of the waist and hip particles nor the trunk arm consisting of the trunk and shoulder particles moves vertically. Therefore, the waist arm and the trunk arm rotate on the horizontal plane only. Figure 4 shows a linearized model of the biped humanoid robot. The terms about the motion of the upper-body particles are put on the left-hand side as unknown variables, and the terms about the moment generated by the low-limbs particles on the right-hand side as known parameters. According to the components of three axes, we write decoupled and linered equations as follows:

$$m_u(\bar{z}_{cs} - \bar{z}_{zmp})(\ddot{\bar{x}}_{cs} + \ddot{x}_q + g_x) - m_u(\bar{x}_{cs}$$
$$-\bar{x}_{zmp})g_z + m_w(\bar{z}_{cw} - \bar{z}_{zmp})(\ddot{\bar{x}}_{cw} + \ddot{x}_q + g_x)$$
$$-m_w(\bar{x}_{cw} - \bar{x}_{zmp})g_z = -M_y(t), \qquad (3)$$

$$m_u(\bar{y}_{cs} - \bar{y}_{zmp})g_z - m_u(\bar{z}_{cs} - \bar{z}_{zmp})(\ddot{\bar{y}}_{cs}$$
$$+\ddot{y}_q + g_y) + m_w(\bar{y}_{cw} - \bar{y}_{zmp})g_z - m_w(\bar{z}_{cw} -$$
$$\bar{z}_{zmp})(\ddot{\bar{y}}_{cw} + \ddot{y}_q + g_y) = -M_x(t), \qquad (4)$$

$$m_s R_s^2 \ddot{\theta}_t + m_h R_h^2 \ddot{\theta}_w = -\tilde{M}_z(t) - M_z(t), \qquad (5)$$

where R_s and R_w are the radius of the trunk and waist, respectively. $\ddot{\theta}_t$ and $\ddot{\theta}_w$ denote the rotational accelerations of the trunk and waist, respectively. The moment caused by the yaw axes of the trunk and waist, $\tilde{M}_z$, can be derived from the pitch and roll motions of the trunk and waist as follows:

$$\tilde{M}_z(t) = m_u(\bar{x}_{cs} - \bar{x}_{zmp})(\ddot{\bar{y}}_{cs} + \ddot{y}_q + g_y)$$
$$-m_u(\bar{y}_{cs} - \bar{y}_{zmp})(\ddot{\bar{x}}_{cs} + \ddot{x}_q + g_x)$$
$$+m_w(\bar{x}_{cw} - \bar{x}_{zmp})(\ddot{\bar{y}}_{cw} + \ddot{y}_q + g_y)$$
$$-m_w(\bar{y}_{cw} - \bar{y}_{zmp})(\ddot{\bar{x}}_{cw} + \ddot{x}_q + g_x). \qquad (6)$$

Transferring the known valuables of Equations (3), (4) and (5) to right-hand side and adding them to the moment produced by the lower-limb motion, we can rewrite the ZMP equations as

$$m_u(\bar{z}_{cs} - \bar{z}_{zmp})\ddot{\bar{x}}_{cs} - m_u \bar{x}_{cs} g_z + m_w(\bar{z}_{cs}$$
$$-\bar{z}_{zmp})\ddot{\bar{x}}_{cw} - m_w \bar{x}_{cw} g_z = -\hat{M}_y(t), \qquad (7)$$

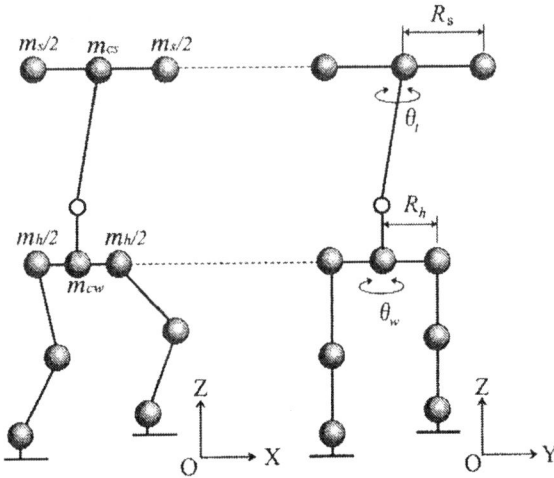

Fig. 4. Linearized model.

$$-m_u(\bar{z}_{cs} - \bar{z}_{zmp})\ddot{\bar{y}}_{cs} + m_u\bar{y}_{cs}g_z - m_w(\bar{z}_{cw} \\ -\bar{z}_{zmp})\ddot{\bar{y}}_{cw} + m_w\bar{y}_{cw}g_z = -\hat{M}_x(t), \quad (8)$$

$$m_s R_s^2 \ddot{\theta}_t + m_h R_h^2 \ddot{\theta}_w = -\hat{M}_z(t), \quad (9)$$

where $\hat{\boldsymbol{M}} = [\hat{M}_x, \hat{M}_y, \hat{M}_z]^T$ is the known valuable term of the biped model, which is derived from the lower-limb motions and the time trajectory of ZMP.

It is difficult to calculate the compensatory motion of the trunk and waist from Equations (7), (8) and (9) due to six unknown variables such as $\bar{x}_{cs}$, $\bar{x}_{cw}$, $\bar{y}_{cs}$, $\bar{x}_{cw}$, θ_t and θ_w. So, the known valuable term $\hat{\boldsymbol{M}}$ is classified into the trunk and waist terms as

$$\hat{M}_y = \hat{M}_{yt} + \hat{M}_{yw}, \\ \hat{M}_x = \hat{M}_{xt} + \hat{M}_{xw}, \\ \hat{M}_z = \hat{M}_{zt} + \hat{M}_{zw}. \quad (10)$$

The linearized euations are

$$m_u(\bar{z}_{cs} - \bar{z}_{zmp})\ddot{\bar{x}}_{cs} + m_u g_z \bar{x}_{cs} = -\hat{M}_{yt}, \\ m_w(\bar{z}_{cs} - \bar{z}_{zmp})\ddot{\bar{x}}_{cw} + m_w g_z \bar{x}_{cw} = -\hat{M}_{yw}, \quad (11)$$

$$-m_u(\bar{z}_{cs} - \bar{z}_{zmp})\ddot{\bar{y}}_{cs} - m_u g_z \bar{y}_{cs} = -\hat{M}_{xt}, \\ -m_w(\bar{z}_{cw} - \bar{z}_{zmp})\ddot{\bar{y}}_{cw} - m_w g_z \bar{y}_{cw} = -\hat{M}_{xw}, \quad (12)$$

$$m_s R_s^2 \ddot{\theta}_t = -\hat{M}_{zt}, \\ m_h R_h^2 \ddot{\theta}_w = -\hat{M}_{zw}. \quad (13)$$

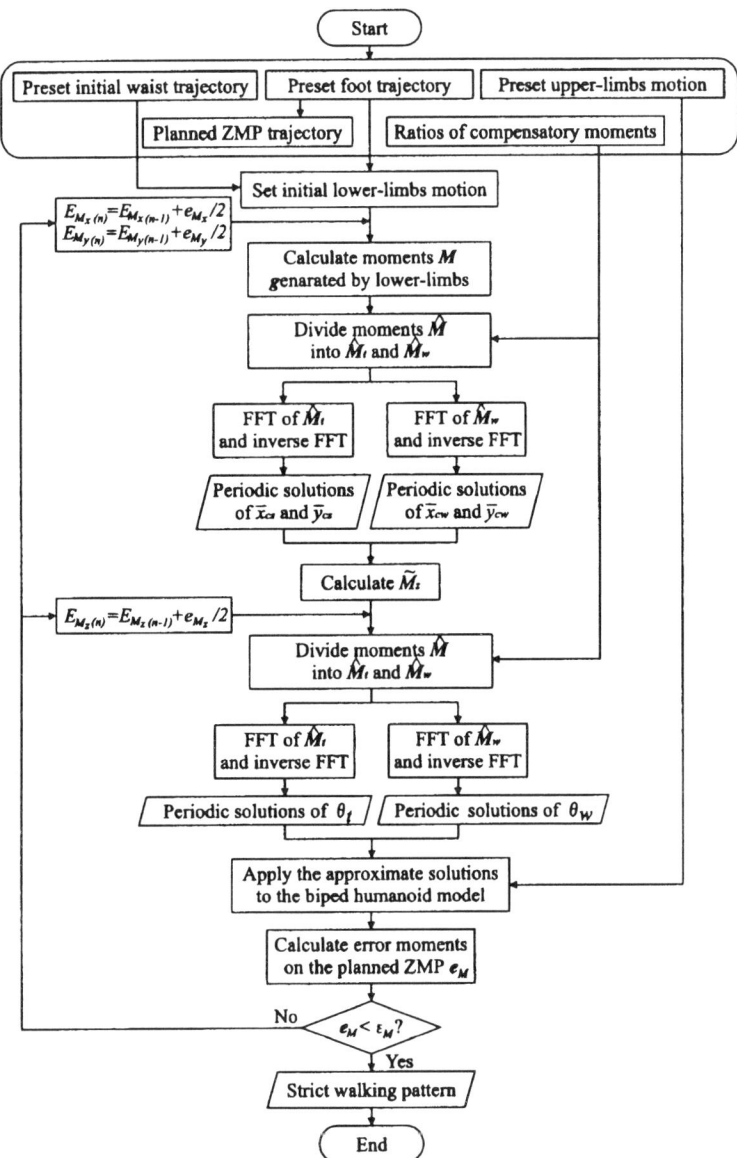

Fig. 5. Iteration method.

In the case of the steady walking, the right-hand terms of the moment equation $\hat{M}$ are periodic functions because each particle of the lower-limbs and the time trajectory of ZMP move periodically with respect to the waist frame $\bar{\mathcal{F}}_O$. Each equation, therefore, can be represented as a Fourier series. Comparing the Fourier transform coefficients from both sides of each equation, the approximate periodic solutions for the motion of the trunk and waist can be easily obtained using Inverse Fourier Transform. To determine the offset terms of the yaw trunk and waist, we take into consideration that the generated yaw-motion angles are in the range of the rotatable region of the yaw-axis actuators. In addition, the above computation is applicable not only to steady walking but also to the complete walking. We regard the complete walking as one walking cycle and make the static standing states before and after walking long enough.

To compute the strict solutions of the trunk and waist, we propose an algorithm that iteratively computes the approximate solutions. Figure 5 shows the block diagram of an iteration method. At first, we substitute the approximate periodic solutions of the linear Equations (7), (8) and (9) for the ZMP equation (2) represented in the waist frame $\bar{\mathcal{F}}_0$, and calculate the errors of moments $e_M = [e_{M_x}, e_{M_y}, e_{M_z}]^T$ according to the planned ZMP. This calculation is repeated until the moment errors drop below a specific tolerance level $\epsilon_M = [\epsilon_{M_y}, \epsilon_{M_x}, \epsilon_{M_z}]^T$.

3 Motion Parameters

In this section, motion parameters are discussed to express emotions. In happiness, the motion parameters are as follows; A: step length, B: foot, C: toe, D: pitch waist, E: roll waist, F: pitch trunk and head, G: roll trunk and head and H: arm. As an example of happy parameters, the biped humanoid robot should (1) walk with a long step, (2) lift its foot and put it down vertically like a M-shaped walking, (3) turn the toes outward, (4) move the waist up and down using knee joints, (5) rotate the waist using the hip joints, (6) greatly shake the head and upper trunk side-to side, (7) greatly move the head and upper trunk front-to-back and (8) largely swing the arms.

To choose the motion parameters having a great effect on three emotions we simulate as follows. First, 32 walking patterns each emotion are preset by the combinations of 8 parameters and 4 walking types. Second, the combined motion patterns of the trunk and waist are set using the balance control method to cancel the moments generated by the motions of lower-limbs. Finally, 3D simulations are videotaped and evaluated as five steps, from 0 point(not like) to 4 point(very much like). The evaluation is done by 17 male undergraduates as shown in Table 1.

Table 2 shows a result of a analysis of variance using Table 1. The motion parameters are determined by the analysis-of-variance table. To calculate the critical value of the F_α statistic, we use the tables of percentage points of the F distribution [9]. In happy walking, the critical values of the F_α statistic ($\alpha 5 = 0.05$ or $\alpha 1 = 0.01$) for $\nu_t = 1$ and $\nu_e = 17$, and $\nu_t = 3$ and $\nu_e = 17$ are as follows:

$$F^{\alpha 5,1}_{\alpha 1,17} = \begin{cases} 4.45 \\ 8.40 \end{cases}, \quad F^{\alpha 5,3}_{\alpha 1,17} = \begin{cases} 3.20 \\ 5.18 \end{cases}$$

where α is the probability that the test statistic will fall in the rejection region. ν_t and ν_e are the degrees of freedom for treatments and the degrees of freedom for errors.

Table 1. Evaluation of happy walking.

Gaits	A	B	C	D	E	F	G	H	Sum
1	1	1	1	1	1	1	1	1	33
2	2	1	2	2	2	1	2	2	6
3	1	1	1	1	2	2	3	2	22
4	2	1	2	2	1	2	4	1	11
5	2	1	1	2	1	3	3	1	10
6	1	1	2	1	2	3	4	2	35
7	2	1	1	2	2	4	1	2	5
8	1	1	2	1	1	4	2	1	37
9	1	2	2	2	1	1	3	2	4
10	2	2	1	1	2	1	4	1	24
11	1	2	2	2	2	2	1	1	16
12	2	2	1	1	1	2	2	2	12
13	2	2	2	1	1	3	1	2	24
14	1	2	1	2	2	3	2	1	12
15	2	2	2	1	2	4	2	1	45
16	1	2	1	2	1	4	4	2	5
17	2	3	2	1	2	1	3	1	37
18	1	3	1	2	1	1	4	2	4
19	2	3	2	1	1	2	1	2	25
20	1	3	1	2	2	2	2	1	13
21	1	3	2	2	2	3	1	1	18
22	2	3	1	1	1	3	2	2	29
23	1	3	2	2	1	4	3	2	13
24	2	3	1	1	2	4	4	1	24
25	2	4	1	2	2	1	1	2	6
26	1	4	2	1	1	1	2	1	30
27	2	4	1	2	1	2	3	1	9
28	1	4	2	1	1	2	3	1	10
29	1	4	1	1	2	3	3	2	36
30	2	4	2	2	1	3	4	1	9
31	1	4	1	1	1	4	1	1	37
32	2	4	2	2	2	4	2	2	2

Table 2. The analysis of variance in happy walking.

Parameter	SS	DOF	MS	F
CM	11362.781	1	11362.781	0
A	69.031	1	69.031	2.269
B	54.094	3	18.031	0.593
C	52.531	1	52.531	1.726
D	3140.281	1	3140.281	103.203
E	11.281	1	111.281	0.371
F	238.844	3	79.615	2.616
G	216.844	3	72.281	2.375
H	504.031	1	504.031	16.565
Error	517.281	17	30.428	0
Total	16167	32	-	-

4 Experiments

To verify the effectiveness of the motion parameters for emotion expression, emotional walking experiments are conducted using a forty-three mechanical degrees of freedom WABIAN-RII with a human configuration. The height of the WABIAN-RII shown in Section 1 is about 1.84[m] and its total weight 127[kg]. The biped robot is controlled by a PC/AT compatible computer PEAK-530 (an Intel MMX Pentium 200[MHz] CPU processor) which are govern by a OS, MS-DOS 6.2/V (16-bit). The computer system is mounted on the back of the waist, and the servo driver modules are mounted on the upper part of the trunk. The external connection is only an electric power source.

Table 3. Evaluation of emotional walking.

Walking style	Percentage of agreement [%]
Happy walking	90
Sad walking	80
Angry walking	30

The schemes of three emotional walking experiments are as follows: (1) the preset walking patterns of the lower-limbs, waist and head are planned based on the motion parameters: (2) the compensatory motion pattern of the waist is changed by the body control method: (3) these motion patterns are commanded to the WABIAN-RII using the program control.

The dynamic complete walking is realized with the step time of 1.28[sec/step] and the step width of 0.15[m/step]. Figure 6 shows the scenes of the happy walking experiment. Table 3 shows the evaluation of the emotional walking. The agreement rate in the happy walking was 90 per cent and that in the sad walking was 80 per cent. However, the agreement rate in the angry walking was 30 per cent due to the joint angle limitation of the WABIAN-RII. In these results, we can imagine that the swing waist and arm and the moving head have a great effect in the happy walking. Also, the bent trunk and drooped head have a great effect in the sad walking while the stuck face and clenched fist have a great effect in the angry walking.

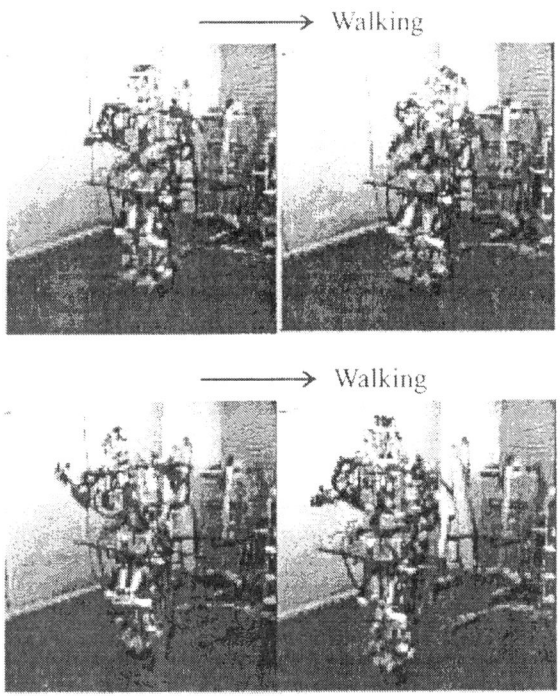

Fig. 6. Happy walking experiment.

5 Conclusions

In this paper, a balance control and parameterization were introduced to explore the emotional walking behavior. The walking motion of a biped robot is parameterized by the analysis of variance and preset as an emotional walking pattern. The balance control method generates the combined motion of the trunk and waist that compensates for the three-axis moments produced by the motions of the lower-limbs and upper-limbs. The experimental results clarify that the parameterization of the biped robot is useful for the emotion expression.

Acknowledgment

This study has been conducted as a part of the humanoid project at Humanoid Research Institute, Waseda University. This study was funded in part by JSPS under Japanese Grant-in-Aid for Scientific Research and Joint Research Project under the Japan-Korea Basic Scientific Cooperation Program and NEDO (New Energy and Industrial Technology Development Organization). The authors would like to thank Okino Industries Ltd., Harmonic Drive Systems Inc. and YKK Corporation for supporting us in developing the hardware.

References

1. A. Takanishi, M. Ishida, Y. Yamazaki, and I. Kato, "The realization of dynamic walking by the biped walking robot," in *Proc. IEEE Int. Conf. Robotics and Automation*, St. Louis, MO, Mar. 1985, pp. 459–466.
2. A. Takanishi, H. O. Lim, M. Tsuda, and I. Kato, "Realization of dynamic biped walking stabilized by trunk motion on a sagitally uneven surface," in *Proc. IEEE/RSJ Int. Workshop Intelligent Robots and Systems*, Tsuchiura, Japan, Jul. 1990, pp. 323–329.
3. A. Takanishi, M. Kumeta, K. Matsukuma, J. Yamaguchi, and I. Kato, "Development of control method for biped walking under unknown external force acting in lateral plane," in *RSJ Annual Conf. Robotics Society of Japan*, Tsukuba, Japan, Nov. 1991, pp. 321–3224. (in Japanese).
4. A. Takanishi, S. Sundo, N. Kinoshita, and I. Kato, "Compensation for three-axes moments by trunk motion: Realization of biped walking under unknown external force," in *RSJ Annual Conf. Robotics Society of Japan*, Kanezawa, Japan, Oct.-Nov. 1992, pp. 593–596. (in Japanese).
5. S. Setiawan, S. Hyon, J. Yamaguchi, and A. Takanishi, "Physical interaction between human and a bipedal humanoid robot realization of human-follow walking," in *Proc. IEEE Int. Conf. Robotics and Automation*, Detroit, Mich., May 1999, pp. 361–367.
6. H. Miura and I. Shimoyama, "Dynamic walk of a biped," *Int. J. Robotics Research*, vol. 3, no. 2, pp. 60–74, Summer 1984.
7. M. Vukobratovic, A. A. Frank, and D. Juricic, "On the stability of biped locomotion," *IEEE Trans. of Biomedical Engineering*, vol. 17, no. 1, pp. 25–36, 1970.
8. K. Hirai, M. Hirose, Y. Haikawa, and T. Takenaka, "The development of honda humanoid robot," in *Proc. IEEE Int. Conf. Robotics and Automation*, Leuven, Belgium, May 1998, pp. 1321–1326.
9. M. Merrington and C. M. Thompson, "Tables of percentage points of the inverted beta (f)-distribution," *Biometrika*, vol. 33, pp. 73–88, 1943.

Design and Control of the Humanoid Robot ARMAR

K. Berns, T. Asfour and R. Dillmann

Forschungszentrum Informatik at the University of Karlsruhe
Haid-und-Neu-Str. 10-14 76131 Karlsruhe, Germany

Abstract. This paper addresses the development and current research work on the humanoid robot ARMAR which has to work autonomously or interactive in cooperation with humans in workshops or home environment. ARMAR is developed to realize interaction with humans and mobile manipulation.

1 Introduction

Robots of the current generation have been used in fields isolated from the human society. They suffer major shortcomings because of their limited abilities for manipulation and interaction with humans. Humanoid Robots are expected to exist together with human beings in the everyday world such as hospitals, offices and homes. An intelligent behaviour of a robot with human-like skills can only be carried out through the permanent interaction between cognition components, the robot system itself and a human operator who demonstrates typical actions. In cooperation with human beings humanoid robots should react human friendly. Therefore, they need a light-weight body, high flexibility, many kinds of sensors and high intelligence. They have to be adaptive to new situations and capable of performing tasks in dynamic environment. Their design requires also a high extent of integration of mechanical, electronical and computational technologies. Tanie (1999) presented the different reseach topics for the development of a humanoid robot and its application.

At the Forschungszentrum Informatik Karlsruhe (FZI) the humanoid robot ARMAR is developed for applications like assistance in workshops or home environment (Asfour et al., 1999). Main focus of our research concentrates on the programming and execution of manipulation tasks of ARMAR by a direct and real-time mapping between the robot and the person, which demonstrates the task (Asfour et al., 1999). In cooperation with human beings humanoid robots should share the same working space and should react human friendly. The manipulation ability of ARMAR has to be comparable to that of a human. In the following the present state of our research is described.

2 Mechatronics

In order to achieve a high degree of mobility and to allow the simple and direct cooperation with humans, the structure (size, shape and kinematics) of the arm and of the torso should be similar to that of a human. Up to now it is not planned to use legs for the locomotion of the machine, since in a workshop environment it is not necessary to have such a flexible locomotion system. In fact one function normally supported by legs is the change of the total height. This influences

 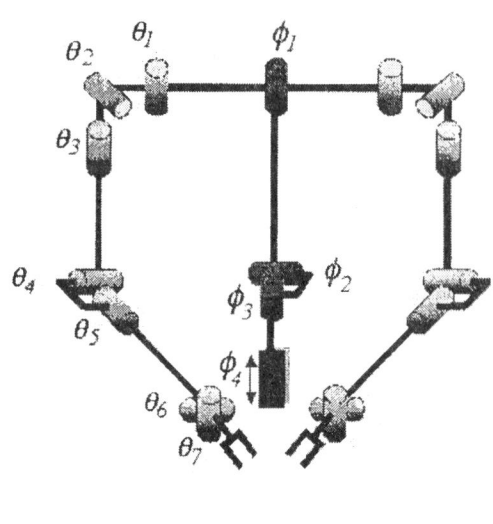

Figure 1. The humanoid robot ARMAR and its kinematic model.

the workspace of the two arm system. However we installed a telescopic joint in the torso of ARMAR to have this degree of freedom, too.

The mechanical concept of the humanoid robot ARMAR consists of an autonomous mobile wheel-driven platform, a body with 4 DOF, a two arm system with a simple gripper and a stereo camera head. The total weight of ARMAR is about 45kg. The mobile platform consists of two active driven wheels fixed in the middle of an octagonal board and another two wheels as passive stabilisers (Fig. 1). The maximum velocity of the platform is about 1m/s. The anthropomorphic body of the robot is placed on the mobile platform and supports a rotation of about 330°. It also can be bended forward, backward and sideward (circa 110°). To adapt the height of the robot (180 cm), a telescopic joint is included in the body. With this joint the total height of the machine can be increased by 40cm.

For the dual arm system of ARMAR, we designed two anthropomorphic arms, each having 7 DOF and a length of 65 cm (including the gripper). At present a simple parallel jaw gripper is implemented. Currently, a new humanoid five-finger lightweight hand with only one actuator with 20 DOF is under construction. The new hand has the ability of performing most of human hand grasp types. Since the robot should support a simple and direct cooperation with the human, the physical structure (size, shape and kinematics) of the anthropomorphic arm is developed as close as possible to the human arm in terms of segment lengths, axis of rotation and workspace. The anthropomorphic arm design is based on a simplified kinematics model, which approximates the kinematic, kinetic and anthropomorphic characteristics of the human arm (see figure 1). Details about the mechanics of the arm of ARMAR are reported in (Berns et al., 1998).

The computer architecture of the robot has been designed to be modular. Currently, the robot is controlled by a cluster of seven C-167 micro-controllers and a standard PC. The mico-controllers are coupled with special power cards, which control 4 motors. The micro-controller boards are connected via CAN-Bus with a maximum transfer rate of 1 Mbit/s to the PC. For real-time requirements the real-time operating system RT-Linux is used. Modules like trajectory planning, data interpretation from every sensor directly coupled with the micro-controllers are running on the PC. The PC is connected via wireless Ethernet to a PC-network. Programs for simulation and special GUIs for the monitoring and the control of ARMAR are running on these external PCs (see Asfour et al., 1999).

The sensor system consists of angle encoders for each joint with a resolution of $0.1°$. Current and the voltage of each motor are determined by the power electronic card. For gripping various kinds of objects an artificial skin is placed on the inner side of the four fingers. It is realized by measuring the electrical resistance of the conducting rubber that is divided in several fields of an array. Additionally, it is planned to include stain gauges on different parts of the body of ARMAR. To detect the environment, a stereo camera system is fixed on ARMAR. The sensor system of the mobile platform includes eight ultrasonic sensors and a laser scanner. Both types of sensors are used for a collision free navigation of the robot.

3 Control

The implementation of full dynamic control on a robot still remains a challenge to robot scientists and researchers today. It is known that the performance of a robot can be improved with the including of robot dynamics into its controller. However, the complexity and, more important, the lack of knowledge about the dynamic parameters of the robot, lead robots to be controlled mostly by PD "poportional derivative" or PID "proportional, integral, derivative" control, where the control is done independently for each joint.

For the control problem of the dual arm system of ARMAR only the kinematic control is considered. The control problem is solved in two stages: first, an inverse kinematic problem is solved to transform task variables into the corresponding joint variables for the arms and body of the robot. The obtained joint variables are input of a suitable joint control scheme. The coordination is then solved at the inverse kinematics level while arm interaction can be considered at joint control level. Since the robot is moving at low speed, traditional position joint controllers are used, because they can better deal with nonlinear friction.

The purpose of a position controller is to drive the motor so that the actual angular displacement of the joint will track the desired angular displacement specified by a preplaned trajectory. The joint-angle measurements of the arms and body of ARMAR are obtained by accurate encoders. A robust robot control requiring only position measurements is easy to implement and increases the dynamic performance of the robot manipulator. Nevertheless, when velocity and force sensors are available, feedback of the velocity and forces can be added to improve the performance of the system. Figure 2 shows the structure of the controller we use. The controller chooses a set of parameters of a classical position joint controller depending on the configuration of the arm. The sets of parameters are established through experiments.

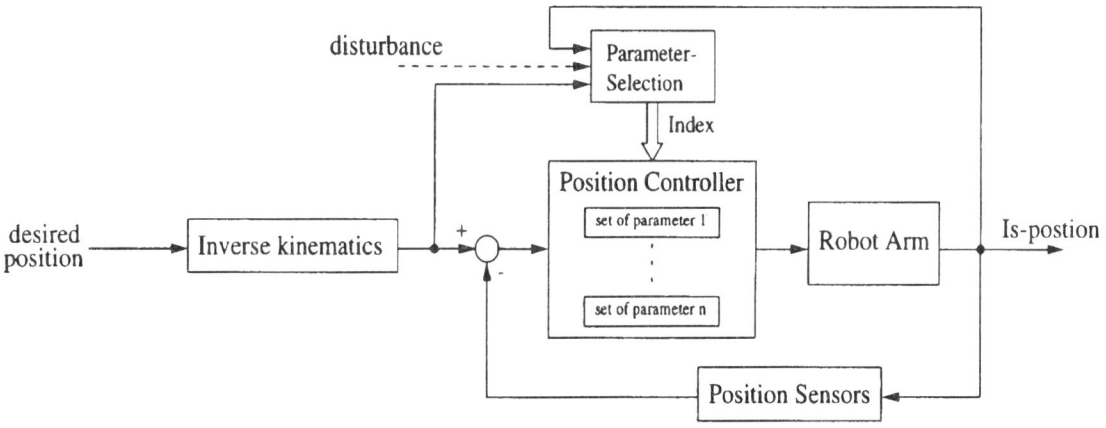

Figure 2. The structure of the controller

4 Programming of manipulation

The problem of human robot interaction, such as copying the motion of a operator by a robot, is addressed by the use of a full model of the human body. An introduction of the modeling of human can be found in (Morecki, 1999). Our concept of programming manipulation tasks is based on motion measurement via position sensors attached to the human arm. Currently, two commercially available position sensors are used to measure the positions and orientations of the elbow and the wrist. Therefore, a mechanical master device for controlling the robot like in (Chang et al., 1999) is not necessary. Our approach in programming of manipulation tasks can be divided into three main parts:

- Measurement and analysis of human arm movements demonstrated by a human operator. The human arm movements are only kinematically represented, and the dynamics for human manipulation tasks is taken into account as a post-processing step. It is not necessary to consider the dynamics unless realistic velocity distribution for manipulation motions are required.
- Transfer of the demonstrated movements to ARMAR using an inverse kinematics algorithm. This is necessary because most manipulation tasks are specified in terms of the object trajectories.
- Adaption of body/arm movements of ARMAR according to the description of the object to be manipulated and the robot environment.

Up to now the first two steps of the programming approach are implemented. Starting from the human arm movement detected by the above mentioned tracking system an adequate 3-D model of the human arm is selected. The two sensors are attached to the elbow and wrist of the human arm. The arm configuration can then be computed from the sensor data using a specialised inverse kinematics algorithm. In order to compute the joint angles of the robot arm corresponding to the operator's current arm configuration, we assume that the shoulder positions are fixed. The proposed method suffers major drawbacks because of the limited range and accuracy of the used

Design and Control of the Humanoid Robot ARMAR

magnetic sensors. Our proposed model of the human arm provides a quantitative description of the mechanical parameters of the bones, the joint kinematics and the coordinates systems required to describe their relative degrees of freedom.

Based on the elbow and wrist trajectories of the human arm model, the joint angles of the arms of ARMAR are calculated via inverse kinematics. The problem when solving the inverse kinematics of the 7-DOF robot arm arises from the under-determination of the inverse kinematics. An analytical-geometrical method for a closed form solution of the inverse kinematics problem is reported in (Asfour et al., 1999. The robot arm redundancy can be described by an analytical curve result from all feasible positions of the elbow (shoulder and wrist joints are fixed). Note that in order to apply the evaluated joint angles to the control of the robot arm it is necessary to take into account the mechanical joint limits.

5 Test Results

Up to now motion planners are developed for two types of motion: point-to-point (PTP) motion and curve-tracking (CT) motion. The inputs of motion planners are specified by a human operator. The outputs of the motion planners are sequences of joint angles, and are executed in the simulation as well as on ARMAR.

A simple dual-arm motion planner for coordinated motion is developed. This motion planner considers the closed kinematic chain of both arms and the object. This method is not powerful enough for the manipulation of arbitrary objects because of the limited accuracy of the mechanics and the noise in the sensor information. Therefore, it is planned to use additionally joint-torque sensors to perform tasks involving contacts and forces. Figure 3 shows typical manipulation task which should be performed by ARMAR.

Figure 3. ARMAR is performing a manipulation task, which was demonstrated by a human supervisor.

6 Conclusion

In this paper we presented how human arm movements of typical manipulation tasks can be transfered to the humanoid robot ARMAR. The mechatronics concept of our humanoid robot ARMAR is also introduced. The transfer of human arm movements is directly done using a closed form solution of the inverse kinematics problem. Early cooperative manipulation tasks of both arms of ARMAR are also performed to demonstrate the capabilities of the humanoid robot.

In the future we intend to install the new five-finger hand. In parallel we have started to evaluate the stereo camera information for the detection of objects and the description of the environment. With the help of this information the last step of our approach – the adaptation of the ARMAR's movements – will be implemented. In Addition, the control system for the navigation of the mobile platform will be implemented. This will enable ARMAR to navigate based on maps of the environment, to avoid obstacles and to integrate mobility in manipulation tasks.

References

Tanie, K. (1999). MITI's Humanoid Robotics Project. In *Proceedings of the Second International Symposium on Humanoid Robots*, 71–76.

Asfour, T., Berns, K. and Dillmann, R. (1999). The Humanoid Robot ARMAR. In *Proceedings of the Second International Symposium on Humanoid Robots*, 174–180.

Asfour, T., Berns, K., Schelling, J. and Dillmann, R. (1999). Programming of Manipulation Tasks of the Humanoid Robot ARMAR. In *The 9th International Conference on Advanced Robotics (ICAR'99)*, 174–180.

Berns, K., Vogt, H., Asfour, T. and Dillmann, R. (1998). Design and Control Architecture of an Anthropomorphic Robot Arm. In *The 3rd International Conference on Advanced Mechatronics (ICAM'98)*,

Morecki, A. (1999). Modelling and Identification of Man's Motion. In *Proceedings of the Second International Symposium on Humanoid Robots*, 1–8.

Chang, S., Kim, J., Kim, I., Borm J. H. and Lee, C. (1999). KIST Teleoperation System for Humanoid Robots. In *Proceedings of the International Conference on Intelligent Robots and Systems (IROS'99)*, 1198–1203.

On Dynamics of Movement of Walking Machines with Gears on the Basis of Cycle Mechanisms

E.S. Briskin, V.V. Chernyshev, A.V. Maloletov and S.V. Sherstobitov

Volgograd State Technical University, Volgograd, Russia

Abstract. Multilegged statically stable walking machines of ground practicability with gears on basis of cycle mechanisms are under consideration. The dynamics of walking machines moving on comparatively flat surface with poor carrying layer was learned out both theoretically and experimentally in environmental conditions. Variations of improving the kinematic characteristics of moving of supporting point of walking mechanism were proposed and analyzed.

1. Introduction

In conditions of poorly facilitated roads while moving along the soils with poor carrying layer or along the layer ecologically injured, any movement of a vehicle can hardly be supplied with a wheeled or tracked moving gear. This makes it necessary to search for new types of gears, among which the most walking gears must be considered advantageous [1].

Figure 1. Walking Support

The basic features of stepping machines designed in Volgograd State Technical University (figures 1-3) moving along relatively flat surface and aiming to solve the problem of soil practicability, are that they might be used with cycled stepping mechanisms combined into gears so that one of such mechanisms will be always set into the phase of interaction with soil. Such approach to the problem makes the machine statically stable and it provides the possibility of ignoring the keeping the gait and affords concentration upon the design of the controlling system which is the analog of the control of machines with traditional types of gears.

Figure 2. Transportation machine

Figure 3. Walking Manipulator

2. Mathematical Model of Movement Dynamics

The touching powers $P_{\tau j}$ of discretely interaction of the j-support with the soil, modeled by swampy and elastic environment, might be determined with the following formula:

$$P_{\tau k} = \begin{cases} P_{\tau k}^* = C_\tau(x_j^* - x_j) - \mu \dot{x}_j, & \text{if } |fP_{\tau j}| \le P_{nj} \\ -fP_{nj}\text{sign}(\dot{x}_j), & \text{if } |fP_{\tau j}| > P_{nj} \end{cases}, \quad (1)$$

where $P_{\tau j}$, P_{nj} — accordingly are the touching and normal reaction of the soil under j support; c_τ, μ — are accordingly toughness and the coefficient of the viscosity resistance of the system "Soil-the mechanism's support"; f — is the coefficient of interdigitation; x^*_k, x_{kj} — are accordingly the coordinate of the point of the j-support on soil and its current coordinate.

Among the powers that affect the stepping machine the power of movement resistance is also taken into consideration which is predetermined with the expenditure of energy for the normal deformation of the system "Soil-Support" in the stepping mechanism [2];

$$F = \sum_{j=1}^{N} \frac{P_{nj\max}^2 \gamma}{2C_n S} = f_e G, \quad f_e = \frac{1}{G}\sum_{j=1}^{N} \frac{P_{nj\max}^2 \gamma}{2C_n S}, \quad (2)$$

where $P_{nj\max}$ — is the maximum normal reaction which affects j-gear; γ — is the coefficient of the regime; C_n — is the normal toughness; S — the step length, f_e — movement resistance coefficient, G — the machine's weight.

The controlling system of a stepping machine might be presented in the form of equations for holonomic nonstationary constrains which put restriction on movements of support ρ_j — of the stepping mechanisms towards the gear as well as on the step-surface orientation ε_j of the gears towards the mounting at different moments of time

$$\theta_k = \theta_j(\rho_1,\ldots,\rho_N,\varepsilon_1,\ldots,\varepsilon_N,t), \quad (k=1,2,\ldots). \quad (3)$$

The description of stepping machines dynamics exploited in marching regimes leads to the application model consisting of the system of combined hard bodies flatly moving along. This system includes the mounting, the gears jointed to in the N-shape by hinges. Each gear consists of minimum two stepping mechanisms that are moving counterphasedly and interlock with the soil discretely.

The equations of movement of the solid bodies system analyzed here are made by means of Lagrange equations that connect each other $2N+3$ generalized coordinates (x_c, y_c, φ, ρ_j, ε_j) as well as their first and second derivatives [3].

Joint solution of the equations received alongside with the equations that are in accordance with the system of control structure (3) gives the opportunity to find some laws that can't be found in the machines with traditional moving gear types. They are based upon the high vibroactivity, which is determined by the inertial power, through the quick-mass bearing of the walking mechanisms as well as through the discrete interaction their supports with the soil. Thus, we found two groups of critical moving speeds V^* that are coinciding with the resonance state of a vehicle

$$V_{cr,n} = \frac{S}{2\pi\gamma n}\sqrt{\frac{\sum C_\tau}{M}}, \quad V_{cr,l} = \frac{S}{2\pi\gamma l}\sqrt{\frac{\sum C_\tau(d_j^2+b_j^2)}{J}}, \quad (n,l=1,2..), \quad (4)$$

where M, J are accordingly the mass and the inertial moment of a vehicle; d_j, b_j — are middle meanings of a coordinate moving gear foot j interaction with the soil within the system of coordinates of the machine strictly jointed with the vehicle base.

3. Dragging-Dynamical Calculation

The machine vibrations upon the ground lead to the increase of power expenditure and additional vibro-movement of the machine. To calculate the dragging dynamics it is convenient to take the power expenditure increase into account with the equivalent coefficient of additional resistance f^* which depends on the speed of movement and the hook force Q. This is why the walking machine dragging-dynamical balance equation differs from the analogous equation for the machine with traditional types of gears by additional item and is presented as follows:

$$G[f + f^*(V,Q)] + Q = \frac{M_g \eta_1 \eta_2}{k_i}, \quad (5)$$

where M_g is the moment upon the driving gear shaft, η_1, η_2 — accordingly efficiency coefficients of transmission and the gear; k_i — the transfer relation upon i-transmission that connects the linear speed of a machine with the angle speed of the gear shaft.

To retain the succession of the movement theory of wheeled and caterpillar machines the evaluation of additional vibro-movement of a walking machine is being made with the assistance of the skidding coefficient δ that also depends on the speed and summary power of the movement resistance $\delta = \delta[V,(f+f^*)G,Q]$.

The features of the dependence of the skidding coefficient δ (figure 4) show that in the vicinity of resonance states skidding lessens and can even be negative. The physical meaning of the negative skidding coefficient is nothing but the walking machine movement with the higher speed than it is supplied by the kinematics of the walking machine.

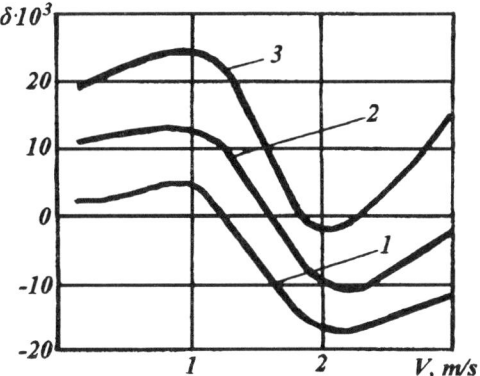

Figure 4. The dependence of the skidding coefficient of the speed. 1 — $Q/G=0$; 2 — $Q/G=0,22$; 3 — $Q/G=0,44$.

Another important feature of the walking machine movement is its high tendency to course instability. The cause of this phenomenon is based on the asymmetry of inertia forces in quickly fetched masses of mechanisms of walking and the forces that make an impact on the machine as a result of discrete interaction of supports with the ground. Transversal and angle vibrations, overlapping the programmed course movement of a machine, cause its moving aside.

4. Experimental Research

To research the dynamics of movement of designed experimental samples, to determine the dragging-joint features and to evaluate the practicability and the ability to maneuver the experiments were carried out in real field conditions [4]. The research of the dynamics of a walking machine was made by means of the method of video shooting of the object movement process followed by the flashback video research that followed. To reach this goal two parallel rows of shields with the scale-coordinated net were installed. Video cameras were installed strictly along the boards of the rear end of the machine, so that they could scan the central part of the scale-coordinated net together with the walking gear lambda-shaped mechanism balance levers. The cameras were working within the mode of seconds' countdown and that gave the opportunity to synchronize by seconds the features on different cameras in the process of experimental data analysis. The length of the shields allowed to record several completed cycles (steps) of a walking machine movement. By using a video the data of the wattmeter installed on the board had also been recorded during the experiments. The wattmeter was included in the electric chain of the electric driving gear, whereas its data characterizing the summary power spent by electric engines of power drive on both boards of a walking machine in the course of the movement.

Some of typical dependencies are displayed in figures 5,6.

The experiments showed that the walking machines are extremely practicable what is supplied by high dragging – joint features. At the same time the analysis of the results of the experiment disclosed a number of unsolved problems that stammer the realization of potential advantages of walking machines over the wheeled and caterpillar ones. The most important of them are non-linearity of the trajectory of the support spot of the mechanism of walking, non-stability of its speed at the stage of interaction with the ground and low ascend of the support spot of the walking mechanism at the phase of removing.

Non-linearity of the trajectory of the support spot affects the periodical ascend and descend of the center of machine-masses and is unfavorable for its energy features. Inequality of the support spot speed upon the ground also leads to slipping of supports on the ground and to higher energy expenditures affecting the course stability of movement. The increase of the height of the support ascend at the removal phase is favorable to increase the profile practicability, whereas the ability to change the orientation of the pads of the walking gears is essential for the kinematically strict turn of the machine organization.

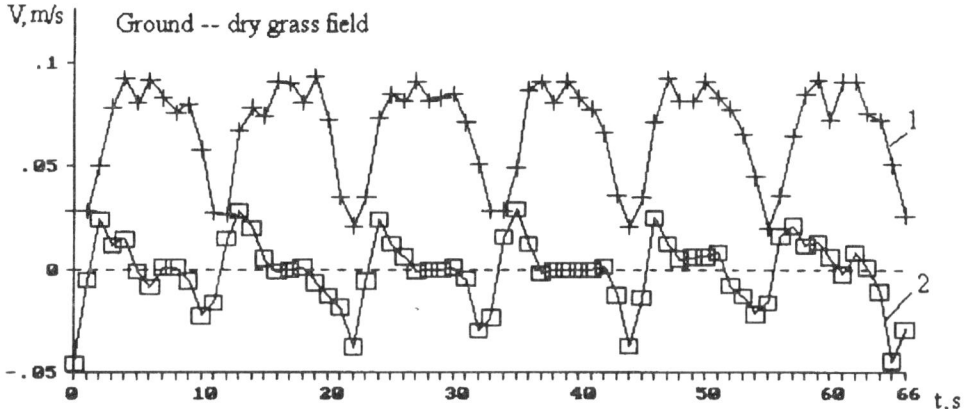

Figure 5. Dependencies of the horizontal 1 and vertical 2 speed of the center of masses of the machine. The cycle period makes 22 seconds.

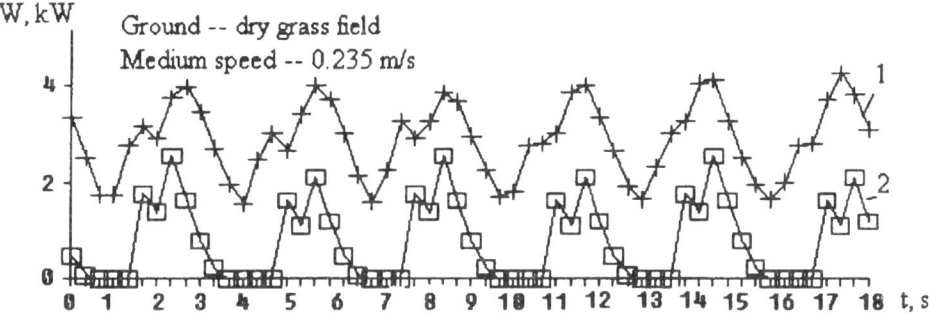

Figure 6. Dependencies of time. 1 – full power, 2 – power upon the vertical movements. The cycle period makes 6,33 seconds.

5. Improvement of the Exploitation Features of the Walking Mechanism

One of the variants of improvement of the rule of support spot movement of the walking mechanism lies in the introduction of unequal angle speeds into the course movement of the reducer-corrector based on Hook's joint, between the outer shaft of the board reducer and the shaft of the crank of the walking mechanism. (figure 7) In this case one can achieve a faster movement of the support spot at the stage of removing and a slower one at the stage of interaction with the ground. This is explained by the non-linear dependence between the angle speeds input ω_{in} and output ω_{out} shafts of the Hook's joint.

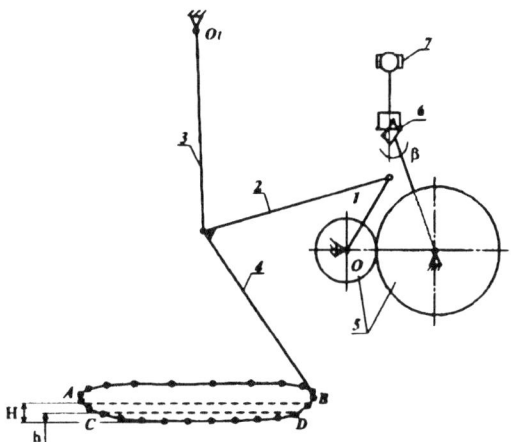

Figure 7. Modificated walking mechanism. 1 — crank, 2 — connecting-rod, 3 — rocker, 4 — element of support, 5 — multiplicator, 6 — gimbals joint, 7— motor

$$\omega_{out} = \frac{1 - \text{tg}(\psi_{in})}{1 - \text{tg}(\psi_{out})} \cos(\beta) \omega_{in}, \qquad (6)$$

where ψ_{out}, ψ_{in} make the angle of the input-output shafts accordingly, β makes the inter-shaft angle of the joint.

Then the change of the supports of both coupled and counterphase-working mechanisms of walking will take place not on line AB, but on line CD, i.e. "the depth of lowering", which is understood as the item of body vibrations of the machine while walking, is getting much lower. The flatness of horizontal speed of the support spot at the stage of interaction with the ground also becomes better. "The depth of lowering" and the flatness of speed may be characterized by non-sized items η and ξ accordingly.

$$\eta = \frac{h}{H}, \qquad \xi = \frac{V_{min}}{V_{max}}, \qquad (7)$$

where h, H — "the depth of lowering", V_{min}, V_{max} — minimum and maximum meanings of the horizontal speed at the support area.

Graph (figure 8) demonstrates the dependencies of these data on inter-shaft angle of gimbal for the investigated lambda-shaped mechanism of walking.

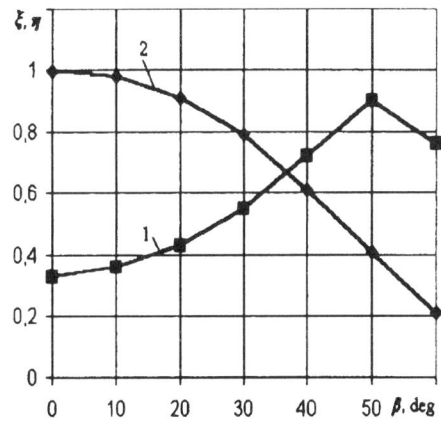

Figure 8. Dependencies $\xi=\xi(\beta)$ (1) and $\eta=\eta(\beta)$ (2)

6. Conclusion

The research undertaken proved the high ground practicability of walking machines stipulated at co-measured parameters by a lower resistance to movement and a larger coefficient of joint compared with wheeled and caterpillar machines.

To complete all potential advantages of walking machines over the machines with traditional types of gears it seems necessary to design walking mechanisms that are kinematically sufficient for the support spot movement.

This work is executed at financial support of the Russian Fund of Basic Researches.

References

1. Д.Е. Охоцимский, Ю.Ф. Голубев Механика и управление движением автоматического шагающего аппарата.—М.: Наука, 1984.—312 с.

2. Е.С. Брискин, В.М. Соболев Тяговая динамика шагающих машин с ортогональными движителями. Проблемы машиностроения и надёжности машин, №3, 1990. С.28–34.

3. Е.С. Брискин Об общей динамике и повороте шагающих машин. Проблемы машиностроения и надёжности машин, №6, 1997. С.33–39

4. Е.С. Брискин, В.В. Чернышев Экспериментальные исследования динамики многоопорной шагающей машины с движителями лямбдаобразного типа. Известия вузов. Машиностроение, №4, 1999. С. 32–37.

Summary

Walking machines of ground practicability moving on comparatively flat surface with low carrying capacity were under investigation. The equations of movement of such machines are made by means of Lagrange equations alongside with equations describing the machine control system. Joint solution of these equations gives the opportunity to find some specific laws of dynamics that can't be found in the machines with traditional types of moving gears. A number of experiments made in real environmental conditions proved the usefulness of elaborated mathematical models.

Development of Walking Machines: Novel Leg Drive Design and Control

Teresa Zielińska[1,2] and John Heng[1]

[1] School of Mechanical and Production Engineering, Robotics Research Centre
Nanyang Technological University, Nanyang Avenue
Singapore 639798
[2] Institute of Aeronautics and Applied Mechanics, Warsaw University of Technology
ul.Nowowiejska 24, 00-665 Warsaw, Poland

Abstract. In this paper, a novel design of a leg drive mechanism, hardware architecture and leg control method is described for a walking machine being developed for the study of various walking gait strategies. The leg mechanism employs an inverse differential gear drive system, to provide a large leg lift and swing sweep angle on a common pivotal point while being driven collectively by a pair of motors. The development platform consists of a pair of legs mounted adjacent to each other on a linear slide. A 3-axis piezo transducer is mounted on the foot to measure the various vector forces in the leg during the walking phase. The description of this unique leg drive system, the hardware architecture used, the position and force control strategy adopted, force sensing results gathered from the 3 axis force sensor, power conservation methods are discussed and presented.

1 Mechanical Structure

The general structure of walking machine LAVA (Legged Autonomous Vehicular Agent) is shown in Fig.1. The thigh section employs a differential gear drive system to achieve both leg swing and leg lift functions. This drive system offers two distinct features that are superior to conventional leg design. Firstly, leg lift and leg swing functions operate from a common geometrical pivot point. This feature will prove beneficial when performing workspace and kinematic modeling. Secondly during leg swing and leg lift motions, both motors are constantly working together to achieve the desired motion. The advantage would be that two smaller lighter motors can be utilized which can be combined to provide a cooperative effort. The result would provide savings in power consumption, weight penalty and size constraints. To provide maximum foot placement flexibility with precise turning functions, full 3 DOF per leg was incorporated in the leg design. The large leg lift and swing angle complements the symmetrical leg design, which enables the walking machine to be invertable.

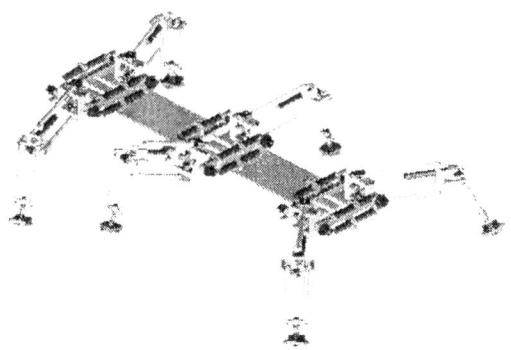

Figure.1. LAVA using multi-purpose leg being developed at the Robotics Research Centre

2. Control System

The functional structure of the control software was decomposed into hierarchically related levels The lowest level includes **joint** control. The angular joint positions are evaluated from the leg-end trajectory shape defined in Cartesian space. Inverse kinematics model is implemented there to evaluate the joints angular position. As the feedback devices, the incremental rotary optical encoders mounted on motor shafts are used. The motor controllers provide the PID algorithm for the angular position control loop.

The upper level of functional structure of the control system - **leg level**, produces the leg-end trajectory according to the proper timing scheme. The next level is the gait **level**. The rhythmic and free gait will then be generated. In the case of pick and place operations, this level will also to generate trajectories of front legs treated as manipulators. The last top level of the control software will be responsible for the generation of the body (**body level**) trajectories according to the user commands or according to the sensory readings. For the gait and body level, the most serious problem is to elaborate the method of free gait generation under assumptions that there are obstacles of different size and density (distance between obstacles) which must be omitted. The transition from one state to the another one is according to the stability conditions, sensory readings, goal of machine motion and motion states and leg-end coordinates of other legs. The planning of free gait is performed in parallel for the legs.

he real time QNX system and Watcom C are being used in the development of the control software. Inter-process cooperation has a typical *client-server* relation. Currently these three processes have been developed into software: *leg* process, *driver* process and s*ensor* process. The *Leg* process is the client while both the *sensor* and *driver* processes are the servers. The *Leg* process is responsible for the generation of motion trajectories according to the rules given by the programmer and the data received from *sensor* process, which reads the data delivered by force sensor. The *Driver* process is responsible for the co-operation with hardware. It receives the both the data and commands from the *leg* process, transforms that data to the format acceptable by hardware (motion controllers) and communicates with the hardware. On the back-paths (from servers to clients) includes the transmission of the send-only sensor data (from *sensor* process), the confirmation of the movement done (from *driver* process) and the information about the errors which can be hardware or software type.

Force control feedback is closed on the leg level of the controller functional structure. Adequate procedures are responsible for calculating maximum velocity and acceleration for each micro-step. On trajectory following movement, to prevent the leg-end vibrations, acceleration must be constant. Proper values of acceleration were obtained experimentally - for each motor separately. Those values are different for leg-end transfer and for support phase. The programme is responsible for proper evaluation of acceleration and velocity. Error in this calculation can destroy the motion time scheme that can result in motor shaft vibrations. For the point-to-point motion it was assumed that the time of one micro-step is long - 4s when compared with 0.03s in continuous path motion. One sixth of this time, motors should accelerate next 4/6 of micro-step, motor speed should be constant and next one sixth - motor must decelerate (Fig.2). It was tested by experiments that for this values and for every possible range of movement inside the working space calculated acceleration and velocity is never above the maximum range.

If the number of samples for one micro step is equal to n, and the distance that must be passed is equal to Δs (in counts) it is easy to calculate that the velocity v must be equal to:

$$v = \frac{6}{5}\frac{\Delta s}{n} \qquad (1)$$

and acceleration:

$$a = \frac{36}{5}\frac{\Delta s}{n^2} \qquad (2)$$

For the trajectory following movement, the motor's acceleration should be constant (for smooth leg-end movement). For this case to reach every possible reference position during the fixed micro-step the time of acceleration/deceleration must be flexible and velocity must be calculated in the proper way. To assume that unknown acceleration time (in samples) is equal to deceleration time and is expressed by x, we can find that the change of position during n samples is equal to:

$$\Delta s = ax^2 + ax(n-x) \qquad (3)$$

From the above, to know that the total acceleration and deceleration time - x must be at least one half shorter that one micro step, we have:

$$x = 0.5n - 0.5\sqrt{n^2 - \frac{\Delta s}{a}} \qquad (4)$$

Analyzing the above relation, it is easy to find that the acceleration must be greater than the certain value to prevent having as a solution an unrealistic complex number. On the other hand, the acceleration cannot be too big, which means very short acceleration/deceleration time and sharp motor motion. Assuming that this time must be longer than $1/12$ of micro step we find:

$$\frac{4\Delta s}{n^2} \leq a \leq \frac{16\Delta s}{n^2} \quad (5)$$

Distance increment Δs can be very different, for this reason it is difficult to calculate acceleration using only (5). In practice proper value of acceleration was found experimentally but paying the attention to (5). For experimental evaluation of a, many movements were observed monitoring the Δs values and extreme values of the acceleration when the fixed velocity profile (rel.(1),(2)) was used. Later considering (5), acceleration was fixed separately for leg-end transfer and for support phase. Transfer phase is usually much shorter than support.

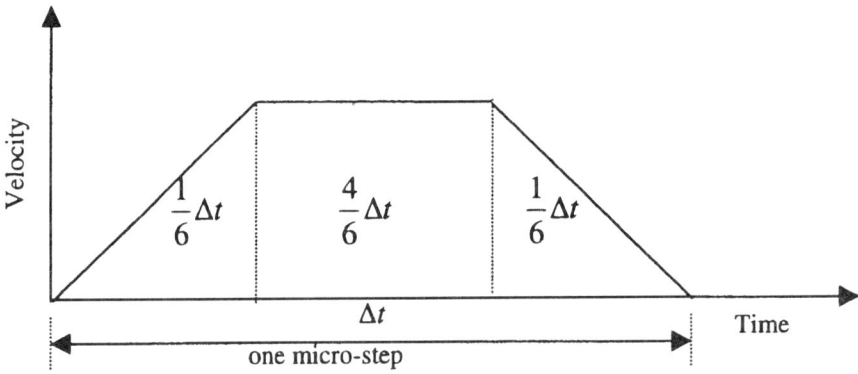

Figure 2. Trapezoidal velocity scheme with fixed time

3. Force Sensing

Force control is needed to raise the adaptability of the machine to irregular terrain and on different types of soil. In locomotion over a complex terrain, the necessity to control the horizontal force components, so that contact forces are within friction cones may arise. In locomotion on soft soil, it is necessary to control the leg loads because of their sinkage into the soil. In locomotion over slightly uneven terrain, the leg sinkage can be determined taking into account leg joints position, readings from inclinometers and loading on the legs read from leg-end force sensors. For loose sandy loam the sinkage is irreversible. Such soil behaves as an absolutely rigid support if the load on the foot becomes less than the maximum value already achieved.

Force controlled walking machine would give additional advantages by increasing energy efficiency by the reduction in internal forces between legs and providing the desired support forces regardless of the behavior of the terrain walked on. Low adaptability to the environment is the problem of position control. A position-controlled leg of walking machine would either move in the air without producing any forces for the body or exert all the forces available in the case of an uneven terrain. The latter possibility happens if there is a position error (due the lack of proper environment model, due to the control method or due to the change of environment properties).

4. Experiments On Force Control

For this experiment the test rig with one-leg mounted with a three components force sensor mounted on the foot was used. The vertical component of the reference Cartesian leg end trajectory and vertical force in relation to the micro-steps (time) are given. The average value of support force is 15N, variance is in range 2N and maximum force peaks are of about 5N (Fig 3). There is no observed peak (impact) at the beginning of the support phase and only the peak at the end. The reaction force variations shows that leg-end control should be improved by better positioning, which means greater density in division of leg-end Cartesian trajectory to the reference points (between this points leg-end movement results only from the leg structure and motors velocity regime). Such a solution results in greater calculation demand (more amounts of control commands and more often inverse kinematics must be solved). Other solution is the application of force control. In this case the leg-end motion between the Cartesian points will be also "uncontrollable", but leg-end position correction can help to smooth the reaction force. In our experiments, body motion was simulated by movement of the "ground" supported by the leg. The friction coefficient between the ground and nonmoving support was very small. Force sensing can be helpful not only for synthesis of force control methods but also for the synthesis of position control.

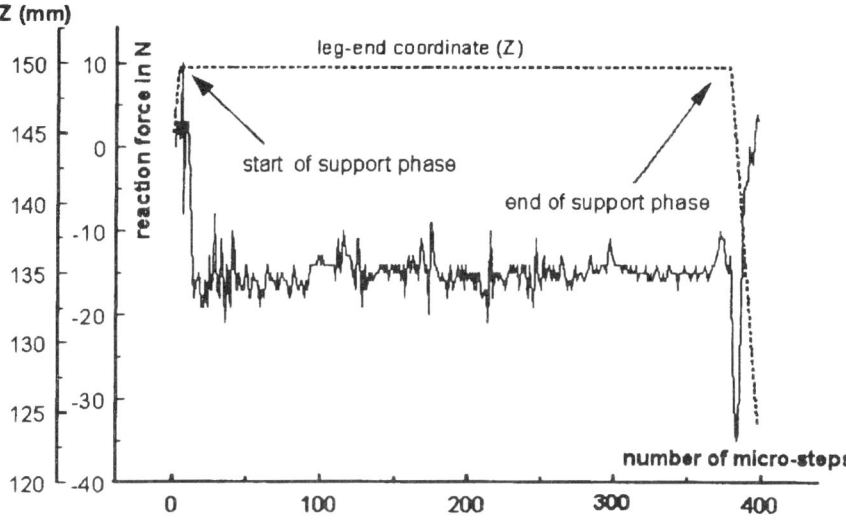

Figure. 3 Force distribution during support phase

The force decomposition is based on the leg positions. Since all leg positions are needed for the calculations, the proper place of these calculations is in the main processor (computer) of the body level of the control system. Therefore there will be a delay between the time of the leg position readings and activation of the corresponding force references. The other drawback is the calculation effort that is needed for proper force control.

Leg force distribution problem is complex from the numerical point of view and it has not a unique solution. There is an indefinite number of solutions to produce the six degrees of freedom (DOF) body force reference, because at least three legs are always used to support the body in the case of static walking and there are at lest nine motors (motor torque's) in the supporting legs. Choosing the best suitable solution is an optimization task called "leg force distribution". Optimality depends on the point of view, because reaction forces have influence on other factors. Any optimal solution can be considered when behaviour is analyzed according to partial goals.

5. Motion Simulator

Motion generator/simulator is implemented in Visual C. Its function is to transform the motion commands, which are written in defined language to the body position and legs trajectories which results in the reference trajectory following movement. Fig.2 shows the screen view of the motion simulator, the top view of the terrain and body trajectories are shown. The arrows mark the body motion history. Moreover, the enlarged view of the body with the legs is displayed for every moment of simulation time. The figure inset in the bottom left corner is the final screen view that shows the final body and leg positions when the body trajectory was completed. The simulator will form part of the motion generator that is to be used in the real control system in future.

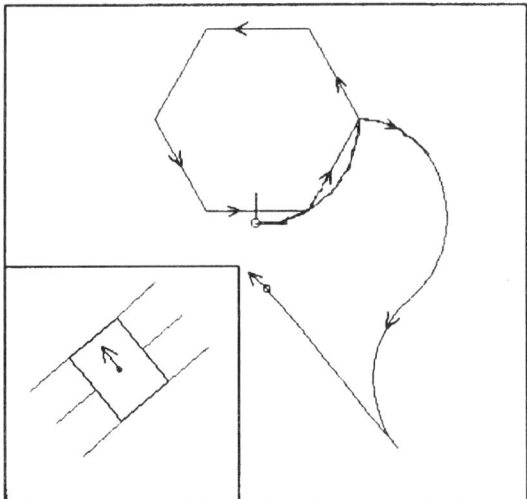

Figure 4. Motion simulator

6. Summary

The development and usability of walking machines can be constantly improved by the aid of proper design. In comparison to the industrial manipulators, the task to build adaptable, autonomous walking machine is more difficult. Walking machines have more active degrees of freedom than industrial robots. All this joints must be controlled properly in real time. It also means that the hardware and software systems must meet more critical demands than those for industrial controllers. Systematic approach to mechanical and control system design can benefit in the flexibility for future development and modifications. Selected problems that meet the design of mechanical and control system of walking machine (*LAVA*) are shown in this paper. The current research focuses on the synthesis of control software and experimental tests of walking behavior.

7. Acknowledgments:

This work was conducted with the support of Robotics Research Centre, Nanyang Technological University, Singapore.

References

Pugh, R.D., Ribble, E.A., Vohnout, V.J., Bihari, T.E., Walliser, T.M., Patterson, M.R., Waldron, K.J. (1990), Technical description of the adaptive suspension vehicle. The Int. Journal of Robotics Research., 9(2), 24-42

Hartikainen, K. (1996). Motion Planning of a Walking Platform Designed to Locomote on Natural Terrain. Helsinki Univ. of Technology.

Pal, P.K., Jayarajan, K. (1991). Generation of free fait - a graph search approach. IEEE Trans. On Robotics and Automation, 7(4), 299-305

Gardner, J.F (1992). Efficient computation of force distribution for walking machines on rought terrain. Robotica.,10, 427-433

Gornievsky, D.M., Schneider, A.Yu. (1990). Force control in locomotion of legged vehicles over rigid and soft surfaces. The Int. J. of Robotics Research., 9(2), 4-22

Klein, Ch.A., Kittivatcharapong, S. (1990). Optimal force distribution for the legs of a walking machine with friction cone constraints. IEEE Trans. On Robotics and Automation., 5(1),3-85

Manko, D.J. (1992). A General Model of Legged Locomotion on Natural Terrain. Kluver Academic Publishers.

Nagy, P.V., Desa, S., Wittaker, W.L. (1994). Energy-based stability measures for reliable locomotion of statically stable walkers: theory and application. The Int. Journal of Robotics Research, 13(3), 272-282.

Song, S. M., Waldron, K.J. (1986). Geometric design of a walking machine for optimal mobility. Transactions of The ASME: Journal of Mechanisms, Transmissions, and Automation in Design, Dec., 1-8

Brooks, R.A., Stein, L.A. (1994). Building brains for bodies. Autonomous Robots., 1, 7-25

Klein, Ch.A., Olson, K.W., Pugh, D.R. (1983). Use of force sensor for locomotion of a legged vehicle over irregular terrain. The Int. Journal of Robotics Research., 2(2), 3-17.

Loh, J., Heng, J., Seet, G., Sim, S.K. (1998). Behavior-based Search using Small Autonomous Mobile Robot Vehicles Proc. 2nd Knowledge-Based Intelligent Electronic Systems International Conference, Adelaide, Australia, Apr. 21st to 23rd, 3, 294 – 301

Heng, J (1997). ALV Design and Developmental Work – Interim Report, Dec. Nanyang Technological University, School of MPE, Nanyang Ave, Singapore 639798

Zielinska, T. (1997). Utilisation of biological patterns in reference trajectories generation of walking machines. IEEE 8th Int. Conference on Advanced Robotics. Workshop II: New Approaches on Dynamics Walking and Climbing Machines. Monterey, California, USA, 92-104.

Autonomous Locomotion of Walking Machines in Rough Terrain

M. Frik, A. Buschmann, M. Guddat, M. Karataş and D.C. Losch

Gerhard-Mercator-University of Duisburg,
Faculty of Mechanical Engineering, Department of Engineering Mechanics,
47048 Duisburg, Germany

Abstract. The paper describes a method of autonomous locomotion control of walking machines in unknown and rough terrain which is based on our software-library for gait generation combined with sensory information on distant and near obstacles. The presented methods are applicable for nearly any walking machine with at least two symmetrical pairs of legs. The application of the developed algorithms and libraries are demonstrated within a simulation as well as with the six-legged walking machine TARRY II.

1 Introduction

To simplify the generation of various gait patterns, a software-library for gait generation for walking machines with at least two symmetrical pairs of legs has been developed. The movement of the central body and the legs is defined by a set of gait parameters. A detailed description of the parameters can be found in Buschmann et al. (1998).

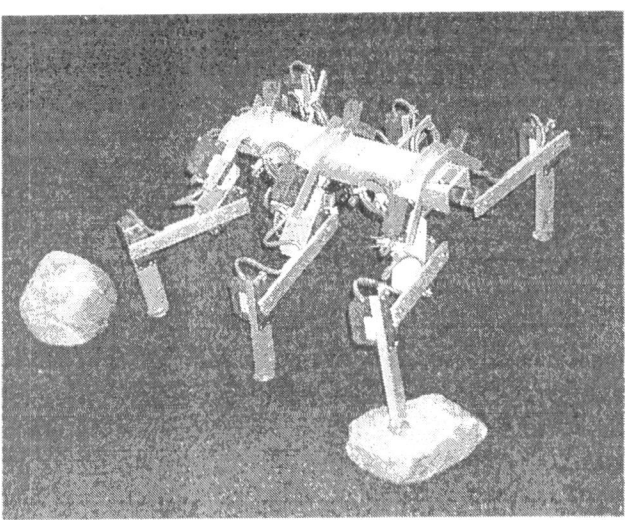

Figure 1. TARRY II.

The trajectories of the feet are stretched cycloids, which minimize the impact when the feet hit the ground. The periodic gait patterns of the simulation are used to train neural networks. Gaits in all directions on planar grounds can be generated due to the ability of the neural networks to generalize. In rough terrain the machine has to adapt to the environmental conditions. To do this the machine reacts on different kinds of sensor output with so called reflex actions. The possible reactions can be divided into local and global reactions. Local reactions manipulate one leg only, whereas global reactions influence two or more legs. Small disturbances such as bumps or dents in the ground are in general adjusted with local reactions. To deal with greater problems such as large obstacles that can not be climbed over, global reactions have to be used. The scheme of the control architecture is shown in fig. 2 (Frik et al., 1999). The components used to drive the real robot are discussed in Frik et al. (1998).

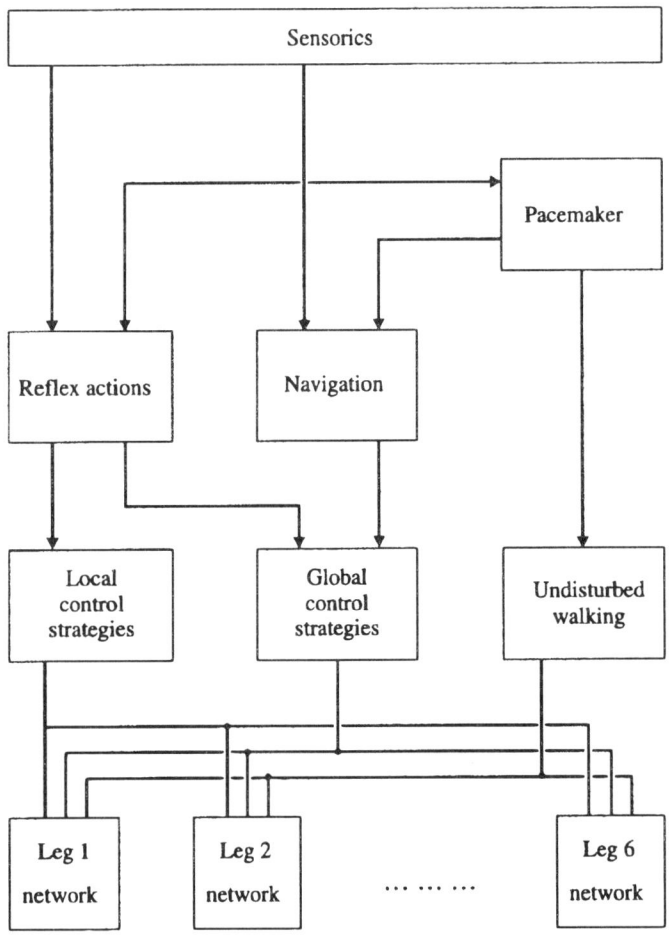

Figure 2. Schematic view of the control scheme.

This paper presents some of the basic reflexes that are used to react on elements in the close environment of the machine to show how adaptive walking works. Additionally a method to deal with the output of a rotatable ultrasonic sensor to avoid collisions with large obstacles by changing the route of the machine as well as some simulation results are presented.

2 Reflex actions

TARRY II has several sensors to initiate reflex actions to deal with different situations. The most important of these reflexes is the reaction to secure a proper standing of the machine. To trigger this reflex the data of the ground contact sensors, one mounted in each leg, are used. To survey the current stance a neural network is implemented. This network has been trained with data that was gained by recording the respective data of the real machine during undisturbed walking. The data is collected together with the output of a central pacemaker that is used for synchronization. This pacemaker signal is used as input for the neural network whereas the sensor signals, forming the stance pattern, are used for comparison. If a difference between the current pattern and the network created pattern has been detected, commands are created to raise or lower the concerned legs.

Another fundamental reflex is the so called Levator-Reflex that is also watched at the walking stick insect *carausius morosus*. If a leg hits an obstacle during its swing phase it is retracted and raised to a higher level. At this new level a new attempt to climb the detected obstacle is taken. This action is repeated until the barrier has been passed with this leg or until the leg reaches its maximum height. To detect collisions during the swing phase the ground contact sensors and the current of the hip-servo motor is used. If it exceeds the normal values, determined during normal walking, a collision is detected and the countermeasures are taken.

In addition to these most basic reflexes an inclination control reflex and the central generation of a ground map are implemented (Guddat and Frik, 2000).

The inclination control is done via a very simple algorithm. To keep the simple neural generation of the walking patterns shown in Frik et al. (1999) no roll or pitch angles are used as input. Instead the inclination is leveled by changing the individual leg heights. The geometric inconsistencies that may occur, such as a leg accidentally loosing ground contact, are intercepted by the other reflexes.

As the machine does not use a completely decentralized approach, it is able to store some data in a central repository. In this way the machine keeps a ground map of the terrain below the machine. Due to the data gathered in this map the robot is able to avoid subsequent collisions with obstacles other legs have already passed. By this it is possible to move much more fluently and reduce the number of collisions. This makes the machine faster and decreases impacts on the material.

3 Navigation

The main goal of the navigation is to avoid collisions with large obstacles in unknown environments. The only sensor information available for distant obstacles is gathered by a rotatable ultrasonic sensor which is attached to the front of the machine. This sensor detects the distances to obstacles in several directions as shown in fig. 3. For each direction the signal is sent to a

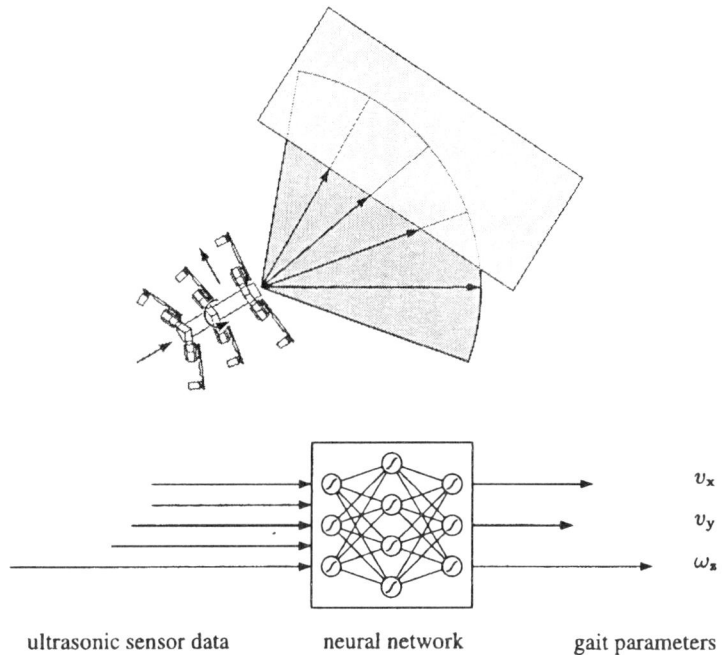

Figure 3. Obstacle detection.

single input of a neural network which generates new gait parameters that lead to a walking path without collision.

The training data for this network was created by steering a machine manually serveral times in a simulated environment through a course with obstacles (Losch, 1999a). The simulated sensor information was recorded with the steering commands of the operator. After the learning process the network was also able to navigate the machine in a similar, unknown environment. In general this method works fine, but in special cases collisions may happen. The problem arises if the training data for the network was not unique. This occurs if the operator reacts in the same situation (same ultrasonic signals) in a different way while recording training data. For example if the machine is heading straight towards an obstacle there are the options of passing to the right or the left side. It is not advisable to train both paths with the same network, because the learning process does not converge and the result becomes unpredictable. The worst imagineable case is that the neural network generates the average solution and moves straight forward into the obstacle.

To avoid the learning of similar input values assigned to different output values the neural network was divided into more specialised networks for different situations. For each situation a reaction scheme exists, illustrated in fig. 4. In this case seven reaction schemes are used: moving towards a wall, turning to the left, turning to the right, passing a gate, wall on the left side, wall on the right side and passing a small obstacle. Each reaction scheme is trained by a single neural network.

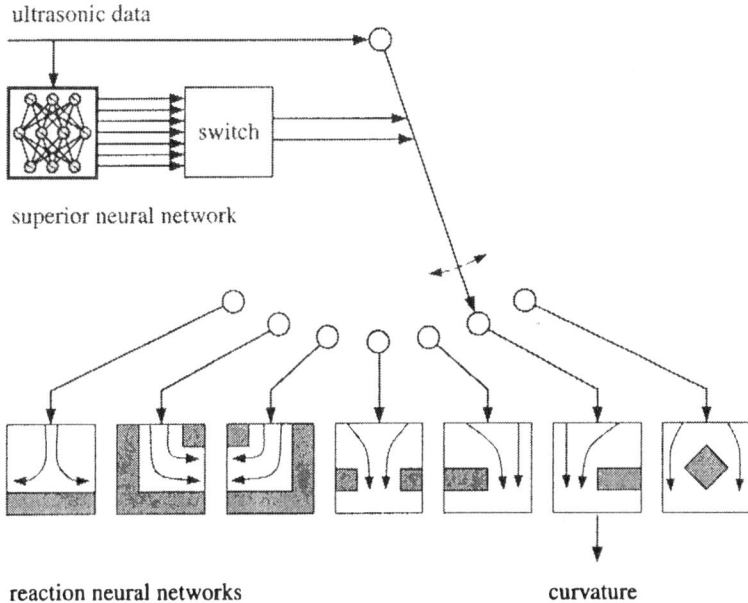

Figure 4. Global control strategy.

3.1 Training of the neural network

To achieve the training data for the reaction scheme networks the desired paths around the obstacles are generated. In the two dimensional simulation this is done by creating polygons which are replaced by b-splines to create the curved paths. For each step of the path the simulated ultrasonic sensor data is recorded and stored together with the curvature of the path. In previous experiments the body-fixed velocities in longitudinal and transversal direction and the rotational velocity were stored instead of the curvature. The advantage of learning the curvature is that the neural network has only one output and the path can be traced with different path velocities. The disadvantage is that no transversal motion is possible.

Because there has to be a decision which network should be used in the current situation of the machine, another neural network must be trained. This superior network has the same number of input units like the others, but as many output units as the number of existing reaction networks. The training data for the superior network consists of all ultrasonic sensor data. For every sensor record the corresponding output unit is activated and the other output units are set to zero. The network learns to compare the current environment with the reaction schemes by evaluating the sensor input.

3.2 Control system

The superior neural network calculates a different signal for each reaction scheme network with the ultrasonic sensor data as input. These signals are forwarded to a switching module. The task of the switching module is to decide which reaction scheme network should be activated. In the simplest case this is the network assigned to the strongest signal. Then the sensor data is sent to the selected reaction network and the curvature of the path is generated. With the desired path velocity the gait parameters, needed to follow this path, can be calculated.

3.3 Simulation

In the two dimensional simulation seven neural feed forward networks with a (11-15-15-15-1) layer structure have been trained by a backpropagation algorithm. The seven reaction schemes are illustrated in fig. 4. The superior network had a (11-15-15-15-7) structure. The goal of the simulation was to move in an unknown environment without collision. The simulation was stopped after 400 seconds. The path velocity was 0,6 m/s and the detection rate was set to 25 Hz. The paths for two different starting points are illustrated in fig. 5.

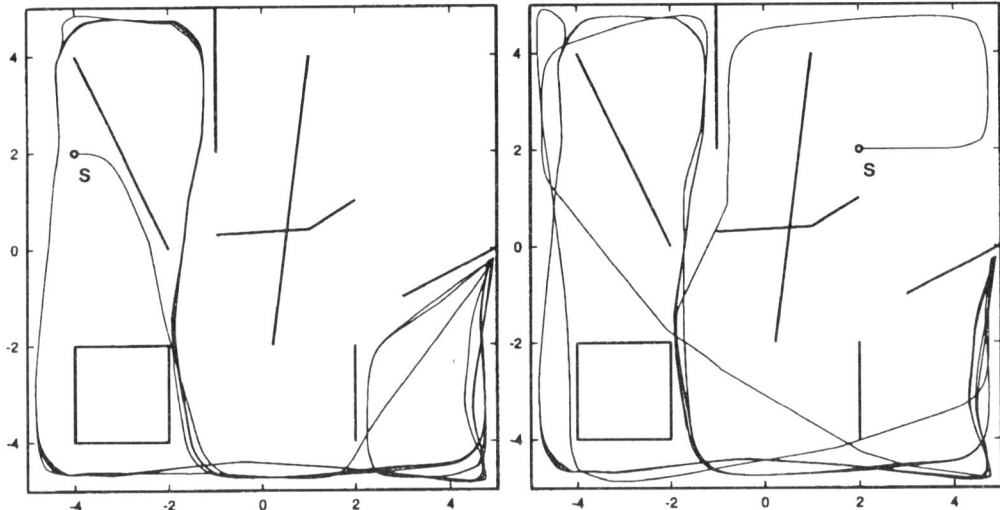

Figure 5. Paths with starting point S(-4;2) and S(2;2).

4 Summary

Due to its system of reflex actions the TARRY II walking machine is able to deal with several kinds of obstacles. The machine can cope with bumps and dents, edges, small obstacles and slopes as well as seesaw like situations. This ability to adapt to adverse situations in immediate vicinity is augmented by a method to detect larger obstacles in advance and to change the route of the machine accordingly. This strategy is based on a set of neural networks, one network decides which approach is chosen and a set of other neural nets creates an appropriate evasive action.

References

Buschmann, A., Frik, M., Guddat, M., Karataş, M., and Losch, D. C. (1998). Modular generation and optimization of gait patterns for walking machines. In *Autonomous Walking 98: Theory and Practical Realisation of Walking Machines*, 7–14. Fraunhofer IFF – Workshop, Magdeburg, Germany.

Frik, M., Guddat, M., Losch, D., and Karataş, M. (1998). Terrain adaptive control of the walking machine TARRY II. In *Proc. European Mechanics Colloquium, Euromech 375 – Biology and Technology of Walking*, 108–115.

Frik, M., Guddat, M., Karataş, M., and Losch, D. C. (1999). A novel approach to autonomous control of walking machines. In Virk, G. S., Randall, M., and Howard, D., eds., *Proceedings of the 2nd International Conference on Climbing and Walking Robots CLAWAR 99, 13–15 September, Portsmouth, UK*, 333–342. Bury St. Edmunds: Professional Engineering Publishing.

Guddat, M., and Frik, M. (2000). Control of walking machines with artificial reflexes. Manuscript submitted for publication.

Losch, D. C. (1999a). Ein Ansatz zur Navigation autonomer Roboter mittels neuronaler Netze. Interner Forschungsbericht 2/99, University of Duisburg, Department of Engineering Mechanics.

Losch, D. C. (1999b). Kollisionsvermeidung autonomer Roboter in unbekannter Umgebung mittels neuronaler Netze. Interner Forschungsbericht 8/99, University of Duisburg, Department of Engineering Mechanics.

Design and Control of a Biped Robot

Klaus Löffler, Michael Gienger and Friedrich Pfeiffer

Institute B for Mechanics, Technical University of Munich, Germany

Abstract. The project of a two-legged walking robot is presented. The control system is based on a nonlinear control scheme using the method of feedback linearization. Special effort has been devoted to the trajectory generation which is computed such that the system remains controllable throughout the entire gait cycle. The controller and simulation algorithm have been implemented parametrically. Parameter studies have been performed which lead to the final design. The mechanical design with particular emphasis on actuators and sensors is presented.

1 Introduction

In recent years, the number of scientists working in the field of two-legged walking machines has increased rapidly. Centers of this research are in Japan and the USA. The realized walking machines range from simple two-dimensional robots to fully self-contained autonomous robots as e.g. introduced by Honda Motor Co., Ltd. (Hirai et al., 1998). Running motion with ballistic phases has been investigated by Raibert (1986).

This paper deals with the control and design of the autonomous biped walking robot "Johnnie" which is currently being developed at the TU München. Particular emphasis has been devoted to the realization of a "dynamically" stable motion which allows for "running" including ballistic phases. The geometry is based on anthropometric data. The joint structure (figure 1) has been chosen to realize a human-like motion, which is characterized by the main determinants of human gait (Saunders, Inman, 1953). The joints provide the required degrees of freedom for walking and running on plain and uneven ground and around curves.

2 Mechanical Design

The robot is supposed to be autonomous in terms of the computational power and the actuators, while the operating power will be supplied by external sources. The size of the machine will be 1.80 m, its geometry is based on anthropometric data (Hahn, 1994). Electric actuation is used to generate the motion of the system, since former research has proven that electric actuation has the highest power-to-weight-ratio in the weight region of interest (Weidemann, 1993). Other reasons for choosing electric actuation are its good controllability and the robustness against overload. The actuation is realized by DC motors with inside rotor design. Commutation with brushes has been found out to be superior to brushless commutation in terms of the choice of the gear components. In order to implement high transmission ratios in a very compact design, Harmonic Drive gears and ball screw drives are used. Figure 1 shows the present state of the design.

The robot has 15 driven degrees of freedom allowing for a natural gait pattern. The hip joint is of particular interest since three degrees of freedom are implemented. Figure 1 shows the final design of this joint. The three axes of rotation intersect in one point. The motors of the α- and β joints are arranged coaxially to the joint axis, the Harmonic Drive gear is directly attached to the motor shaft. The γ joint is driven by two DC motors via toothed belt. To achieve a light weight design, the actuation of the ankle joint is realized with two linear drives based on ball screws. This way two motors can be used in parallel to realize one motion. The design has been tested in an experimental setup and turned out to be very efficient. The foot will contain a damping mechanism as impacts have to be reduced and the time gap between the sensor input and motor response has to be bridged.

3 Sensors and Electronics

The sensor system can be subclassified into joint sensors, vestibular sensors and force sensors. Each joint will be equipped with an incremental encoder which is attached to the motor shaft. The encoder allows for the measurement of the exact joint position as well as the joint angular velocity by numerical differentiation of the position signal. Reference points are detected by electronic switches. Experiments have shown that the signals are very accurate. In addition, the digital signal leads to a high robustness against electromagnetic interference. The vestibular sensor system consists of gyroscopic sensors and inclinometers. The spacial orientation of the upper body will be measured by numerical integration of the velocity signal of the gyroscopes. The inclination sensors are used for drift compensation. The ground contact forces will be measured by six-axes force/torque sensors which will be placed in the feet.

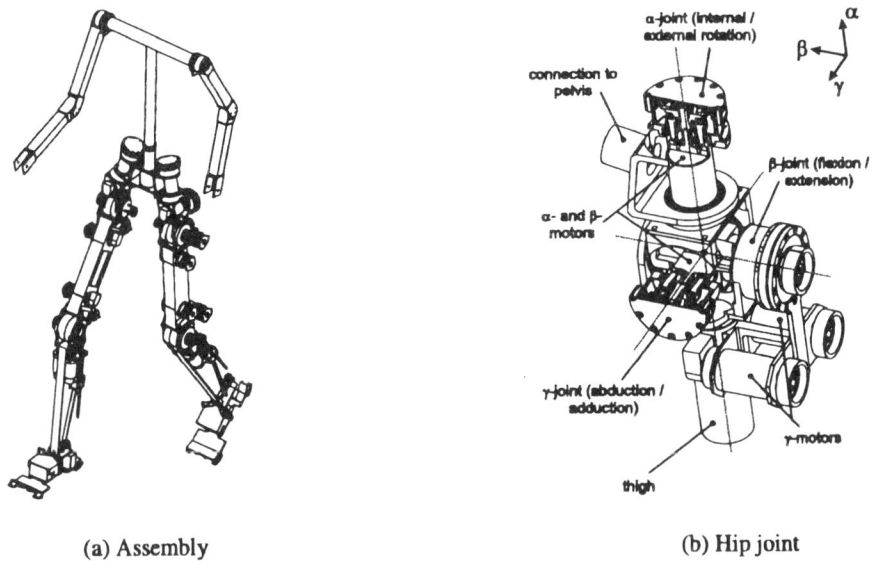

(a) Assembly (b) Hip joint

Figure 1: Design of biped robot "Johnnie".

4 Control of the Robot

The mathematical model of the plant corresponds to the design of the robot and consists of a system of 16 bodies connected by $m = 15$ rotational joints (figure 2). Then the overall system has $n = m + 6 = 21$ degrees of freedom. When the robot is standing on one leg, the 6 degrees of freedom of the upper body correspond to a specific orientation of the 6 rotational joints of the supporting leg. Furthermore there is one joint in the pelvis and one joint for each arm, which are used to compensate dynamic forces.

Each foot has 4 rubber elements that ensure that the coefficient of friction between foot and ground is high and that impacts are reduced that occur when a foot hits the ground. These foot elements are modeled as three dimensional springs.

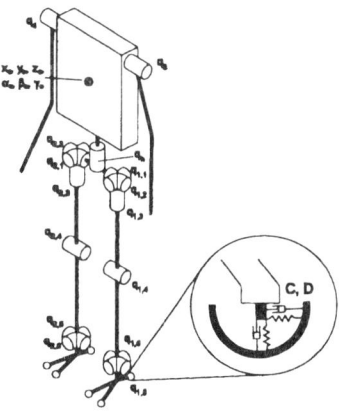

Figure 2: Plant model.

4.1 Gait Pattern

The gait pattern of the biped robot is generated on the upper control level. As shown in figure 3, the gait cycle is divided into five different phases. In the support phase one leg is standing on the ground while the other one is swinging to its next stance point. At the end of the step first the heel lifts off from the ground. This strategy is known from the human gait pattern and allows to start swinging the knee of the supporting leg forward before the foot actually leaves the ground. In the lift off phase the normal forces to the gound are reduced until the foot finally looses contact. Then both feet have no contact to the ground during the air phase. The subsequent step begins when the other foot has touched the ground and has taken over the whole weight of the robot. During each phase a set of m variables can be controlled by the m driven joints, while $n - m = 6$ degrees of freedom result from kinematic contstraints. In general the trajectory of the robot can be prescribed in any m coordinates, as long as there exists a unique transformation to the generalized coordinates. Since these constraints are varying, a different set of controlled variables is chosen in each phase of the gait pattern. The vector of controlled variables, $x_{c,i}$ consists of variables x_p that are controlled in all phases and a vector $x_{v,i}$ that has different elements in each phase i. The vector x_p consists of the angular position of the upper body around the x- and

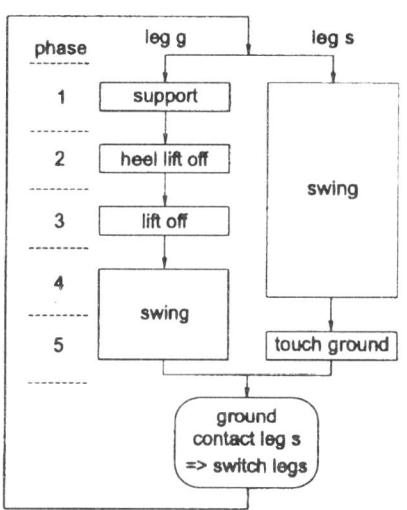

Figure 3: Phase sequence.

y-axis α_u, β_u, the orientation of the pelvis joint x_p, the rotation of the arms x_a and the spacial orientation of the swinging foot $\alpha_s, \beta_s, \gamma_s$.

$$x_{c,i} = (x_p^T \; x_{v,i}^T)^T \quad x_p = (\alpha_u \; \beta_u \; x_p \; x_a^T \; \alpha_s \; \beta_s \; \gamma_s)^T \tag{1}$$

In the support phase the angular position of the upper body around the vertical axis γ_u, the position of the center of gravity x_{cg} and the position of the swinging foot x_s, y_s, z_s are included in $x_{v,1} = (\gamma_u \; x_{cg} \; y_{cg} \; z_{cg} \; x_s \; y_s \; z_s)^T$. The position and rotation of the stance foot result from the ground contact conditions. In phase 2 the heel lifts off from the ground. Then the horizontal velocity of the center of gravity is not controlled any more, but it is replaced by the control of the flexion of the thigh. Then $x_{v,2} = (\gamma_u \; \phi_{thigh} \; y_{cg} \; z_{cg} \; x_s \; y_s \; z_s)^T$.

In phases 3 and 5 one foot touches ground or lifts off from the ground. Therefore all six degrees of freedom of this leg and the height of the other leg are included in $x_{v,3}$ and $x_{v,5}$, respectively. When the feet do not have ground contact in the air phase, the center of gravity follows a ballistic curve and its position is not controllable. Therefore the height and orientation of the previous stance foot are included into the vector of controlled variables. In this case $x_{v,4} = (z_g \; \alpha_g \; \beta_g \; \gamma_g \; x_s \; y_s \; z_s)^T$.

4.2 Feedback Linearization

The control of the robot is based on the method of feedback linearization (Slotine, Li, 1991, Rossmann, 1998), which starts out from the equations of motion:

$$M\ddot{q} + W_1\lambda = h + Q_e \tag{2}$$

Here M is the mass matrix, $q \in \mathbb{R}^n$ is the vector of generalized coordinates and h and Q_e are the gyroscopic and external forces, respectively. The vector $\lambda \in \mathbb{R}^m$ comprises the applied torques of the joints that are projected on the generalized coordinates with the Jacobian W_1. Given a set of control inputs λ, a linear behavior can be prescribed in m coordinates x_c that are selected according to the phase in the gait pattern. A linear PD control law is prescribed in x_c with constant matrices C and D. The deviation from the reference trajectory is denoted $\Delta\dot{x}_c = (\dot{x}_c - \dot{x}_{c,ref})$ for the velocity and $\Delta x_c = (x_c - x_{c,ref})$ for the position, respectively: $\ddot{x}_c = \ddot{x}_{c,ref} - D\Delta\dot{x}_c - C\Delta x_c$. When the system is in a non-singular configuration, the linear control law can be projected on the vector of generalized coordinates:

$$\dot{x}_c = J\dot{q} \; ; \quad \ddot{x}_c = J\ddot{q} + \dot{J}\dot{q} \quad \Rightarrow \quad J\ddot{q} = \ddot{x}_{c,ref} - D\Delta\dot{x}_c - C\Delta x_c - \dot{J}\dot{q} \tag{3}$$

With abbreviation $w = \ddot{x}_{c,ref} - D\Delta\dot{x}_c - C\Delta x_c - \dot{J}\dot{q}$ we obtain the equations of motion in standard form:

$$M\ddot{q} + W_1\lambda = h + Q_e \tag{4}$$

$$J\ddot{q} + w = 0 \quad \Rightarrow \quad \lambda = (JM^{-1}W_1)^{-1}(w + JM^{-1}(h + Q_e)) \tag{5}$$

4.3 Foot Placement

As long as the mathematical model of the plant is close to the real system, stability of the linearized system can be achieved by placing the closed loop poles with matrices C and D. However problems arise from the limitation of the maximum forces and torques that can be transmitted to the ground. Defining the forces and torques that act between foot and ground as F_x, F_y, F_z and T_x, T_y, T_z, the constraints are denoted:

$$|T_x| \leq 0.5\, F_z l_y \quad |T_y| \leq 0.5\, F_z l_x \quad |T_z| \leq \mu_d F_z \quad \sqrt{F_x^2 + F_y^2} \leq \mu_t F_z \quad F_z \geq 0 \qquad (6)$$

Here l_x and l_y are the length and width of the foot, μ_d and μ_t represent the coefficients of friction. The contstraints on T_z can be satisfied by an adequate motion of the arms. Also the tangential forces can be kept within their limits when phases 3 and 5 are long enough. However, the constraints on the torques of the ankle joint T_x and T_y cannot be satisfied directly. When these torques reach their limits, the system becomes underactuated. Practically this means that the robot starts tipping over in the sagittal or lateral direction. Therefore the reference trajectory has to be calculated such that the maximum torques are not exceeded in some area around the ideal trajectory. In order to obtain a trajectory that satisfies these constraints, the position of the support foot x_F has to be optimized at each step. Choosing the maximum value of the torques $|T_x|$ and $|T_y|$ as cost functions, it is possible to calculate x_F such that $max\{|T_x|\}$ and $max\{|T_y|\}$ are minimzed.

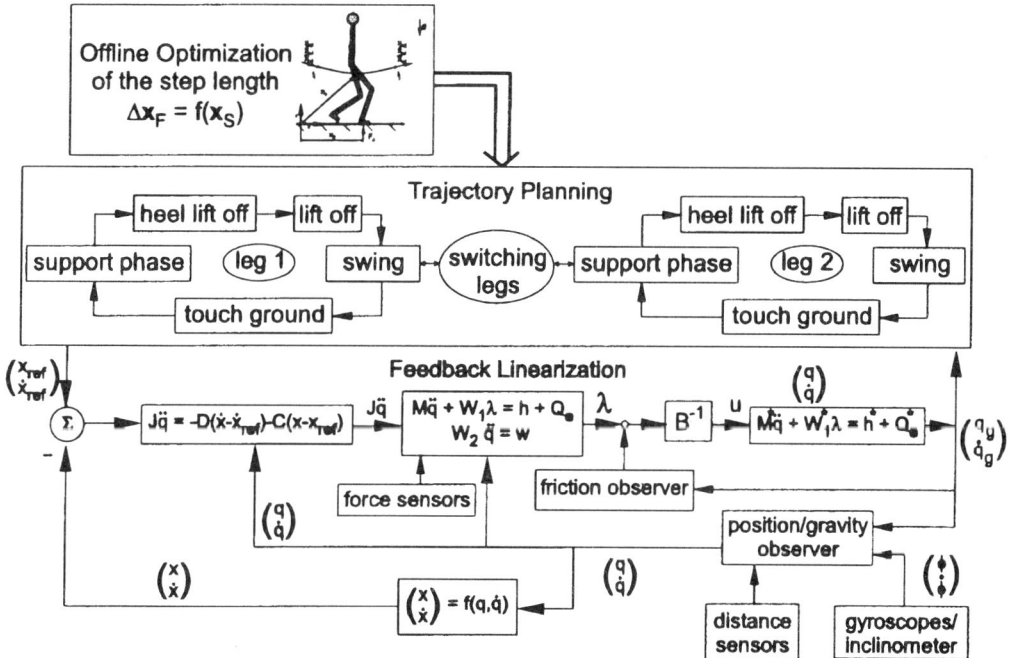

Figure 4: Control system.

This procedure is theoretically correct, but it requires intensive computation since the cost functions can only be determined by integration over one step period. Practically the method cannot be used for real time applications. Therefore simplified models are mostly suggested in the literature. Typically inverted pendulums or inverted double pendulums are used, with a lumped mass attached to the upper body. These models can be integrated with low computational effort or it is even possible to find analytical solutions for the optimal stance point x_F.

Simulations show, however, that these models are insufficient for a robot with a realistic mass distribution. In particular, the motion of the swinging leg influences the overall system behavior significantly. We therefore suggest to use a comprehensive model of the multibody system for the optimization of the foot placement, and to calculate the results in simulations. As long as the state of the system is close to the reference trajectory, the optimal position of the foot can be calculated in advance for a certain range of walking speeds and is then included into the controller as a lookup table.

This way the controller can operate in real time and ensure that the robot is controllable throughout the gait cycle. Figure 4 shows a simplified block diagram of the overall control system with friction, gravity and position observers.

5 Simulation

Multibody simulations are used to test the performance of the controller and to optimize the mechanical design of the robot. The mathematical model that is used in the simulations is similar to that of the controller, but it includes additional effects to obtain a realistic system behavior.

The nonlinear friction of the Harmonic Drive Gears and the dynamics of the DC-motors are considered. Add to this the contact between foot and ground is of particular importance. The compliance of the feet is modeled with spring-damper elements, while the contact itself is modeled as a rigid body contact and included into the multibody simulation as a linear complementarity problem.

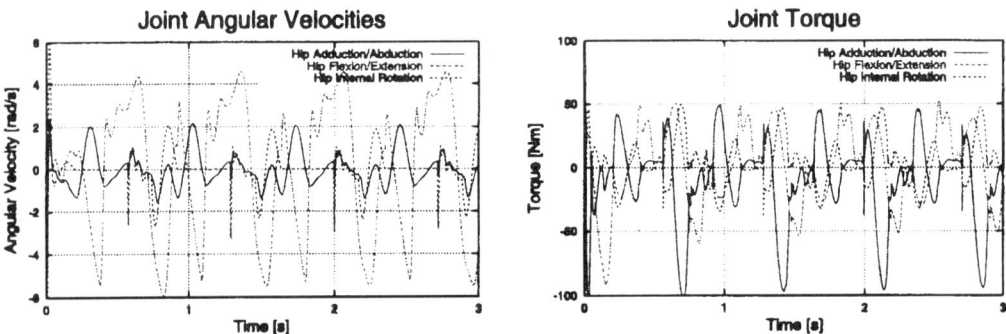

Figure 5: Joint velocities and torques in the hip joint.

6 Summary and Conclusions

The control and design of a humanoid biped robot are presented. Dynamically stable motion is achieved by using the method of feedback linearization. The control problem can be solved in real time and is suitable for implementation on the real robot. The parametric simulation and control algorithm have been the basis for parameter studies. Based on these results, the mechanical design was performed. Particular emphasis has been devoted to a lightweight design. In the near future, the robot will be manufactured. With its completion, the control strategy will be verified practically.

References

Cruse, H. (1976). The Function of the Legs in the Free Walking Stick Insect, Carausius morosus. *Journal of Comparative Physiology*, 112.

Cruse, H. (1990). What mechanisms coordinate leg movement in walking arthropods?. *Trends in Neurosciences 13*, 15-21.

Glocker, C. (1995). Dynamik von Starrkörpersystemen mit Reibung und Stößen. *Fortschritt-Berichte VDI*, Reihe 18, Nr. 182, Düsseldorf: VDI-Verlag.

Hahn, U. (1994). Calculation of Anthropometric Data for Human Body Segments. *Implementiert in Programm "Calcman3d"*.

Hirai, K., Hirose, M., Takenaka, T. (1998). The Development of Honda Humanoid Robot. *Proceedings of the 1998 IEEE International Conference on Robotics and Automation*, Leuven, Belgium, 1321–1326.

Raibert M.H. (1986). Legged Robots that Balance. Cambridge: MIT Press.

Rossmann, Th. (1998). Eine Laufmaschine für Rohre. *Fortschritt-Berichte VDI*, Reihe 8, Nr. 732, Düsseldorf: VDI-Verlag.

Saunders, J. B., Inman, V. T. (1953). The Major Determinants in Normal and Pathologican Gait. J. Bone Jt. Surg., 35-A, 543–559.

Slotine, J.-J. E., Li, W. (1991). Applied Nonlinear Control. Englewood Cliffs, NJ: Prentice Hall.

Weidemann, H.-J. (1993). Dynamik und Regelung von sechsbeinigen Robotern und natürlichen Hexapoden. Fortschritt-Berichte VDI, Reihe 8, Nr. 362, Düsseldorf: VDI-Verlag.

Six Link Mechanisms for the Legs of Walking Machines

A.P. Bessonov, N.V. Umnov, V.V. Korenovsky, E.E. Silvestrov and S.V. Khoborkov

Mechanical Engineering Research Institute, Moscow, Russia

1 Introduction

When designing walking robots one should separate the walking robots proper from the robots mounted on a walking chassis. It is clear that a task for the first ones is to achieve a maximum cross-country capability and therefore to get a maximum adaptation, maximum profile and soil cross-country capability, manoeuvrability etc. In order to deliver an object to the given point "at any rate". These are transport robots with very high performance. They should possess a developed flexible system of control, system of sensors but mechanical part — a propeller should not limit their potentialities. Moreover, mechanical characteristics of such robots are calculated on extreme conditions and their potentialities exceed those with which robot encounter normally in its functioning. Besides, when analysing robot characteristics the maximum values of the obstacles being overcome appear on the foreground meanwhile such characteristics as velocity, load-carrying capacity, effectiveness deviate on the background and are not very important for the achievement of the goal.

Unlike transport robots for extreme conditions the robots of the second group are common ones on the mobile base in which a walking chassis is used. The object of such symbiosis is of course, the increase of a cross-country capability giving the chance for the robot to function in such conditions in which it would not do it being on wheels or even on tracks chassis. This, first of all, the ground — swamp, tundra, snow or rather flat surface but with a lot of obstacles or finally rugged firm surface as a result of natural or technological catastrophes. It is important to note that the main work of the robot occurs *after* reaching the given place and is weakly connected with the process of overcoming obstacles on the way to this place. For such robots the velocity of walking, load-carrying capacity are more important than the sophisticated software permitting to plan the overcoming of obstacles in such a way in order to reach "the other side".

2 Regimes of Walking

It would not be out of place to observe that, as is known, the value of vertical robot adaptation decreases its velocity and static stability. It is connected with the fact that for the adaptation increasing it needs "to lift legs" high, time spent for this operation decreases the part of the supporting phase time, with which the supply of stability decreases. For the increase of the supply up to the safe limits one should limit the velocity of the vehicle. The way out of this contradiction is the division of the regimes of the robot work to cruising and adaptive regimes distinguishing by the value of adaptation. At the cruising regime the height of legs lifting when transferring is minimum that gives real opportunity to increase the machine velocity while at the adaptive regime after a considerable decrease of velocity one can increase largely the legs lifting keeping the necessary supply of static stability. It is important to note that all this occurs only in the phase of the leg transfer and do not produce an effect on the interaction of the leg and ground by which in full measure keeps the perfect ground passability of the walking propeller.

All said above gives the ground to conclude that for walking robots of the first group the use of the mechanisms with constant cycle for propellers is hardly advisable while for the second group it is not just possible but can be proved rather effective. This conclusion is based on the use of a linkage for the fulfilment of the main walking cycle on cruising regime without involving an additional adaptation mechanism that permits for the robot to travel with a considerable velocity limited only by dynamics of the latter. The leg lifting is a part in the main walking cycle and is determined by the mechanism used. Emphasize once again that the adaptation mechanism for additional leg lifting-lowering in this regime does not work. Only at the necessity of overcoming the obstacle higher than the height of leg lifting, determined by the mechanism of the main walking cycle, the vertical adaptation mechanism begins to work. The machine velocity at this can be reduced to the necessary safe value up to its full stop for making a decision in choosing further strategy. From this, by the way, follows the conclusion that economical orthogonal propellers do not fit much for such type of robots since the regime of lifting - lowering coincides with the regime of vertical adaptation of the propeller as the both are realised by one of the same drive. Of course, for cruising regimes one can use the regime of not full adaptation and some technical tricks that however, deprives the orthogonal propeller of its tempting simplicity and casts doubt on the advisability of its use for such type of robots.

3 Synthesis of Mechanism of Propeller

The use for the plane linkages working in vertical plane is normally rejected immediately by designers mainly because of a constant cycle of the mechanism work, difficulties of the rearrangement of its parameters and practically full lack of adaptation. These charges mainly are correct since rare instances of the designed adaptive walking mechanisms demonstrate rather complicated kinematic chains [Ibraev, 1999].

Simultaneously, the combination of a plane mechanism for the realisation of the main walking cycle with additional adaptation mechanisms for which a mechanism of orthogonal type can be used seems the solution removing all contradictions. On this promising way there is just one obstacle — lack of a good mechanism.

4 Basic Conditions

Let us consider it in detail. Synthesis of the mechanism propeller of the walking machine is connected with the design of such chain of crank mechanism which should reproduce a certain closed trajectory. Proceeding from the designated purpose this trajectory is a trajectory of the propeller supporting point and should satisfy the following *basic* conditions:

1. There should be straight section in a trajectory
2. Duration of this straight section of a trajectory by the crank mechanism turn angle should not be less than 180 degrees
3. Supporting straight section of the trajectory should be outer as for the trajectory of the supporting point and for the trajectories of all other points belonging to the mechanism. The rest part of the trajectory corresponding to the phase of the supporting point transfer should be on the inner branch and should not cross the supporting part of the trajectory.

As is known [Bessonov, Umnov, 1991] the plane four link mechanisms do not satisfy in full these *basic* conditions — normally the mechanisms with outer position of the straight part of the trajectory have no necessary extent and in the straight-line mechanisms with a trajectory of superb quality the straight section of the trajectory either is not "outer" or "cross" the mechanism that eliminates direct use of the reproducing point of the mechanism as a supporting point of the walking propeller. As is known a four-link mechanism was successfully realised for the propeller of the real walking robot for working in swamps constructed at 1996 in Volgograd by Briskin E.S. However, performance of its propeller is far from being ideal that reduces the values of the machine on the whole. The picks of power consumption because of the too short duration of the supporting part and higher slip of the legs because of variable velocity of the propeller in a supporting phase — these are its main disadvantages.

It testifies that for the use of the four link mechanism as an effective one of the main step cycle it needs substantial updating and first of all duration of its supporting phase. The most simple way is the refusal from constant velocity of a crank and passage to variable. One should slow velocity in the supporting phase and accelerate it in the phase of transfer. Of course, the direct control of the drive velocity in different phases of motion using electricity is absolutely irrational. It is easier to do it if to place between evenly rotating drive and the crankshaft any mechanism of variable rotation with a mean ratio equalling to one. As such one there can be used, for instance, an elliptical gear mechanism, plane double crank four-bar mechanism or its spatial analogue with higher potentialities — Bennett mechanism; inversion of slider-crank mechanism, gear- lever mechanism and other any solutions that may be not so obvious. Evident disadvantage of such way of solving the problem is the increase of variability of travelling in the supporting phase that increases, in addition, substantial slip of the supports on the ground. Justifiably, for soft swampy grounds for the work on which the Volgograd walking machine was designated, this disadvantage is not very substantial.

Another, also natural approach to the creation of effective propeller concludes in the use, as mechanisms of the main step cycle, of more complicated crank mechanisms with many links for the propeller mechanism. We shall consider further the problems connected with the use of only six link mechanisms for the walking systems propellers satisfying the *basic* conditions introduced above.

The difficulty concludes in the fact that such mechanisms are unknown, they should be synthesized and the problems involved will be discussed. Traditionally, synthesis of mechanisms is executed in some stages. In the first stage normally the choose of a structure synthesis is executed [Levitsky, 1979]. Our task in this stage is to list all possible schemes of plane six link mechanisms useful for using as a propeller. We limit ourselves with the mechanisms with revolute pairs.

All plane six link mechanisms with one degree of freedom can be taken from the analysis of all six-link chains. From six moving links ($n = 6$) and seven lower pairs ($p = 7$) one can form only two closed kinematic chains of Watt and Stephenson [Hain, 1967]. Sequentially fixing in each of them different links one can take 5 structural chains of six link mechanisms. Choosing different links as initial link one can define all 9 known six-bar mechanisms.

In each of these mechanisms there are links making a complicated motion, trajectory points of which can be analysed for their fitness for walking propellers. However, one should analyse not all links but only those that make more complicated motion than the coupler of the four link mechanisms and therefore have coupler curves of higher degree. Such contraction of the area of analysis connected with the fact as it was said [Bessonov, Umnov, 1991], that the coupler of four link mechanisms do not give a total solution of the problem formulated.

The analysis showed that only twelve structural chains of six-link mechanisms represented in Fig. 1 can be considered as candidates for synthesis of coupler curves, more complicated than four-link ones. It is seen in Fig. 1 that among these twelve mechanisms there are seven mechanisms of the second class (schemes 1-7), three mechanisms of the third class (schemes 8-10) and two mechanisms of the fourth class (schemes 11 and 12). All schemes are shown uniformed with the maximum unification of symbols. On all schemes the point G symbolised the supporting point of the mechanism the trajectory of which is subjected to analysis. Besides, the points of suspension of the propeller to the walking machine body (linkage point of the fixed member of the six link mechanism) are points O and A for two supporting mechanisms and the points O, A and E for three supporting ones (schemes 7 and 8). In all schemes the link AB — crank. The origin of the co-ordinates — point O. The complete kinematic cycle of any mechanism is executed for one revolution of the crank AB. For our purposes only crank assemblies of the mechanism are of interest. The table of the schemes (Fig. 1) is taken from the special computing complex designated for the automatized synthesis of propellers since the possibilities of universal program means for these purposes are limited.

In the second stage of the walking propellers synthesis for each scheme (Fig. 1) it is necessary to define dimensions of the mechanism performing pointed earlier main conditions to the trajectory of the supporting point G. Besides, one can introduce some additional conditions the performance of which influences the quality of the mechanism work but not its fitness. In order to use the existing methods of synthesis of approximate guiding mechanisms [Levitsky,1979] to which the schemes are belonging presented in Fig. 1 of the scheme it is necessary to find the trajectory of the supporting point G depending on the turn angle φ of the crank AB. The effectiveness of the synthesis procedure will depend on the effectivity of the algorithm of defining the co-ordinates of a supporting point. If for the schemes 1-7 (Fig. 1), schemes of the mechanisms of the second class, the procedure of defining the co-ordinates of the coupler curve is rather traditional, well-known by the literature [Levitsky,1979] and has no features, but the description of the trajectory for the rest schemes of the mechanisms of the third and fourth classes is rather laborious and consists of the whole row of unobvious features complicating the synthesis procedure.

5 Features of the Synthesis Procedure

The second stage of synthesis in its turn consists of two substages. At first it is necessary to choose a certain initial set of parameters, approximately satisfying the formulated task and then by any way to specify the parameters improving the final result. We should make some remarks. It is clear that the nearer initial parameters to the parameters of hypothetical optimal mechanism the more probability of taking it. Being far from optimal solution it is practically impossible, using common methods of optimisation, to reach it mainly because of the problem of local minimums. At the same time too many parameters (15 in our case) practically eliminate the possibility of their direct exhaustion with a sufficient discreteness and in reasonable time. Here the system of tests performed in the earliest stages of synthesis can be of aid. Let us describe the approach to an automatic rejection of variants of mechanisms in the process of the work of the above mentioned automatized system of synthesis. In this system a trajectory of the supporting point of the mechanism-candidate is subjected to automatic analysis. The procedure of such analysis is as follows. Having taken the massive of the co-ordinates of the trajectory points (48 in the program) we choose consequently each point of the trajectory as initial point of the supporting phase. The fact is that if one does not see a mechanism it is difficult to foresee the optimal position of the straight section but a visual control of the program work in interval stages of synthesis was principally excluded. If the supporting phase consists of half cycle, the final point of the supporting phase can be found easily. Having drawn a straight line through the co-ordinates of these two points, we will consider it a line of the supporting plane, supporting straight line. Now we should check whether all points of all mechanism links are on one side from the supporting straight line. This will be done easily analysing only a sign of the distance from the point under check till the supporting straight line. Also the rest point of the trajectory corresponding to the phase of transfer should be checked. If any point under check does not stand this test then it means that the link to which corresponding point checked crosses the supporting straight line. Therefore, the point of the trajectory chosen as an origin of the supporting phase is not fit and it does not provide outer position of the supporting section of the trajectory and should be rejected. Now the next point should be checked. It is enough to check half of all points of the trajectory (24 in our case). If none of the points of the trajectory does not satisfy the check for the outer position the mechanism should be excluded.

In case, if the results of the check for "outer" of the chosen point are positive, the accuracy of reproduction of the trajectory in the supporting phase and this value is stored as a feature of the chosen initial point. Also, there occurs the transfer to the analysis of the next point. After the analysis of all points of the trajectory the accuracy of those initial points of mechanism for which the check was positive is compared. The best accuracy among all checked points is recognised as accuracy feature of the mechanism with these parameters.

If the accuracy of the mechanisms not worse than the given one, parameters are recognised. The point of the trajectory with the highest accuracy of reproduction of the straight line as a point of origin of the supporting phase of the mechanism with the analysing set of parameters. Final point of the supporting phase is shifted relative to initial by half cycle and is determined from the massive of co-ordinates of the trajectory points. By this two points "horizon" of the propellers is formed that normally does not coincide with the axes of co-ordinates taken at the calculation of the co-ordinates of the trajectory of the supporting point.

One can try to improve the taken mechanism by a common optimizational procedure. It turned out useful to estimate visually the taken mechanisms in this stage in order to specify their fitness for some weakly formulized parameters. Let us make some practical remarks generalising the experience of the working with the program.

We refused a direct exhaustion of their parameters in order to reduce a number of preparatory variants of the mechanisms. On the other hand methods of random choice widely using in practice of synthesis of a mechanism seem not to be for because of uncertainty in full coverage of the whole range of the area of existence of parameters. We used LP-τ-sequences permitting with the aid of so-called points of Sobol [Sobol,Statnikov,1981] to choose the value of parameters uniformly distributed in a space of parameters. Such approach permits substantially to reduce a number of test points and at the same time guarantee that the area of parameters will be looked through. Use of this method permitted to reduce a number of the variants looked through up to some thousands comparing with some billions at the overall exhaustion that finally gave the possibility to make a complete analysis in reasonable time.

This method is widely used at the automatization of experiments permitting substantially to increase their effectiveness.

However, in our case it has a methodical disadvantage connected with the fact that the points of Sobol give sets of parameters the values of which are different. In other words, if the parameters are lengths of links, only mechanisms with different lengths of links are subject to analysis. Simultaneously majority of classical guiding mechanisms normally has some links of the same length. We assumed, however, that at the next correction of the links lengths in optimizational procedure they will become equal if it permits to increase a purpose function. Wonderfully but this did not occur.

Another feature of the optimisation procedure is the choice of the purpose function. The fact is that traditional criteria connected with the increase of accuracy only on the straight line of the trajectory normally at the optimisation, bring to degeneration — either the trajectory of the mechanism contracts to a point or the links increase unfoundedly. We used full trajectory of the supporting point including not only the supporting phase but the phase of transfer as a standard. (Fig. 2). Then we chose the deviation of points of the trajectory of the mechanism from the points of a standard curve on the full cycle of travelling as a criterion. This has permitted to avoid the common degeneration.

Substantial difficulties appeared with the mechanisms of high classes (schemes 8-12, Fig. 1). These difficulties were connected mainly with the computation of points of the trajectory of the mechanism. The fact is not only that the mechanisms of the third and fourth classes have more number of cognates (6 in this case), but also with difficulties of taking constrained consequence of points regarding to the mechanism of the same cognate. This last feature occurs at the analysis or roots for any neighbour values of the crank position. Comparing two sets from six numbers it is not always easy immediately to determine which number is the continuation of the sequence. Big difficulties arise when the quantity of number in the set changes. This requires substantial additional time for analysis. Perhaps, this was a reason of the fact that we failed to have found any fit mechanism of the third class of sufficiency accuracy satisfying *basic* conditions.

6 Results

Two mechanisms of the second class shown in Fig. 3 and Fig. 4 were taken as a result of working the synthesis program. The mechanism in Fig. 4 seems more preferable since there is no double crank in it. The mechanism in Fig. 3 is the mechanism from scheme 5 (Fig. 1), and the mechanism in Fig. 4 is the mechanism from scheme 7. Linkages dimensions of those mechanisms are: for Fig. 3 — OA = 1.05, AB = 0.15, AD = 0.18, BAD = 75 GR., BC = 1.27, OC = 0.51, COE = 202 GR., OE = 1.26, EF = 1.67, DF = 0.53, FEG = 219 GR., EG = 1.19, beginning point = 28, length of supporting fase = 1.03, accuracy = 0.45%.

for Fig. 4 — OA = 0.23, AB = 0.13, BC = 0.65, OC = 0.59, BCD = 208 GR., CD = 0.47, AOE = 296 GR., OE = 1.26, EF = 1.43, DF = 0.86, DFG = 208 GR., FG = 0.98, beginning point = 41, length of supporting fase = 1.03, accuracy = 0.39%.

It is necessary also to note that other 5 variants of schemes of the mechanisms of the second class and 5 variants of schemes of the mechanisms of the higher class have not yet given any acceptable mechanisms.

Six Link Mechanisms for the Legs of Walking Machines

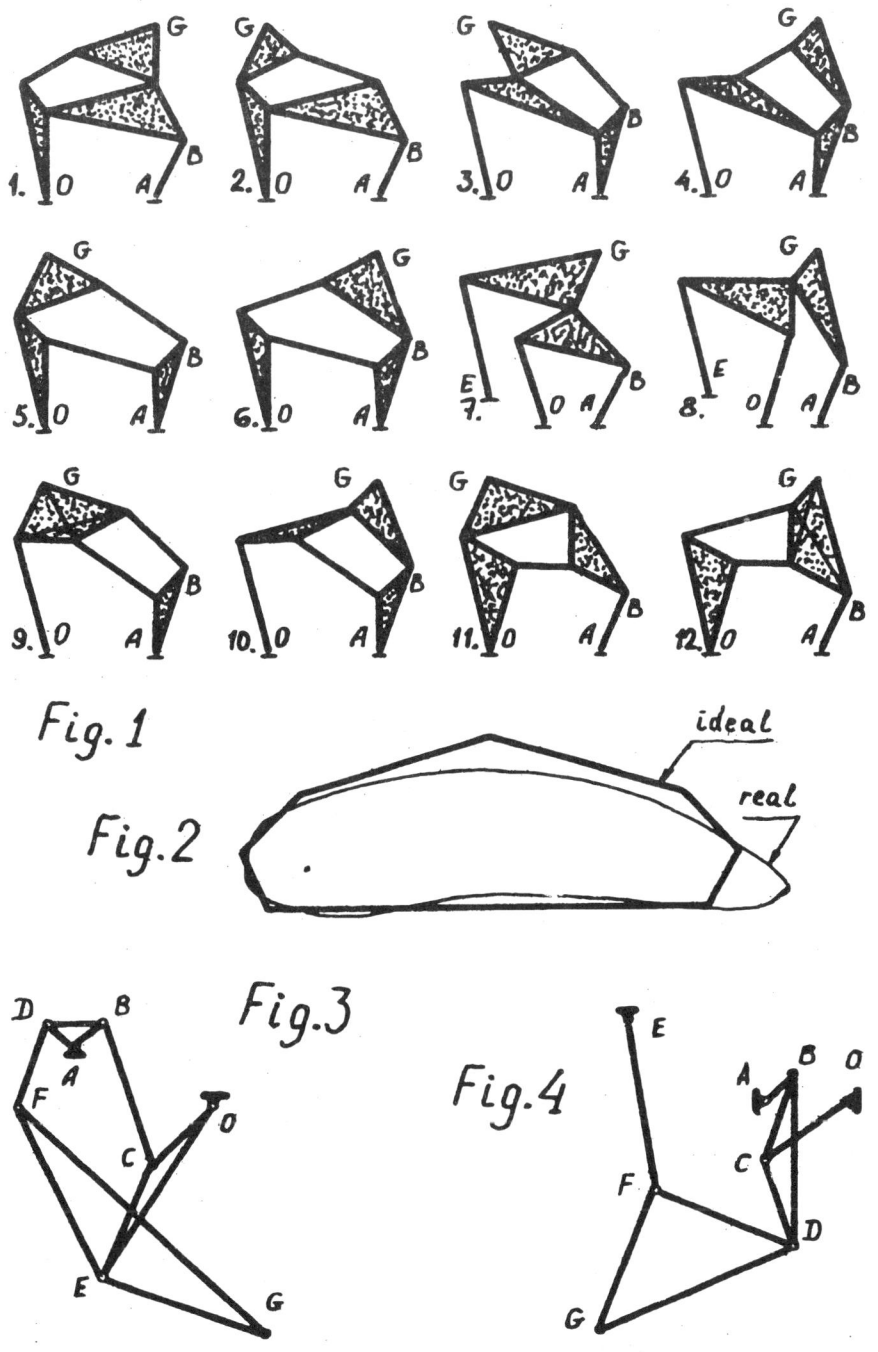

Figures 1–4.

7 References

1. Ibraev S.M., Synthesis of Walking Vehicles Leg Mechanisms with Decoupled Foot-Point Motion, In: Tenth World Congress on Theory of Machines and Mechanisms, 1999, Oulu, Finland, Vol. 3, pp. 1188-1195.
2. Bessonov A. P., Umnov N. V., Application of Mechanisms with Translational Motion of Moving Elements for Self-propelled Machines. In: VIII World Congress on Theory of Machines and Mechanisms, 1991, Prague, Czechoslovakia, Vol. 6, pp. 1610-1631.
3. Levitsky N.I., Theory of Machines and Mechanisms, Nauka Press, 1979, 576p. (in Russian)
4. Hain K., Applied Kinematics, 1967 MacGrawHill, N.Y.
5. Sobol I.M., Statnikov R.B., Vybor oprtmal'nyh parametrov v zadachah so mnogimi kriterijami, Moscow, Nauka Press, 1981, 107p. (in Russian)

Design, Analysis and Measurements Problem of Mili-Walking Machines Using Multi-Body System Formulation

Janusz Fraczek and Adam Morecki[1]

[1] Institute of Aeronautics and Applied Mechanics, Warsaw University of Technology, Poland

Abstract. The paper presents the generalised approach to design, kinematic and dynamic modelling of walking robot based on multi-body formalism implemented in commercial package. The impact forces were simulated using FEM approach and reaction forces were calculated. Numerical results were validated using reaction forces measurements with tensometer sensors.

1 Introduction

The design and analysis of legged robots capable of moving over surface with complicated profile is of interest both practically and theoretically. They are essentially complex due to the mechanical problems involved in the design of such a system which has, as a rule at least four to six legs with several degrees of freedom in each. Moreover the control system synthesis and force control algorithms require in many cases the proper choice of dynamical model.

The paper presents the general multi-body system approach to design, motion planning and kinematic and dynamic analysis of walking machines. The subject was focused on the small machines (having small dimensions). In the literature several multi-body formalisms are presented, based mainly on Newton-Euler, Lagrange, Hamilton equations in symbolic, numerical or mixed implementations. For the sake of simplicity of formulation the classic Lagrange formalism was used based on theorem of virtual power. This approach leads to differential algebraic equations of index 3. The developed ideas were used to simulate motion of HERMES^{TM} (Robotics IS, 1994) mili-walking robot, which incorporates high-powered processing, rich sensing and high mobility into a compact body. To develop the kinematics and dynamics of machine the model of real insect locomotion was used.

The contact leg - ground model, having the greatest influence on results of dynamic simulation was investigated using nonlinear Finite Element Method. The results obtained were implemented in multi-body model of the contact interactions. The reactions forces obtained from computer simulations were validated by direct experimental measurements (tensometer sensors) on the machine legs during gait. Majority of calculations based on developed algorithms was carried out using commercial packages (ANSYS^{TM}, ADAMS^{TM}) (ANSYS manual, rel. 5.4, ADAMS manual, rel.10).

2 Design and Systematics of Mili-Walking Machines

Kinematic synthesis of the walking machine legs and actuators is usually one of the first step in the walking machine design. The common source being used in the kinematic synthesis of the mecha-

Table 1. Comparison of existing mili-walking machines.

Nr	NUM. OF LEGS n =	NAME OF THE MACHINE	DATA						
			YEAR	COUNTRY	HEIGHT m	WIDTH mm	LENGTH mm	MASS Kg	VELOCITY cm/s	DOF
1.	2	BAASILICS	1992-98	MEXICO	360	40	460	3	30	-
2.	2	ROMA	1995-98	SPAIN	344	400	736	92	100	-
3.	2	WALKING GYROSKOPE	1980	USA	200	25	50	0,45	-	1+4p
4.	3	PLIF	1996	BELGIUM	20	20	20	-	20	-
5.	3	ROBINSPEC	1995	BELGIUM	300	300	400	-	-	-
6.	4	ARL SCOUT	1996	CANADA	270	230	260	2,3	10	16(4x4)
7.	4	BEAST	1993-94	FRANCE	130	170	170	1,0	14	-
8.	4	EXPLORATES	1997	HUNGARY	270	400	600	-	-	-
9.	4	MENO	1995-97	USA	300	500	500	12	-	-
10.	4	POLYPOD	1992-94	USA	6	6	6	-	-	-
11.	4	RITHMO	1990-93	GERMANY	344	719	736	67	1	-
12.	4	ROBO TRAC		GERMANY	150	270	550	1,3	8	8 (2m)
13.	4	ROBUG II	1989-90	U.K.	250	100	500	15	1	3
14.	4	TITAN VIII	1997	JAPAN	250	400	660	19	-	12(2m)
15.	4	THING	1994	USA	250	-	-	1,6	-	12(4x3)
16.	4	I.N.-4	1994	POLAND	120	120	300	0,04	1	4
17.	4	AIBO	1999	JAPAN	266	156	274	1,6	-	18
18.	6	HERMEST™II	1990	USA	140	220	290	2,5	2,8	12(6x2)
19.	6	BOADICEA	1992-95	USA	-	-	500	2,5	10,0	
20.	6	GORIBURI	1995	JAPAN	100	76	200	0,6	30cm/min	-
21.	6	CWRU ROB. II	1992-95	USA	200	460	760	-	-	18+6p
22.	6	CYCLON	1998	GERMANY	200	150	350	-	-	-
23.	6	IOAN	1994-96	BELGIUM	150	100	400	1,2	4	-
24.	6	LAULON	1992-95	GERMANY	300	700	800	12	-	-
25.	6	LAURON II	1995-	GERMANY	300	700	700	16	-	-
26.	6	MASCHA	1971-94	RUSSIA	210	700	700	20	14	-
27.	6	LEONARD	1994-95	SWITZERLAND	200	200	240	1	10	-

28.	6	MAX	1993-96	U.K.	160	300	300	0,762	10	-
29.	6	KARLA	1991-95	GERMANY	400	400	700	-	-	24+4p
30.	6	TARRY	1992-	GERMANY	150	300	400	-	-	-
31.	6	SILEX	1991-	BELGIUM	-	-	-	15	10	-
32.	6	LAMI (R.H.))	1989	SWITZERLAND	130	80	300	0,4	20	-
33.	6	ROBOTY	1995-	USA	200	180	380	-	-	12+6p
34.	8	NERO	1990-91	USA	160	350	800	15,6	10	-
35.	8	P.C.R.	1992-95	GERMANY	150	12-30(0)	300	3,1	30	-
36.	8	LOBSTER ROB.	1998-	USA	450	150	600	-	-	7+9p.
37.	7	WALKING BEAM	1989-90	USA	300	760	1500	-	-	-

nisms are the knowledge bases concerning existing mechanisms. In the table 1 the various types of walking machines were gathered (limited to 0.5m height) systematized according to the number of legs and other criteria. At the design stage the kinematic structures of the legs and the whole machine structure can be chosen from existing prototype structures e.g. described by table 1 (details can be found in references). The kinematics and the dynamics of the machine having investigated structure of driving mechanisms and actuators can be further analyzed using general multi-body formalism and formulas proposed below. The results of simulations of this "virtual prototype" become source of information for control system synthesis and motion planning. In case the kinematic and dynamic parameters of the machine does not satisfy the design assumptions the different structure can be analyzed using conception of virtual prototyping with multi-body formalism. It is obvious that some factors, like leg-ground contact-friction phenomena, which are difficult both to measure and model, can significantly affect obtained results.

3 Kinematics of the Machine (Generalized Co-ordinates, Constraints Equations)

The kinematics of the machine can be described in absolute (Cartesian coordinates). Generalized vector for each body is given by Cartesian co-ordinates vector (Haug, 1984) $\mathbf{r}_k = [x, y, z]_k^T$ of any point on the body (e.g. mass center) and by the Euler parameters vector determining the orientation of each link in the inertial global co-ordinate system. The vector of generalized co-ordinates q describing the whole mechanism is *6n-dimensioned* (*n* - number of the links). The mutual geometric relation between the generalized co-ordinates imposed by the kinematic pairs is described by holonomic constraints. For the whole machine they can be presented in a vector form:

$$\mathbf{F}^P(\mathbf{q}) = [(\mathbf{F}_1^P(\mathbf{q}))^T, (\mathbf{F}_2^P(\mathbf{q}))^T, ..., (\mathbf{F}_n^P(\mathbf{q}))^T]^T \tag{3.1}$$

If the number of independent constrains equation is l, then the number of DOF is equal to $n - l$. For kinematic analysis it is necessary to define $n - l$ additional equations, called driving constrains (Haug, 1984):

$$\mathbf{F}^D(\mathbf{q}) = [(\mathbf{F}_1^D(\mathbf{q},t))^T, (\mathbf{F}_2^D(\mathbf{q},t))^T, ..., (\mathbf{F}_{n-l}^D(\mathbf{q},t))^T]^T = 0. \tag{3.2}$$

The solution of kinematic problem can be obtained in a numerical way by solving the system of non-linear equations (3.1) and (3.2) simultaneously.
Equations (3.1) and (3.2) for velocity and acceleration can be written in the form:

$$\begin{bmatrix} \mathbf{F}_\mathbf{q}^P \\ \mathbf{F}_\mathbf{q}^D \end{bmatrix} * \dot{\mathbf{q}} = \begin{bmatrix} 0 \\ -\mathbf{F}_t^D \end{bmatrix}, \tag{3.3a}$$

$$\begin{bmatrix} \mathbf{F}_\mathbf{q}^P \\ \mathbf{F}_\mathbf{q}^D \end{bmatrix} \ddot{\mathbf{q}} = \begin{bmatrix} -(\mathbf{F}_\mathbf{q}^P \dot{\mathbf{q}})_\mathbf{q} \dot{\mathbf{q}} \\ -(\mathbf{F}_\mathbf{q}^D \dot{\mathbf{q}})_\mathbf{q} \dot{\mathbf{q}} - 2\mathbf{F}_{\mathbf{q}t}^D \dot{\mathbf{q}} - \mathbf{F}_{tt}^D \end{bmatrix}. \tag{3.3b}$$

a)   b)

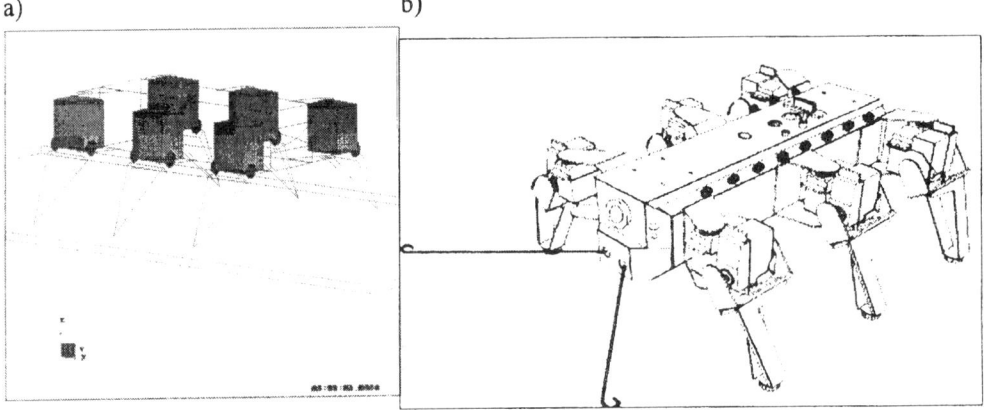

Figure 1. HERMES™ machine - computer view a) and real scheme b).

The algebraic equations (3.1), (3.2) can be solved using modified Newton-Raphson algorithm using sparse matrices techniques. Equations (3.3) are linear systems and can be solved with traditional numerical schemes. The starting point $\mathbf{q}_0$ was chosen using optimization techniques DFP (Haug, 1984) or null space method for the function:

$$\psi(\mathbf{q}, t_0, \mu) = [\mathbf{q} - \mathbf{q}_0]^T [\mathbf{q} - \mathbf{q}_0] + \mu \mathbf{F}^T(\mathbf{q}, t_0) \mathbf{F}(\mathbf{q}, t_0) \tag{3.4}$$

where μ - the weight multiplier.

In HERMES machine each of the six legs is two-degree of freedom system with two-revolute kinematic pairs. The driving constraints were found from relative angles in revolute pairs, which were programmed for each of the machine with trapezoidal profile of velocities. Motion planning (sequence of steps) was based on a real model insect locomotion (see earlier Fraczek et al. 1998). The intervals between steps of the hind and middle and between middle and foreleg were constant.

4 Equations of Dynamics

Lagrange's equations with multipliers (of I kind) were used to describe the problem of dynamics.

$$\begin{cases} \mathbf{M\ddot{q}} + \mathbf{F}_q^T \lambda = \mathbf{Q} \\ \mathbf{F}(\mathbf{q},t) = 0 \end{cases} \quad (4.1)$$

Where $\mathbf{F} = \begin{bmatrix} \mathbf{F}^P \\ \mathbf{F}^D \end{bmatrix}$, $\mathbf{M}$ – mass matrix, λ - Lagrange multiplier.

Equation (4.1) constitutes DAE equation (differential algebraic equation) with index equal to 3. Generally DAE equations are stiff and require special numerical algorithms, which are based usually on BDF methods (Haug, 1984).
In the paper Gear algorithm was used implemented in ADAMS (ADAMS manual, rel.10). as the GSTIFF method.

5 Contact Forces With Friction

The approach of classical dynamics simulation is to consider this problem as a piecewise differential algebraic equation, using the sign of the constraint force as a test to decide if the respective contact will break or not (Haug et al., 1986) Another approach consists in formulating the multi-body rigid body contact problem as a complementarity problem (LCP). For problems without friction it is known that LCP can always be solved. Unfortunately for problems involving Coulomb friction, there is no guarantee that a solution exists (Pfeiffer et al., 1996).

In the approach presented in the paper contact forces were modelled using the simple impact function acting outside the geometry of the legs and ground with respect to the geometry normal orientation and involving the impact stiffness (with exponent w fixed between 1 and 3). The impact functions are presented in Figure 2.

The contact parameters depend on kind of contact surfaces and kind of materials. In case of HERMES machine each of the legs is ended with solid hemisphere made of soft rubber. The ground was modelled as flat, rigid metal-type material. The impact function, given in Figure 2 can be described by four parameters - stiffness coefficient K, stiffness force exponent w, the damping coefficient C and damping ramp-up distance d. Two methods were applied to identify these parameters.

In the first method stiffness coefficient K and exponent w were found using classical Hertzian contact (Nikravesh, 1990, Johnson, 1985) theory.

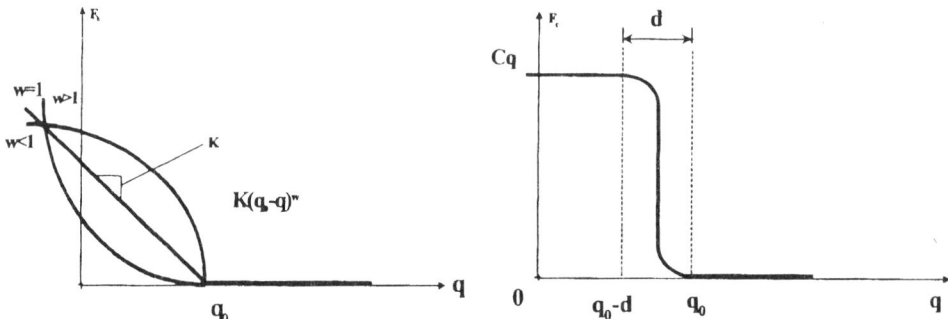

Figure 2. Stiffness and damping force-penetration dependency.

Using well known hertzian formulas stiffness K and exponent w can be expressed in the form (for plane surface $R_2 = 0$):

$$K = (\frac{16RE^2}{9})^{\frac{1}{2}}, \quad \frac{1}{R} = \frac{1}{R_1} + \frac{1}{R_2}, \quad \frac{1}{E} = \frac{1-v_1^2}{E_1} + \frac{1-v_2^2}{E_2}, \quad w = \frac{3}{2}, \quad (5.1)$$

Damping coefficient C can be expressed in the hysteresis form proposed by (Nikravesh, 1990):

$$C = \mu d^n \quad (5.2)$$

where μ is called hysteresis damping factor.
In the second method the relation between stiffness force and penetration was found using nonlinear (large deformation) Finite Element Method. The tip of the machine leg was modeled using hemisphere contacting with rigid ground (see Figure 3) (the friction effect was taken into account). For soft rubber sphere the Mooney-Rivlin (ANSYS Manual, 1998) material with coefficients found experimentally was applied. Numerical calculation were carried out using ANSYS^{TM} universal package for FEM modeling and analysis.
For both method of contact modeling the friction and stiction force law were based on continuous approximation of Coulomb friction.

6 Results and Conclusion

The computer analysis of the machine motion was performed in time of 18 sec of real motion. The commercial package ADAMS^{TM} was used in substantial part of the calculation. Characteristics of the contact forces, driving forces and the trunk mass center were registered. For the measurements of the contact force between the leg and the ground special force transducer (strain gauges) were built and tested.
For exemplary comparison of numerical calculations to experimental results the reaction force were chosen. Figure 5 shows a comparison between results obtained from computer simulations of normal reaction forces with contact coefficient taken from ANSYS calculations and registered from force sensors attached to the foot (sensor 1) and to the trunk of the walking robot

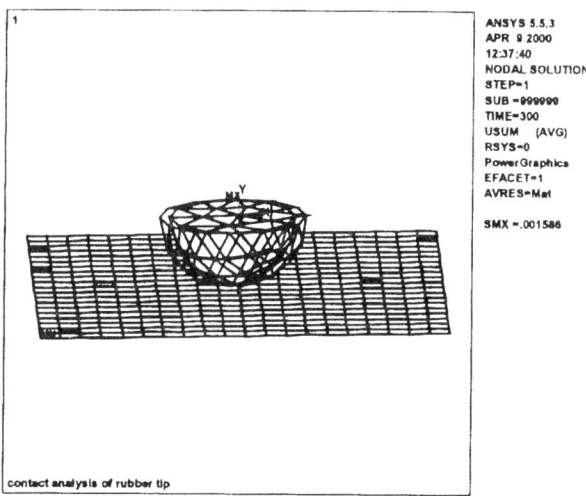

Figure 3. Finite element analysis of the machine leg tip (natural rubber).

(sensor 2). The contact forces for two legs are shown: for left-foreleg and left-middle in time of 5s. The contact forces are of discrete type and are characterized by sudden jumps of values in the contact phase.

The experimental results (sensor 1) show a good correspondence to numerical results for left-foreleg. For the other legs the correspondence of absolute values is some worse, although the courses of experimental and numerical results are similar.

References

HERMES™, *the compact walking robot*, (1994) IS Robotics Inc. Twin City Office Centre, Sommerville, HA 02143, USA.

ANSYS - Theory Manual rel. 5.4 (1998), SAP.

ADAMS - rel. 10, MDI, (1999).

Haug E.J (1984). Elements and Methods of Computational Dynamics. In *Computer Aided Analysis and Optimisation of Mechanical System dynamics in NATO ASI Series*, vol. F9, ed. by E.J. Haug. Springer-Verlag Berlin Heildelberg.

Frączek J., Morecki A., Malec A., Ober J., Cieślak R. (1998). Geometry, kinematic and dynamics control of six-legged miliwalking machine (HERMES™II) combine with mini-manipulator, *Euromech 375*, Munich, Germany.

Haug E.J., Wu C.S, Yang S.M. (1986). Dynamics of Mechanical System with Coulomb Friction, Stiction, Impact and Constraint Addition-Deletion-I. *Mechanical and Machine Theory*.

Pfeiffer F., Glocker H (1996). *Multibody Dynamics with Unilateral Constraints*. Wiley and Sons, Inc.

Nikravesh P.E. (1990). A Contact Force Model With Hysteresis Damping for Impact Analysis of Multi-body Systems, *ASME Journal of Mechanical Design*, 369-376.

Johnson K. (1985). *Contact mechanics*. Cambridge University Press.

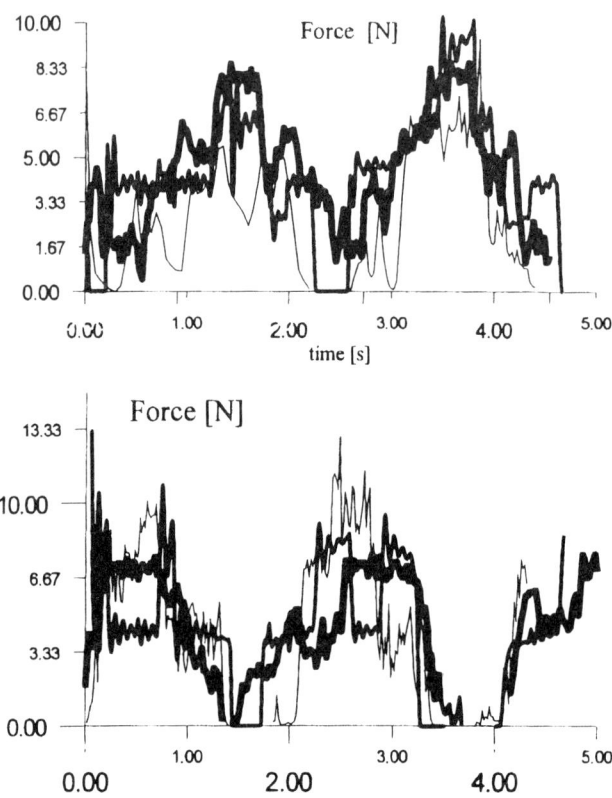

Figure 4. The comparison between the contact forces measurements and computers simulation for two legs of the walking machine.
Legend:
Computer simulation
Measur. (sensor1)
Measur. (sensor 2)

Modelling, Simulation and Nonlinear Control of a Combined Legged and Wheeled Vehicle

Jörg Müller and Manfred Hiller

Gerhard-Mercator-Universität Duisburg
Fachgebiet Mechatronik, Lotharstraße 1, 47057 Duisburg, Germany

Abstract. During the design process of the walking robot ALDURO *virtual prototyping* is used to test the design studies in a computer simulation, where the different parts can be tested under various conditions. It is not sufficient to have separate models of the mechanical, hydraulic and electronic subsystems. A single detailed *mechatronic design approach* that incorporates all subsystems, and their influence on each other, is required. For the ALDURO conventional linear controllers do not provide satisfactory results due to the highly nonlinear kinematics and dynamics of the legs. Therefore a nonlinear control concept is under development using a detailed model of the mechanical structure and the method of the exact input/output linearization.

1 Introduction

The principal aim of modelling and simulating walking machines is to make available a nonlinear dynamic simulation model describing the complex system, with all relevant coupling effects. These models can be used as virtual prototypes for the development of a walking machine, replacing the real prototype and showing the overall behavior of the system. In this case the models are also required as a basis for calculation of the output variables for a nonlinear controller. The model of the overall system, including the mechanical system, a model of the ground contact and the hydraulic system, is implemented in the object-oriented programming environment MOBILE (Kecskeméthy, 1993) using the language C++.

2 The ALDURO

The system under investigation is the hydraulically driven autonomous large-scale combined legged and wheeled vehicle ALDURO (<u>A</u>nthropomorphically <u>L</u>egged and Wheeled <u>Du</u>isburg <u>Ro</u>bot) which is being built in the Mechatronics Laboratory at the University of Duisburg. The machine will have a platform length of 2.2 m and a width of 2.0 m, a weight of approximately 1200 kg and a payload capacity in the region of 300 kg. It can be used as a quadruped walking machine (Figure 1), while, by replacing the two hind feet with wheels, it also can be used as a combined legged and wheeled vehicle (Figure 2). The latter combines the advantages of a walking machine – its mobility – with the stability of wheeled vehicles.

An essential feature of this walking machine is the spatial leg mechanism with anthropomorphic properties which is a modification of the leg presented in Kecskeméthy (1994), realized

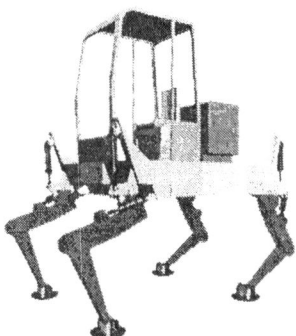

Figure 1. Walking Machine

Figure 2. Combined Legged and Wheeled Vehicle

using a spherical joint with three degrees of freedom at the hip, a revolute joint at the knee and a pivotally suspended foot. The leg can be treated as a spherical parallel manipulator, with four coupled kinematic loops specifying the orientation of the thigh, and a planar subsystem with one kinematic loop taking care of the motion at the knee.

3 The Simulation Model

The dynamic model used in a model based motion controller includes the kinematics and the dynamics of the system so that the actuator forces necessary to fulfill a desired motion can be calculated. Here, the system consists of a central platform with four identical legs with four hydraulic cylinders each. An efficient method of generating the equations of motion for such complex multibody system models is the method of the kinetostatic transmission elements, which uses the formulation in minimal coordinates. Beyond that a detailed foot ground contact model and a model of the hydraulic circuit are required for a overall simulation model that can be used as a virtual prototype.

3.1 Kinematic Structure of the Overall System

The mechanical structure of the ALDURO can be represented as a multibody system consisting of several rigid bodies and joints, including coupled kinematic loops. The topology of the overall system in shown in Figure 3. For free movement of the central platform the fixed inertial system is connected to the platform with an additional virtual joint with six degrees of freedom (DOF) – three translational joints r_{cb} and three rotational joints φ_{cb} (i. e. as BRYANT angles). In the case of the direct kinematics no additional kinematic loops occur, because the foot ground contact is modelled as a force element with no kinematic constraints. This leads to an increase in the overall number of DOF from 10 up to 16 for the walking machine and from 12 up to 18 DOF for the combined legged and wheeled vehicle, depending on the number of legs with contact to the ground.

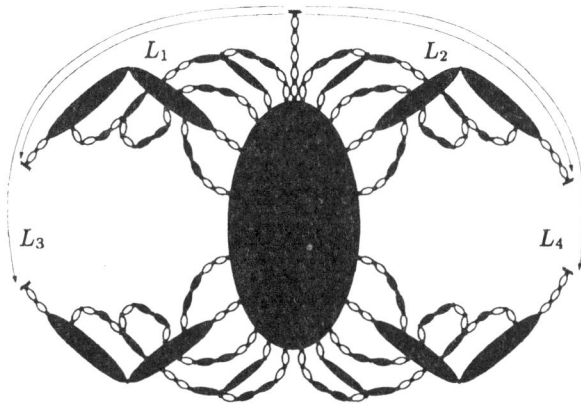

- ○ joint
- • generalized coordinate
- ● rigid body
- ○ rigid body without mass

Figure 3. Topology of the ALDURO

The kinematics of such a multibody system can be solved in two steps: first the *relative kinematics* are solved, where a relationship between the independent kinematic inputs q and the dependent joint variables β is established. In the second step, solving the *absolute kinematics*, the absolute position of the bodies is determined as a function of the joint variables β. For the direct kinematics the input vector consists of the linear motions of the four hydraulic cylinders in each leg and of the virtual joint described above:

$$q_d = \left[s_{\mathrm{fl}}^\mathrm{T}, s_{\mathrm{fr}}^\mathrm{T}, s_{\mathrm{rl}}^\mathrm{T}, s_{\mathrm{rr}}^\mathrm{T}, s_{\mathrm{cb}}^\mathrm{T}\right]^\mathrm{T} ; \tag{1a}$$

$$s_i = \left[s_{i,1}, s_{i,2}, s_{i,3}, s_{i,4}\right]^\mathrm{T}, \quad \text{with} \quad i = \mathrm{fl, fr, rl, rr}, \tag{1b}$$

$$s_{\mathrm{cb}} = \left[r_{\mathrm{cb}}^\mathrm{T}, \varphi_{\mathrm{cb}}^\mathrm{T}\right]^\mathrm{T} . \tag{1c}$$

The design and calculation of the anthropomorphic leg is explained in detail in Müller et al. (1998).

3.2 Dynamic Model

To calculate the dynamics of a system formulated in minimal coordinates a coupled system of nonlinear ordinary differential equations has to be solved. These equations of motion are:

$$M(q)\,\ddot{q} + g^c(q,\dot{q}) = g^a(q,\dot{q}) . \tag{2}$$

The number of DOF f in the system is equal to the number of independent joint coordinates. The inertia matrix $M \in \mathbb{R}^{f \times f}$ is symmetric and positive definite and $g^c \in \mathbb{R}^f$ is the vector of the generalized centrifugal forces. The vector of the generalized applied forces $g^a \in \mathbb{R}^f$ can be split up into

$$g^a(q,\dot{q}) = g^{a'}(q,\dot{q}) + G(q)\,u, \tag{3}$$

where $g^{a'}$ is the vector of the generalized applied forces without the generalized driving forces $Q_q^* = G\,u$. The generalized driving forces are required for the exact input-output linearization embedded in the model based controller. The matrix $G \in \mathbb{R}^{f \times p}$ describes the mapping of the p driving forces and torques $u \in \mathbb{R}^p$ to the generalized forces (see e. g. Hiller and Schneider, 1997).

3.3 Foot-Ground Contact

A model of the complete walking machine ALDURO has to include a detailed model of the foot-ground contact because the forces that arise when setting the foot on the ground are very high compared to the forces in the static system. Here, a three dimensional regularized contact model with spring-damper elements was developed based on the planar contact model described in Schuster (1999) for automotive suspensions. This model compares the force tangential to the ground plane with the maximum static friction force, which depends on the force normal to the ground plane. The foot starts sliding if the tangential force exceeds the static friction force. A schematic view of the regularized contact model is shown in Figure 4.

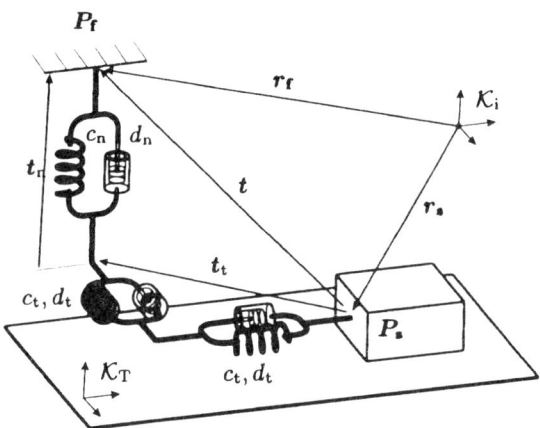

Figure 4. Three Dimensional Regularized Contact Model

The model consists of three orthogonal spring-damper elements and a sliding shoe with a friction plane. The tangential force F_t can not exceed the static friction force $F_{\mu,\max}$, so that in the model the coordinates of the sliding shoe have to be corrected. In the following equations the algorithm for the calculation of the regularized contact will be derived.

As shown in Figure 4, the vector $t = r_f - r_s$ leads from the sliding shoe point P_s to the foot point P_f. This vector can be split up in a vector normal to the plane

$$^{\mathrm{T}}t_n = \begin{bmatrix} 0, 0, ^{\mathrm{T}}t_z \end{bmatrix}^{\mathrm{T}} \tag{4}$$

and a vector in the ground plane

$$^T t_t = \begin{bmatrix} ^T t_x, ^T t_y, 0 \end{bmatrix}^T, \tag{5}$$

both described in components of $\mathcal{K}_T$. At $^T t_z = 0$ the foot is on the ground plane, for all $^T t_z > 0$ the foot can move freely over the ground. In a first step the normal force F_n acting on the foot can be calculated as:

$$F_n = \begin{cases} 0 & \text{for } ^T t_z > 0, \\ -(c_n t_n + d_n \dot{t}_n) & \text{for } ^T t_z \leq 0, \end{cases} \tag{6}$$

where c_n and d_n are the spring and damping coefficients in the direction normal to the plane. After this the maximum static friction force is $F_{\mu,\max} = \mu F_n$. In the same manner the tangential forces in the x and y directions can be calculated. As long as the static friction force exceeds the total tangential force the foot remains fixed on the ground Ouezdou et al. (1998). Therefore, the sliding shoe stays at the same position between time step (0) and time step (1).

$$(\|F_{\mu,\max}\| > \|F_t\|) \wedge (^T t_z \leq 0) :$$

$$r_s^{(1)} = r_s^{(0)}. \tag{7}$$

If the static friction force is smaller than the calculated tangential force the foot starts sliding over the ground and the sliding shoe moves in the direction of the tangential force, so that the tangential force equals the maximum static friction force.

$$(\|F_{\mu,\max}\| \leq \|F_t\|) \wedge (^T t_z \leq 0) :$$

$$r_s^{(1)} = -\frac{1}{c_t} \|F_{\mu,\max}\| \left(\frac{1}{\|F_t^{(0)}\|} \right) F_t^{(0)} + r_f. \tag{8}$$

The new vector $t^{(1)} = r_f - r_s^{(1)}$ is used to calculate the corrected regularized tangential force:

$$F_t^{(1)} = \begin{cases} 0 & \text{for } ^T t_z^1 > 0, \\ -\|F_{\mu,\max}\| \left(\frac{1}{\|F_t^{(0)}\|} \right) F_t^{(0)} & \text{for } ^T t_z^1 \leq 0. \end{cases} \tag{9}$$

If the foot point P_f is above the ground plane, there is no foot ground contact and the position of the sliding shoe is set to the position of the foot point in the ground plane.

3.4 The Hydraulic Subsystem

The state variables z_{hydr} of hydraulic subsystems are the pressures p in the system, in contrast to the mechanical subsystems where the position and velocity in generalized coordinates are used to build the state vector. These pressures can be calculated with the help of the pressure build-up equation:

$$\dot{z}_{hydr} = \dot{p}, \quad \text{with} \quad \dot{p}_i = \frac{1}{C_{h_i}(p_i)} \cdot \sum Q, \quad i = 1, \ldots, n, \tag{10}$$

where n is the number of hydraulic degrees of freedom and $C_{h_i}(p_i)$ is the hydraulic capacity. The required volume flows Q can be calculated for each hydraulic component as a function of the actual pressures and the geometric and electrical properties of the component.

The coupling of the mechanical and hydraulic subsystems described in Hiller and Schneider (1997) has to be carried out in four steps in order to calculate the state vector for the complete system (Figure 5). First the *global kinematics* of the mechanical subsystem have to be solved as

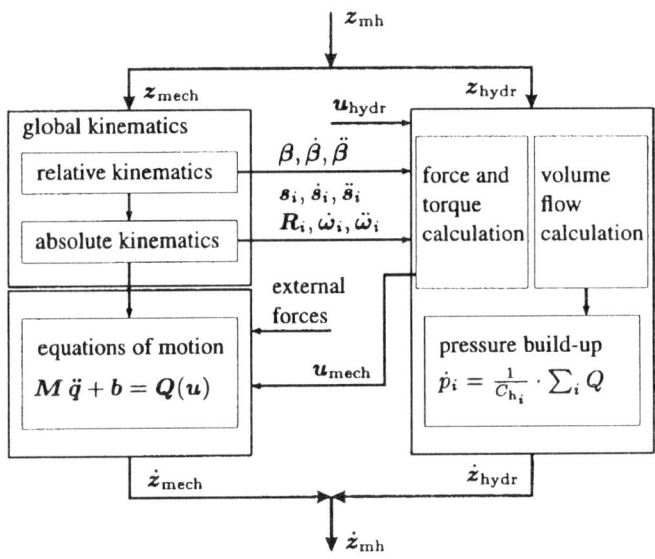

Figure 5. Coupling of Mechanical and Hydraulic Subsystems

described in section 3.1. These calculations can be carried out without knowledge of the hydraulic subsystem. The volume flows within, and the forces u_{mech} generated by, the hydraulic elements – cylinders, pumps, etc. – are calculated. The input vector u_{hydr} gives in this case the electrical inputs for the servo valves. The pressure build-up in the hydraulic system is calculated next. It depends on the actual pressure, the volume flow rates and the kinematics of the multibody system. The equations of motion are solved in a final step. These depend on the global kinematics and the forces provided by the hydraulic subsystem, as well as the hydraulic differential equations.

4 Force Based Control

The control of four legged walking machines needs more complex controllers than six or eight legged machines of the same size (e. g. in Song and Waldron (1989)) because four legged systems can become unstable more easily. The ALDURO is steered by the driver with the help of a joystick. Therefore only the platform position r_{cb}, orientation φ_{cb} and their time derivatives are specified Müller et al. (2000). The coordination of the legs and the trajectories r_s of the legs in the swing phase are calculated using a walking pattern generator. This leads to a set of control output variables, that are different for case A with all legs on the ground and case B with one leg in the swing phase:

$$y^{(A)} = \left[r_{cb}^T, \varphi_{cb}^T\right]^T \in \mathbb{R}^{m^{(A)}}; \quad m^{(A)} = 6 \text{ (case A)}, \tag{11a}$$

$$y^{(B)} = \left[r_{cb}^T, \varphi_{cb}^T, r_s^T\right]^T \in \mathbb{R}^{m^{(B)}}; \quad m^{(B)} = 9 \text{ (case B)}, \tag{11b}$$

These control variables can also be used as *minimal coordinates q*, because the number of control variables equals the number of minimal coordinates:

$$f^{(A)} = m^{(A)}, \quad \text{and} \quad f^{(B)} = m^{(B)}. \tag{12}$$

In contrast to equation (1) the cylinder coordinates build the vector of the dependent coordinates $\beta = \beta(q)$. The time derivative leads to

$$\dot{\beta} = J_\beta \dot{q}, \quad J_\beta = \frac{\partial \beta}{\partial q}. \tag{13}$$

The Jacobian J_β is also used in transposed form to describe the mapping of the driving forces u to the generalized forces Q_q:

$$Q_q = J_\beta^T u. \tag{14}$$

The purpose of the input/output linearized controller used is the calculation of the desired driving forces u^* with respect to the given acceleration $\ddot{y}^* = \ddot{q}^*$. Therefore the desired generalized driving forces are calculated as:

$$Q_q^* = M \ddot{q}^* + g^c - g^{a'}. \tag{15}$$

The *linear input/output channels* have to be stabilized by using *linear* control laws Schneider (1999). When solving equation (14) with respect to u there are more unknown variables than equations, so additional equations are needed for a unique solution. Therefore the classical optimization problem

$$Z(u^*) = \frac{1}{2} u^{*T} P u^* - p^T u^* \stackrel{!}{=} \min, \quad \text{with} \quad P \in \mathbb{R}^{p \times p}, \quad p \in \mathbb{R}^p \tag{16}$$

has to be solved by minimizing the LAGRANGE function

$$L = Z + \lambda^T (J_\beta^T u^* - Q_q^*) \stackrel{!}{=} \min \tag{17}$$

with the LAGRANGE multipliers $\lambda \in \mathbb{R}^f$. The necessary conditions for a minimum are

$$\frac{\partial L}{\partial u^{*T}} = 0 \quad \text{and} \quad \frac{\partial L}{\partial \lambda^T} = 0, \tag{18}$$

which lead to the linear set of equations:

$$\begin{bmatrix} P & J_\beta \\ J_\beta^T & 0 \end{bmatrix} \begin{bmatrix} u^* \\ \lambda \end{bmatrix} = \begin{bmatrix} p \\ Q_q^* \end{bmatrix}. \tag{19}$$

Here the Matrix P and the vector p can be used to influence the calculated solution. These equations can be solved using solvers for systems of linear algebraic equations. The driving forces have to be controlled using decentralized force controllers (see e. g. Schneider, 1999).

References

Hiller, M., and Schneider, M. (1997). Modelling, simulation and control of flexible manipulators. *European Journal of Mechanics, A/Solids* 16, special issue:127–150.

Kecskeméthy, A. (1993). *Objektorientierte Modellierung der Dynamik von Mehrkörpersystemen mit Hilfe von Übertragungselementen*. Fortschritt-Berichte VDI, Reihe 20 Nr. 88. Düsseldorf: VDI-Verlag.

Kecskeméthy, A. (1994). A spatial leg mechanism with anthropomorphic properties for ambulatory robots. In Lenarčič, J., and Ravani, B., eds., *Advances in Robot Kinematics and Computational Geometry*. Dordrecht, Boston, London: Kluwer Academic Publishers. 161–170.

Müller, J., Schneider, M., and Hiller, M. (1998). Modelling and simulation of the large-scale hydraulically driven ALDURO. In *Proceedings of the Euromech Colloquium 375*, 116–123.

Müller, J., Schneider, M., and Hiller, M. (2000). Modelling, simulation and model based control of the walking machine ALDURO. *IEEE Transactions On Mechatronics* 5(2).

Ouezdou, F. B., Bruneau, O., and Guinot, J. C. (1998). Dynamic analysis tool for legged robots. In Schiehlen, W., ed., *Multibody System Dynamics*, volume 2. Dordrecht, Boston, London: Kluwer Academic Publishers. 369–391.

Schneider, M. (1999). *Modellbildung, Simulation und nichtlineare Regelung elastischer, hydraulisch angetriebener Großmanipulatoren*. Fortschritt-Berichte VDI, Reihe 8, Nr. 756. Düsseldorf: VDI-Verlag.

Schuster, C. (1999). *Strukturvariante Modelle zur Simulation der Fahrdynamik bei niedrigen Geschwindigkeiten*. Berichte aus der Fahrzeugtechnik. Aachen: Shaker-Verlag.

Song, S., and Waldron, K. (1989). *Machines That Walk: The Adaptive Suspension Vehicle*. Cambridge, Massachusetts; London, England: The MIT Press.

A New Local Path Planner for a Nonholonomic Wheeled Mobile Robot in Cluttered Environments

Gabriel Ramírez and Saïd Zeghloul

Laboratoire de Mécanique des Solides, Université de Poitiers, France

Abstract. This paper presents a new local path planner based on distance information, for mobile robots with nonholonomic constraints. The nearby obstacles are mapped as linear constraints over the robot's velocities to form a Feasible Velocities Polygon. This polygon represents the set of velocities that the robot can use without collision with the obstacles. The planner, composed by two modules, uses the FVP representation to ensure the collision-free navigation. The first module allows the robot to continuously approach the goal position, avoiding the obstacles and following a stable reference trajectory, obtained from an exponential control law. When a deadlock situation occurs, the second module allows the robot to follow the obstacle's boundary in order to escape the deadlock. The presented results demonstrate the capabilities of the proposed method for solving the collision-free path-planning problem.

1 Introduction

The path-planning problem consists in finding a collision-free trajectory, which leads the mobile robot from its initial position to a desired location. Motivated by its several applications, this problem has been the center of attention of the scientific community in recent years. The two principal points of view are the global and the local approaches:
(a) Global methods need a complete representation of the robot's environment, such as the grid of visits [1], the numeric potential fields [2], the global graphs (of visibility [3], of tangents [4], Voronoï [5]), behavior-based models [6] and genetic algorithms [7], among others. Generally, these methods find a solution if it exists. Their main disadvantage is the expensive calculation time, which makes them prohibitive for use in real-time applications.
(b) In local methods, the robot reacts to the local obstacle configuration. Simpler and faster than the global methods, the local methods are used for real-time path planning, especially in sensor-based navigation. Examples of this approach are the potential fields [8], the Bug algorithm [9], the behavior-based models [10], the probabilistic models (VFH [11], certainty grids [6]), the fuzzy logic [12][13], etc. However, these methods do not guarantee a solution.

For the most of these methods, the mobile robot is considered as an holonomic point. Recently, the nonholonomic properties (nonintegrable constraints over the velocities) of mobile robots have been included into the path-planning problem.

For the local path planner proposed in this paper, the robot and the obstacles are modeled by convex polygons and the nonholonomic constraints are considered. This work is focused to differential mobile robots (robots equipped with two parallel driving wheels and no steering wheel). In order to avoid the collisions, a similar approach to the one proposed in [15] is used: the objects in the robot's influence zone are mapped as linear constraints in the robot's velocity space, forming the **Feasible Velocities Polygon** (FVP). The collision-free trajectory is obtained by minimizing the deviation of the robot's current trajectory from the reference trajectory, under the obstacle constraints. This minimization problem is solved by a minimal distance calculation between the FVP and the point describing the goal.

The planner is composed by two modules, both based on FVP analysis. The first module continuously approaches the robot to the goal, using the optimization approach. The second module uses the FVP and the goal distance to escape from the deadlock situations.

2 Kinematic Model of a Differential Mobile Robot

In order to define the position of a mobile robot in a plan, we can use the following state vector:

$$\mathbf{q} = [x \ y \ \theta]^T$$

where (x, y) is the position of a fixed point R on the robot and θ the orientation of the frame linked to the robot with respect to the X axis. The linear velocity v and the angular velocity ω of the robot are given by:

$$v = \frac{v_d + v_g}{2}, \quad \omega = \frac{v_d - v_g}{A}$$

where v_d and v_g are respectively the right wheel and the left wheel velocities. The kinematic model of the mobile robot can be written as follows:

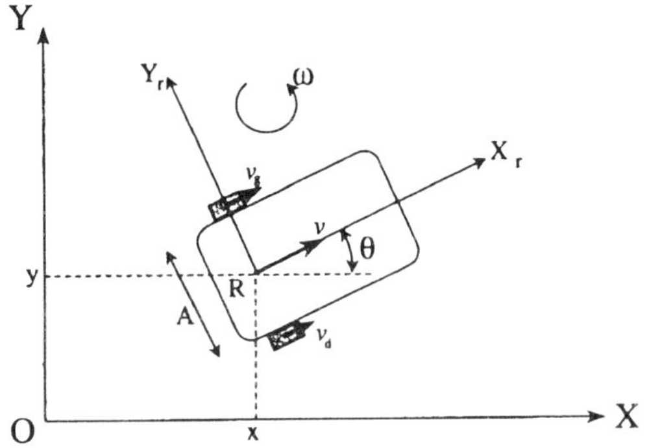

$$\begin{bmatrix} \dot{x} \\ \dot{y} \\ \dot{\theta} \end{bmatrix} = \begin{bmatrix} \cos\theta & 0 \\ \sin\theta & 0 \\ 0 & 1 \end{bmatrix} \begin{bmatrix} v \\ \omega \end{bmatrix}; \quad \text{thus } \dot{\mathbf{q}} = \mathbf{J}(\mathbf{q})\,\mathbf{u}$$

Figure 1. A differential mobile robot

where $\mathbf{J}(\mathbf{q})$ is the robot's Jacobean matrix, and $\mathbf{u} = [v, \omega]^T$ is the input vector (or control vector). For this system, the nonholonomic constraint is given by $\dot{y} - \dot{x}\tan\theta = 0$.

3 Exponential Control Law

The nonholonomic systems cannot be stabilized to an equilibrium point using a time-invariant, smooth (or even continuous) feedback control law [16]. Thus, most existing results from linear and nonlinear systems theory are not directly applicable in this case. In [14], we have proposed a piecewise smooth control law, which makes the final position globally exponentially stable. The final robot's orientation is not taken into account for the construction of the control law.

Let $\mathbf{q}_f = [x_f, y_f, 0]^T$ be the desired goal position (regardless of the final orientation) and $\mathbf{q} = [x, y, \theta]^T$ the robot's current position. Using the polar coordinates (a, α) of the goal with respect to the robot's frame, the control law is established as follows (see Figure 2):

$$v = k_1 a \cos\alpha$$
$$\omega = k_2 \alpha + k_1 \sin\alpha \cos\alpha$$

with $k_1, k_2 > 0$. This control law makes the origin (which represents the goal) globally exponentially stable, as proved in [14].

Using the control law described above, we obtain the path shown in Figure 3. The initial position is the point $q_i = [6, 3, \pi/4]^T$ and the goal position is located at the origin $q_f = [0, 0, 0]^T$. The coefficients k_1, k_2 are set to 0.6.

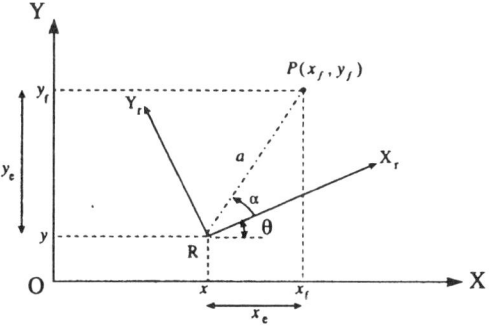

Figure 2. Error definition

4 Path Planning Algorithms

Faverjon and Tournassaud have proposed a substitute for the potential field method [15], for a manipulator robot with a high number of degrees of freedom in a cluttered environment. In this approach, called *constraints method*, the obstacles are mapped as linear constraints over the robot's joint velocities. The collision-free trajectory is obtained by minimizing the deviation of the current robot's trajectory from the reference trajectory. The minimization problem can be stated as follows:

$$\text{Minimize } f = \frac{1}{2}\|\dot{q} - \dot{q}_{goal}\|^2 \quad \text{with} \quad \dot{q}_{goal} = \frac{q - q_f}{\max_{i=1,...,n}\|q^i - q_f^i\|}$$

subject to the obstacle constraints given by the *velocity damper* model:

$$\dot{d} \geq -\xi \frac{d - d_s}{d_i - d_s}$$

where q and q_f are respectively the current and final configuration vectors, $\dot{q}_{goal}$ is desired joint velocities vector, calculated in order to obtain a straight line in the joint space, d is the minimal distance between the robot and the considered object, d_i is the influence distance from which a constraint becomes active, d_s is the security distance that must be respected, and ξ represents a coefficient in order to adapt the convergence speed.

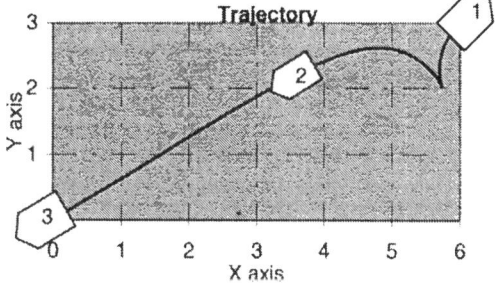

Figure 3. Robot's trajectory

4.1 "Reaching the Goal" Module

The first module of our method, *"reaching the goal"*, uses a similar approach. The optimization problem can be stated as follows: find a vector $u = [v, \omega]^T$ that minimizes the function

$$f_r = \frac{1}{2}\|u - u_{goal}\|^2$$

where $\mathbf{u}_{goal}$ is obtained from the exponential control law proposed in section 3, as:

$$\mathbf{u}_{goal} = \begin{bmatrix} k_1 a \cos\alpha \\ k_2 \alpha + k_1 \sin\alpha \cos\alpha \end{bmatrix}$$

For any obstacle at a distance d less than d_i, we can express the *velocity damper* constraint as a function of the control vector $\mathbf{u}$ (see Fig. 4):

$$(v\mathbf{m} + \omega\hat{k} \times \overline{RP}) \cdot \mathbf{n} \leq \xi \frac{d - d_s}{d_i - d_s}$$

where $\mathbf{m}$ is the unit vector along the robot axis X_r and $\mathbf{n}$ is the unit vector along the segment PQ of minimal distance between the robot and the obstacle.

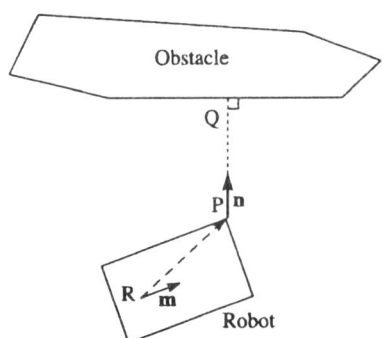

Figure 4. Establishing the obstacle constraints

The set of obstacle constraints and the velocity limit constraints are linear in (v, ω) and define a convex subset in the (v, ω) space. This subset, that we call the *Feasible Velocities Polygon*, represents the bi-dimensional set of velocities that the robot can use without collision with the obstacles, while respecting the velocity limits (Figure 5).

The optimization problem

$$\min_{\mathbf{u}} f_r = \frac{1}{2}\|\mathbf{u} - \mathbf{u}_{goal}\|^2$$

subject to the linear constraints

$$(v\mathbf{m} + \omega\hat{k} \times \overline{RP}) \cdot \mathbf{n} \leq \xi \frac{d - d_s}{d_i - d_s} \quad , \quad d \leq d_i \quad \text{and} \quad |v| \leq v_{max} \quad , \quad |\omega| \leq \omega_{max}$$

is a minimal distance calculation problem between the FVP and the reference point $\mathbf{u}_{goal}$. The solution to this problem is represented by the point $\mathbf{u}^*$, which is the closest point on the polygon to $\mathbf{u}_{goal}$. To solve this problem, we use a fast minimal distance calculation algorithm, proposed by S. Zeghloul and P. Rambaud [18].

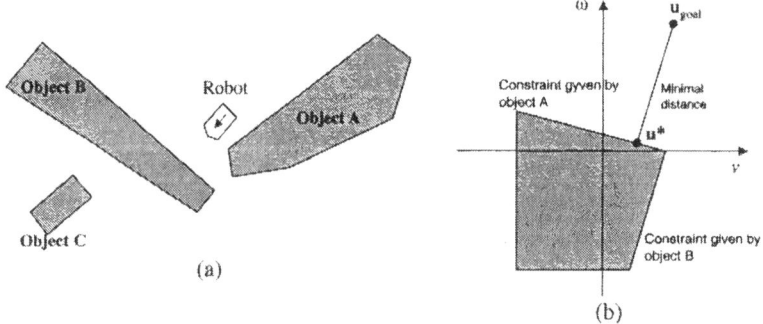

Figure 5. (a) Obstacle configuration, (b) Feasible velocities polygon

The first module, "*reaching the goal*", is defined by the following iterative process:
(i) compute the reference velocity $\mathbf{u}_{goal}$, using the exponential control law ;
(ii) build the FVP, by mapping the obstacles in the robot's influence zone;
(iii) solve the optimization problem by minimal distance calculation ;

(iv) apply the velocities represented by the vector **u***.

This module allows the robot to solve simple situations where the obstacle configuration does not create local minima points. An example is shown in Figure 6. While the robot uses the solution **u***, the *"distance"* function:

$$V(\mathbf{z}) = \frac{1}{2}a^2 + \frac{1}{2}\alpha^2$$

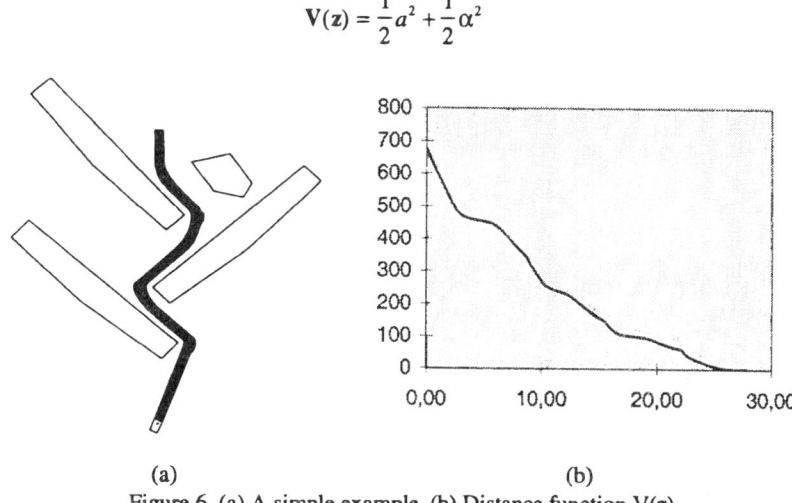

(a)　　　　　　　　　　　　　(b)
Figure 6. (a) A simple example, (b) Distance function V(**z**)

is strictly decreasing, since the robot is continuously approaching to the goal (Figure 6(b)). This property is used in the second module to solve the deadlock situations.

When the solution of the optimization problem, **u***, is located at the origin of the plan (v, ω), i.e. **u*** is a null vector, the robot cannot continue to move using the *reaching the goal* module (see Figure 7). This condition corresponds to a deadlock situation. In this case, the second module, called *boundary following*, is applied.

4.2 "Boundary Following" Module

The *"boundary following"* module and its interaction with the *reaching the goal* module have been inspired by the Bug algorithm [9]. The module analyzes the FVP configuration and the value of the *distance* function V(**z**) to allow the robot to follow the obstacle's boundary and to escape from the deadlock.

When a deadlock situation is detected by the first module (**u***=0), the current value V_{block} of the *distance* function V(**z**) is recorded and the two adjacent vertices to **u***, s_R and s_L, are identified (see Figure 7). The local configuration of the blocking obstacles defines which obstacle constraint will be chosen to surround them, and which of the vertex s_R and s_L will be used. Thereafter, the *boundary following* module tracks the evolution of the chosen constraint on the FVP and applies the velocities represented by the chosen vertex. This allows the robot to follow the boundary of the blocking obstacle at a distance $d \geq d_s$.

While the robot is surrounding the obstacles, the value of the distance function V(**z**) is continuously compared with V_{block}. When the value of the distance function is less than V_{block}, the robot has found a point on the obstacle's convex boundary that is closer to the goal than the deadlock point. At this moment, the planner switches from the *boundary following* module to the *reaching the goal* module in order to continue to the goal position.

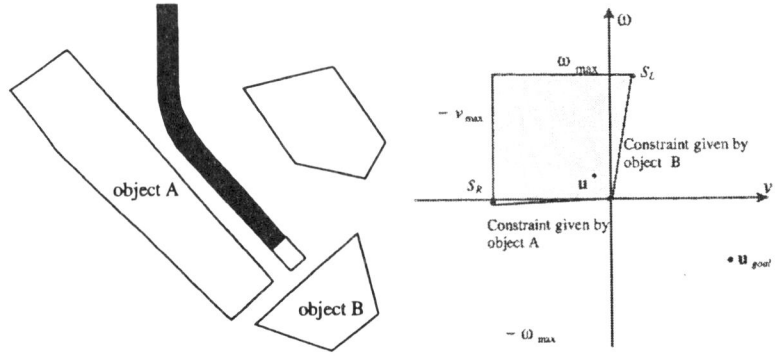

Figure 7. Example of a dead-lock situation

When a new obstacle prevents the robot from following the obstacle boundary, the chosen vertex converges to the origin of the (v, ω) plan. In this case, the method simply switches from the followed constraint to the new one. Therefore, the robot will surround the new obstacle.

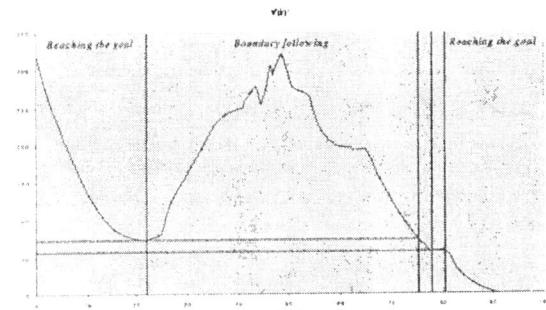

Figure 8. Distance function

The resulting trajectory is a sequence of *reaching the goal* and *boundary following* intervals. By construction, for every *boundary following* interval the final value of the distance function is less than the initial one (V_{block}). Furthermore, the distance function is strictly decreasing in the *reaching the goal* intervals (Fig. 9). If the *boundary following* intervals are finite, we can consider the *reaching the goal* sections as a sequence of strictly decreasing positive values. Since the origin is the only equilibrium point of the kinematic system, we can affirm that the distance function converges to zero. Thus, the robot reaches the final position.

Both modules are based on the FVP analysis only, which is built from the distance information between the robot and the obstacles, thus the method is well adapted for sensor-based navigation. The proposed path planner has been implanted on a real mobile robot (RoboSoft Robuter) equipped with ultrasonic sensors and the results are very satisfactory.

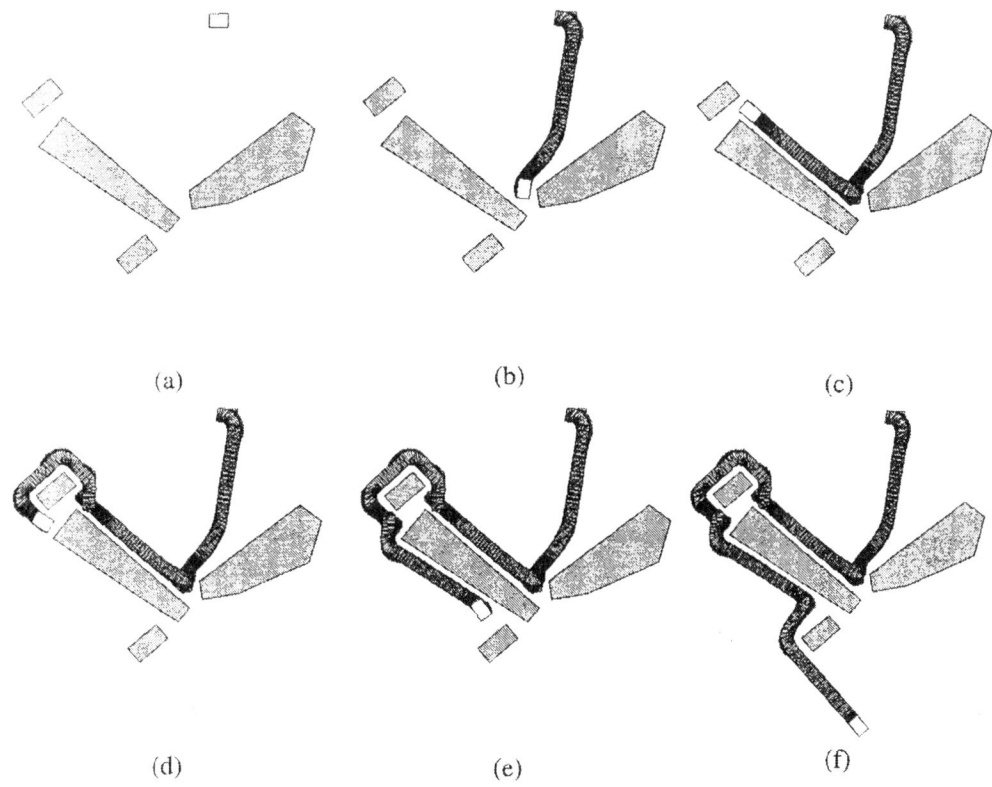

Figure 9. Path planning solution

5 Conclusion

In this paper, a new collision-free path planner for nonholonomic mobile robots has been presented. Two modules compose this planner. The first module allows the robot to continuously approach the goal position, avoiding the obstacles and following a stable reference trajectory obtained from an exponential control law, which considers the nonholonomic constraints. In order to avoid the collisions, a similar approach to the one proposed in [15] is used: the objects in the robot's influence zone are mapped as linear constraints in the robot's velocity space, forming the *Feasible Velocities Polygon*. The collision free trajectory is obtained by minimizing the deviation of the robot's current trajectory from the reference trajectory, under the obstacle constraints. This minimization problem is solved by a minimal distance calculation in the velocity space, between the FVP and the point describing the goal velocity.

The second module is activated when the first module detects a deadlock situation. The module uses the FVP representation to escape from the blocking condition, by following the obstacle's boundary and evaluating the distance to the goal. The results show that the proposed path planner gives a solution in almost all cases where the classical local methods fail. The computing times are very short, which allows the implantation of the proposed method for real-time application. The path planner has been implanted on a commercial mobile robot and experimental tests have been successfully performed.

References

[1] T. Balch and R. Arkin, "Avoiding the past: a simple but effective strategy for reactive navigation," *IEEE*, pp. 678-685, 1993.

[2] J. Barraquand et al., "Numerical potential field techniques for robot path planning," *Procs. Inter. Conf. Adv. Robotics*, pp. 1012-1017, Pisa, Italie, 1991.

[3] F. Garcia and R. Mampey, "Mobile robot planning by reasoning both at itinerary and path levels," *IEEE Int. Conf. Adv. Robotics*, pp. 1074-1080, Pisa, Italie, 1991.

[4] Y. H. Liu and S. Arimoto, "Proposal of tangent graph and extended tangent graph for path planning of mobile robots," *Procs. Robotics and Automation*, pp. 312-317, 1991.

[5] S. S. Iyengar et al., 'Robot navigation algorithms using learned spatial graphs," *Robotica*, pp. 93-100, 1986.

[6] D. W. Cho et al., "Experimental investigation of mapping and navigation based on certainty grids using sonar sensors," *Robotica*, pp. 7-17, 1993, vol. 11.

[7] T. Shibata and T. Fukuda, "Coordinative behavior by genetic algorithm and fuzzy in evolutionary multi-agent system," pp. 760-765, *IEEE*, 1993.

[8] W. Tianmiao and Z. Bo, "Time-varing potential field based „ perception-action " behaviors of mobile robot," *Procs. IEEE Int. Conf. Rob. Autom.*, pp. 2549-2554, 1992.

[9] T. Skewis and V. Lumelsky, "Experiments with a mobile robot operating in a cluttered unknown environment," *Procs. Int. Conf. Rob. Autom.*, pp. 1482-1487, 1992.

[10] M. Wolfensberger and D. Wright, "Synthesis of reflexive algorithms with intelligence for effective robot path planning in unknown environments," *SPIE Mobile Robots*, pp. 70-81, 1993.

[11] J. Borenstein and Y. Koren, "The vector field histogram - Fast obstacle avoidance for mobile robots," *IEEE Trans. Robot. Autom.*, pp. 278-288, 1991, vol. 7.

[12] Y. Maeda, "Collision avoidance control among moving obstacles for a mobile robot on the fuzzy reasoning," 1990.

[13] B. Beaufrere and S. Zeghloul, "A mobile robot navigation method using fuzzy logic approach," Robotica, pp. 437-448, 1995.

[14] G. Ramirez and S. Zeghloul, "Path planning for a nonholonomic wheeled mobile robot in cluttered environments," *Proc. of the 4th Japan-France Congress on Mechatronics*, volume 1, pp. 337-342, 1998.

[15] B. Faverjon and P. Tournassoud, "A local based apporach for path planning of manipualtors with high number of dregrees of freedom," *IEEE Procs. Int. Conf. Robot. Autom.*, pp. 1152-1159, 1987.

[16] R. W. Brockett, "Asymptotic stability and feedback stabilization," *Differential Geometric Control Theory*, pp. 181-208, Birkhauser, 1983.

[17] O. J. Sordalen and C. Canudas, "Exponential control law for a mobile robot: extension to path following," *Procs. IEEE Int Conf. Rob. Autom.*, pp. 2158-2163, 1992.

[18] S. Zeghloul and P. Rambeaud, "A fast algorithm for distance calculation between convex objects using the optimization approach," *Robotica*, pp. 355-363, 1996.

Three-Dimensional Simulation of Walk of Anthropomorphic Biped

Fabrice Gravez[1], Olivier Bruneau[2] and Fethi Ben Ouezdou[1]

[1] Laboratoire de Robotique de Paris, France
[2] Laboratoire de Vision Robotique, France

Abstract. The work presented in this paper seeks the development of a new approach of three-dimensional walk simulation of anthropomorphic biped robots. Our approach is based on a three dimensional extension of previous work which introduced a reference structure and a reference gait modified by spatial and temporal transformations of the ankle trajectory (Bruneau et al., 1998). The three main points explained in this paper are the following: in a first part, the 3D-biomechanical data of the angular joints and the main geometric and inertial characteristics of a standard human being are described. Secondly, in order to carry out simulations of the dynamic behavior of an anthropomorphic biped, inverse kinematics of one leg with six degrees of freedom are given. Thus, angular joints can be obtained by the knowing of the positions and orientations of the foot relatively to the hip. This method permits an interactive change of these positions and orientations to modify the locomotion cycle. Finally, the first steps of a three-dimensional walk is easily obtained with our method.

1 Introduction

Today, a specific class of robot, the biped robots, are the subject of a lot of research programs. The technological posibility to build advanced anthropomorhic bipeds such as Honda robots P2 and P3 (Hirai et al., 1998), or WABIAN (Yamagushi et al., 1997), able to move in environments well-adapted to human beings, opens new potential applications in the fields of cooperations between humans and robots. One of the most critical point in the development of this kind of structures is their quick displacements with dynamic gaits. Furthermore, this displacement can be improved by the design of the mechanical structure. Thus, we present in this paper a method to produce easily the dynamic three dimensional start of walk for anthropomorphic bipeds having flexible feet.

2 Reference Structure and Biomechanical Data

First of all, statistical biomechanical studies give the main geometric and inertial characteristics of a standard human being which has a total mass of 75 kg and a total height of 1.78 m (Bouisset et al., 1995), (Winter, 1990), (Seward et al., 1996). Each leg has six dof (degrees of freedom) broken down as follow (fig. 1(a) and 1(b)) : 3 d.o.f. for the hip, 1 d.o.f. for the knee, and 2 d.o.f. for the ankle.

Each segment of the structure is approximated by a composition of solid primitives such as ellipsoid of revolution for the head, parallelepipeds for the torso and the pelvis, truncated cones for the arms and the legs. Geometric and inertial parameters of the biped are given in table 1. We

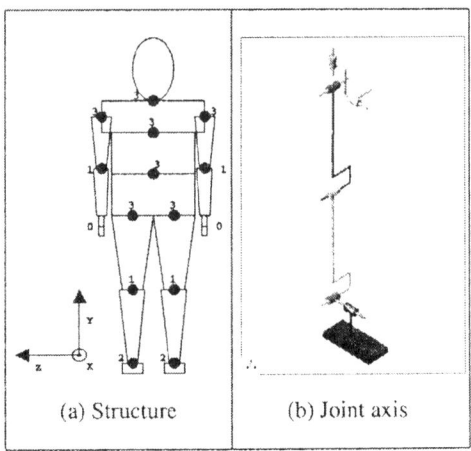

Figure 1. Solid primitives and degrees of freedom.

have to note that the center of mass of all rectangular parallelepipeds and ellipsoids of revolution are located at their geometric center. The geometric characteristics of the truncated cone are calculated such that its center of mass is located at 43 % of link length from the bottom radius which is a law given by biomechanicians. The most interesting data on producing velocity of the biped is the vector of the three positions and three orientations of each foot with regard to a frame attached to each hip. These data obtained by cubic-spline interpolations of the recorded motions are given in figures 2(a), 2(b), 3(a) and 3(b). At the beginning, the biped stands up on the two legs with an initial velocity equal to zero. These data are related to the beginning of the walk, that is to say the transition from standing to walking during 1s. After this phase, the curves give the periodic regime of walk with a time periode of 1s. In order to simulate the dynamic behavior of the robot, we want to use the set of parameters $q_i(t)$ (i=1,..6) which are the articular joint motions imposed in the inverse dynamic model. They are obtained by using the inverse explicit kinematic model of one leg, described in the next section.

3 Inverse kinematics

The inverse explicit kinematic model permits to obtain the evolution of the articular joint motions as analytical functions of the feet motions with regard to the platform. The six degrees of freedom of one leg are given in figure 1(b). The homogeneous matrix T which gives the foot position and orientation with regard to the platform is the following :

$$T = \begin{pmatrix} A & B & C & X \\ E & F & G & Y \\ I & J & K & Z \\ 0 & 0 & 0 & 1 \end{pmatrix}$$

X, Y and Z are the positions of the foot directly given by curves 2(a) and 2(b).
A, B, C, E, F, G, I, J, K are the compositions of the three rotations ry,rx,rz successively around

Three-Dimensional Simulation of Walk of Anthropomorphic Biped

Table 1. Table of physical parameters

body part	shape	length (m)			mass (kg)	Inertia (kg.m^2)		
		Rx	Ry	Rz		Ixx	Iyy	Izz
head	ellipsoid of revolution	0.189	0.336	0.189	5.400	4.015E-02	1.926E-02	4.015E-02

body part	shape	length (m)			mass (kg)	Inertia (kg.m^2)		
		Lx	Ly	Lz		Ixx	Iyy	Izz
torso	rectangular parallelepiped	0.182	0.178	0.365	12.540	1.721E-01	1.738E-01	6.788E-02
high pelvis	rectangular parallelepiped	0.212	0.208	0.303	14.172	1.596E-01	1.618E-01	1.042E-01
low pelvis	rectangular parallelepiped	0.212	0.208	0.303	14.172	1.596E-01	1.618E-01	1.042E-01
foot part 1	rectangular parallelepiped	0.055	0.053	0.089	0.289	2.599E-04	2.651E-04	1.428E-04
foot part 2	rectangular parallelepiped	0.110	0.046	0.089	0.499	4.181E-04	8.405E-04	5.991E-04
foot part 3	rectangular parallelepiped	0.083	0.031	0.089	0.255	1.896E-04	3.155E-04	1.679E-04
foot part 4	rectangular parallelepiped	0.027	0.020	0.089	5.535	3.846E-05	4.007E-05	5.460E-06

body part	shape	length (m)			mass (kg)	Inertia (kg.m^2)		
		L	R	r		Ixx	Iyy	Izz
forearm	truncated cone	0.277	0.030	0.045	1.400	9.053E-03	1.056E-03	9.053E-03
arm	truncated cone	0.326	0.036	0.056	2.400	2.168E-02	2.755E-03	2.168E-02
thigh	truncated cone	0.434	0.056	0.087	7.500	1.227E-01	2.059E-02	1.227E-01
leg	truncated cone	0.434	0.039	0.060	3.700	5.790E-02	4.919E-03	5.790E-02
hand	extrusion			0.015	0.450	3.576E-04	2.025E-04	5.523E-04

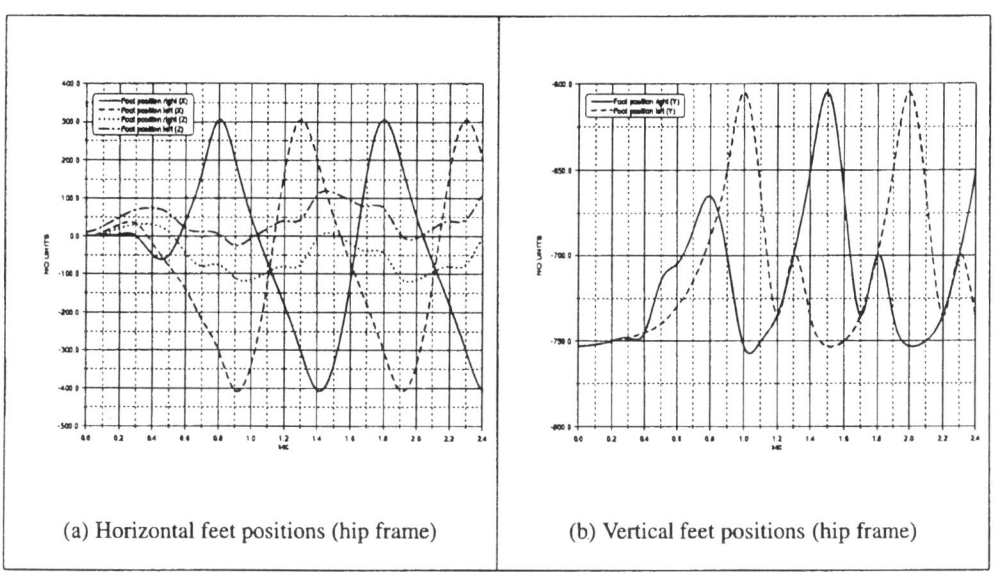

(a) Horizontal feet positions (hip frame) (b) Vertical feet positions (hip frame)

Figure 2. Positions of the feet.

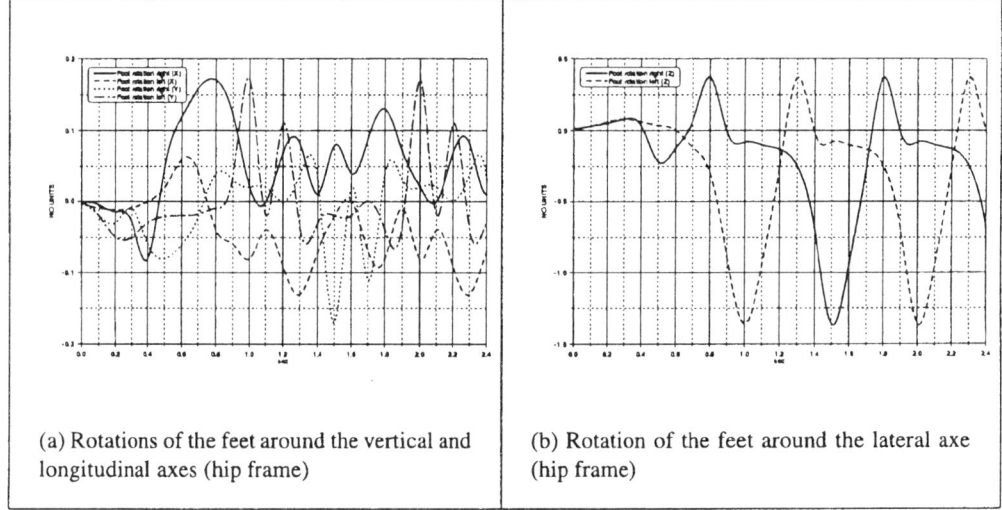

(a) Rotations of the feet around the vertical and longitudinal axes (hip frame)

(b) Rotation of the feet around the lateral axe (hip frame)

Figure 3. Orientations of the feet.

Y,X and Z-axis given by the curves 3(a) and 3(b) with :

$A = cos(ry)cos(rz) + sin(rx)sin(ry)sin(rz)$ $\quad B = -cos(ry)sin(rz) + sin(rx)sin(ry)cos(rz)$
$C = cos(rx)sin(ry)$ $\quad E = cos(rx)sin(rz)$ $\quad F = cos(rx)cos(rz)$
$G = -sin(rx)$ $\quad I = -sin(ry)cos(rz) + sin(rx)cos(ry)sin(rz)$
$J = sin(ry)sin(rz) + sin(rx)cos(ry)cos(rz)$ $\quad K = cos(rx)cos(ry)$

This homogeneous matrix is equal to the composition of the homogeneous matrix from the hip to the foot :

$$T = T_{01}T_{12}T_{23}T_{34}T_{45}T_{56} \tag{1}$$

with :
$T_{01} = Rot(\mathbf{y}, q_1)$ (from the pelvis frame to the first hip joint frame)
$T_{12} = Rot(\mathbf{x}, q_2)$ (from the first hip joint to the second hip joint)
$T_{23} = Rot(\mathbf{z}, q_3 - \frac{\pi}{2})$ (from the the second hip joint to the third hip joint)
$T_{34} = Tr(\mathbf{x}, l_1)Rot(\mathbf{z}, q_4)$ (from the hip to the knee)
$T_{45} = Tr(\mathbf{x}, l_2)Rot(\mathbf{z}, q_5 + \frac{\pi}{2})$ (from the knee to the ankle)
$T_{56} = Rot(\mathbf{x}, q_6)$ (from the first ankle joint to the second ankle joint)

By identification of the two members of equation 1, we obtain (notation $c_{ij} = cos(q_i + q_j)$):

$X = (l_1 * s_3 + l_2 * s_{34}) * c_1 - (l_1 * c_3 + l_2 * c_{34}) * s_1 * s_2$ $\quad Y = -(c_3 * l_1 + l_2 * c_{34}) * c_2$
$Z = -(l_1 * c_3 + l_2 * c_{34}) * c_1 * s_2 - (l_1 * s_3 + l_2 * s_{34}) * s_1$
$A = c_1 * c_{345} + s_1 * s_2 * s_{345}$ $\quad B = (s_1 * s_2 * c_{345} - c_1 * s_{345}) * c_6 + s_1 * c_2 * s_6$

$$C = (c_1 * s_{345} - s_1 * s_2 * c_{345}) * s_6 + s_1 * c_2 * c_6 \qquad E = c_2 * s_{345}$$
$$F = c_2 * c_{345} * c_6 - s_2 * s_6 \qquad G = -c_2 * c_{345} * s_6 - s_2 * c_6 \qquad I = c_1 * s_2 * s_{345} - s_1 * c_{345}$$
$$J = (c_1 * s_2 * c_{345} + s_1 * s_{345}) * c_6 + c_1 * c_2 * s_6 \qquad K = c_1 * c_2 * c_6 - (s_1 * s_{345} + c_1 * s_2 * c_{345}) * s_6$$

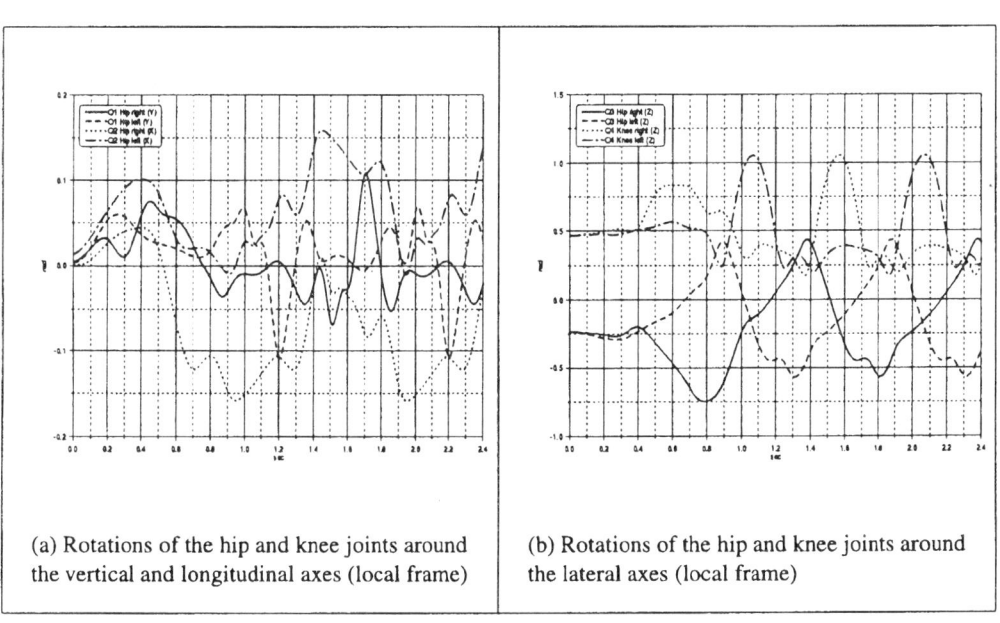

(a) Rotations of the hip and knee joints around the vertical and longitudinal axes (local frame)

(b) Rotations of the hip and knee joints around the lateral axes (local frame)

Figure 4. Angular joint motions of the knees and the hips.

The complete inverse kinematic model gives the six following articular joint parameters :

$$q_1 = \arctan\left(\frac{I * Y - E * Z}{E * X - A * Y}\right)$$

$$q_2 = \arctan\left(\frac{\sin(q_1) * A + \cos(q_1) * I}{E}\right)$$

$$q_3 = \arctan\left(\frac{t_x * (l_1 + l_2 * \cos(q_4)) + t_y * l_2 * \sin(q_4)}{t_x * l_2 * \sin(q_4) - t_y * (l_1 + l_2 * \cos(q_4))}\right)$$

with $t_x = \cos(q_1) * X - \sin(q_1) * Z$ and $t_y = \frac{Y}{\cos(q_2)}$

$$q_4 = -\arccos\left(\frac{X^2 + Y^2 + Z^2 - l_1^2 - l_2^2}{2 * l_1 * l_2}\right)$$

$$q_5 = \theta_5 - q_3 - q_4$$

with $\theta_5 = \arctan\left(\frac{E}{\cos(q_2) * (\cos(q_1) * A - \sin(q_1) * I)}\right)$

$$q_6 = \arctan\left(\frac{\cos(q_1) * C - \sin(q_1) * K}{\sin(q_1) * J - \cos(q_1) * B}\right)$$

The evolution of the six angular motions versus time are thus obtained. The curves 4(a),4(b),5(a) and 5(b) give the joint motions of the hips, the knees and the ankles for the first step during 1 s and the periodic walk with a cycle time of 1 s. The distibuted and elastic contact model used for the simulation is described in a previous work (Bruneau et al., 1999).

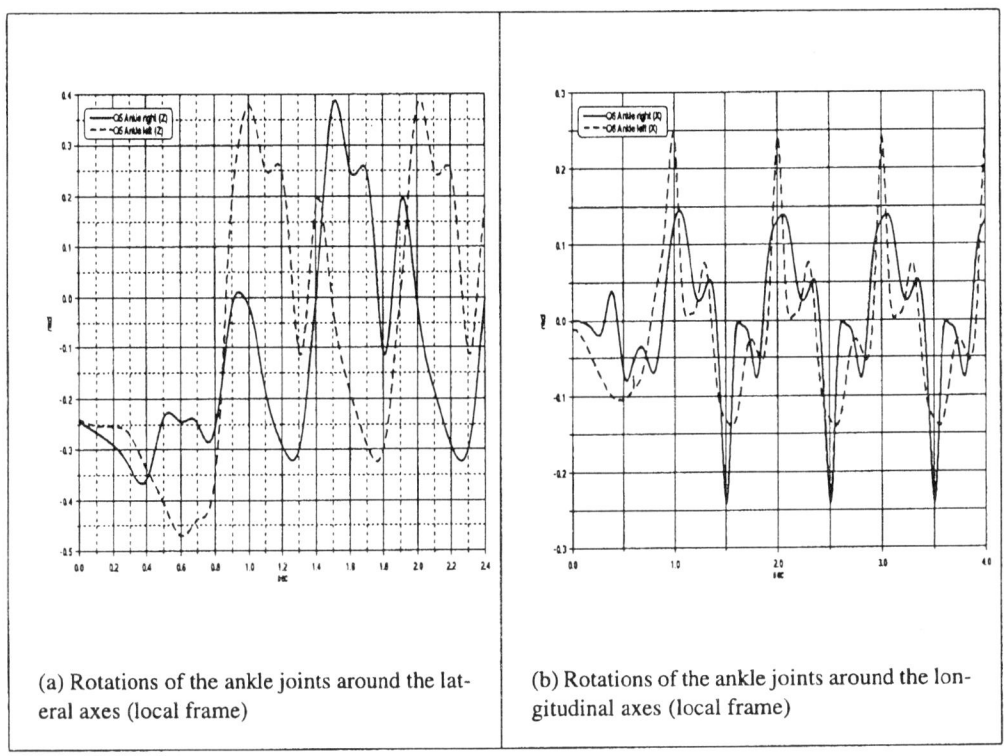

(a) Rotations of the ankle joints around the lateral axes (local frame)

(b) Rotations of the ankle joints around the longitudinal axes (local frame)

Figure 5. Angular joint motions of the ankles.

4 First Steps in 3D with flexible feet

The results of the first steps of the three-dimensional walk of a biped robot having the particularity to walk with flexible feet are given in this section. Each foot is composed by four elements related by three rotary joints. The following torsion couple is applied on each rotary joint i :

$$\tau_{torsion\,i} = k_i(q_i^o - q_i(t)) - c_i\dot{q}_i(t) \quad (2)$$

where k_i and c_i are stiffness and damping coefficients, q_i^o is an equilibrium position for each torsion spring-damper, $q_i(t)$ and $\dot{q}_i(t)$ are articular positions and velocities.

The two major effects of the introduction of flexible feet are the diminution of the intensity of normal force when the heel touches the ground and the increasing of the phase of double support allowing a better stability of the system (Bruneau et al., 1999b). Since the structure is different from a real human being, modifications of the initial curves have to be made. These modifications are carried out by an interactive change of the controlling points (ten per locomotion cycle) of curves 2(a), 2(b), 3(a) and 3(b). Furthermore, the locomotion cycle duration is modified and equal to 1.2 s. A significant advantage given by the analytical inverse kinematics is the direct change of the positions and orientations of the feet relatively to the hips which is more intuitive and easier than a modification of the articular joint time evolution. Some snapshots of the simulation are shown on figure 6. The duration of the simulations is 2.8 s. During 1.2 s, the biped stands up on the two legs with an initial velocity equal to zero. The two first steps of the walk are then carried out.

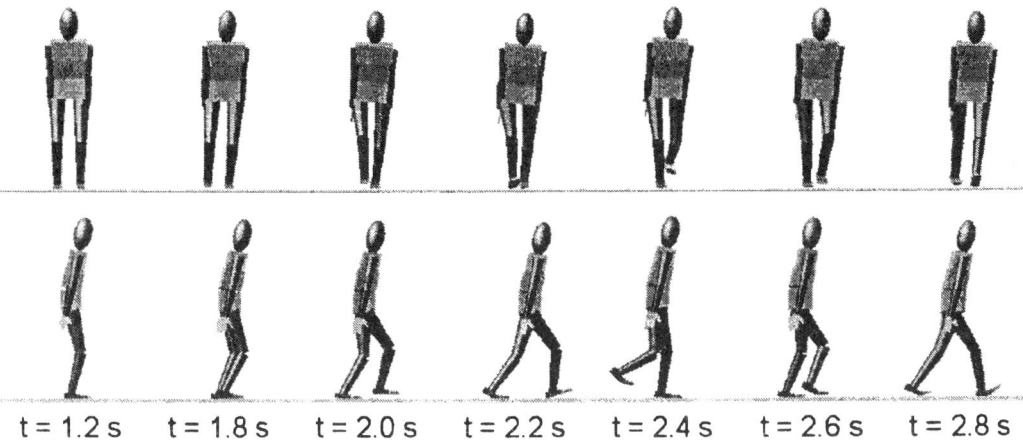

t = 1.2 s   t = 1.8 s   t = 2.0 s   t = 2.2 s   t = 2.4 s   t = 2.6 s   t = 2.8 s

Figure 6. Start of walk.

The curve 7(a) gives the normal resultant applied by the ground on each foot. The double-support phase allowed by the flexible feet can be observed. Furthermore, for the second step, except at the instant of impact, the shapes of the curves are similar as those recorded for a real human being : two maximas greater than the weight and one minimum between these two maximas smaller than the weight. The curve 7(b) gives the time evolution velocity of the pelvis which goes from 0 to 1.25 m/s. It has to be noticed that this value is similar to the one obtained by a human being for a nominal regime of walk.

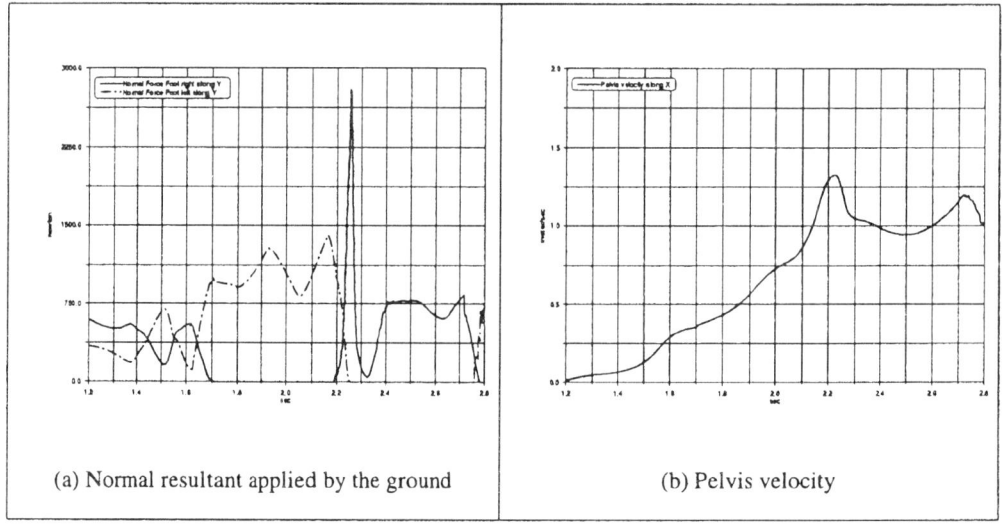

Figure 7. Simulation data curves

5 Conclusion

We have explained a method to generate the start of the three-dimensional walk of an anthropomorphic biped robot. After a description of the 3D-biomechanical data giving the angular joints and the main geometric and inertial characteristics of a standard human being, we have given the inverse kinematics of one leg with six degrees of freedom in order to carry out simulations of the dynamic behavior of an anthropomorphic biped with flexible feet. We have then explained how to create an interactive change of the positions and orientations of the feet relatively to the hips to modify the locomotion cycle and to obtain easily the first steps of a three-dimensional walk. One of the further developments is to establish, thanks to elementary transformations of the reference locomotion cycle, the transition between the three-dimensional walk and the three-dimensional run. Another development is the complete design and construction of a prototype.

References

O. Bruneau, F.B. Ouezdou, J.C. Guinot. (1998). Dynamic Simulation Tool for Biped Robots. In *Symposium on Theory and Practice of Robots and Manipulators, Paris, France*.

K. Hirai, M. Hirose, Y. Haikawa. (1998). The Development of Honda Humanoid Robot. In *Proceedings of the 1998 IEEE, Inter. Conf. on Robotics and Automation, Leuven, Belgium*.

J. Yamagushi, A. Takanish. (1997). *Development of a biped walking robot having antagonistic driven joints using nonlinear spring mechanism*. In *IEEE Proc. of Int. Conf. on Rob. & Aut.*

S. Bouisset, B. Maton. (1995). Muscles, posture et mouvement. In *Hermann 1995*.

D.A. Winter. (1990). Biomechanics of human movement. In *A witley-interscence publication, New York*.

D.W. Seward, A. Bradshaw, F. Margrave. (1996). The anatomy of a humanoid robot. In *Robotica, Vol. 14, Cambridge University Press*.

O. Bruneau, F.B. Ouezdou. (1999). *Distributed ground/walking robot Interactions* In *Robotica, Vol. 17, Cambridge University Press, 1999.*

O. Bruneau, F.B. Ouezdou. (1999). Marche dynamique d'un bipède anthropomorphique muni de pieds flexibles In 14ème *Congrès Francais de Mécanique, Toulouse.*

Chapter V

SENSING AND MACHINE INTELLIGENCE

Cooperative Micro Object Handling by Dual Micromanipulators under Vision Control

Antoine Ferreira [1] and Shigeoki Hirai [2]

[1] Laboratoire de Vision et Robotique, Ecole Normale Supérieure d'Ingénieurs de Bourges, France.
[2] Micro-Robotics Lab., Electrotechnical Laboratory, Tsukuba, Japan.

Abstract. We proposed a newly Automated Micro Assembling System (AMAS) concept for a desktop micro device factory composed of two micromanipulators equipped with micro tools operating under a light microscope. First, we consider the transportation of micro objects by controlling the manipulator push of an micro object on a flat surface with point contact to a desired position. A manipulator control method for the micro object to follow a planned trajectory in pushing operation is proposed under vision based-position control. Then, we present the cooperation control strategy of the micro handling operation under vision-based force control. Finally, different simulation and experimental results show the effectiveness of the proposed controllers.

1. Introduction

The design of advanced products moves toward smaller components thus requiring higher accuracy. The realization of micromachines made of micro-mechanical parts is however still limited, essentially due to the lack of dedicated tools to assemble them. The approach to assembly and manipulation tasks in the micro-world is different to the one followed in the real world. Due to the small dimensions, the operator has no direct access to the objects and a general view of the working space is usually not available. The mass production of micromachines will soon require fully automatic operations with very limited human interaction. Building three-dimensional components often requires different operating tasks, i.e. manipulating the objects in space, applying forces during the handling process, and also assembling micro-sized objects with high-accuracy. Several concepts have already been proposed in the literature by Kasaya (1999), Coudourey (1995), Fatikow (1996), and so on. Most of them are based on the micro-teleoperation of micro-robots under vision-based control. In view of the emerging applications in this field, a project named *"Automated Micro Assembling System"* has been started at the ElectroTechnical Laboratory (ETL). The goal is to realize an automated watch device assembling process which integrates different functions, i.e. handling, manipulation and assembly of micro-sized objects.

Previously, we investigated some important issues in the design of automatic micro object handling by intelligent sensor-based control under light microscope (Ferreira *et al.*, 1998). The purpose of this paper is to study a cooperative micro handling control system performed by two planar micromanipulators. To solve this problem, we introduce a decoupling strategy of the position control and handling force control operating under limited control and observation freedom. Several simulations and experimental results are shown to confirm the validity of the proposal approach.

2. The Desktop Micro Hand-Eye System

A view of the current configuration of the desk-top micro hand-eye system designed at ETL is shown in figure 1(a). The system is composed of two micromanipulators operating under a light microscope equipped with a motorized zoom. To provide the user with visual information, the microscope is mounted with one camera. A second camera mounted on the side of the microscope provides the user with a side view.

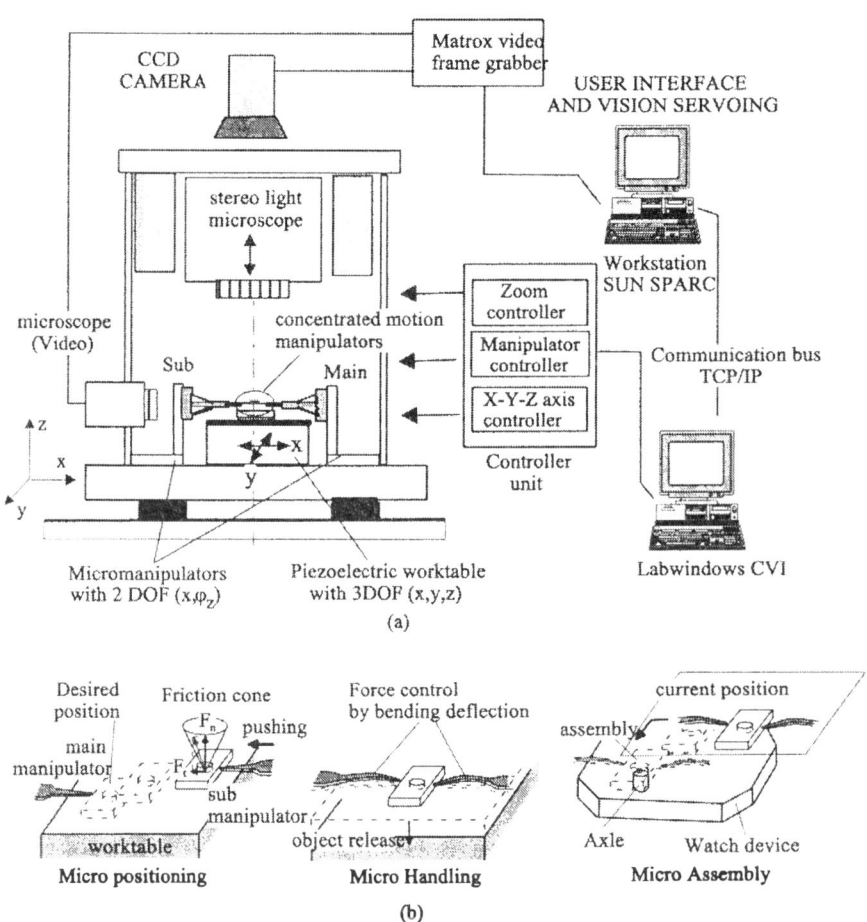

Figure 1. a) Set-up of the desktop micro hand-eye system, and b) operating principle of micro-object assembly for watch device manufacture under vision-based control.

2.1. Mechanical Configuration

The current configuration of the micro handling system consists of two concentrated motion micromanipulators (main and sub) equipped with a micro tool holder, a X-Y-Z precise piezoelectric worktable and an optical vision system (a stereo microscope and a video camera). The microscope is the dominant device in micro-handling systems for the following reasons: 1) it is the largest and heaviest system component, 2) its field of view is much smaller than the system, 3) higher is the magnification ratio, smaller is the working distance. Hence, to ensure the task execution under microscope monitoring, the tool tips must always stay within the field of view of the microscope. Considering these design requirements, our system has been designed so that: 1) the degrees of freedom (d.o.f.) are divided between two arms, allowing better operation and flexibility under microscope, 2) the concentrated motion micromanipulators are so designed that all rotational centers of the degrees of freedom coincide each other at the tool tips.

The operating principle of the micro assembly task is decomposed in different steps as it is shown in Fig.1(b). The right arm (main micromanipulator) is the *carrier object* allowing 2 planar degrees of freedom (X, φ_z). It allows to carry a micro-sized object, using stepping pushing operation, to its final position before to be handled. Variations in the friction force at the object-support interface are the main cause of disturbances that appears particularly during the translational push. This results in a unknown change in object orientation. In order to correct its orientation, cooperative commands are sent to the main micromanipulator and the 3 d.o.f. (X,Y,Z) piezoelectric worktable (resolution of 10.10^{-9}m). The left arm (sub micromanipulator) is the *tool holder*, and has the same number of d.o.f. as the former. It allows to handle the micro-object with a stable gripping. Once the micro-object is handled, micro assembly task can thus be initiated (Figure 1(a)). This configuration offers actually redundancy between the micromanipulators and the worktable. This is needed to allow each tool to be moved to or retracted from the field of view of the microscope.

If excessive contact force is applied to small objects which are light and fragile, they can flip away or be broken easily. A force sensor should therefore monitor the minute micromanipulation force during handling operations. A specific micro-endeffector integrating a buckling-type force sensor has been developed and settled at both endeffectors. The proposed force measurement methodology is based on a non-contact technique. More precisely, it consists to measure the out-of-plane bending deflection of the endeffector by an optical measurement system. When a normal pushing force is applied to the endeffector, the stress on the flexible buckling beam results in an out-of-plane deflection. The amplitude measured on the top of the buckling beam is used to measure the magnitude of the applied force at the contact interface between the object and both manipulators. The detail drawing of the proposed endeffector design is shown in Figure 2(b). In the design of the endeffector, we have to consider also that the surface tension forces are dominant for object dimensions ranging from several hundred micrometers to few millimeters. We adopted a multi-contact geometry of the endeffector's tip for adhesive force reduction as proposed by Arai (1996). Polysilicon micro-contacts have been realized by etching process and glued onto the endeffector's tip surface. Experimental results (Figure 2(c)) shows that the force sensor presents good performances (linearity, sensibility and accuracy) in the linear behavior of the endeffector structure. Hysteresis is negligible owing to the elastic properties of the silicon material. Accuracy of this force sensor is less than few 10^{-3} gf order. We can improve its accuracy by reducing the structural rigidity of the prototype endeffecor.

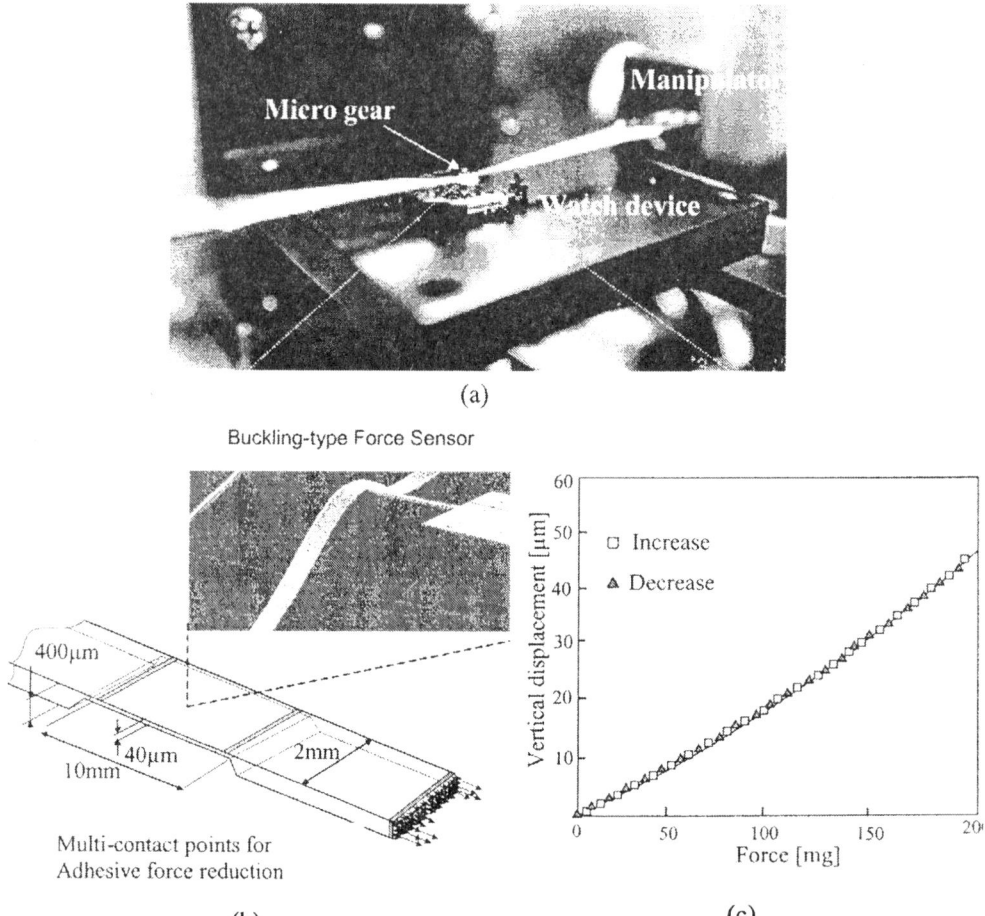

Figure 2. (a) Image of the micro-handling operation, (b) geometry of the micro-endeffector with integrated buckling-type force sensor and multi-contact points adhesive force reduction, and (c) calibration of the integrated force sensor.

2.2. Cooperative Position and Handling Force Control System under Visual Servoing

For high precision movements and for micro assembling task oriented teleoperation, the manipulation system has to be controlled automatically. In our approach, we have chosen to control the different steps of the micro assembling process by means of a vision system looking at the micro-sized object through a microscope linked to a stereo camera. To obtain operation information of three dimensional (3D) location, multi-directional location and multi-directional imaging is very effective. To obtain an accurate 3D position in the working space, one microscope should be located perpendicularly to the other. To observe in the same space at the same time, all fields of view are concentrated. Thus, all optical axis of the microscope intersect at one point in space in a concentrated visual field of the manipulation system. The proposed control structure is illustrated in Figure 3. It is imperative to use a such vision system for the following reasons: 1) as mentioned above, most of the d.o.f. have a good resolution but a poor repeatability, 2) during the pushing operation, friction forces at the micro-object/support interface induces errors in the orientation and positioning of the micro-object which implies the use of a vision-based position control system, and 3) to control the contact force at the micro-object/manipulator tips interface, a vision-based handling force control must be used. For highly accurate manipulations the parameters describing the relationship between the image frame and the frame attached to manipulators must be continuously and accurately updated. These parameters are unknown a priori and are not directly measurable, but can be deduced by observing the scene change during motion. Because our simple vision system does not process images in real time, the algorithm is based on a "look and move" strategy, i.e. the motion is realized in steps, during which the manipulators and the worktable are controlled in a closed-loop. In addition, we calculate predictions of pattern positions on-line for optimal trajectory control. This is done using an Extended Kalman Filter (EKF).

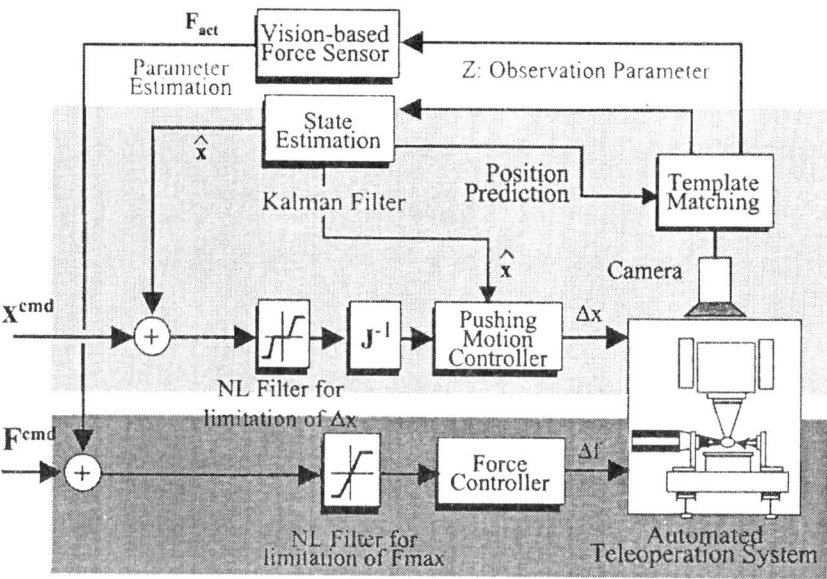

Figure 3. Block diagram of cooperative position and force control system for two-manipulators operating under a visual servoing.

2.3. System Integration Computer

A workstation (SUN Sparc) with a Matrox video frame grabber is used as the system integration computer. The images are then analyzed and presented to the operator. A dedicated interface allows to pilot the micromanipulators with high precision and high flexibility.

3. Vision-Based Position Control

3.1. Pushing Motion Controller

In order to handle small objects in the range of size from several hundred μm to few millimeters, the main manipulator (carrier tool) is controlled in such a way to transport the micro-object to its final location using pushing operation. When pushing an object, it is difficult to control its motion for the following reasons; 1) there is slip between the object and the pusher due to the frictional forces, (2) there is dynamic frictional forces at the object/support interface and 3) there is a non-holonomic constraint in the motion of pushed object with point contact condition. It implies the use of an appropriate position and trajectory control scheme. However, since there are uncertain factors, which are not considered in planning, such as changes of the friction distribution during actual position, we need a vision feedback control method for pushing the object along the planned trajectory. A control method previously investigated by Kurisu (1996) has been implemented in the micro-world. We consider that the desired trajectory of the object is given by the position $^{U}p_{Od}(t)$ and velocity $^{U}\dot{p}_{Od}(t)$ of the center of friction expressed in Σ_U fixed on the worktable, and the orientation $\theta_d(t)$ and rotational velocity $\dot{\theta}_d(t)$ of the object (see Figure 4).

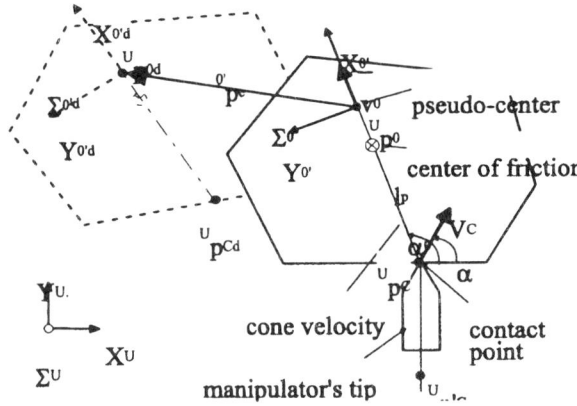

Figure 4. Definition of the parameters.

We also assume that we can measure the position $^{U}p_C$ and velocity $^{U}v_c$ of the object at the contact point expressed in Σ_U, and the orientation θ and rotational velocity $\dot{\theta}$. The motion of the pushed object can be regarded as the motion of the micromanipulator $^{U}p'_C$. Let $\Sigma_{0'}$ denote the coordinate frame with its origin at the pseudo-center and its $X_{0'}$ axis parallel to the line which joins the contact point and the center of friction. Then, we consider the position error of the pseudo-center expressed in $\Sigma_{0'}$ with respect to the desired trajectory. The error is denoted as $^{O'}p_e = [^{O'}x_e, ^{O'}y_e]^T$. Denoting the actual position of pseudo-center expressed in Σ_U as $^{U}p_{0'}$, the desired position as $^{U}p_{0'd}$, and the rotational matrix from $\Sigma_{0'}$ to Σ_U as $^{U}R_{0'}$, the error $^{O'}p_e$ is given by:

$$^{O'}p_e = ^{U}R_{0'}^{T}\left(^{U}p_{0'd} - ^{U}p_{0'}\right) \quad (1)$$

Denoting the orientation error as $\theta_e = \theta_d - \theta$, and adopting a differential feedback rule given by:

$$\begin{bmatrix} \bar{v}_0 \\ \dot{\theta} \end{bmatrix} = \begin{bmatrix} \bar{v}_{0'd} \cos\theta_e + K_x {}^{0'}x_e \\ \dot{\theta}_d + \bar{v}_{0'd}(K_y {}^{0'}y_e + K_\theta \sin\theta_e) \end{bmatrix}, \qquad (2)$$

it is guaranteed that $e_{0'} = [{}^{O'}p_e{}^T, \theta_e]^T$ converges uniformly asymptotically to 0 as far as $\bar{v}_{0'd} > 0$. The terms K_x, K_y, and K_θ are feedback parameters. The velocity ${}^O v_c$ can be expressed as:

$$^O v_c = B^{-1}[\bar{v}_{0'}, \dot{\theta}]^T \quad \text{with} \quad B = \begin{bmatrix} \cos\alpha_0 & \sin\alpha_0 \\ \dfrac{1}{l_p}\sin\alpha_0 & -\dfrac{1}{l_p}\cos\alpha_0 \end{bmatrix}. \qquad (3)$$

we can expect to make the object follow the desired trajectory. The control variable, that is, the manipulator tip velocity ${}^U v_c$ is given by ${}^U v_c = {}^U R_0 {}^O v_c$. As it has been reported before, our vision system is based on a "look and move" strategy. After every step motion, a new image is acquired by the camera. The orientation and the rotational velocity of the micro-object can be measured from the pattern's pose change by detecting its geometry. The orientation of the object, the position of center of friction and pseudo center are computed from the angle of the manipulator tip. New cooperative command signals are computed from Eq.(2) and sent to the main manipulator and the worktable in order to match with the desired position.

3.2. Simulation and Experiments

In this section, we will show the effectiveness of the proposed control method by simulation and experiments. In the simulation, the object is assumed to have a rectangular base with 5×3×2 [mm], and have a friction distribution uniform. We want to move this object from ${}^U p_{0'd}(0) = [2,0]^T$ ([mm], [mm]), $\theta(0)_d = 0$[deg] to ${}^U p_{0'd}(15) = [0,30]^T$ ([mm], [mm]), $\theta(15)_d = 0$[deg] in the time interval of 15 secondes. The contact is the foot of the perpendicular from the center of friction to the side of the object. It is assumed that there is no slip or detachment at the manipulator's tip/object interface. The coefficient of friction at the contact point is assumed to be $\mu = 0.2$ in the simulation. Note that the simulation described before is the case that the manipulator tip velocity lies in the velocity cone. Then, to observe the convergence of the errors with respect to the desired position and trajectory, we have chosen the feedback parameters as $K_x = 18$, $K_y = 66$ and $K_\theta = 15$ (critical damping case). Figure 5(b) shows a result of simulation. As can be seen from the figure, the actual trajectory converges to the desired one quickly. This result shows the effectiveness of the proposed control method.

Experimental results are shown in Figure 5(a) and (c). In the Figure 5(a), the desired trajectory of the reference point is shown by the dotted line and its actual trajectory is shown by the solid line. The rest information is shown in the same way as Figure 5(c). Experimental results have shown that the object follows the desired trajectory in x-y coordinates with good performance, i.e., typical error values (${}^U x_{0'd} - {}^U x_{0'}$) and (${}^U y_{0'd} - {}^U y_{0'}$) are less than 10μm although the tracking performance of the orientation is slightly degraded; i.e. $(\theta_d - \theta) < 2$ degrees. The smooth damping and error position are due to the fact that the actual friction distribution is supposed to change according the movement of the object due to the slight unevenness of the support. In practice, it has been observed that any slip between the object and the manipulator tip occurred. The result shows the effectiveness of the proposed method in normal conditions of manipulation of the proposed desktop manipulation system.

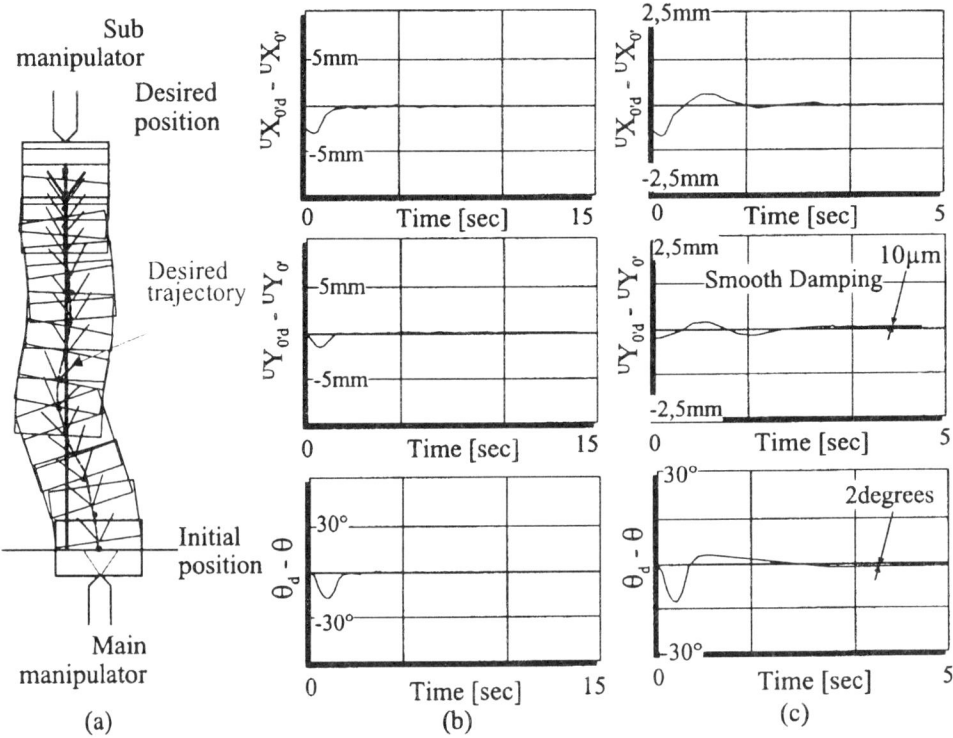

Figure 5. (a) Experimental results of position and trajectory tracking, (b) simulation and (c) experimental results of tracking errors.

4. Vision-Based Force Control

This section describes the image processing techniques used for the vision-based force measurement. It is constituted by an optical interferometer combined with a vision system looking at the micro-manipulators through an optical microscope.

4.1. Graphic Image Processing of the Deformation Shape

After the calibration procedure realized through an optical interferometer, bidimensional deflection measurements of the elastic structure is investigated in this paragraph in order to estimate the contact forces. The grey-level data vizualized through the optical microscope and acquired by the high-definition CCD camera are sampled by the video frame grabber. The data are then transferred to the workstation. We have to take into account several problems related to graphic image processing of the deflection distribution. First, the distorsion problem cannot be neglected because we observe the task at a relatively low magnification ratio of 1,400. It can be noticed at the circumferential part in the field of view. Therefore, the left and right sides of the image are cut away in order to keep the central area. Images issued from the optical microscope are generally noisy. Noise reduction is implemented by applying a Gaussian filter. This filter reduces the sharpness of the image but do not affect image recognition and definition performance. Detection of edges from smoothened images is then initiated. The number of detected edges are limited to the sensing part (rectangular shape). It allows to do not degrade the reliability and speed of the optical image processing since the structure shape is quasi-static.

Afterwards, from the sampled image data of the buckling-type force sensor, the grey-level distribution of the out-of-plane beam deflection is calculated. However, the visual control of the beam deflection give information about qualitative but not quantitative displacements. Precise quantification of the micrometric out-of-plane deflection acquired by the vision system is realized using the combination of an additional optical interferometer integrated in the vision system. The latter measures the maximal out-of-plane displacement that occurs on the top of the buckling beam. Figure 6(a) shows experimental bidimensional distribution of the vertical displacement for two values of force contact. A three-dimensional graphic representation (Figure 6(b)) is reconstructed. In order to obtain the smooth wire-frame curve line from small numbers of sampled data, a cubic Spline function was applied (Ahlberg, 1967). Based on the deformation shape of the force sensor, we can estimate the maximal contact force occurring at the object/manipulator tip interface.

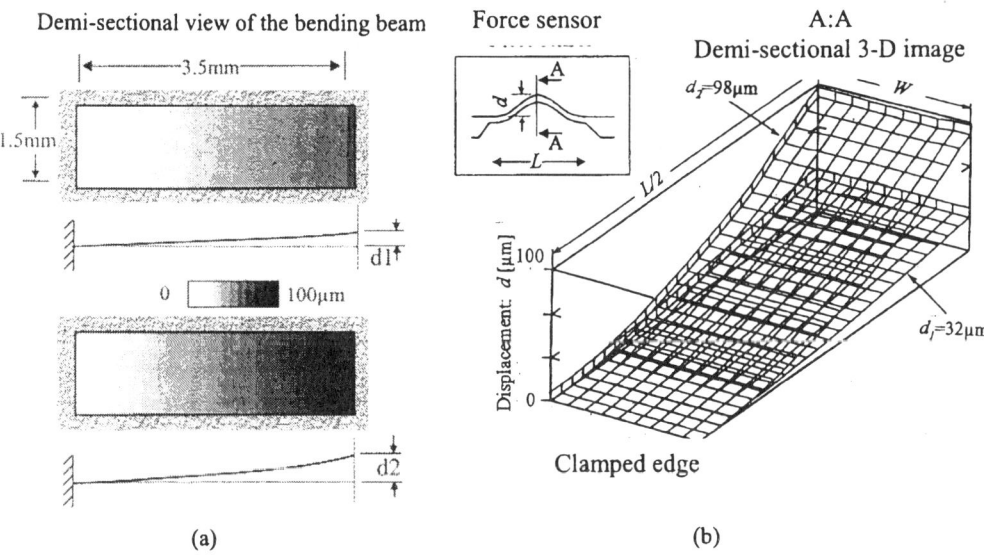

Figure 6. (a) Experimental results of position and trajectory tracking, (b) simulation.

5. Conclusion

We have presented the design of a new desktop micro handeye system with vision-based position and handling force control for handling, manipulating and assembling micro-sized objects. The dimensions of the manipulated object are ranging from one hundred μm to several millimeters. The implementation of this system was described and the very first test results were presented. It features a micrometric precision of object positioning within a workspace of 3cm^3 and the control of gripping forces less than 100.10^{-9} Newtons. An automatic execution of the task is currently being programmed. There are however some handling (object sticking, stability of gripping,) and control (3D shape recognition, real time processing,) problems in the micro world that prevent us from a fully automatic execution task.

6. References

Ahlberg, J-H., Nielson, E-N. and Walsh, J-L. (1967). The theory of Spliens and their applications. New York : Academic.

Arai, F., Andou, D., Nonoda, Y., Fukuda, T., Iwata, H. and Itoigawa, K. (1996). Micro endeffector with micro pyramids and integrated piezoresistive force sensor. *Proceedings on IEEE Int. Conf. on Intelligent Robotics and Systems*, 842-849.

Codourey, A. *et al.* (1995). A robot system for automated handling in micro-world. *Proceedings IEEE of the International Conference on Intelligent robots and Systems*, 170-175.

Fatikow, S. and Rembold, U. (1996). An automated microrobot based desktop station for micro assembly and handling of microobjects. *Proceedings IEEE conference on Emerging Technologies and Factory Automation*.

Ferreira, A. and Hirai, S. (1998). Development of desk-top micro hand-eye system for automated watch device assembling process. *12th CISM Symposium. on RoManSy 98*.

Kasaya, T., Miyazaki, H., Saito, S. and Sato, T. (1999). Micro object handling under SEM by vision-based automatic control. Proceedings of IEEE Conference on Robotics & Automation, 2189-2196.

Kurisu, M. and Yoshikawa, T. (1996). Tracking control for an object in pushing operation. *Proceedings IEEE of the International Conference on Intelligent robots and Systems*, 729-735.

A Comparative Study of Torque Control Using a Wrist or a Base Force/Torque Sensor

F. Geffard[1], C. Andriot[1] and G. Morel[2]

[1] Service de Teleoperation et Robotique, CEA, Fontenay-Aux-Roses, France
[2] Université L. Pasteur Strasbourg I, ENSPS, LSIIT, France

Abstract. In teleoperation like in conventional Robotics, accurate control of joint torques is essential to obtain good performance in fine motion task and high resolution contact task. In the case of manipulators not equipped with joint torque sensors, some external sensor solutions must be considered. This paper proposes a theoretical and an experimental comparison between a solution using a base force/torque sensor, and a solution using a wrist force/torque sensor.

1 Introduction

Manipulator (in particular industrial robots) exhibit disturbances such as joint friction, backslash, and motor cogging, at their actuator/transmission systems. This affects their force resolution and leads to difficulties in executing precise and slow motion. However, accurate joint torque control can reduce the effect of these perturbations. For example, Pfeffer & al. [Pfeffer et al, 1989] have equipped a PUMA 500 with joint torque sensors, and have experimentally shown that the effective friction torque had been reduced by 97 %. Avoiding this equipment, Williams and Khatib [Williams and Khatib, 1995] have mounting a 6 axis force/torque sensor at robot's wrist. The joint torques are estimated multiplying the measured wrench by the extremity jacobian transpose. Later, Morel and Dubowsky [Morel and Dubowsky, 1996] proposed an other external solution using a base force/torque sensor for estimating the joint torque. They have shown that this solution significantly improved the motion precision for small and slow displacement.

The aim of this paper is to compare in a one axis case the WFTS (Wrist Force/Torque Sensor) solution and the BFTS (Base Force/torque Sensor) one. First of all, we give theoretical limitations of both solutions. And these limitations are then verified on an industrial manipulator (RX90 from Stäubli).

2 Problem position

The CEA has equipped an industrial manipulator with two force/torque sensors. The first one is mounted on the wrist of the robot, and the second is located at the base of the manipulator. In static, both wrist and base force/torque sensors measure the same wrench (apart from the gravity wrench). And in both cases it is possible to calculate the joint torque with the measured wrench using the manipulator jacobian transpose J^T. We have :

$$\tau_{est} = J_B^T F_B = J_W^T F_W \quad \text{with} \quad J_B = Ad_{g_{BW}} J_W$$

where F_B represents the wrench measured by the base force/torque sensor and F_W is the wrench measured by the wrist force/torque sensor. Moreover, g_{BW} maps the base coordinates into the wrist coordinates. And $Ad_{g_{BW}}$ is the adjoint transformation associated with g_{BW} [Murray et al, 1994].

Then, the joint torque control problem is solved for each joint. Therefore, in the joint space, both solutions lead to control SISO systems which are more easy to modelize and to analyse than the initial MIMO one. So the problem is then equivalent to design a control law in a one-axis case.

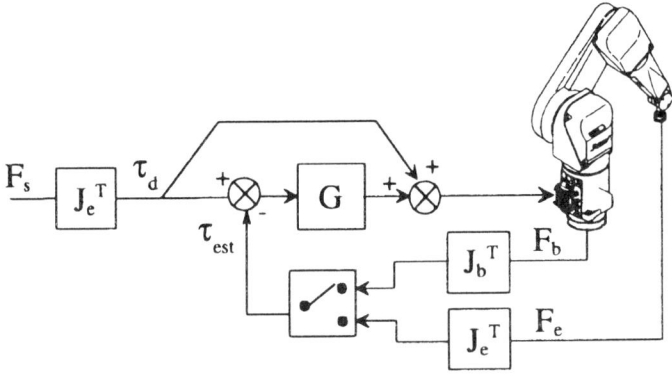

Figure 1 : Control Scheme with BFTS and WFTS

3 Theoretical comparison between WFTS and BFTS

Considering the resulting SISO case, comparison will be done here in a one axis linearized flexible case. Both solutions will be compared in the sense of the passivity criteria described in [Colgate and Hogan, 1989]. It must be noted that the model we use here is a simplified model of a collocated robot's axis.

The benefit action of integral controller in force control is widely recognized [Visher and Khatib, 1995] [Volpe and Khosla, 1993]. Moreover, adding a proportionnal term allows more important integral gains [Newman and Zhang, 1994]. Therefore, the compensator we have used for this study was a PI compensator with a feedforward term (f_{est} being the estimated torque):

$$f = \left(1 + K_p + \frac{K_i}{s}\right) f_d - \left(K_p + \frac{K_i}{s}\right) f_{est}$$

3.1 Modeling

Consider the simplified system shown on Figure 2. We suppose here, that the measured force is perfect for both sensors, and we neglect the dynamic of the sensors. Moreover, we will neglect all viscous damping but B_m (B_m can also represent an active damping term).

On Figure 2 M_m is the motor inertia, M_l the link inertia, K_t the transmission stiffness, and n is the gear ratio. f is the command force/torque, f_e the force/torque exerted by the environment on the link, and f_{NLm} is the motor non-linear friction force.

In order to compare both solutions, we propose to introduce some variations on the measured force. Thus, the measured force will be able to represent either the BFTS measure, or the WFTS measure.

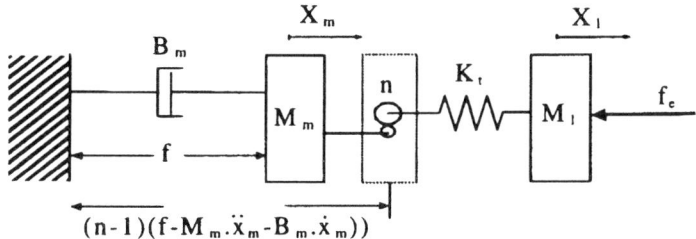

Figure 2

Suppose that we take for the measure:

$$f_{est} = f_{mes\_b} + M'_m \ddot{x}_m + M'_l \ddot{x}_l \quad \text{with} \quad f_{mes\_b} = f_e + M_m \ddot{x}_m + M_l \ddot{x}_l$$
$$= f_e + \Delta M_m \ddot{x}_m + \Delta M_l \ddot{x}_l$$

The different cases we can modeled are:

- $\Delta M_m = \Delta M_l = 0$ stands for the WFTS case,
- $\Delta M_m = M_m$, $\Delta M_l = M_l$ stands for the BFTS case.

Then, the closed-loop admittance of the environment interaction port is given by:

$$Y_s(s) = \frac{v_a(s)}{f_e(s)} = \frac{b_3 s^3 + b_2 s^2 + b_1 s + b_0}{(a_3 s^3 + a_2 s^2 + a_1 s + a_0)s} \tag{1}$$

with :

$a_3 = (M_m n^2 M_l + \Delta M_m n M_l K_p)$ $a_2 = (B_m n^2 M_l + \Delta M_m n M_l K_i)$ $b_1 = (K_t K_p + K_t)$

$a_1 = (\Delta M_m n K_t K_p + K_t M_l + M_m n^2 K_t + K_t \Delta M_l K_p)$ $b_2 = (\Delta M_m n K_i + B_m n^2)$

$a_0 = \Delta M_m n K_t K_i + K_t \Delta M_l K_i + B_m n^2 K_t$ $b_3 = (\Delta M_m n K_p + M_m n^2)$ $b_0 = K_t K_i$

Positivity

The admittance of the interaction port will be positive if :

$$\frac{B_m K_p}{M_m} \leq K_i \leq \frac{n^2 B_m (1+K_p)}{n^2 M_m - n \Delta M_m + M_l - \Delta M_l}$$

and $(M_l - \Delta M_l - n \Delta M_m) K_p \leq n^2 M_m$

Note that if we take $n=1$, $\Delta M_m = 0$, and $\Delta M_l = 0$, we find the Newman's limit gains with a WFTS [Newman and Zhang, 1994].
Anyhow, it is straightforward to observe on this positivity conditions that *the BFTS solution allows greater gains than the WFTS solution*. Indeed, in the case of the BFTS solution, there is no theoretical limits on K_p, and the ratio $\frac{K_i}{1+K_p}$ is greater than for the WFTS case. This last point is an important result for disturbance rejection (see next section).

Closed-loop behavior

At low frequencies, the closed-loop equations become:
- « WFTS » case :

$$\dot{x}_e = I_W(s)\left(f_e - \frac{s}{(1+K_p)s + K_i} n f_{NLm}\right) \quad \text{with} \quad I_W(s) = \frac{(1+K_p)s + K_i}{(n^2 B_m)s} \tag{2}$$

- « BFTS » case :

$$\dot{x}_e = I_B(s)\left(f_e - \frac{s}{(1+K_p)s+K_i}nf_{NLm}\right) \text{ with } I_B(s) = \frac{(1+K_p)s+K_i}{(nM_mK_i+M_lK_i+n^2B_m)s} \quad (3)$$

$I_B(s)$ and $I_W(s)$ represent passive equivalent admittances of the environment interaction port. And when we use the maximum value for K_p and K_i, the admittances become:

$$I_W(s \to 0) = \frac{1}{M_l s} \quad \text{and} \quad I_B(s) = \frac{(1+K_p)s+K_i}{(nM_mK_i+M_lK_i+n^2B_m)s}$$

Therefore, both solutions reveal an inertial behavior at low frequencies seen from the extremity port. And, even with an important K_i, the BFTS behavior will be more inertial than the WFTS one. Another interesting thing is that in both cases, the motor friction effect on these inertias is identically filtered. It means that in both cases a larger ratio $\dfrac{K_i}{1+K_p}$ leads to a better attenuation of motor friction. And the positivity conditions show that the BFTS is better than the WFTS for this purpose (Cf. Figure 3).

Now we have to determine if it is more interesting to have a larger inertia with less friction than the opposite. The answer is visible on Figure 4, where we have simulated the response of the manipulator end-effector velocity (see Equations (2)(3)) after having applied an external force step and a ten times more important friction step. The applied friction force in the opposite direction of f_e explains why the sign of $\dot{x}_e$ is initially negative. Obviously on a real system the sign would be always the same as for f_e.

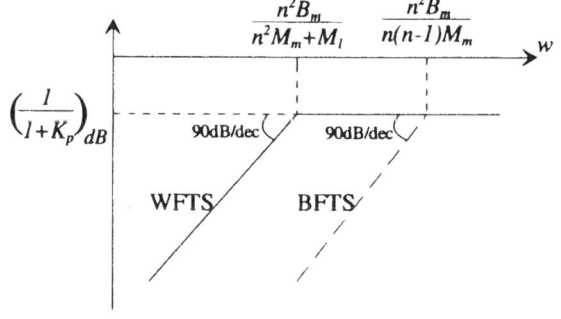

Figure 3: Motor friction attenuation.

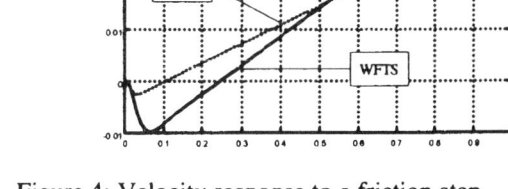

Figure 4: Velocity response to a friction step.

Nevertheless, we see on Figure 4 that the manipulator controlled with the WFTS is less inertial (greater slope), so after few seconds it is more transparent than with the BFTS solution. But, as the motor friction is better compensated with the BFTS solution, the manipulator starts to move earlier (at 0.1s) than with the WFTS (at 0.25s). This is an important result for force sensitivity.

$M_m = 1e^{-4} Kg.m^2$	$M_l = 19 Kg.m^2$	$n = 101.09$
$B_m = 7e^{-2} N.m.rad^{-1}.s^{-1}$	$K_{pb} = 0.94$	$K_{ib} = 93.0$
$K_l = 57000 N.rad^{-1}$	$Kpw = 0.36$	$K_{iw} = 38.0$

Table 1: RX90's first axis parameters values

For this simulation, inertia, viscous damping, and gains, result from experimental measures on the first axis of the industrial manipulator RX90. These parameter values are given in Table 1, where K_{pb}, K_{ib}, K_{pw}, K_{iw} are the proportional and integral gains obtain experimentally when using respectively the BFTS and the WFTS.

4 Experimental comparison

4.1 Experimental setup

The experimental tests have been done on an industrial manipulator RX90 (Staübli). It is mounted on an ATI force/torque sensor (FT3807) and another ATI force/torque sensor (FT3538) is mounted at its wrist. The description of the range and resolution of both sensors is given in Table 2.

Sensor model	FT3538	FT3807
Sensing Ranges :		
±Force X and Y (N)	130	2600
±Force Z (N)	260	5300
±Torque (N.m)	10	750
Resolution :		
Fx, Fy, (N)	0.10	2.25
Fz, (N)	0.20	4.5
Tx, Ty, Tz, (N.m)	0.005	0.02

Table 2 : Force/Torque sensors resolution

4.2 PI Tuning

In order to compare both solutions, we have chosen a tuning method which was warranting the same margin and performance whichever the process was. As we needed to tune a PI controller, we have adopted the Aström tuning method [Aström et al, 1998].

In accordance with [Eppinger and Seering, 1987], we have obtained more important gains with the collocated solution (BFTS measure) than with the non-collocated solution (WFTS measure). This also directly confirms the theoretical results of section 3.

4.3 Backdrivability tests

In order to verify results of section 3, we have done some backdrivability trials on the RX90's first axis (see Figure 5).

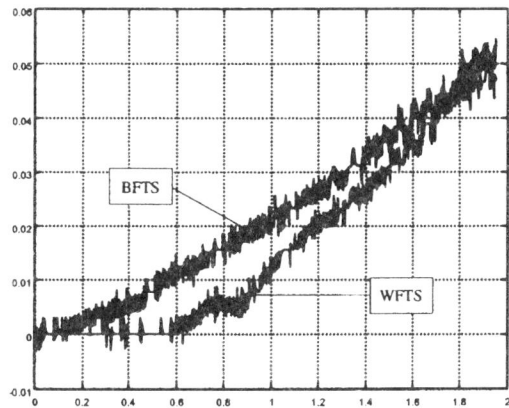

Figure 5: Velocity (rad/s) response at an end-effector applied force step (0.66N).

These results have been obtained by applying an external force step of 0.66N at the extremity of the manipulator. This step has been applied using a known load and a pulley. The manipulator was firstly hold in a precise position, and it was then released precisely at t=0s. As it was expected, the WFTS solution appears to be more inertial, but it better rejects the motor friction. Thus, the robot starts to move earlier with the BFTS solution (0.1s) than with the WFTS solution (0.6s). It is an interesting thing for teleoperation, that the slave reacts rapidly to an external applied force. Indeed, with a position/position control scheme, the operator will have a better sensitivity to the environment with the BFTS solution, than with the WFTS solution.

5 Conclusion

This paper compares two external solution for torque control. It has been theoretically shown that torque control with the use of a base force/torque sensor is better for backdrivability than using a wrist force/torque sensor. These results have been experimentally verified on the first axis of an industrial manipulator by applying a small external force at its extremity.

6 References

[Pfeffer et al, 1989] : Pfeffer L. E., Khatib O., Hake J., « Joint Torque Sensory Feedback in the Control of a PUMA Manipulator », IEEE Trans. Robotics and Automation, Vol. 5 (4), 1989.
[Williams and Khatib, 1995] : Williams D., Khatib O, « Improved Force Control for Conventional Arms Using Wrist-Based Torque Feedback », Prep. Fourth Intern. Symp. Exper. Rob., ISER'95, p323-328.
[Morel and Dubowsky, 1996] : Morel G., Dubowsky S., « The Precise Control Of Manipulators With Joint Friction : A Base Force/Torque Sensor Method », IEEE Intern. Conf. Robotics and Automation, 1996.
[Murray et al, 1994]: Murray R.M., Li Z., Sastry S.S, « A Mathematical Introduction to Robotic Manipulation », CRC Press, 1994.
[Eppinger and Seering, 1987]: Eppinger S.D., Seering W.P., « Understanding Bandwith Limitation in Robot Force Control », IEEE Intern. Conf. Robotics and Automation, 1987.
[Colgate and Hogan, 1989] : Colgate E., Hogan N., « The Interaction of Robots with Passive Environments : Application to Force Feedback Control », 4th Int. Conf. Advanced Robotics, 1989.
[Visher and Khatib, 1995]: Visher D., Khatib O., "Design and Development of High-Performance Torque-Controlled Joints", », IEEE Trans. Robotics and Automation, Vol. 11 (4), 1995.
[Volpe and Khosla, 1993] : Volpe R., Khosla P., « A theorical an Experimental Investigation of Explicit Force Control Strategies for Manipulators », IEEE Trans Rob. and Automation, Vol. 38 (11), 1993.
[Newman and Zhang, 1994] : Newman W. S., Zhang Y., « Stable Interaction Control and Coulomb Friction Compensation Using Natural Admittance Control », Journal of Robot. Systems, 11 (1), 1994.
[Aström et al, 1998] : Aström K.J., Panagopoulos H., Hägglund T. « Design of PI Controllers Based on Non-Convex Optimisation », Automatica, Vol. 34, No5, pp. 585-601, 1998.

Chapter VI

APPLICATIONS

Quality Feature Based Adjustment of Robot Programs Exemplified for the Welding Process – MAGROB

R. D. Schraft, J. Neugebauer, and W. Schaaf

Fraunhofer Institute Manufacturing Engineering and Automation (IPA), Stuttgart, Germany

Abstract. This paper presents a new concept for the process-oriented adjustment of robot programs. The approach has been successfully implemented in the software MAGROB for welding a fillet weld. The result of the production process is visually assessed and determined in the manner of the quality features which are the input of the proposed system. Using fuzzy logic rules the quality features are translated into the process parameters which are then the basis for the calculation of the robot parameters.

1 Introduction

Robot programs often need specific adjustment in order to meet the quality demands of the product, e.g. during the comissioning when the program is tested under real conditions at the first time or during production when the conditions slightly change and, so, quality decreases. During this adjustment and testing the robot stands still. Therefore, this time must be minimized. Program adjustment can be avoided by using sensors for a closed-loop control of the process. However, sensors do not detect the final overall quality and reduce the availability of the robot system due to their own low availability.

For this reason we propose in this paper an approach to support the quality feature based iterative adjustment of robot programs by an user-friendly knowlegde-based tool. This approach will be described for the process of metal-arc active gas welding (MAG) of a thick plate fillet weld of which the quality parameters are defined in the norms (DIN EN 25817, ISO 5817).

2 State of the art

Knowledge Based Systems, i.e. Expert Systems, have been widely introduced in welding technology (Hartfuss 1996, Park 1993). For this purpose, fuzzy logic was accepted as the basis for such systems in the nineties. For the closed loop control of the welding process sensors can be used (Roosen 1997, Starke 1995, Scheller 1994). For the development of a tool for iterative and offline adjustment of robot programs can be found that the fuzzy classification of the welding parameters and the correlations of these parameters and the resulting quality have been intensively investigated.

Basic approaches for the reduction of the time for the adjustment of robot programs are only in the field of the easy programming of positions and orientations (AMIRA 2000,

FANUC 1998). An iterative offline, i.e. open loop control on the basis of the fuzzy classification of the process parameters have not been investigated so far.

3 Analysis and Requirements

The objective is to facilitate the process of the adjustment of robot programs with respect to the product quality by using an assisting tool. Therefore we propose the following requirements: userfriendly interaction, hardware independence as far as possible, quality feature based input parameters and expandable concept of the system. Userfriendly interaction demands a process-oriented icon-based interface. Hardware independence means that the system should be as far as possible independent from specific robot control systems, i.e. should be widely manufacturer independent.

The requirement of a quality feature based system means that the quality parameters are the input of the system and that they are the basis for the program adjustment calculated by using fuzzy logic. The expandability is given if we apply an object-oriented software contruction in which we can easiliy add further quality features as well as we can easily adopt the system to any robot control hardware.

4 Concept of the System

In order to meet the requirements of a quality feature based and process oriented system we propose the classification of the **paramters X** into **quality parameters Q** and **execution parameters E**. The execution parameters are calculated by the system using the quality parameters Q. In order to meet wide hardware independence we further subdivide the execution parameters E into **process parameters P** and **robot parameters R**. The process parameters P descibe the process control system independently whereas the robot parameters R are used in the robot control system. A parameter can be defined as an **absolute parameter** $^{Absolute}X$, i.e. can be used directly, or as a **corrective parameter** $^{Corrective}X$, i.e. must be added to the actual parameter in order to calculate the desired parameter. The **actual parameter** $_{Actual}X$ refers to the actual production result at the **iteration i** and the **desired parameter** $_{Desired}X$ is used to perform the **next production iteration (i+1)**. Corrective parameters can be subdivided in **relatively corrective parameters**, which are given as a percentage, and **absolutely corrective parameters**, which are given in the matching physical unit.

In this way we have a set of parameters Q, P and R at each iteration step i and, so, we can describe the optimization procedure using the number q of quality parameters, the number p of process parameters, the number r of robot parameters and the number t of iterations by the following matrices:

$$\underline{\underline{Q}} = \begin{bmatrix} Q_1^1 \cdots Q_1^t \\ \vdots \\ Q_q^1 \cdots Q_q^t \end{bmatrix}, \underline{\underline{P}} = \begin{bmatrix} P_1^1 \cdots P_1^t \\ \vdots \\ P_p^1 \cdots P_p^t \end{bmatrix}, \underline{\underline{R}} = \begin{bmatrix} R_1^1 \cdots R_1^t \\ \vdots \\ R_r^1 \cdots R_r^t \end{bmatrix}.$$

So, we propose to divide the system into three main modules: the man machine interface, the knowlegde base and the modification module. In this way only the modification module is hardware dependent and can easily be adopted to any robot system for a given process. We can now describe the iterative optimization procedure at iteration step i as follows (see Figure 1):

Quality Feature Based Adjustment of Robot Programs

1. determination of the vector $_{Actual}\vec{Q}^i$ $\vec{Q}$ of the quality parameters by interaction with the interface of the system,
2. calculation of the vector $_{Desired}\vec{P}^i = f_{Fuzzy}(_{Actual}\vec{Q}^i)$ of the process parameters using the fuzzy logic knowledge base,
3. calculation of the vector $_{Desired}\vec{R}^i = f_{Analytical}(_{Desired}\vec{P}^i, _{Actual}\vec{R}^i)$ of the absolute robot parameters using analytical formulas and
4. production of the iteration (i+1) using the parameters $_{Desired}\vec{R}^{i+1} \equiv _{Desired}\vec{R}^i$.

The object-oriented class diagram and the approach for data gathering is shown in Figure 2 and 3. For the conception of the knowledge base we considered the three variants of Figure 4 and found that the variant 3 where a knowlege base is assigned to each process parameter is most suitable.

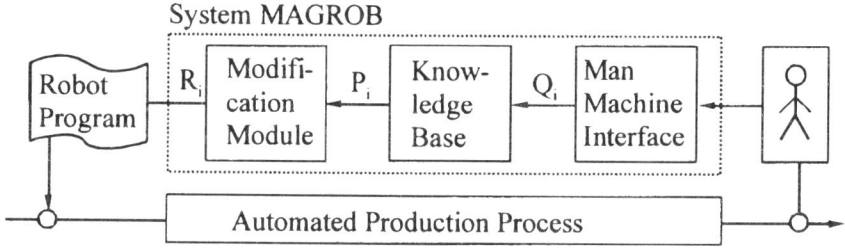

Figure 1. Iterative optimization procedure using the proposed system

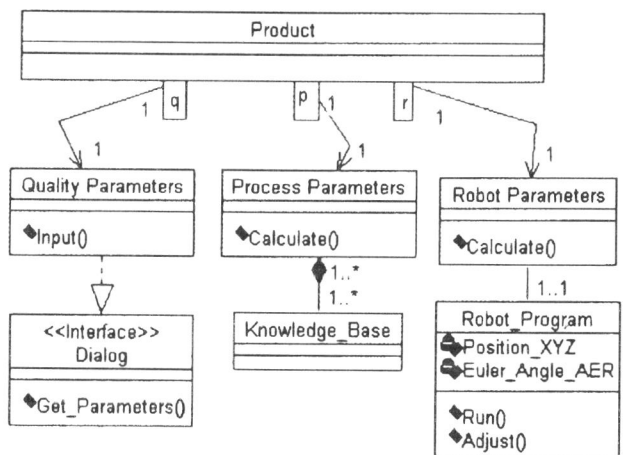

Figure 2. Object-oriented class diagram for the system

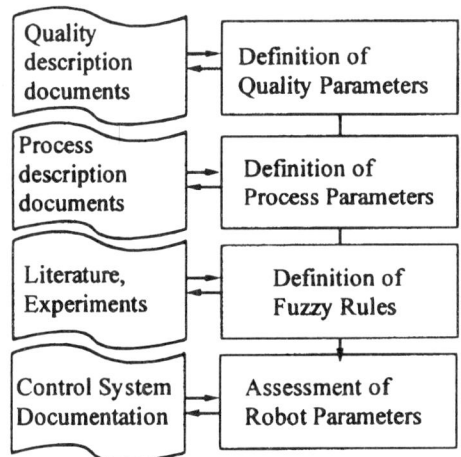

Figure 3. Approach for data gathering.

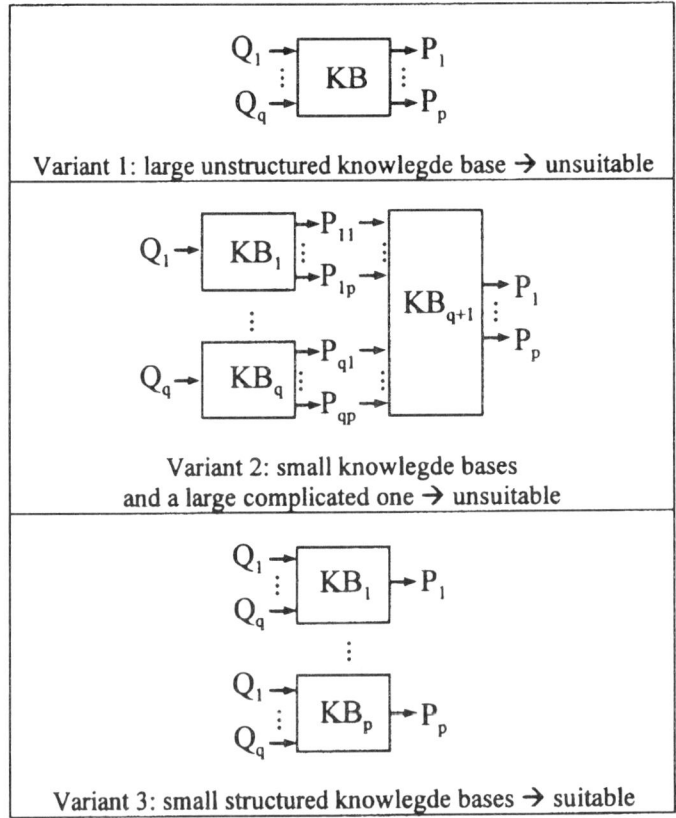

Figure 4. Variants for the knowledge base concept.

Input Parameters Q		Knowledge base / Rules	Output parameters P	
Q6	Unequal sides	P2>45°: Q6 ↓ ⇒ P3 ↓	P3	Weld torch angle β
P2	Actual weld torch angle β	P2<45°: Q6 ↑ ⇒ P3 ↑		
Q1	Weld reinforcement	Q4 = 0:	P4	Weld torch angle α
Q2	Undercut	Q1 ↑ ⇒ P4 ↓		
Q3	Spatter	Q2 ↑ ⇒ P4 ↑		
Q4	Seam thickness deviation	Q3 ↑ ⇒ P4 ↓		
Q1	Weld reinforcement	Q1 ↑ ⇒ P5 ↑	P5	Voltage
Q2	Undercut	Q2 ↑ ⇒ P5 ↓		
Q3	Spatter	Q3 ↑ ⇒ P5 ↓		
Q1	Weld reinforcement	Q1 ↑ ⇒ P6 ↓	P6	Wire feed speed
Q4	Seam thickness deviation	Q4 ↑ ⇒ P6 ↓		
Q3	Spatter	Q3 ↑ ⇒ P6 ↑		
Q1	Weld reinforcement	Q4 = 0:	P8	Weld speed
Q2	Undercut	Q1 ↑ ⇒ P8 ↑		
Q3	Spatter	Q2 ↑ ⇒ P8 ↓		
Q4	Seam thickness deviation	Q3 ↑ ⇒ P8 ↓		
Q5	Longitudinal offset	Q5 ↑ ⇒ P9 ↓	P9	Longitudinal offset
Q6	Unequal sides	Q6 ↑ ⇒ P10 ↓	P10	Horizontal offset
Q6	Unequal sides	Q6 ↑ ⇒ P11 ↑	P11	Vertical offset
Q1	Weld reinforcement	Q4 = 0:	P7	Contact tube distance
Q3	Spatter	Q3 ↑ ⇒ P7 ↓		
Q4	Seam thickness deviation	Q1 ↑ ⇒ P7 ↑		
Q7	End crater depth	Q7 ↑ ⇒ P1 ↑	P1	Hold time at end position
Legend: ↑...increasing, ↓...decreasing, ⇒...inference "then"				

Table 2. Knowledge base for thick plate fillet welding.

The Fuzzy Controller now performs as shown in Figure 5. The robot parameters were calculated for the robot control system "Comau C3G" of the company "Comau Robotics". This calculation was done using offsets, factors and kinematic formulas (Craig 1989).

5 Application of the Concept to a Fillet Weld

As the first step in our approach we define the parameters, their range and their fuzzification. As fuzzifcation of the quality parameters and the process parameters we propose to use five fuzzy terms (very low, low, medium, high, very high) and simple membership functions such as the Z-type and the Λ-type. The quality features of a weld seam track are called weld imperfections or discontinuities and are described in the norms (DIN EN 25817, ISO 5817). To build up the knowledge base we firstly define the parameters in Table 1 and, secondly, define the fuzzy rules according to literature data and experiments in Table 2 and, thirdly, calculate the robot parameters.

	Parameters	Classification			Range	
		Absolute Parameter	Corrective Parameter			
			Absolutely	Relatively		
Q_1	Weld reinforcement		X		0	+5 mm
Q_2	Undercut		X		0	+3 mm
Q_3	Spatter		X		0	+1
Q_4	Seam thickness deviation		X		-5 mm	+5 mm
Q_5	Longitudinal offset		X		-5 mm	+5 mm
Q_6	Unequal sides		X		-5 mm	+5 mm
Q_7	End crater depth		X		-4 mm	+4 mm
...	Expandable					
P_1	Hold time at end position		X		-1 s	+1 s
P_2	Actual weld torch angle β	X			0°	90°
P_3	Weld torch angle β		X		-15°	+15°
P_4	Weld torch angle α		X		-60°	+60°
P_5	Voltage			X	-50 %	+50 %
P_6	Wire feed speed			X	-90 %	+90 %
P_7	Contact tube distance		X		-4 mm	+4 mm
P_8	Weld speed			X	-50 %	+50 %
P_9	Longitudinal offset		X		-5 mm	+5 mm
P_{10}	Horizontal offset		X		-4 mm	+4 mm
P_{11}	Vertical offset		X		-4 mm	+4 mm
...	Expandable					
R_1	Position X	X			-	-
R_2	Position Y	X			-	-
R_3	Position Z	X			-	-
R_4	Eulerangle A	X			-	-
R_5	Eulerangle E	X			-	-
R_6	Eulerangle R	X			-	-
R_7	Weld speed	X			-	-
R_8	Energy	X			-	-
R_9	Weld voltage	X			-	-
R_{10}	Hold time at end position	X			-	-
...	Expandable					

Table 1. Definition of parameters for welding a fillet weld.

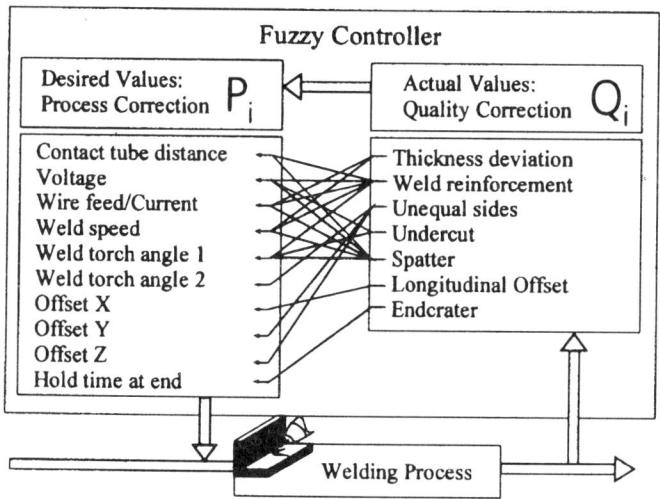

Figure 5. Fuzzy controller for a fillet weld.

6 Experiments and Discussion

The software MAGROB was tested by applying it to intentionally "bad" robot programs. An example of the optimization and the man machine interface is shown in the figures 6 and 7. It could be verified that "low" imperfections of the weld seam track were compensated after a few iterations, i.e. three to five, in an easy way. The automatic program adjustment could be done much faster than to adjust it manually. However, when "high" imperfections or several imperfections occur the number of iteration steps increases to five to ten. This is due to the visual "unsharp" determination of the quality features which can lead the fuzzy logic controller to an overshooting. Another uncertainity depicts the complexity of the welding process and its rough fuzzy logic model which results in an mutual impact of the parameters.

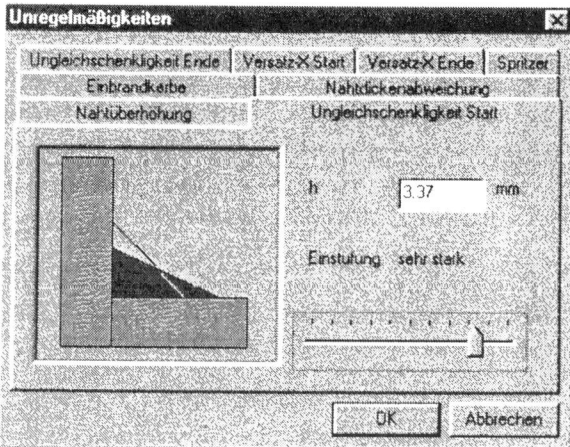

Figure 6. Example for a process optimization using the software MAGROB.

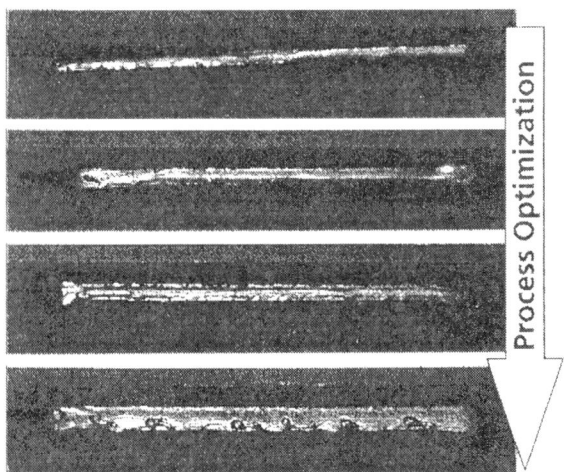

Figure 7. Process optimization.

7 Summary and Outlook

This article describes a method for a quality feature based adjustment of robot programs emplified for the fillet welding of thick plate steel. Therefore, a classification of parameters, the system concept and an object-oriented framework is presented. The implementation of the method in the software MAGROB as well as the successful testing using a welding robot is outlined.

Further work can be carried out on the transfer of the software to other processes. Moreover, existing expert systems can be connected and the knowledge base can be enhanced on using the parameters of the preceding iterations in the sense of a learning system.

8 Acknowledgements

This reserch project was funded by the German Ministry for Economy and Technology (BMWi) through the "Arbeitsgemeinschaft industrieller Forschungsvereinigungen 'Otto von Guericke' e.V. (AiF)". The company Comau Deutschland GmbH in Leonberg, Germany, has kindly delivered a robot welding system for the practical verification of the technique.

9 References

AMIRA (2000): Homepage des Projekts AMIRA URL: http://www.cee.etnoteam.it/amira/, January.

Craig, J. J. (1989): *Introduction to Robotics. Mechanics and Control.* Reading, MA: Addison-Wesley, 2nd edition.

DIN EN 25817 (1992): Deutsches Institut für Normung: *Lichtbogenschweißverbindungen an Stahl. Richtlinie für die Bewertungsgruppen von Unregelmäßigkeiten.* Berlin: Beuth Verlag.

FANUC (1998): "Arc Smart and Easy Package" from FANUC. In *Jara Robot News*, 10/6: 3.

Hartfuss, C. (1996): *Wissensbasierte Programmierung von Industrierobotern zum Schutzgasschweißen im Stahlhochbau.* Berlin, Heidelberg, New York: Springer.

ISO 5817 (1992): International Standardization Organization: *Lichtbogenschweißverbindungen an Stahl; Richtlinie für die Bewertungsgruppen von Unregelmäßigkeiten.*

Park, J.-Y. (1993): *Fuzzy-Logic-basiertes Beratungssystem zur Prozeßoptimierung und Fehlerdiagnose beim MAG-Schweißen.* Aachen: Verlag Shaker (= Aachener Berichte Fügetechnik 93/6).

Roosen, S. (1997): *Online-Prozeßoptimierung beim MAG-Schweißen mit Hilfe eines Expertensystems.* Aachen: Verlag Shaker (= Aachener Berichte Fügetechnik 97/4).

Scheller, W. (1994): *Einsatz künstlicher neuronaler Netze in Schweißkopfführungssystemen für das Metallschutzgasschweißen.* Aachen: Verlag Shaker (= Aachener Berichte Fügetechnik 94/1)

Starke, G. et al. (1995): Schweißprozeßparameteroptimierung mit Fuzzy-Logik. In *Schweißen und Schneiden* 47/6: 494-503.

Application of the RNT Robot to Milling and Polishing

K. Mianowski[1], K. Nazarczuk[1], M. Wojtyra[1], W. Szynkiewicz[2], C. Zieliński[2] and A. Woźniak[2]

[1] Institute of Aeronautics and Applied Mechanics, Warsaw University of Technology
[2] Institute of Control and Computation Engineering, Warsaw University of Technology

Abstract. A new type of robot with six DOF invented by K. Nazarczuk and K. Mianowski (1992) is presented in the paper. The robot arm has serial-parallel structure, high stiffness - comparable to that of parallel manipulators and very large workspace – comparable to that of serial robots. Lately the RNT robot has been adapted to milling of soft materials and polishing large surfaces. The controller of the robot is based on the MRROC++ library/language described by Zieliński at all. (1997, 1998, 1999). Due to the modularity of the software and its open structure, it is especially well suited to investigative tasks. It enables easy incorporation of any sensors. The information gathered by sensors can be used for on-line trajectory generation, monitoring the progress of machining or for subsequent analysis of results. The MRROC++ control system can generate any trajectories, and for the purpose of machining it interprets APT program CL-files produced by a UNIGRAPHICS CAD system. The power of the controller lies in the possibility of combining the off-line generated trajectory coded in APT with sensor-gathered information to generate an on-line resultant trajectory. The RNT robot has very good position repeatability (±0.02mm). For achieving good absolute accuracy a special procedure of local kinematic model correction has been developed and implemented. Large workspace and high manoeuvrability of the RNT robot enables it to machine work-pieces that no NC-machine can produce - any 3D surfaces can be obtained. The robot has successfully polished large metal surfaces and milled soft materials, e.g. wood and plastic.

1 Manipulator of the RNT robot

In this paper we present some results of investigations of a Robot of New Type (RNT) designed, and built by the authors at Warsaw University of Technology. The manipulator of this robot is shown in Fig. 1. First DOF, i.e. rotation of the skew bracket 1 of the arm around the vertical axis of the column 0 is driven by electric motor M1 and a four-bar mechanism with screw-nut gear located inside the base. The outrigger 3 of the arm is connected to the skew bracket 1 by link 2 with kinematic pairs 1-2 with vertical axis and 2-3 with horizontal axis. Second and third DOF of the arm are realised by a mechanism in the form of movable spatial truss consisting of an outrigger 3 with constant length, parallely driven by two actuators L2 and L3 with screw-nut mechanisms driven by motors M2 and M3. In fact, the motion of the end of the arm is a combination of spherical movement of outrigger 3 in the frame of bracket 1 and rotation of this bracket around the vertical axis of a column 0. At the end of the outrigger 3 a spherical wrist is mounted. It consists of serially connected pitch 4, yaw 5 and roll 6 elements driven by motors M4-M6 with gears located at the back of the outrigger. Weight of these motors acts as a counterbalance and the spring S compensates the weight of outrigger 3 combined with the wrist.

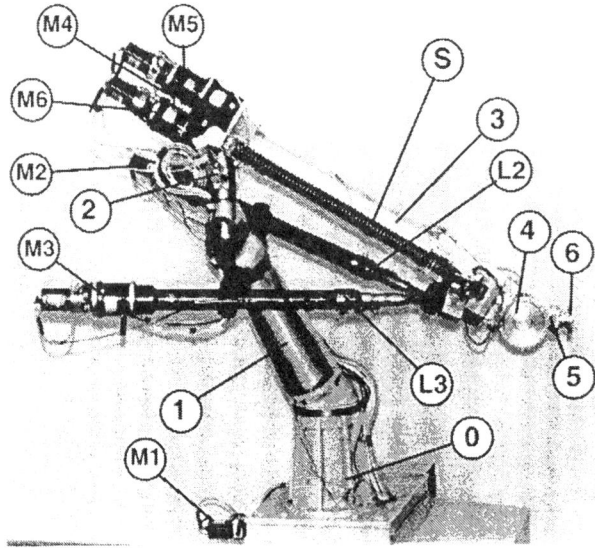

Fig. 1. General view of the RNT manipulator

Serial-parallel structure of the arm and ball screws used as gears ensures very high stiffness and relatively small mechanical hysteresis of the arm. The stiffness measured at the end of the arm depending on the orientation of acting forces is $2*10^5$ - $1.5*10^7$ N/m, while hysteresis is not greater than ±0.02mm. The performance of the arm is similar to that of typical parallel manipulators, while the workspace, which is a fragment of a torus is similar to that of serial ones. Utilisation of toothed gears renders the performance of the wrist inferior to that of the arm. It is worth noting, that minimal frequency of natural vibrations is approximately 20-25 Hz. Some of the experimental investigations of the mechanical performance of the arm were reported by the authors in articles (1995, 1999).

2 Structure of the MRROC++ controller

The controller of the robot is based on the MRROC++ library/language described in details by Zieliński (1997, 1998, 1999). Due to its modularity and an open structure, it is especially well suited to investigative tasks. It enables easy incorporation of any sensors. The sensor-gathered information can be used both for on-line trajectory generation and monitoring of progress of machining.

The MRROC++ system has a hierarchical structure (see Fig. 2). The overall structure is divided into three parts. The first part is hardware dependent, the second is task dependent and the third composes the operator interface and is constant. In this way the modifications due to hardware or task changes are disjoint. MRROC++ is capable of controlling many robots, but in this case it was used to control a single robot system. It usually runs on PC computers (Pentium based are preferred) connected by an Ethernet network, but a single computer configuration can also be used if the computational power is sufficient. It is supervised by a real-time operating system QNX-4. A single process co-ordinating the operation of the whole system is called the Master Process (MP). Each effector (either a robot or a co-operating device) has two processes controlling it: the Effector Control Process (ECP) and the Effector Driver Process (EDP). The former is responsible for the execution of the user's task assigned to this effector, and the latter for direct control of this effector. EDP is supervised by ECP. In this way the user's task and the effector specific control have been separated and are independent of each other.

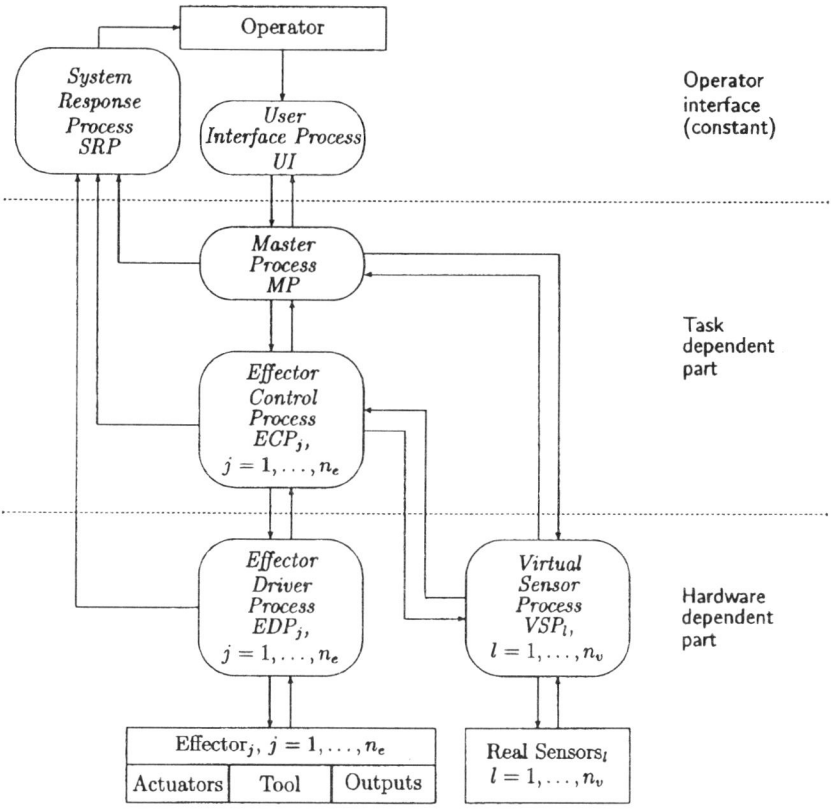

Fig. 2. MRROC++ hierarchical structure.

Data obtained from real (i.e. hardware) sensors usually cannot be used directly in robot motion control. The process of extracting meaningful information for the purpose of motion control is named data aggregation and is performed by a virtual sensor. Data aggregation is done by Virtual Sensor Processes (VSPs).

Moreover, the system contains two processes dedicated to the interaction with the operator. User Interface Process (UI) handles operator commands. System Response Process (SRP) displays all the system status and error messages on the screen of the monitor. Both processes perform in a windows environment, so operator commands such as: initiation of execution of the user's program, its termination or pausing and resuming are done by clicking certain icons.

The EDP is commanded by its ECP, and so its code usually is not be modified by the user. It varies only when the effector hardware changes. If the type of the robot is changed, a new EDP has to be supplied, but the remaining part (i.e. the task dependent part) is not altered. To fulfil this requirement a standard EDP communication protocol has been devised. EDP is treated as a server interpreting commands issued by the other parts of the system described by Zieliński (1999).

Possible applications of the RNT robot such as milling, grinding or contour following require a high-performance motion control system. Good command tracking, fast dynamics, and high disturbance rejection are the basic requirements in the servo level control. Implementation of very sophisticated control algorithms is possible within the framework of the proposed open structure controller. The servomechanism structure for a single joint is shown in Fig. 3.

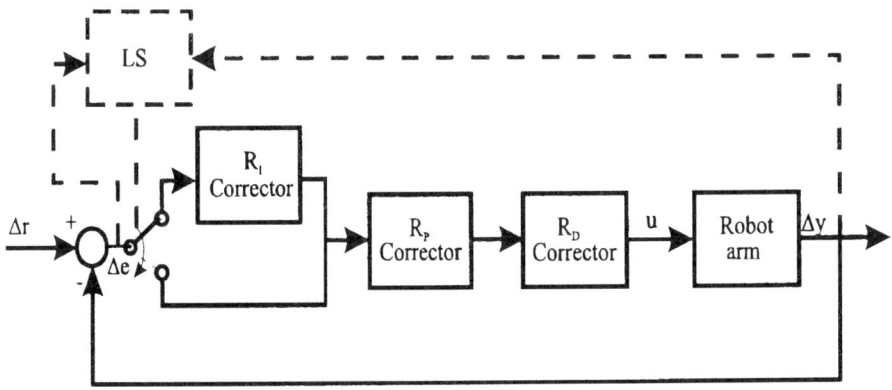

Fig. 3. Servo controller structure

Discrete PI+PD type of algorithm was used:

$$u(z) = R_D(z) R_I(z) \Delta e(z)$$

$$R_I(z) = \frac{(1+c)z - (1-c)}{z-1}, \quad c > 0,$$

$$R_D(z) = \frac{b_0 z - b_1}{z - a}, \quad a, b_0, b_1 > 0,$$

where: $\Delta e = \Delta y - \Delta r$ is an error increment, and Δr is the reference input increment. Parameters c, a, b_0, b_1 have been chosen basing on the classical frequency-response method using the W-transform [2]. It is well known that oscillations may occur in position control systems with integral action. In order to reduce undesirable "hunting" and "hangoff" effects, the integral action is turned off, when the current steady-state positional error Δe is near zero and the joint is at rest (i.e. $\Delta y=0$). The moment when the integral action should be turned off is obtained from an empirically chosen condition, which is calculated in the LS block.

2.1 Controller hardware

Each DC electric motor actuated DOF of the arm is controlled by a separate axis controller described previously by Zieliński (1998). The axis controller consists of: a control microcomputer based on the single-chip MCS-51 family 80552 microcomputer, a position measurement circuit, a PID regulator and an interface circuit (see Fig. 4). Axis controllers are connected to a PC type master computer through a parallel interface in such a way that interfacing circuit registers are mapped into the input-output address-space of the master computer. The microcomputer communicates with it and interprets its commands.

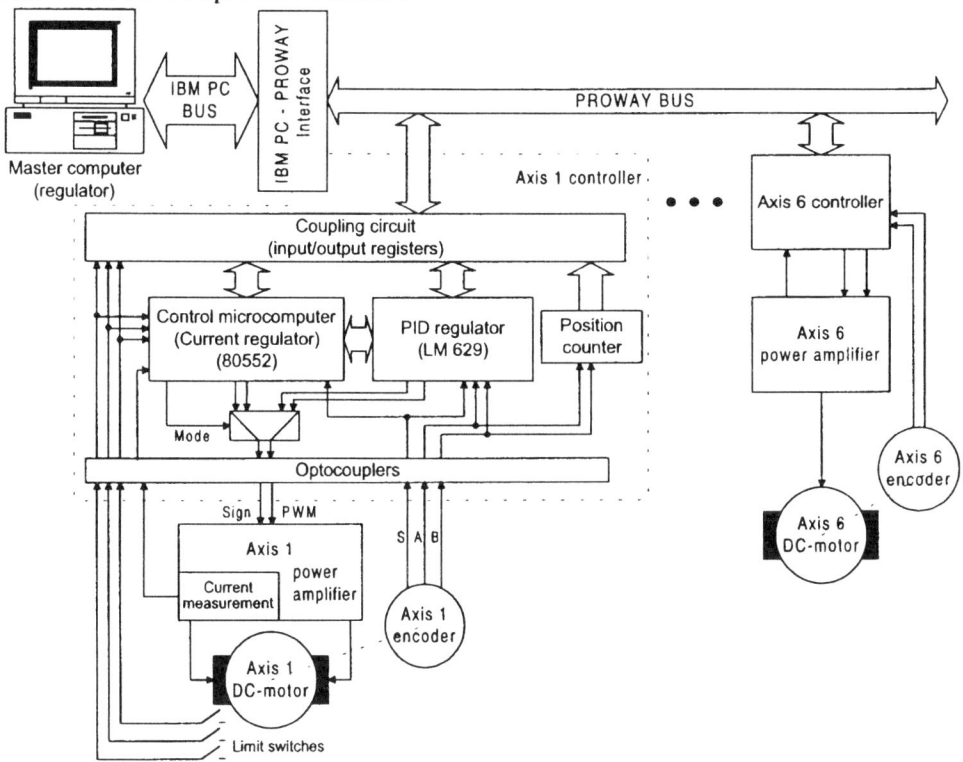

Fig. 4: Hardware structure of the axis controller.

The axis controller has two modes of operation. In the first one, the servo algorithm is implemented in the master computer, so the microcomputer generates a PWM signal and measures the current position by using a specialised 32-bit reversible counter. In the second mode, besides performing the above functions it also serves as a servo-regulator. For this purpose it uses National Semiconductors LM629 PID controller. In this case the master computer does not have to execute the servo algorithm, so it saves the computation power for other tasks. Obviously, in this case the algorithm is limited to PID control only, although its parameters can be modified.

In both modes of operation the microcomputer continuously monitors the state of limit switches and the value of the DC motor current. Activation of a limit switch or transition over the threshold current in the motor causes immediate motor stop and an error message being sent to the master computer.

The RNT manipulator in conjunction with the MRROC++ based control system attains position repeatability of ±0.02mm without overshoots for typical PTP motions. Due to this a decision has been made to apply it to a machining processes, such as grinding, polishing of smooth surfaces or milling small objects.

4 Some typical applications

Application of the RNT robot to a machining processes such as milling or grinding, needs accurate information about positions and orientations of its end-effector. The absolute error generated by the control system and introduced by the mechanical part of the manipulator should be as low as possible. Real mechanical systems are subjected to friction exhibit elasticity and mechanical hysteresis (Lost Motion) even without the backlash. A nominal model of kinematics of the robot arm introduces additional errors due to inadequate knowledge of link parameters, and their elasticity. Friction in the joints aggravates the situation.

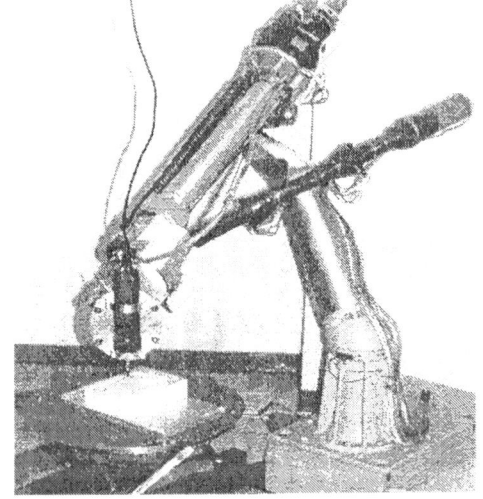

Fig.5a Fig.5b

Fig. 5. Experimental stand with the RNT robot equipped for the investigations of milling.

In the case of milling, the positions and orientations of the end-effector have to be corrected. For the correction of some of the kinematic characteristics of the robot, the procedure reported by K. Mianowski in the paper (1999) was used. General view of the RNT robot equipped for milling wood is shown in Fig. 5. A small machine tool is mounted on the wrist. The linear gauge shown in Fig. 5a has been used to measure real distances between two sets of complementary sockets located on the end-effector and the table attached to the base of the robot. Measurements were taken for a number of manipulator configurations. In this way the parameters for the correction of the nominal kinematic model were obtained for a small area of the robot workspace in which the milling task was to be executed. The obtained parameters form a matrix correcting the nominal kinematic model. Both the nominal kinematic model and the correction matrix have been embedded in the controller. Whenever the end-effector is in the vicinity of the space for which the correction is valid the corrector is switched on. The implementation of this procedure resulted in the improvement of the absolute positioning accuracy. The errors have been reduced to 1/4 of the initial value and for typical machining tasks they did not exceed 0.5mm.

Another machining task that the RNT robot has been put to was polishing of smooth surfaces. For this application a special machine head, shown in Fig. 6b, was used. In this task it is not very important to control the tool trajectory accurately.

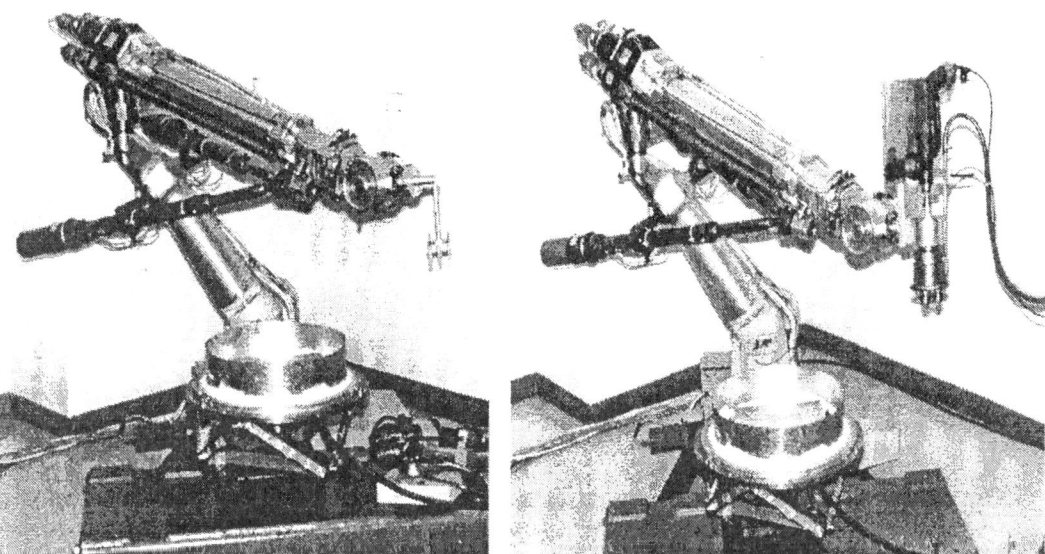

Fig.6a   Fig.6b

Fig. 6. The RNT robot used for polishing smooth surfaces.

The force between the tool and the machined surface is generated by pneumatic actuator. The nominal tool trajectory was taught-in by using a special mechanical substitute element shown in Fig. 6a. To measure the reaction force between the tool and the machined surface, a truss-type table with a structure similar to that of a Stewart platform was designed. This table contains strain-gauges. It is worth noting, that such a construction enables the measurement of six components of reaction force and torque directly in a Cartesian coordinate frame. Traditional polishing machines realise their task without changing the orientation of the tool. The RNT robot can adapt the orientation of the tool to the shape of the polished object. With the use of the RNT robot the quality of the machined surfaces has improved. The machining process requires cooling fluid and abrasion powder. Those harmful to humans substances have to be removed. The table is equipped with a device for that purpose. Experiments were conducted in which the RNT robot was applied to machining big elements of car bodies produced by an automotive industry stamping press.

5 Conclusions

The obtained results show that implementation of local correction of kinematic model can considerably improve the precision of arm motions. This widens the possible applications of the RNT robot. The manipulator of the RNT robot exhibits very high stiffness. Its other mechanical parameters are also relatively good. To improve the mechanical performance of the robot its wrist have to be modified. The modularity of the MRROC++ system enables quick and simple transformation of the control system required by various tasks carried out by the RNT robot. Any sensors can be incorporated into the control system, so there was no problem in connecting the strain gauges located in the table. The information gathered by sensors can be used both for on-line trajectory generation and monitoring progress of machining. The same capability was used to gather data for subsequent analysis of control of the milling process. The precision of machining can be influenced by many factors, so it was of utmost importance that the structure the control system was such that any problems arising in the investigations could be dealt with by simple programming means. For example, problems posed by friction and the inaccuracy of the kinematic model could be successfully resolved.

The unconventional design of the RNT robot arm resulted in good mechanical performance. The manipulator can be used by small companies as a machine tool to machining various elements. Lately the robot has been equipped with a coupler enabling quick changeover of tools mounted on the wrist. Robot can work providing some various operations in a number of stands. It is worth noting, that the RNT robot can machine workpieces that no NC-machine can produce - any 3D surfaces can be obtained. Moreover, it has successfully polished large metal surfaces and milled soft materials, e.g. wood and plastic.

6 Acknowledgments

The authors would like to thank A. Rydzewski for creating the hardware of the control system. This work was partially supported by Program of Control, Information Technology and Automation (PATIA) of Warsaw University of Technology.

References

Bidziński J., Mianowski K., Nazarczuk K., Słomkowski T., A manipulator with an arm of serial-parallel structure, The Archives of Mechanical Engineering, Vol. XXXIX, Warsaw 1992.

Franklin G. F., Powell J. D., Emami-Naeini A., Feedback Control of Dynamic Systems, Reading Ma., Addison-Wesley, 1994

Mianowski K., Nazarczuk K.: Parallel Drive of Manipulator Arm, Proc. Ro.Man.Sy'8, Kraków, Poland 1990.

Mianowski K.: Parallel and Serial-Parallel Robots for the Use of Technological Applications, Proc. PKM'99, Milano, Italy, 1999.

Nazarczuk K., Mianowski K., Olędzki A., Rzymkowski C: Experimental investigation of the robot's arm with serial-parallel structure, Proc. IX World Congress on the Theory of Machines and Mechanisms, Milano 1995, pp. 2112-2116

Zieliński C.: "Object-Oriented Programming of Multi-Robot Systems", Proc. 4th International Symposium on Methods and Models in Automation and Robotics MMAR'97, 26-29 August 1997, Międzyzdroje, Poland, pp.1121-1126.

Zieliński C., Rydzewski A., Szynkiewicz W.: "Multi-Robot System Controllers", Proc. 5th International Symposium on Methods and Models in Automation and Robotics AR'98, 25-29 August 1998, Międzyzdroje, Poland, Vol.3, pp.795-800.

Zieliński C.: "The MRROC++ System", Proc. 1st Workshop on Robot Motion and Control, RoMoCo'99, 28-29 June, 1999, Kiekrz, Poland. pp.147-152.

Path Planning in Complex Environments for Industrial Robots with Additional Degrees of Freedom

Francisco Valero[1], Vicente Mata[1], and Marco Ceccarelli[2]

[1] Department of Mechanical Engineering, Universidad Politécnica de Valencia, Spain
[2] Dipartimento di Meccanica, Strutture, Ambiente e Territorio, Università di Cassino, Italy

Abstract. In this paper a path planning among obstacles is presented as applied to a robot of PUMA 560 type with mobile base. From two given configurations - the initial and goal ones - a configuration space is calculated. The robot configurations are expressed in terms of fully Cartesian co-ordinates and are obtained by solving non-linear optimisation problems between adjacent configurations, a variety of constrains is considered in order to take into account different real operation problems. The path is selected from a weighted graph associated to the map of feasible robot configurations. A search algorithm has been used to minimise an objective function in order to obtain a sequence of robot configurations between the initial and goal ones.

1 Introduction

The path planning problem for a robot that must operate among static obstacles is a question, which has addressed the attention of numerous researchers in the last twenty years. The reason for such interest is due to the increasing number of possible applications in industrial environments. It is recognised that industrial robots operate more and more in environments with increasing complexity, such as, for example, robots in industrial automotive welding, in assembling or inspection operations. In addition, new fields of robotic applications may also require a path planning problem among obstacles, such as robots aided surgery.
Proposed algorithms from several authors have been based on global methods for path planning (Lozano-Pérez and Wesley, 1979), on local methods (Khatib, 1986, Barraquand and Latombe, 1991), and mixed methods (Kavraki, Svestka, Latombe, and Overmans, 1996). However, all of them operate in the robot's space of joints. This requires an important computational effort to express the obstacles, which are inside of the workspace in the above-mentioned space of joints. In a previous work (Valero et al. 1997) a formulation has been proposed for a path planning algorithm that operates exclusively in the Cartesian Space.
In this paper, the proposed algorithm works in two stages. In the first one, a discrete configuration space is generated as specified in Section 3, by using the model of the robot and its environment presented in Section 2. In the second stage the configuration space is transformed into a weighted graph, which allows to obtain the sequence of configurations that define the trajectory, as it is indicated in Section 4. It must be pointed out that in order to obtain a new configuration free of collisions an optimisation problem is considered which is subject to the first degrees of freedom of the robot. In case that no solution is obtained, all the degrees of the robot are considered and

the problem is reformulated. In this way, the large movements, that don't use unnecessary degrees, are boosted.

Although the proposed path planning algorithm could be applied to any manipulator, in this paper it has been investigated on a industrial robot model with 6 degrees of freedom taking into account additionally a movable base. The three-dimensional environment in which the robot operates has been assumed with static obstacles. An illustrative example is obtained by applying the proposed procedure and is shown in Section 5.

2 Environment and Robot Modelling

The robotic system is modelled by means of several significant points, which describe the characteristic constraints of its structure. The variables, which model the robotic system, are the Cartesian co-ordinates of the significant points.

The minimum number of points is considered through significant points that can define the configuration of the system without ambiguity. For example, in the system shown in Figure 1, the points A, C, D, F and P are the chosen significant points. The rest of the points complete the description of the system and their co-ordinates can be obtained from those of the significant points.

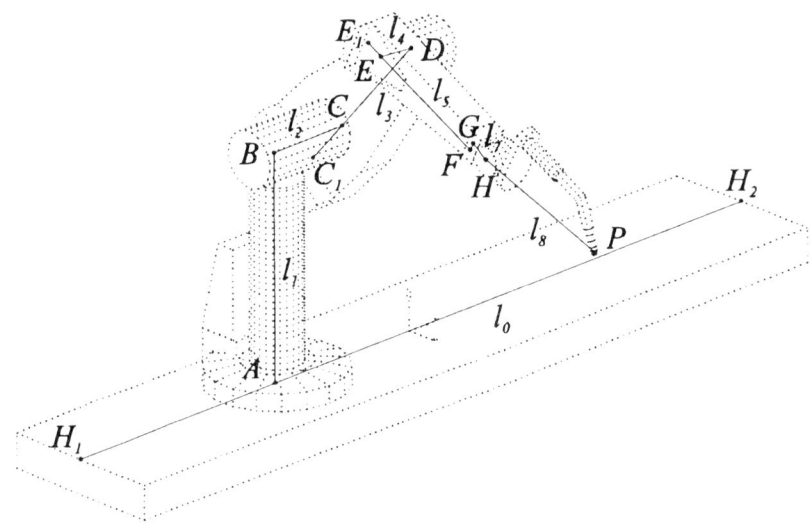

Figure 1

The Cartesian co-ordinates of significant points will determine a robot configuration C^j. In the robotic system of Figure 1 the significant points are indicated with the co-ordinates as A (x_1,x_2,x_3), C (x_4,x_5,x_6), D (x_7,x_8,x_9), F (x_{10},x_{11},x_{12}) and the extremity point P (x_{13},x_{14},x_{15}). Therefore, C^j will be determined by 15 variables in the form $C^j(x_1, x_2,...,x_{15})$.

Assuming the point co-ordinates and the system geometry (type and location of kinematic joints and lengths of links) as given, it is possible to calculate the other points B, E, G, and H.
The significant points are subjected to constraints that make feasible the setting of a configuration. These constraints are:

a) Geometrical constraints of the robot structure;
b) Constraints on the mobility of robot joints;
c) Collision avoidance within the robot workspace;

Obstacles avoidance between adjacent configurations.
The obstacles contained in the workspace are modelled by a set of spheres with radius r_j and centre Q_j. In order to obtain the real dimension of the robot, the technique of growing obstacles of (Lozano-Perez and Wesley, 1979) has been used.
Constraints of type (a) are related with the lengths of links, distances between interesting points, location and orientation of the kinematic joints in order to describe orthogonal or parallel geometry between segments defined by significant points.
Constraints of type (b) formulate both possible movements for the robot's base and the structure constraints that may affect the mobility of the robot's joints. In the case of Figure 1 point A can move along the segment H1H2. An example of how the angular constraints can be considered is shown in Figure 2, that describes the constraint for first rotational joint.

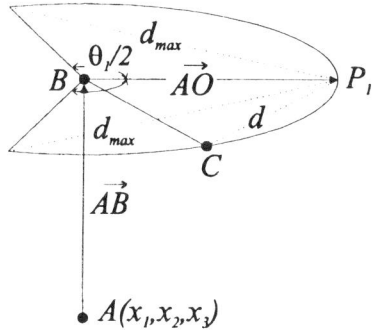

Figure 2

For a feasible joint angle the constraint equation is based on distances by using an auxiliary vector $\overrightarrow{AO}$, which indicates the middle point of the permitted arc from the movement of the joint, as shown in Figure 2. The constraint equation is given as

$$d_{max} \geq d \qquad (1)$$

where d is the distance from the end of the auxiliary vector $\overrightarrow{AO}$ to point C. d_{max} is the maximum value that such distance can have, in agreement with the angular constraint characteristic of the joint.

Constraints of type (c) avoid collisions between links of the robot and obstacles. For the case of the robot of Figure 1 eight constraints can be formulated for every sphere modelling an obstacle. This type of constraints can be formulate as

$$d_{ij} > r_j + w_i \tag{2}$$

where d_{ij} is the distance from sphere j to link i; r_j is the radius of the sphere and w_i is the radius of the smallest cylinder that contain the link i.

Constraints of type (d) avoid the location of obstacles between feasible adjacent configurations of the robot. Due to the fact that a discrete space of feasible configurations is considered, it is necessary to guarantee that the movement between adjacent configurations is possible, so that it must be checked that there are not obstacles between them. These constraints can be formulated as

$$\left|\overrightarrow{A_p A_k}\right| \leq 2 \cdot min(r_j) \qquad \left|\overrightarrow{C_p C_k}\right| \leq 2 \cdot min(r_j)$$
$$\left|\overrightarrow{D_p D_k}\right| \leq 2 \cdot min(r_j) \qquad \left|\overrightarrow{F_p F_k}\right| \leq 2 \cdot min(r_j) \tag{3}$$

where r_j is the radius of the smallest sphere in the modelled workspace. The subscripts p and k denote the considered adjacent configurations.

The workspace can be considered as a rectangular-based prism with parallel edges to the Cartesian System frame $OXYZ$. The end points of its diagonal have been assumed as the robot terminal element points corresponding to the initial and final configurations for the given the trajectory. The set of points that can be reached by the end point of the terminal element are restricted by the prism volume discretization, which has been previously generated by using the following increments

$$\Delta_x = \frac{|y_f - y_i|}{N_x} \qquad \Delta_y = \frac{|y_f - y_i|}{N_y} \qquad \Delta_z = \frac{|z_f - z_i|}{N_z}$$

where the values of Δ_x, Δ_y and Δ_z are established by the user with the condition that they are smaller than the smallest sphere diameter modelling obstacles in the workspace.

3 Configuration Space Generation

Configuration space generation is performed on the discrete environment and locations of the robot terminal element correspond to points of this space.
The proposed procedure is based on the following assumptions:

1. The user gives the initial C^i and C^j configurations, which are used by the procedure. This will affect to the associated configuration space.
2. The robot terminal element motion will be parallel to the axes of the fixed Cartesian Reference System.

A definition for adjacent configuration, applied to the robotic system is needed in order to obtain the configuration space. Given C^k configuration, a new one C^p is said to be adjacent to it if, at least, one of the three following equations is satisfied:

$$\left|x_{13}^p - x_{13}^k\right| = \Delta_x \qquad \left|x_{14}^p - x_{14}^k\right| = \Delta_y \qquad \left|x_{15}^p - x_{15}^k\right| = \Delta_z \tag{5}$$

C^p will be obtained by using C^k as an initial estimation for solving an optimisation problem with the expression of the distance between the configurations

$$\|C^p - C^k\| = \sum_{i=1}^{12}(x_i^p - x_i^j) \tag{6}$$

as objective function.

The constraints for the optimisation problem are the four types that have been illustrated in the previous section.

In the usual working of the robots in industry, the movements of the robots are clearly distinguished between the large displacements performed with the first degrees of freedom and the small movements by the spherical wrist. In the proposed optimisation problem, two alternative strategies have been considered to handle two different groups of constraints. The first one corresponds to the movement of the robot without using the movement of the wrist (actuators 4 and 5). If no solution is obtained for this problem then a second set of constraints are used in which all actuators can move including the wrist.

The generation of a configuration space is based on the idea of generating adjacent configurations from the initial to the final configuration by using an optimization problem with the above-mentioned objective function. In particular, C^j can generate six configurations C^1, C^2, C^3, C^4, C^5 and C^6 which come from ramifying with positive and negative increments along parallel axes of the Cartesian Reference System. Next, the ramification from the C^1 configuration is repeated to all possible directions, where previous adjacent configuration doesn't exist yet, and so on until the configuration space is completed.

This procedure has as main advantage that do not establish a strict order to locate the terminal element for the generation of the configurations, and allows the access to difficult areas from more adequate configurations.

4 Obtaining the path

The path consists of a sequence of configurations, which are between the initial and final ones. This sequence will be obtained by the optimisation procedure that is presented next.

A weighed graph is associated with the generated configuration space, where the nodes correspond to the robot configurations and the arcs are related to joint displacements between adjacent configurations. The weight corresponding to the arc that goes from node K (C^k robot configuration) to node P (C^p robot configuration), can be given as

$$a(k,p) = \sum_{i=1}^{12}(x_i^p - x_i^k)^2 \tag{7}$$

when that C^k and C^p are adjacent. In addition C^k and C^p must satisfy:

1. Constraints type (d) avoiding the obstacles between configurations;
2. The angle increase from C^k to C^p must be smaller than the magnitude of the forbidden zone for that joint, so that large displacements are avoided for movement between adjacent configurations, (See Figure 3).

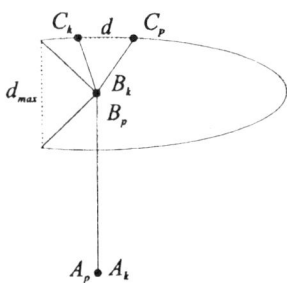

Figure 3

In case the points above mentioned are not met, then we consider that

$$a(k, p) = \infty \tag{8}$$

Finally, the searching is started in the weighed graph with the path that joins the node corresponding to the initial configuration to the node corresponding to the goal configuration. Since the arcs meet that $a(k,p) \geq 0$, the Dijkstra's algorithm is used in order to obtain the path that minimises the distance between the initial and goal configurations. If this path exists, it is easy to get a sequence of robot configurations.

5 Application example

In the Figure 4 the results of an illustrative example is presented as a practical application. The robot environment includes 3 obstacles, which have been modelled trough 47 spheres. The trajectory traced by the robot end-effector and some selected configurations are shown together with some robot configurations.

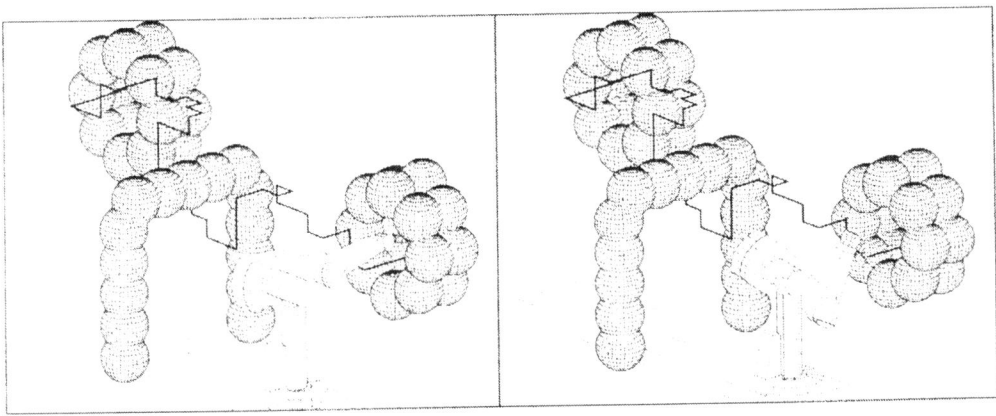

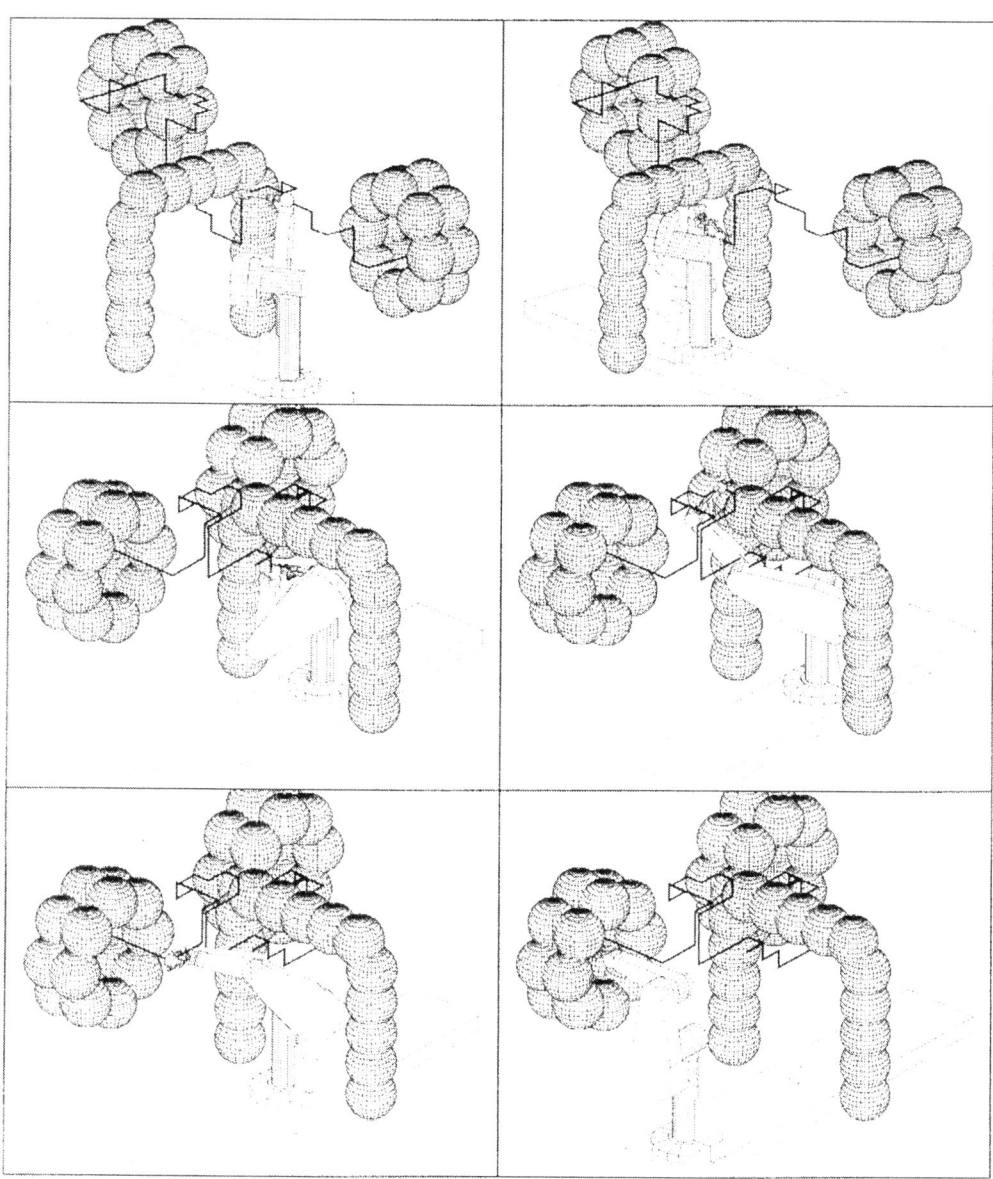

Figure 4

6 Conclusions

The proposed procedure has been used to solve a practical application. But it must be pointed out the flexibility of the formulation, which allows an easy adaptation to other types of robots or systems, even if they are highly redundant, and it can be done without increasing the computational complexity excessively.

In this algorithm, the process of detection of collisions is substituted (since it is computationally expensive) by a sequence of optimisations problems, which take the constraints into account in order to, avoid collisions.

In comparison with global algorithms, the proposed procedure can solve complex problems in reasonable computation time and without high storing needs. Comparing it with local algorithms, it has not local minimum problems.

References

Barraquand, J., and Latombe, J-C. (1991). Robot Motion Planning: A Distributed Representation Approach. *International Journal of Robotics Research* 10(6): 628-649.

Faverjon, B. (1984). Obstacle Avoidance Using an Octree in the Configuration Space of a Manipulator. *Proceedings of the IEEE International Conference on Robotics and Automation*. Atlanta, 504-512.

Joseph, D.A., and Plantiga, W.H. (1985). On the Complexity of Reachability and Motion Planning Questions. *Proceedings of the First ACM Symposium on Computational Geometry*. Baltimore, 62-66.

Kavraki, L.E., Svestka, P., Latombe, J-C., and Overmans, M. (1996). Probabilistic Roadmaps for Planning in High-Dimensional Configuration Spaces. *IEEE Transactions on Robotics and Automation* 12(4): 566-580.

Khatib, O. (1986). Real-Time Obstacle Avoidance for Manipulators and Mobile Robots. *International Journal of Robotics Research* 5(1): 90-98.

Lozano-Pérez, T., and Wesley, M.A. (1979) An Algorithm for Planning Collision-Free Paths Among Polyhedral Obstacles. *Communications of the ACM* 22(10): 560-570.

Tournassoud, P. (1988). *Géometrie et intelligence artificielle pour les robots*. Paris: Ed. Hermes.

Valero, F., Mata, V., Cuadrado, J.I., and Ceccarelli, M. (1997). A formulation for path planning of manipulators in complex environments by using adjacent configurations. *Advanced Robotics* 11 (1): 33-56.

Zhu, X. Y Gupta, K. (1993). *On local minima and random search in robot motion planning*. Available as Technical Report, Simon Fraser University, Burnaby, Columbia Britannica, Canada.

Robotic Deburring Using a Fuzzy Force Controller

Robert Bicker[1] and Kevin Burn[2]

[1] Department of Mechanical, Materials & Manufacturing Engineering, University of Newcastle, UK
[2] School of Computing, Engineering & Technology, University of Sunderland, UK

Abstract. Most industrial robots are designed to operate as position servo-controlled manipulators, which is appropriate if the robot is to simply follow a predefined trajectory in free space. However, if the robot's end-effector comes into contact with the environment, then position control generally will not suffice. Stability becomes a major issue when the robot end-effector becomes highly constrained, and the success of the operation is largely dependent upon the correct selection of the gains in the force control algorithm. These gains are themselves dependent upon the compliance at the task interface, and another major difficulty occurs when this is unknown or varying. This paper describes the application of fuzzy logic in the development of an industrial robot force controller, primarily used for robotic deburring operations. Experimental results illustrate the effectiveness of the controller's self-adjusting gain strategy, and the ability of the robot to successfully carry out a deburring operation.

1 Introduction

Although most robots have traditionally been designed for accurate position control, the demand for them to be able to perform more complex tasks has led to an increasing number of applications where forces must be directly controlled, usually at the end-effector/task interface. An example is the use of a robot to guide a tool across a surface using force feedback, in order to maintain a specified contact force, e.g. in processes such as grinding and polishing. To fulfil these extra demands, an important area of robotics research is the implementation of stable and robust force control strategies. This is often difficult to achieve in practice and no single solution to the force control problem has been found particularly where robots are operating in unpredictable or disordered environments. Most current schemes only provide adequate force control when the controller is tuned to specific task requirements, and performance rapidly degrades if there is any significant variation in the robot tool/task environment characteristics, such as compliance.

Two major problems in the implementation of practical force controllers are stability and robustness. Stable force control is particularly difficult to achieve in 'hard' or 'stiff' contact situations. To improve stability various methods have been proposed (Ganwen & Ahmad,1997), one solution being to employ so-called 'active compliance' filters, where force feedback data is digitally filtered to emulate a passive spring/damper arrangement (Kim et al,1992). Robustness is a problem where environmental uncertainty exists, and effective force control can only be achieved by using environment stiffness detection with smooth switching of controller gains (Ow, 1997). This can slow down task execution, and result in unstable contact where the effective stiffness at the robot/task interface rapidly changes. Also the force-control scheme is constrained by the actual system, e.g. whether direct or indirect methods must be employed. In addition, controller tuning is often done in situ, since the difficulties of estimating a robot's dynamic parameters makes accurate modelling extremely difficult. A further complication is that for many tasks it cannot be assumed that the robot is 'attached' to the environment, i.e. always in contact.

A number of attempts have been made to combine fuzzy logic and robot force control, with a view to improving performance when the robot manipulator is in contact with an environment whose parameters are either unknown or rapidly changing (Burn & Bicker,1999;Tarokh & Bailey,1997). This paper describes a fuzzy logic controller designed to augment a conventional controller in a force control loop, to perform a task in which the stiffness of contact is continually changing. The effectiveness of the new controller is initially demonstrated using a range of experiments involving a six-axis robot contacting a cantilever beam. It is then employed in a deburring application. The method is shown to perform well despite wide variations in environment stiffness.

2 Contact Force Control Problem

Prior to examining the fuzzy logic approach, it is useful to briefly outline the force control problem under consideration and describe a conventional solution. The combined stiffness at the end effector/task interface in the direction of the applied force is K_e. This varies between a minimum value, determined by the objects in the environment with which the robot is in contact, and a maximum value, limited by the stiffness of the arm and torque sensor. Designing a fixed-gain conventional controller to meet a chosen specification for a specific value of K_e is, in principle, a relatively straightforward task. A problem arises when K_e is unknown or variable. Consider, for example, the case where the system is tuned to maintain a desired performance at a low value of K_e. If high K_e contact is then encountered, significant overshoot and oscillatory behaviour will occur, with a danger of instability. Conversely, if the system is tuned for high K_e, at low K_e the system is often sluggish and overdamped.

In view of this potential for instability, conventional force controllers normally require some form of environment stiffness detection technique to enable the controller gains to be switched accordingly. The main problem with this process is that it is time-consuming, often involving 'guarded moves' to contact in order to enable sufficient data to be collected for the algorithm to work. Such methods can also be unreliable in the presence of transducer noise and are not very effective in situations where K_e is variable or rapidly changing. To address these problems, a fuzzy logic approach was adopted.

A fuzzy inference system (FIS) can be considered as a rule-based expert system employing linguistic rules and, in control, can facilitate a mathematical formulation of the uncertainty and imprecision associated with certain processes. This enables non-linear controllers to be devised which would be difficult to design using conventional methods. Fuzzy logic control has been successfully applied in a number of applications where conventional model-based approaches are impractical, particularly in processes which are complex, non-linear or imprecisely defined (Hirota & Sugeno, 1995).

The main objective here was to develop a fuzzy system that could improve the performance of the conventional force controller, with two important criteria. Firstly, it was imperative that the system remained inherently safe, with no increased risk of transmitting unpredictable force demands to the robot. Secondly, the design had to be as simple as possible, both to avoid the problems of dimensionality, and to facilitate implementation in software on the experimental system. Rather than replace the conventional PD controller with an equivalent non-linear fuzzy force controller, the existing controller was retained and a FIS used to modify the controller gains depending upon the response of the robot. The FIS was designed to detect either oscillations or sluggish behaviour, and increase or reduce the control effort accordingly.

3 Experimental Facility

A Puma 762 industrial robot with a modified controller fitted with a wrist mounted force-torque sensor and grinding attachment was used for evaluating the proposed force-control scheme. Bi-directional communication allows path modification data to be transmitted in real-time to an external computer via a dedicated Ethernet link. The host computer handles the communications with the F/T sensor controller. Forward and inverse kinematic solutions are established within structure of the controller. The C matrices determine which axes are to be force or position controlled in a combined force/position scheme. The system architecture is shown in Figure 1, and can be described as a position controller with an add-on outer force control loop.

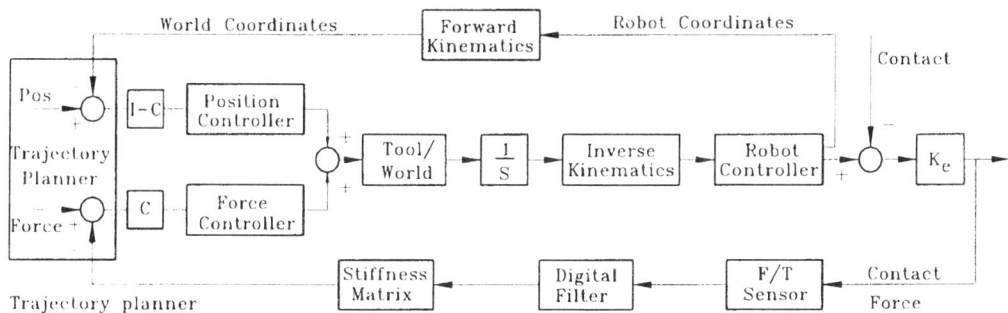

Figure 1. Robot Controller Architecture

To evaluate the dynamic performance of the system during contact tasks, a cantilevered flexible steel beam was mounted in a rigid vice, nominally along the robot's x-axis. The robot was programmed to apply a contact force of 10N on the beam at a position located some distance from the built-in end. The robot was then instructed to move towards the built-in end of the beam at a speed of 10mm/s, whilst maintaining this force, as shown in Figure 2.

The results of the test with a conventional proportional plus derivative (PD) controller employed are shown in Figure 3 (proportional gain $K_p=0.006$, $K_d=0.002$). From the resulting force-time history the instability problem that occurs when the effective K_e increases as the robot approaches the vice is clearly apparent, although the output has been rate-limited.

Figure 2. Experimental Set-up

4 Design of the FIS Force Controller

Preliminary tests with the cantilever revealed that K_p varied between 0.002 and 0.006 when the system was tuned for either 'stiff' or 'flexible' contact with the beam. It was therefore decided to devise a FIS to vary this gain between these limits, using the absolute values of force error (F_e) and force error difference (ΔF_e) as input data. To achieve this, K_p was fixed at 0.006, corresponding to the gain tuned for flexible contact. The numerical (crisp) output from the FIS is then a real number in the range 0.33 to 1.00, referred to as *factor*, and used as a multiplying factor of K_p. An advantage of this approach is that the output from the FIS is always zero or positive, and only a positive input space is required so that additional MFs are not necessary for processing either positive or negative F_e and ΔF_e.

To summarise the design of the FIS in more detail, two input variables (F_e and ΔF_e), and one output variable (*factor*) were assigned. The ranges for each input were determined by analysing experimental data obtained from tests with the conventional controller. This was necessary since a FIS relies on valid input data in order to assign degrees of membership to relevant input MFs during the fuzzification process. Three Gaussian MFs were assigned to each input, labelled *small* (S), *medium* (M) and *big* (B) in each case. Initially, the three MFs for each input were placed symmetrically in their respective input spaces. Three MFs were also assigned to the output, labelled S, M and B. A zero-order Sugeno system was developed (where each output 'set' is a constant), initially with *small*=0.33, *medium*=0.67 and *big*=1.00.

A simple rule base of nine rules was then devised aimed at detecting oscillatory behaviour leading to a reduction in *factor*, or sluggish behaviour leading to an increase in *factor*. For safety reasons, it was decided to initialise *factor* to a low value at the start of each experiment.

The FIS was implemented in real-time using software written in C++. However, prior to this it was tested using experimental data recorded earlier (such as that shown in Figure 3). This data was fed into a Matlab model of the FIS and the output examined. It was apparent from this that the FIS was adversely affected by noisy data, and the output oscillated at high frequency. Also, it was felt that the even distribution of the input MFs was too simplistic, since the relatively large values of F_e and ΔF_e that occur during oscillation were not reflected by the FIS, and it was too sensitive to small fluctuations of F_e. Thus two further modifications were made to the FIS. Firstly, rate limiting was added to the output to prevent large swings in *factor*, effectively acting as an low-pass filter. Secondly, the relative widths of the input MFs were modified. Figure 3 also illustrates the effect upon *factor* of the data obtained during a conventional test, illustrating the ability of the fuzzy system to detect the onset of instability.

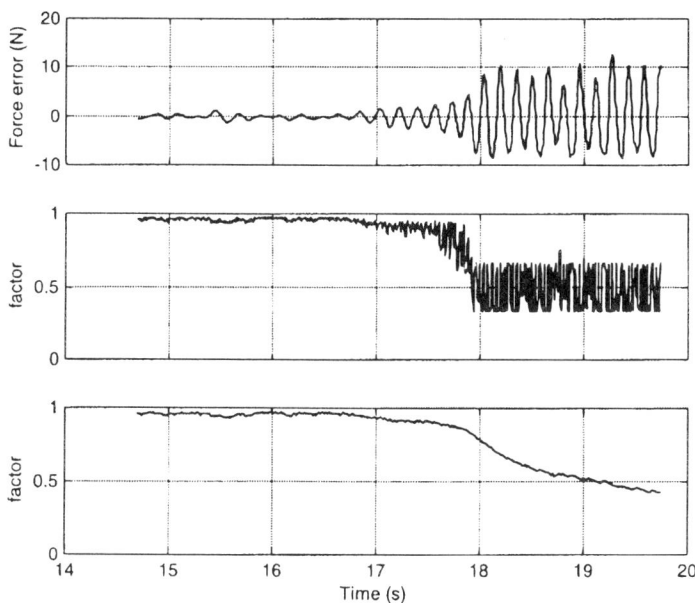

Figure 3. Variation of Force Error using conventional force controller

5 Experimental Test Results

A comprehensive series of rolling contact test using the beam were undertaken. Typical results are shown in Figure 4, where the robot was required to move from the stiff-to-flexible end of the beam whilst maintaining a contact force of 10N. The control effectiveness is clearly illustrated, with good performance being evident throughout the test. The gain factor increases automatically as the robot approaches the flexible end of the beam, however some fluctuation occurs due to the oscillation of the force error with structural resonances. Figure 5 illustrates the results of an actual deburring test along a steel plate. Once again good regulation of force was demonstrated even though the plate stiffness was somewhat higher than in previous tests, although there was an initial overshoot in the contact force due to the dynamics of the tool when switching on.

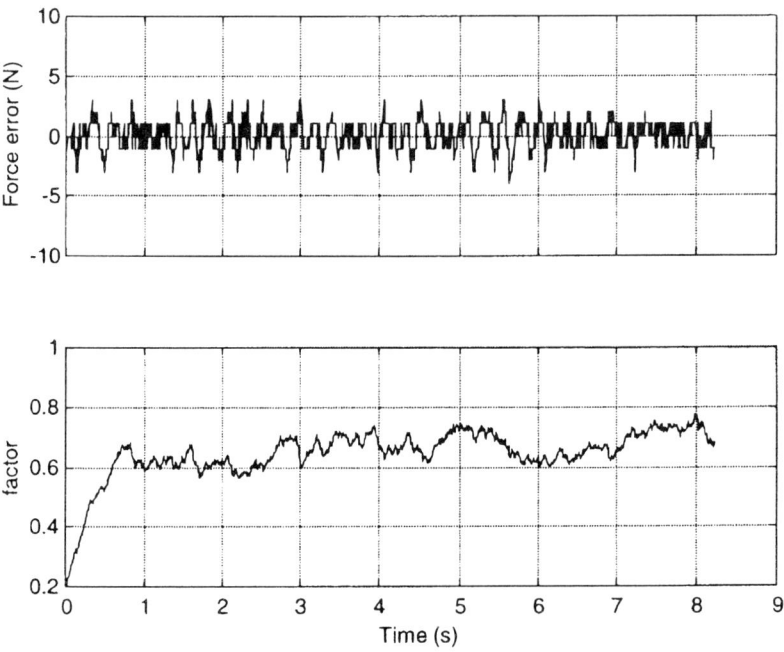

Figure 4. Test results – Fuzzy Controller (Rolling)

6 Conclusions and Further Work

An experimental evaluation of a non-linear fuzzy logic controller in a robotic force control loops has been presented, which combines the advantages of conventional PD control with the intuitive properties associated with fuzzy logic. The strategy is a form of gain scheduling, where the system reacts to changes in the force error, and rate of change of force error. The controller maintains a given level of performance when considerable environmental uncertainty exists, without the necessity of a contact stiffness detection routine. The next stage in the validation of the method described here will be to implement a multi-axis version of the controller in an industrial deburring robot cell.

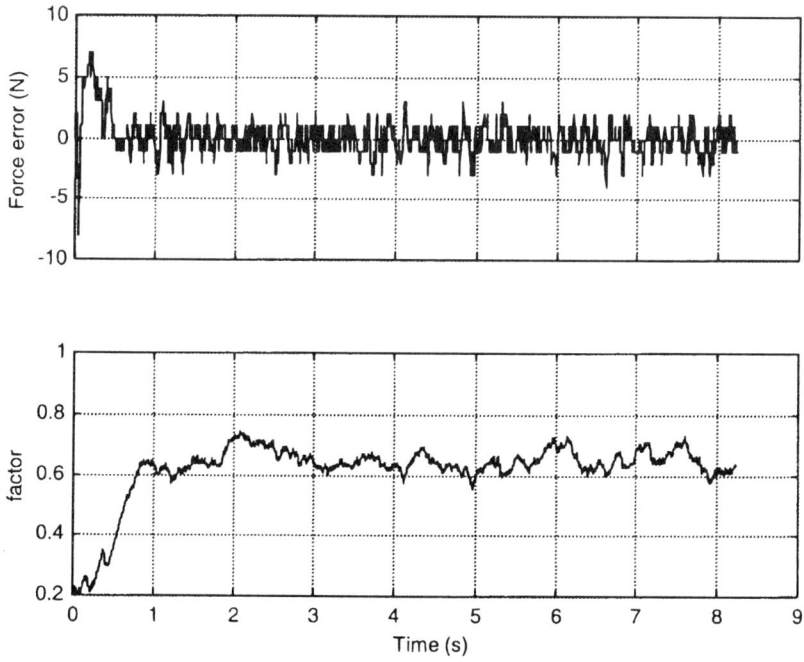

Figure 5. Test Results – Fuzzy Controller (Deburring)

References

Burn K., Bicker, R.(1999), Development of a non-linear force controller using fuzzy logic techniques, *IMechE Journal of Systems and Control Engineering*, Accepted for publication.
Ganwen Zang and Ahmad Hemami (1997) An Overview of Robot Force Control. *Robotica*, Vol 15, 473-482.
Hirota K., Sugeno M. (1995) *Advances in Fuzzy Systems, Applications and Theory Volume 2: Industrial Applications of Fuzzy Technology in the World*, World Scientific Publishing Company.
Kim W.S., Hannaford B. and Bejczy A.K. (1992) Force Reflection and Shared Compliant Control in Operating Telemanipulators with Time Delay. *IEEE Transactions on Robotics and Automation*, Vol 8, No 2, 176-185.
Ow S.M. (1997) *Force Control in Telerobotics*, Ph.D. Dissertation, University of Newcastle upon Tyne, UK.
Tarokh, M., Bailer S. (1997) Adaptive fuzzy force control of manipulators with unknown environment parameters, *Journal of Robotic Systems*, Vol 14, No 5, 341-353.

A Distributed SMA Actuator System and Associated Self-Guiding Control Strategy for a Scalable Endoscope Steering Device

Philippe Bidaud, Jérôme Szewczyk, Nelly Troisfontaine
and Jean-Claude Guinot

Laboratoire de Robotique de Paris, Université de Paris 6, France.

Abstract. This paper describes an original active steering device for endoscopes and boroscopes. Its mechanical structure is based on a tubular hyper-redundant mechanism. Distributed SMA actuators with their own local controller are integrated in this structure for producing bending forces in reaction to the interaction detected between the instrument and its environment. The SMA actuators are two thin NiTi springs in an antagonist configuration. Joint actuation relies on martensite/austenite phase transformation in NiTi alloys. The global behavior of the endoscope is controlled through a multi-agent approach.

1 Introduction

Current instruments for endoscopy suffer from limitations mainly caused by the lack of mobility and ability to perform maneuvers into very small and geometrically complex 3D spaces.

An endoscope is a long thin tubular device for non-invasive inspection in interior cavities, canals, vessels, etc... inserted through a natural or surgically produced orifice. A typical outer diameter of endoscopes is 10 mm and their length varies from 70 to 180 mm. The endoscope body contains several light guides (typically 2), tool channels (biopsy grippers, snare, cytology brush) and optics or electronics for the image transmission. Endoscopes can be rigid or flexible.

A steerable tip can be mounted on most of these instruments. The change of the tip orientation facilitates the endoscope progression in cavities and also modifies the viewing direction. This passively bendable part is generally deflected by one or two pairs of cables (depending upon the number of planes of bending) connected to a remote control mechanism located close to the headset.

These devices, while highly flexible, have limited steering ability. They cannot traverse tight bends nor negotiate complexe interior structures. Moreover, one of the major risk with these instruments is the perforation of the patient's tissues due to their substantial stiffness. There are also problems which results from excessive stresses applied on the operating cables, they frequently break or acquire a permanent strain.

In addition, the trend is to move towards smaller and smaller diameter endoscopes required by applications such as neuro-surgery, cardio-vascular-surgery or obstetrical procedures and this can not be tackled by a scale reduction of the current technologies.

This paper describes the design of a mechanical system and a distributed actuation system with a self-guiding control strategy for a scalable steering device able to dexterously maneuver through small and geometrically complex 3D structures.

Very few devices of this kind can be found in the scientific literature (Müglitz and Schönherr, 1999), (Yusa, 1998), (Dario et al., 1997) but some are described in patents (US Pat $N°5679216$) , (US Pat $N°5482029$) , (US Pat $N°5405337$). They relate generally on shape memory alloy (SMA) distributed films integrated control drivers deposited on a flexible substrate and integrated using VLSI techniques.

2 Design Principles and System Description

The device has been designed to give the user more dexterity in endoscopic procedures than with current instruments.

Here, technology is a key issue. It is clear that cables is no longer a solution for getting tight bends in 3D space. The outer diameter has to be as small as possible considering that the room needed for optical fibers bundles and chanels for surgical tools and fluids, which defines the inner diameter, can not be reduced. Another important feature is that the bending force required for a given stiffness of the inner components increases when reducing the endoscope diameter. Moreover, the technology selection must withstand the sterilization process ($140°C$ during 20 minutes). It has to make the system as simple as possible and to facilitate its manufacturing at small scale.

The controllably bendable portion of the instrument must be able to adapt its local curvature to the interior geometry by a spontaneous reaction to the interactions with the environment while the viewing tip follows a track in a vessel or in a cavity.

Figure 1 schematically illustrated the endoscopic system we designed. The mechanical structure of the device can be viewed as an hyper-redundant manipulator which embrace the endoscope components (optic bundle, light guides, tool chanels). It is a serial arrangement of tubular segments articulated to each other by pin joints. This design is modular, the number of segments can be adjusted to the application and is in theory infinite. On the actual design, the segment length is 4 mm, the inner diameter 5.4 mm and the outer diameter (including the outer elastomer cover) is 8 mm.

This system is usually protected by a metallic sheath in industrial endoscopes

A Distributed SMA Actuator System and Associated Self-guiding Control Strategy

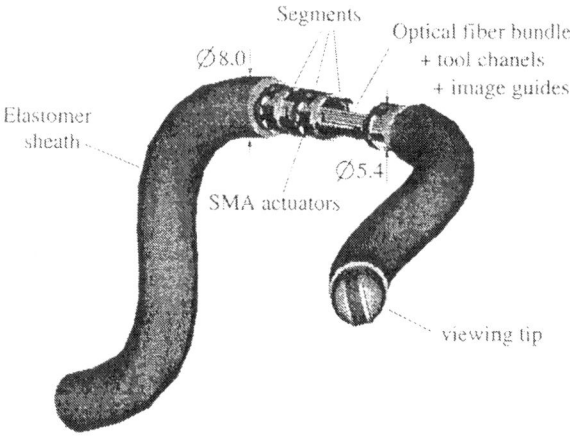

Figure 1: View of the steering system

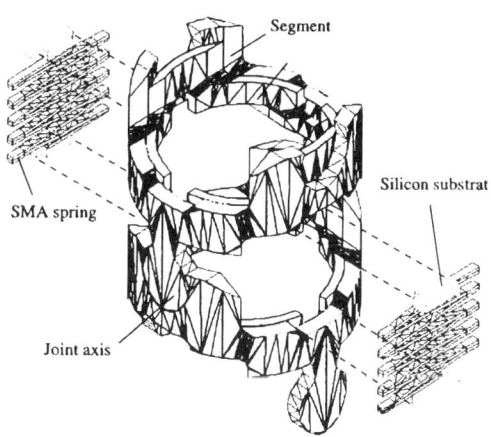

Figure 2: Joint description

and/or by a flexible polymer sheath in medical endoscopes. Notice that this sheath significantly increases the strength requirements for the bending actuators.

3 A Tubular Hyper-Redundant Manipulator

The hyper-redundant manipulator is composed of identical modules (iron rings) linked together by pin joints whose axes are alternatively oriented at $90°$ in the same plane. This mechanical design allows to bend the endoscope body along complex curves in the 3-D space.

The modules are obtained by an electro-erosive processing technique. By using this substractive manufacturing method, joints and links are made in one piece. The pin and the hinge are respectively the positive and the negative cutting in the cylindrical shell of the segment. The relative translation of two consecutive modules along the joint axis is suppressed by inserting a very thin internal ring has shown on figure 2. The steering mechanism is assembled by simply plugging these segments whose length can be reduce to 4 mm such that a 15 mm curvature radius can be achieved.

Two spring-like actuators are integrated with their own control circuit in each module to change the relative orientation of two consecutive segments. These actuator are Shape Memory Alloy (SMA) springs mounted in an antagonist configuration.

The actuator is controlled by the electrical power supplied to the SMA. A control circuit is associated to each actuator. It is integrated on a substrate of alumina by using hybrid electronic technologies.

4 SMA Actuator Design

The SMA actuator elements are springs cut out in a NiTi (Nickel-Titanium) ribbon (Figure 3-a).Basically, these SMA actuators undergo a micro-structural transformation from their austenite phase to their martensite phase (Bidaud et al., 1999). This phase transformation can be activated by heating and cooling the material or by applying an external stress.

For a 50%-50% NiTi alloy, as the one used here, the transition temperatures are $A_s = 40°$ and $M_s = 70°$ for a null applied constraint to the material. This phase transformation also induces a large modification in the material Young's modulus (Figure). E varies from $1GPa$ to $70GPa$ for temperatures between $20°C$ to $120°C$. This property is exploited here to create unballanced pulling forces applied by the two antagonist SMA springs.

Resistance to deflexion, due to the endoscope body and the outer elastomer

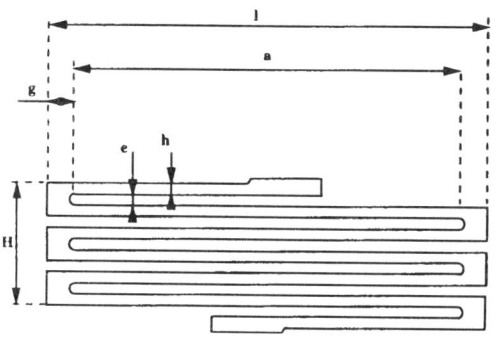
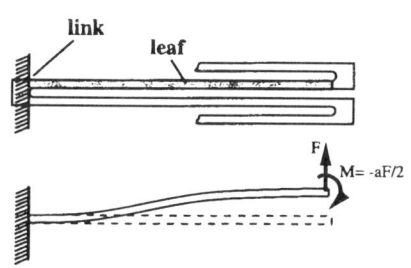

a- Geometry of a SMA spring

b- Deflexion model of one leaf

Figure 3: SMA spring description

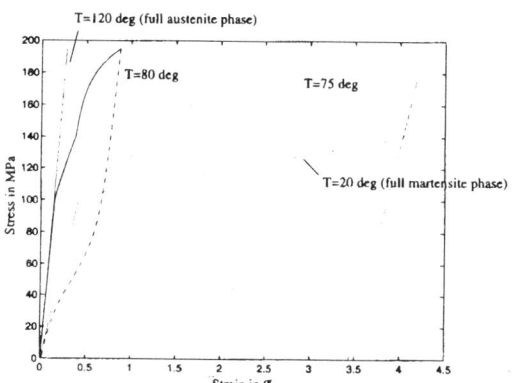
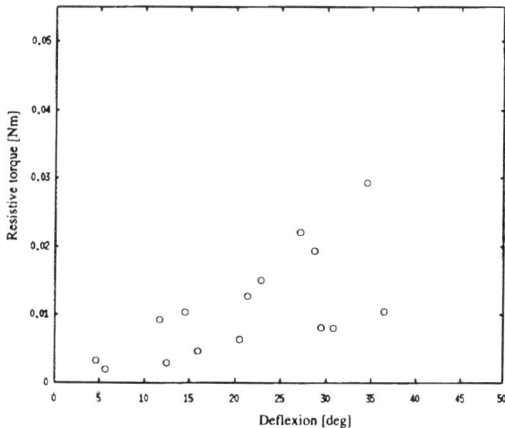

Figure 4: Strain-stress diagram of the NiTi (50%-50%) alloy

Figure 5: Resistive torque [Nm] vs deflexion [deg]

sheath, was experimentally evaluated. On Figure , we represented the necessary output joint torque to bend a rotoïd axis in an existing endoscope. For deflexions smaller than $15°$, the resistive torque is about $0.01 Nm$.

The SMA spring on Figure 3-a can be approximated by an assembly of flexible parts (leaves of length a) linked by rigid parts.

Thus, a leaf can be modelized as a flexible beam rigidly fixed at one extremity and submitted to a combination of a force F and a moment $M(F)$ with $M(F) = -a F/2$ such that a null inclination remains at the other extremity (Figure 3-b).

The desired maximal flexion is $15°$. In this configuration, for $g = 0.135mm$, $e = 0.25mm$ and a number of leaves set to 6, we represent on Figure 6 the normalized output joint torque (the reference value is $0.01 Nm$) and the normalized maximal stress in the material (the reference value is $135 MPa$) ith respect to l and H which are the external dimensions of the springs.

The best trade-off corresponds to $l = 2.35mm$ and $H = 2.0mm$ (i.e. $h = 0.125mm$). In this case, the output joint torque is greater than $0.008 Nm$ and maximal stress is less than $145 MPa$.

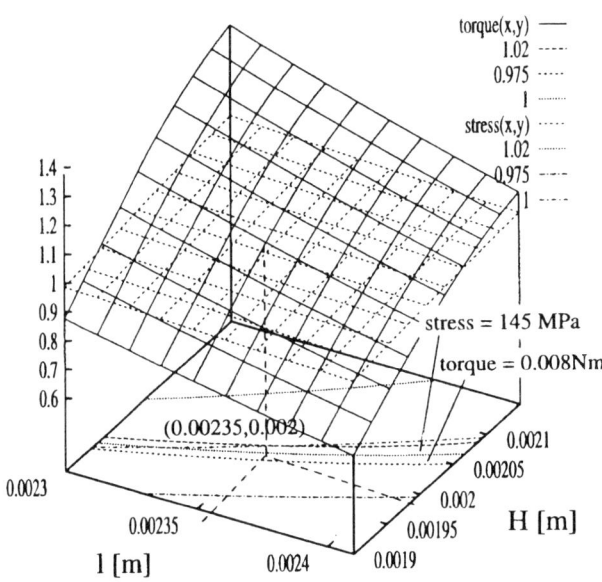

Figure 6: Normalized joint output torque and maximal internal stress for different SMA spring geometries

5 Joint Control

The endoscope configuration is self-guided in such a way that the local interactions between the instrument and its environment are minimized.

Changes in configuration are controlled at joint level by switching between position and a temperature control loops. The resulting controller for the antagonist actuators is represented on Figure 7. Only one spring is actuated at once for producing a displacement in the desired direction. When the error is large, position feedback is used (see below case 1 or 2). A switching on the temperature loop occurs when the static error is less than ϵ (see below case 3 or 4). The temperature input is the one memorized just before switching. If $\tilde{\theta}$ is the desired joint position and $\Delta\theta = \tilde{\theta} - \theta$ the joint position error, then the swiching rules are the following : Case 1 : $\Delta\theta \geq 0$ & $|\Delta\theta| \geq \epsilon$, Case 2 : $\Delta\theta \leq 0$ & $|\Delta\theta| \geq \epsilon$, Case 3 : $\tilde{\theta} \geq 0$ & $|\Delta\theta| \leq \epsilon$, Case 4 : $\tilde{\theta} \leq 0$ & $|\Delta\theta| \leq \epsilon$.

Figure 8 shows experimental results obtained in a position step response of a SMA actuator using this kind of switching controller.

6 Configuration Behavior

Controlling the endoscope configuration aims at positionning and orienting correctly the tip of the structure while limiting forces coming from interactions with the environment.

An algebraic resolution for this problem is extremely complex and the explored environment is a-priori unknown so a reactive resolution method is preferable.

The solution we propose relies on the virtual split of the steering mechanism into independent sub-systems and by considering them as agents (Duhaut, 1993). Each agent is able to detect a contact with the environment and to accordantly modify the endoscope local configuration.

This is a very simple and modular solution indepedent from the length of the structure. Moreover, it is a strictly distributed approach minimizing the quantity of informations exchanged between the agents.

Three different kinds of local behaviors are described on Figure 9. They correspond to sub-systems composed of 1, 2 or 3 segments. The first behavior is very simple but it significantly disturbs the global configuration of the endoscope. The last one requiers information exchanges over three consecutive segments but preserve the global configuration.

The global behavior of a planar structure (20 segments) progressing into a pipe and guided by the local reaction of the agents only, has been tested in simulation. As shown on figure 10-b, ultra-local approch (agent $\equiv$ 1 segment), leads

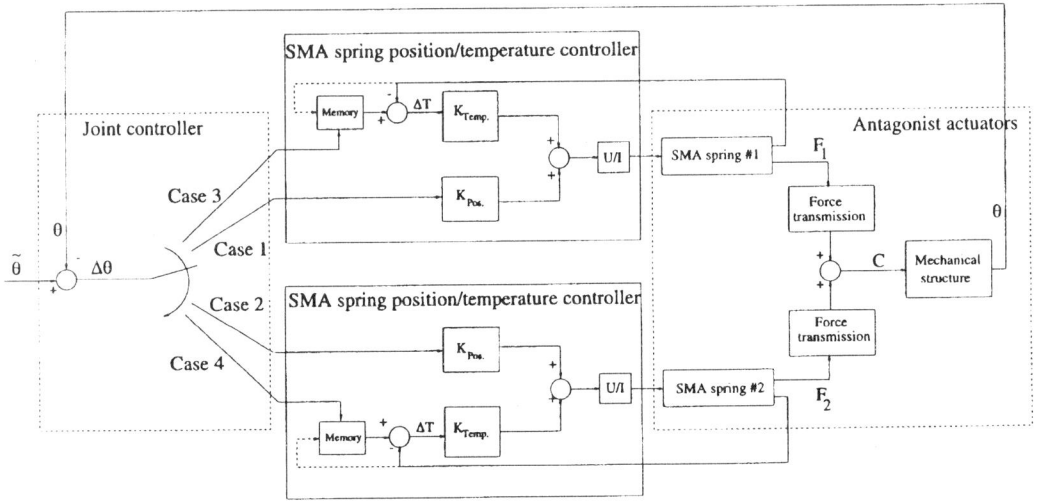

Figure 7: Joint position control

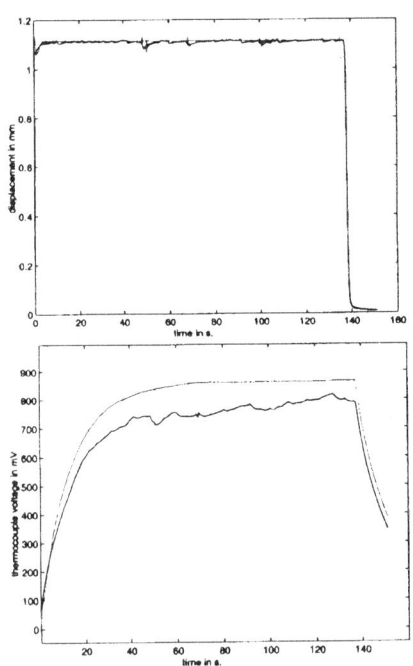

Figure 8: Step response with the position feedback controller (filled line) and the controllers combinaison (dotted line) (on the left-hand side) and the thermocouple output signal (on the right-hand side)

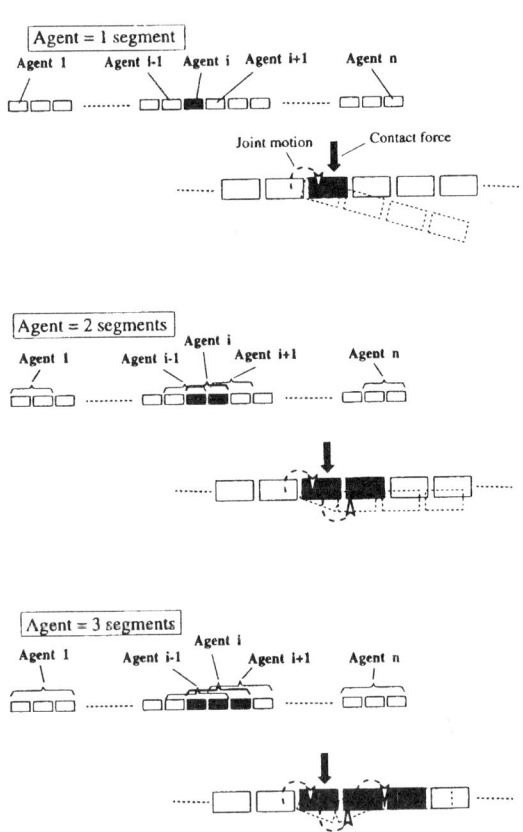

Figure 9: Three different kind of local behavior

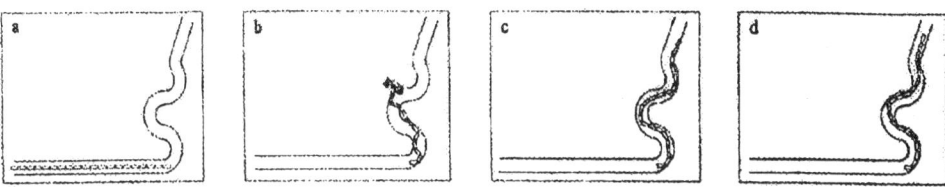

Figure 10: Multi-agent behavior simulation

third solutions (agent ≡ 2 or 3 segments) keep the endoscope stable while minimizing interactions (figures 10-c and 10-d).

7 Conclusion

This paper proposed a new design concept for actively guided steerable endoscope. At this point, the mechanical structure and the associated actuators have been manufactured. The local controller has to be experimentaly improved and the whole integration has to be done for testing the proposed behavior control strategies.

References

[1] P. Dario, C. Paggetti, N. Troisfontaine, E. Papa, T. Guicci, T. Carrozzon, and M. Marcacci. A miniature steerable end-effector for application in an integrated system for computer-assisted arthroscopy. In *International Conference on Robotics and Automation*, pages 1573-1579. IEEE, 1997.

[2] D. Duhaut. Using a multi-agent approach to solve the inverse kinematics. In *Intelligent Robot and System Conference*, IROS, pages 2002-2007, 1993.

[3] J. Müglitz and J. Schönherr. Miniaturized mechanisms - joint design, modeling, example. In *Tenth World Congress on Theory of Machines and Mechanisms*, pages 848-855, 1999.

[4] N. Troisfontaine Ph. Bidaud and M. Larnicol. Optimal design of micro-actuators based on sma wires. *Smart Materials and Structures*, 8: 197-203, 1999.

[5] A. Yusa. Tubular manipulator with multi-degrees of freedom. *Micro Machine Center Journal*, 19, 1998.

Chapter VII

BIOMECHANICAL ASPECTS

Low-Energy Biped Locomotion

Stefan Gruber and Werner Schiehlen

Institute B of Mechanics, University of Stuttgart
Stuttgart, Germany

Abstract. The simulation and realization of legged locomotion robots is an important research topic. A common approach to synthesize a desired walking pattern for a walking machine is to use inverse dynamics techniques. Thus, the nominal control of a walking machine is generated according to fully or partly prescribed and preprogrammed nominal trajectories of legs and body. The equations of motion are solved to obtain the required control forces and torques. These torques are relatively large so that autonomous walking, i.e. walking with energy supply on board, is only possible for a short period. Our aim is to reduce the control torques in the joints of a biped walking model and the total energy consumption of the actuators.

The paper presents the application of the passive walking principle to a biped model. A walking model with knees capable of passive dynamic walking is designed to which small actuators in the joints are added. Active control is used to maintain the passive walking motion compensating the energy losses. The simulation results show that the power consumption during walking is low compared to other machines.

1 Introduction

The motivation of this work was based on the fact that there exist major differences in the energy input which is required to generate the walking motion of existing walking machines. On the one hand there is the well known Humanoid Robot by Honda (2000) which has an average battery operation time of about 25 minutes only. On the other hand, there exist some passive walking machines which were first built by McGeer (1990) and later by Garcia et al. (1998). These passive machines can walk endlessly if their operation conditions are satisfied. They need only a shallow slope to walk on and some appropriate initial conditions to start the walking cycle. In the later case the walking motion can be generated by the passive dynamics of the legs under the influence of gravity only, whereas, in the first case it is generated by actuators which continuously consume a great amount of electric energy by following specified trajectories. Due to this huge difference in energy consumption it is obvious that walking can be made more energy efficient if it is possible to exploit the passive walking principle for active walking.

2 Passive Dynamics

The unforced motion of a mechanical system is commonly called its passive dynamics. Raibert (1986) exploited the passive dynamic principle to generate the vertical motion for a series of running robots with telescopic legs. Further, Ahmadi and Buehler (1999) exploited the passive dynamics of a one-legged machine for the vertical hopping as well as for the leg swing motion.

Once started, even a pair of two-dimensional legs is capable of walking down a shallow slope in a stable gait without any active control or energy supply. This natural walking motion generated by the passive interaction of gravity and inertia is called passive dynamic walking. The phenomenon works for bipeds having straight legs or knees, respectively. The principle of passive dynamic walking for biped machines was pioneered by the work of McGeer (1990) who himself was stimulated by the ballistic walking model described by Mochon and McMahon (1980). Further contributions to passive biped walking were by Garcia et al. (1998) and Goswami et al. (1996).

In this work a model with knees which is capable of passive walking is combined with some energy input and control to exploit the principle for walking on level ground.

3 Biped Model

A biped model with knees was developed in two dimensions consisting of five rigid bodies, namely two lower legs, two upper legs, and one upper body as shown in Figure 1. The foot which is part of the lower leg has circular shape allowing the foot to roll on the ground. The model's position is determined by the two Cartesian coordinates r_1 and r_2 and the absolute joint angles ϕ_i with $i = 1, 2, \ldots 5$ where the numbers correspond to the five rigid bodies. The system has 7 degrees of freedom. A total of 4 control torques is applied to the right and left knee between lower and upper leg and to the hip between right and left upper leg and upper body denoted by u_m with $m = 12, 25, 35$ or 43. The total mass of the system is $20.0\,\mathrm{kg}$ where $13.0\,\mathrm{kg}$ are concentrated in the upper body remaining $2.5\,\mathrm{kg}$ for one upper leg and $1.0\,\mathrm{kg}$ for one lower leg.

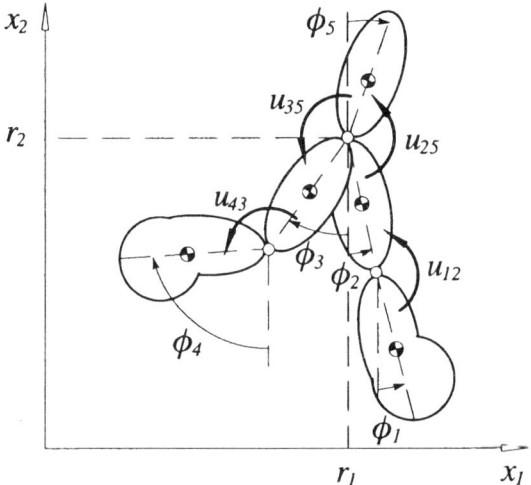

Figure 1. Planar five-link biped model.

3.1 Minimal Formulation

With the 7×1-vector of generalized coordinates

$$y = [r_1, r_2, \phi_1, \phi_2, \phi_3, \phi_4, \phi_5]^T \tag{1}$$

the equations of motion of the multibody system with tree topology can be written as ordinary differential equations in minimal form as

$$M(y)\ddot{y} + k(y,\dot{y}) = q(y,\dot{y}) + Hu, \tag{2}$$

where M is the 7×7 symmetric positive definite mass matrix, k the 7×1-vector of generalized Coriolis forces and q the 7×1-vector of generalized applied forces (e. g. Schiehlen, 1997). The matrix H is the 7×4 control input matrix and the 4×1-vector

$$u = [u_{12}, u_{25}, u_{35}, u_{43}]^T \tag{3}$$

contains the control torques as shown in Figure 1. The equations of motion (2) for the model were derived symbolically using the multibody formalism NEWEUL (Schiehlen, 1990).

3.2 Varying Constraints

The periodic walking cycle consists of four phases denoted by $j = A, B, C$ and D. In order to model the four different phases also four cyclic sets of constraints are imposed on the system as shown in Table 1.

Table 1. Active constraints during the four phases of the walking cycle.

phase j	A	B	C	D
left foot/ground contact x_1	✓	✓		
left foot/ground contact x_2	✓	✓		
right foot/ground contact x_1			✓	✓
right foot/ground contact x_2			✓	✓
left knee locked	✓	✓		✓
right knee locked		✓	✓	✓
no. of active constraints q	3	4	3	4

During all phases the contact between stance foot and ground is modelled constraining the motion in the x_1- and x_2-directions. The ground surface is assumed to have a sufficient coefficient of friction so that no slipping may occur. The knees can be considered to have a mechanical stop preventing them from hyperextension. Therefore, the stance knee can be specified to remain locked straight throughout the step by a torque against that stop.

In phases A and B the knee joint of the left stance leg is locked in straight position. The knee joint of the right swing leg remains free until its lower and upper leg angles coincide indicating the end of phase A. In that position the knee locks and is straight by switching to constraints B In that phase both knees are locked and the model continues rolling about the left stance leg until the step is completed with a collision between right swing leg and ground at the end of phase B. At that instant the swing and stance legs switch instantaneously and the constraint of the knee of the new swing leg is released. As the previous stance leg takes off instantaneously at the moment of the collision between the old swing leg and the ground no double support phase exists. Then, in phase C and D corresponding constraints are active for the legs.

During the phases of the walking cycle different sets of constraints become active when constraining events occur. On the other hand these constraints itself can also be considered as indicator for the occurrence of a constraining event. These four sets of scleronomous holonomic constraints can be implicitly written as

$$g_j(y) = 0 \quad \text{with} \quad j = A, B, C, D, \tag{4}$$

where g_j is a $q \times 1$-vector of at least twice differentiable functions and $q = 3$ for $j = A, C$ or $q = 4$ for $j = B, D$. For the constrained system the number of degrees of freedom remains $f = 7 - q$. Thus, the system at hand has a time-varying number of degrees of freedom and therefore a time-varying topology during walking.

The equations of motion (2) have to be supplemented by the constraint forces resulting in

$$M(y)\ddot{y} + k(y, \dot{y}) = q(y, \dot{y}) + H u - G_j^T(y) \lambda, \tag{5}$$

where $G_j = \partial g_j/\partial y$ is the $q \times 7$ constraint matrix during phase j, and λ the $q \times 1$-vector of Lagrangian multipliers which may be interpreted as generalized constraint forces. The equations of motion of the constrained system consist of a set of differential-algebraic equations (DAEs) formed by equations (4) and (5). The described formulation has the advantage that the simulation can be performed with only one set of differential equations by switching between the four sets of algebraic constraint equations.

3.3 Collision treatment

All of the collisions are treated as inelastic impacts. Thus, the swing and stance legs are switched within an infinitesimal short time where a velocity jump occurs. At that instant a part of the kinetic energy of the system is lost. The new generalized velocities immediately after the collision, $\dot{y}^+$, can be calculated from the previous velocities, $\dot{y}^-$, immediately before and from the generalized coordinates, $y^- = y^+ = y$, in that instant. The angular momentum about the hip and about the point of collision just before and after the impact can be written as $h_0^- = M^-(y)\dot{y}^-$ and $h_0^+ = M^+(y)\dot{y}^+$, respectively, where M^+ and M^- are 7×7 inertia matrices (McGeer, 1990). Conservation of angular momentum at the time of support transfer yields

$$M^+(y)\dot{y}^+ = M^-(y)\dot{y}^-. \tag{6}$$

Equation (6) can be solved for the unknown generalized velocities $\dot{y}^+$.

Once the walking cycle is started energy is required to maintain a periodic motion. The energy will be provided by controls decoupled in several tasks.

The upper body pitch is controlled to remain at a constant desired absolute angle ϕ_{5des} by applying the torque u_{25} or u_{35} against the current stance leg. That torque in the stance hip actuator consists of two parts $u_{stat} + u_{dyn}$ where $u_{stat} = m_5 \, g \, c_5 \, \cos(\phi_5)$ is the torque needed in the static case to hold the upper body at an angle ϕ_5. Further, m_5 is the mass of the upper body and c_5 the distance of the upper body's center of mass from the hip joint. A PD-controller is used to compute the dynamic part as

$$u_{dyn} = k_{p5} \left(\phi_{5des} - \phi_5 \right) + k_{d5} \left(\dot{\phi}_{5des} - \dot{\phi}_5 \right). \tag{7}$$

The desired upper body pitch angle has a value of $\phi_{5des} = -11.0° = -0.2 \, \text{rad}$ which means that it is slightly leaning forward. The torque in the stance hip joint not only keeps the upper body upright but also moves the stance leg backwards. Thereby the torque has the effect of propelling the whole model forward.

The forward motion of the swing leg's upper part is feedforward controlled by a torque in the hip joint. This torque is increasing as a linear function of time until the knee lock occurs. The knee of the swing leg remains passive until the knee locking event occurs. However, if the foot-ground distance is approaching zero a torque is applied in the knee-joint to move the lower leg backwards increasing that distance and ensuring foot clearance. After the knee locking event the straight swing leg's hip angle is PD-controlled to achieve a desired steplength as

$$u = k_p \left(\phi_{des} - \phi \right) + k_d \left(\dot{\phi}_{des} - \dot{\phi} \right), \tag{8}$$

where ϕ_{des} is the desired relative angle calculated from the desired steplength and ϕ is the relative joint angle for the right or left hip, respectively.

4 Simulation

Due to the collisions the model features state dependent discontinuities causing jumps in the generalized velocities. These discontinuities resulting in implicit switching functions need to be localized and handled by the integration code. The numerical integration of the DAEs is performed using the simulation code MBSSIM (Schwerin, 1999) which can deal with such discrete events.

The simulation results of biped walking on level ground are shown in Figure 2. Coinciding lines indicate that the corresponding knee is locked and upper and lower angle are equal.

The mechanical energy added to the system is equivalent to the sum of the work done by the four actuators. The work of the actuators can be calculated as

$$W(t) = \int_{t_0}^{t_1} | u_m \, \omega | \, dt, \tag{9}$$

where u_m is the applied torque, ω the relative angular velocity of the joint and t the time. It is pointed out that $P = | u_m \, \omega |$ is the mechanical power output of the actuator but does not include actuator and transmission efficiency. The active energy input to the system is shown in Figure 3 according to steps and actuators. The total energy cost for one step is about $W_{step} = 1.1 \, \text{J}$. The main part of the consumed energy results from the hip torque which is necessary to hold the upper body upright against the stance leg.

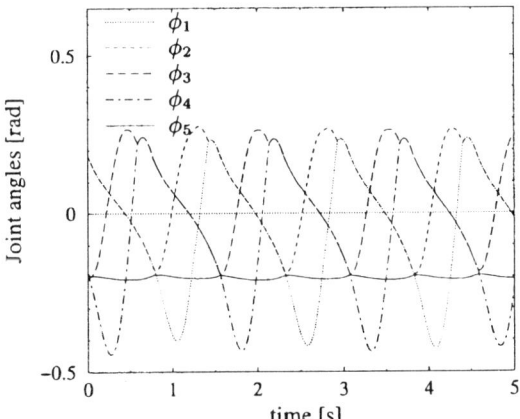

Figure 2. Joint angles during walking.

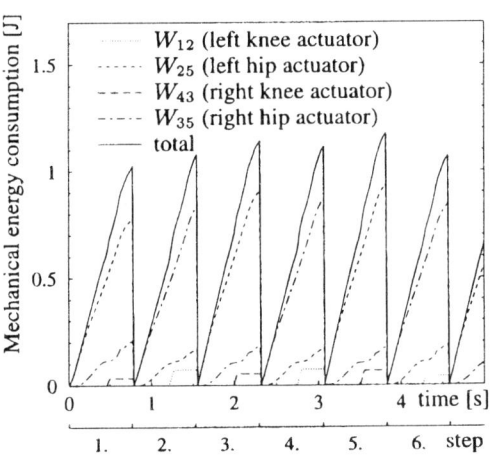

Figure 3. Energy consumption.

A measure to compare the energetic performance of locomotion machines is the specific resistance. The specific resistance at different speeds (Gregorio et al., 1997) can be calculated as the ratio of power, P, and the product of vehicle weight, $m\,g$, and speed, v, as

$$\varepsilon(v) = \frac{P(v)}{m\,g\,v}\,. \tag{10}$$

For the described model with a weight of $m\,g = 20\,\text{kg} \cdot 9.81\,\text{m/s}^2 = 196.2\,\text{N}$ walking at an average speed of $v = 0.31\,\text{m/s}$ while consuming $P_{step} = W_{step}/t_{step} = 1.1\,\text{J}/0.77\,\text{s} = 1.43\,\text{W}$ equation (10) yields $\varepsilon = 0.024$. In comparison with the specific resistance of other walking machines given by Ahmadi and Buehler (1999) as $0.22 < \varepsilon < 11$ this value is small showing the efficiency of the described method. The specific resistance of the same model when walking down a slope of 0.01 rad without any actuation is $\varepsilon = 0.01$. In that passive case the motion is generated by gravity only and the upper body hangs down vertically from the hip. Snapshots from an animation of active biped walking are shown in Figure 4.

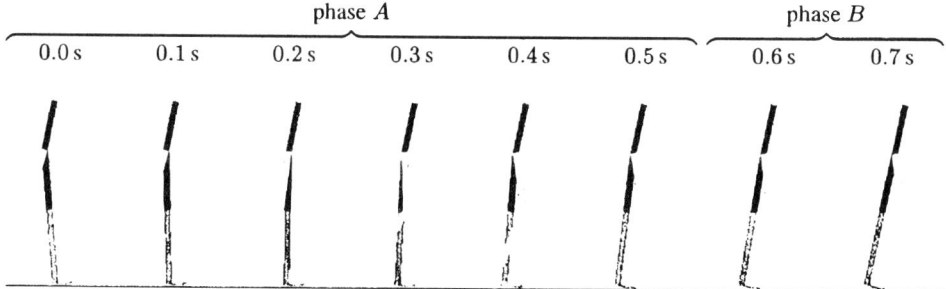

Figure 4. Snapshots from animation of biped walking.

5 Discussion and Conclusions

The principle of passive dynamic walking was exploited for an active five-link biped model with knees. The use of the passive dynamics together with active control showed to be very robust during the simulations. The described approach of extending the passive dynamic principle not only simplifies control but also enables energy savings, therefore making autonomous walking more feasible. The energy consumption of the model compared with other walking machines is very low. Its specific resistance is by a factor of 2.4 higher than in the passive walking case.

It has to be mentioned that in order to apply the passive walking principle in combination with active energy input in practice suitable actuators are required. Standard actuators like electric motors with gears result together with the moving links in large moments of inertia so that the passive swing of a leg might not be possible. This problem will be addressed in future work.

References

Honda Motor Co., Ltd. [web page] (Feb. 2000). Humanoid Robot - Specifications. URL: http://www.honda.co.jp/english/technology/robot/spec1.html.

McGeer, T. (1990). Passive dynamic walking. *The International Journal of Robotics Research* 9(2):62–82.

Garcia, M., Ruina, A. and Coleman, M. (1998). Some results in passive-dynamic walking. In *Proceedings of the Euromech 375 - Biology and Technology of Walking*. Munich, Germany, 268–275.

Raibert, M. (1986). *Legged Robots That Balance*. Cambridge, MA: The MIT Press.

Ahmadi, M. and Buehler, M. (1999). The ARL Monopod II Running Robot: Control and Energetics. In *IEEE International Conference on Robotics and Automation*, Detroit, Michigan.

Mochon, S. and McMahon, T. A. (1980). Ballistic walking: an improved model. *Mathematical Biosciences* 52:241–260.

Goswami, A., Espiau, B. and Keramane, A. (1996). Limit cycles and their stability in a passive bipedal gait. In *Proceedings of the IEEE Conference on Robotics and Automation*.

Schiehlen, W. (1997). Multibody system dynamics: roots and perspectives. *Multibody System Dynamics* 1:149–188.

Schiehlen, W. (1990). *Multibody Systems Handbook*. Springer-Verlag, Berlin.

Schwerin, R. v. (1999). *MultiBody System SIMulation: Numerical Methods, Algorithms, and Software*. Springer-Verlag, Berlin.

Gregorio, P., Ahmadi, M. and Buehler, M. (1997). Design, control, and energetics of an electrically actuated legged robot. *IEEE Transactions on Systems, Man, and Cybernetics* 27B(4):626–634.

Jumping Motion of an Object Controlled by Muscle Contraction

Janis Viba[1], Igors Tipans[1], Olga Kononova[1] and Jean-Guy Fontaine[2]

[1] Institute of Mechanics, Riga Technical University, Riga, Latvia
[2] Centre de Robotique Integree en Ile de France, Velizy, France

Abstract. In the design of robots, manipulators, vibrodrivers and start - stop systems it is very important to have a source of energy with internal interaction. One type of such energy sources may be found in a human or animal muscle contraction process. To synthesize new control systems for excitation or damping of a motion in start - stop systems the idea with a similar principle of action may be used. To determine internal forces, created during the contraction process, we used the Hill's model, numerical values of corresponding parameters were taken from known biomechanical experiments. Such level of modelling allowed us to link processes of electric control and mechanical production of a force during physical exercise. The implementation of optimal control law in real control action which imitates a muscle contraction is proposed in which the discrete control signal switches in the phase plane.

1 Introduction

For a quantitative determination of the generated muscle force contractile component of the Hill's model described by Woledge et al. in 1985 was described as a function of actomyosin cross-bridges, able to develop a force in a given position of a muscle element (Fig.1.). Numerical values for model parameters were taken from the Goubel's (1998) descriptions of corresponding biomechanical experiments. The active or passive phase of a cross-bridge is depending from the attachment of calcium ions to binding sites of the cross-bridges, corresponding mathematical description of that process was given by Tipans in 1998. The concentration of calcium ions in muscle cells is depending from electrophysiological characteristics of nerve impulses, reaching the muscle, and characteristics of internal calcium stores inside muscle cells.

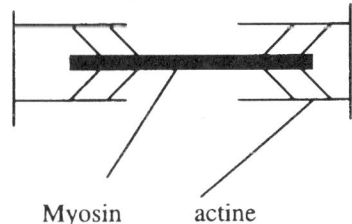

Figure 1. Muscle element.

By composing these elements into a muscle with joint points to a bone we get a scheme to calculate a force (Fig. 2. - Fig. 3.).

Figure 2. Fragment of muscle elements.

For a mathematical modelling of muscle contraction we can use the Hill's model (Fig. 4.). The model with two degrees of freedom (q1, q2) includes elastic (Ela1, Ela2), viscous (Vis1) and contractile (Con) properties of muscles.

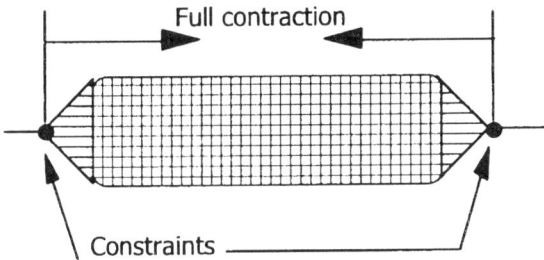

Figure 3. Muscle.

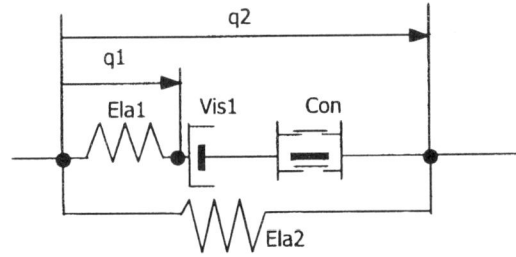

Figure 4. Hill's model.

If the model is used for synthesis of the analogous interaction for a control in mechanical systems, the additional viscous properties (Vis2) may be added (Fig. 5.). In this report, to simplify the calculation and to provide clear results of analysis, the model with one degree of freedom is used (Fig. 6).

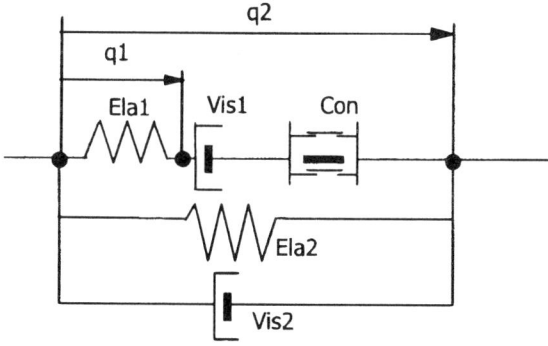

Figure 5. Model with additional damping.

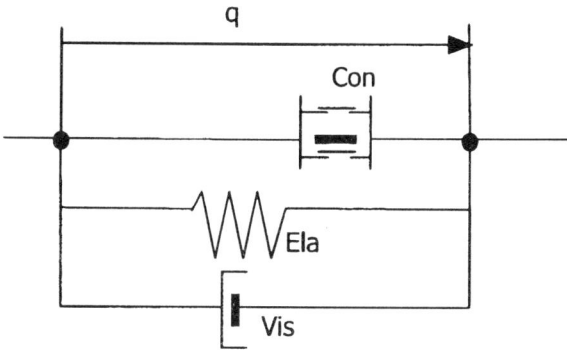

Figure 6. Simplified model.

2 Description of a jumping object

Symmetrical object consists of three rigid bodies 1, 2 and 3 (Fig. 7.). The central body 1 moves only in translation motion without overturn in the vertical direction. Two symmetrical legs 2 and 3 are connected by smooth pins to a central body 1. Legs move in a plane motion. Two indirect control actions (Con1, Con2) are imposed around each smooth pin, imitating a muscle contraction (Fig. 8.). Due to that the excitation of a system is internal and may be made by various moments around the pins. The system has external interaction with a foundation by viscous and elastic forces or rigid rough surface.

If the legs do not contact with a foundation, the system has two degrees of freedom. In that case equations of motion are:

$$\ddot{y}_C = -g; \quad (1)$$

$$A_0(\varphi)\ddot{\varphi} + B_0(\varphi)\dot{\varphi}^2 + C_{01}\sin\varphi = M. \quad (2)$$

where y_C - vertical co-ordinate of mass centre; g - free fall acceleration; $\varphi; \dot{\varphi}, \ddot{\varphi}$ - correspondingly angle, angular velocity and angular acceleration of legs; C_{01} - constant; M - control action, including elastic and viscous properties. Coefficients $A_0(\varphi), B_0(\varphi)$ are

$$A_0(\varphi) = A_{01} + A_{02}\sin\varphi;$$
$$B_0(\varphi) = B_{01}\sin\varphi\cos\varphi,$$

where

A_{01}, A_{02}, B_{01} - constants.

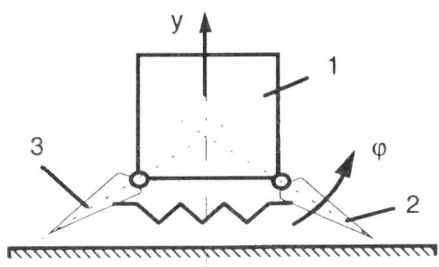

Figure 7. Jumping object.

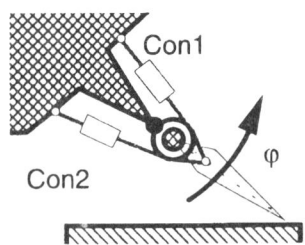

Figure 8. Leg with imitation of muscle action.

From the equation (1) we can conclude that if the system jumps from a foundation, it can not be controlled because the acceleration of the mass centre is constant.

When legs hit against the foundation and stick to it, the system has only one degree of freedom. The equation of motion is:

$$A(\varphi,\dot\varphi)\ddot\varphi + B_2(\varphi,\dot\varphi)\dot\varphi^2 + B_1\dot\varphi + C_1\varphi + D(\varphi,\dot\varphi) = M. \qquad (3)$$

For example, if external interaction is only dry friction, the coefficients are:

$$A(\varphi,\dot\varphi) = A_1 + A_2 \sin^2\varphi + (A_3 + A_4 sign\dot\varphi)\sin\varphi\cos\varphi;$$
$$B_2(\varphi,\dot\varphi) = A_2 \sin\varphi\cos\varphi + (A_3 + A_4 sign\dot\varphi)\cos^2\varphi;$$
$$D(\varphi,\dot\varphi) = D_1 \sin\varphi + D_2 sign\dot\varphi\cos\varphi,$$

where $A_1 - A_4, C_1, D_1, D_2$ - constants.

In equation (3) the moment $M = Con$ includes only control action without elastic and viscous properties (see fig. 6.). To found new principles of control action, the maximum principle of Pontryagin may be used [4]. For that reason we took the boundary value of the control action in the region $-u_0 \le M \le u_0$, where u_0 is constant. The Hamiltonian is

$$H = \psi_0 + \psi_1\dot\varphi + \psi_2[\frac{M - B_2(\varphi,\dot\varphi)\dot\varphi^2 - B_1\dot\varphi - C_1\varphi - D(\varphi,\dot\varphi)}{A(\varphi,\dot\varphi)}] \qquad (4)$$

Corresponding differential equations are

$$\dot\psi_1 = -\frac{\partial H}{\partial \varphi}; \quad \dot\psi_2 = -\frac{\partial H}{\partial \dot\varphi}. \qquad (5)$$

The maximum principle of Pontryagin can be applied in a usual way, and equation (4) gives

$$u_{opt}(t) = u_0 sign\frac{\psi_2}{A(\varphi,\dot\varphi)}. \qquad (6)$$

Therefore the optimal control action $M = u_{opt}(t)$ must be constant during switches from one level to another:

$$\frac{\psi_2}{A(\varphi,\dot\varphi)} > 0, \quad u_{opt}(t) = +u_0; \quad \frac{\psi_2}{A(\varphi,\dot\varphi)} < 0, \quad u_{opt}(t) = -u_0.$$ Three main types of a simple motion control are considered: - motion downwards and stopping (positioning) in a minimum time; - jumping motion from the rest to the maximum height; - stationary jumping motion with a maximum amplitude in a fixed period. For all kinds of motion theoretically the optimal control law for a small angle (f - coefficients of friction) $arctg(f) < \varphi < \frac{\pi}{2}$ is found.

The control action in the time domain has boundary value with one or more switching points.

2.1 Simulation of the stop motion

The motion of the mass centre during the jumping cycle can not be controlled. Therefore the main efforts to simulate the motion were made when legs stick to the foundation. Analysis of the control law (6) in the region $arctg(f) < \varphi < \dfrac{\pi}{2}$ allows to find a new nearly optimal control action like dry friction in the form

$$u_{opt}(t) = u_0 sign \dot{\varphi}. \tag{7}$$

As it is shown by Viba et al. in 1988 and in 1998, such control is very simple, one needs to know only the sign of angular velocity.

Example of stop motion in the time domain and phase plane in a case without control is shown in the Fig. 9. – Fig. 10. The main damping effect is caused by dry friction of legs against foundation.

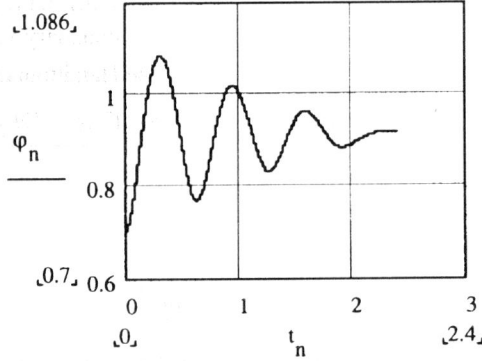

Figure 9. Angle in time domain.

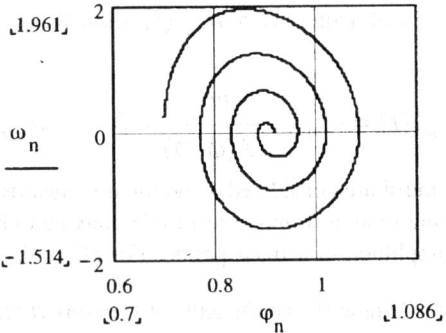

Figure 10. Motion in phase plane.

When a small control action (7) is added, the time of stopping decreases (Fig. 11. - Fig. 12.). In this example the control action has six switching points. If the boundary value increases, the stopping motion may end with one or two switching points (Fig. 13. - fig. 14.).

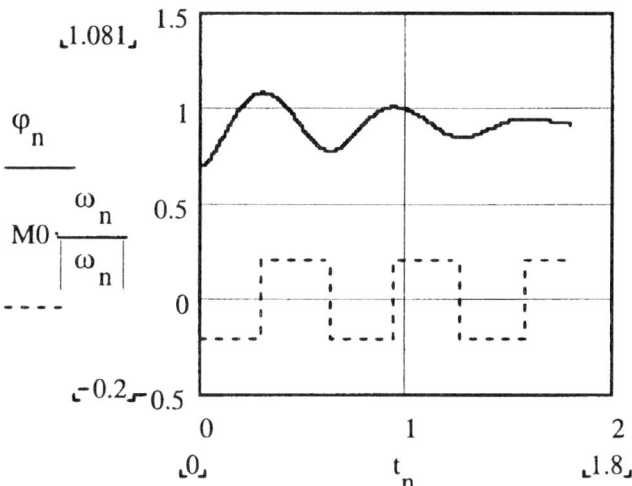

Figure 11. Angle in time domain with control.

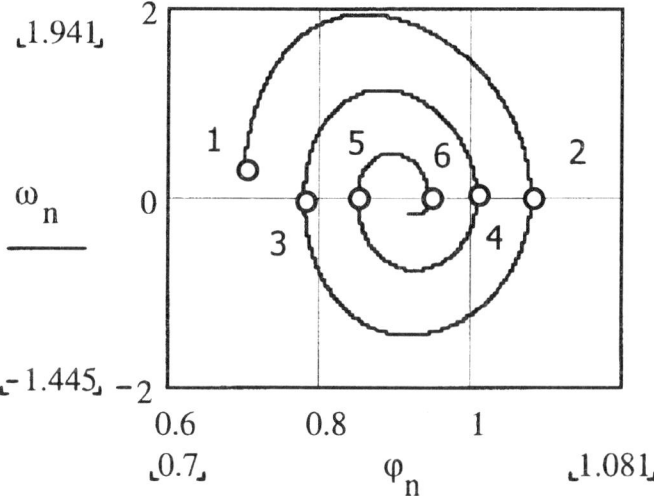

Figure 12. Motion in phase plane with control.

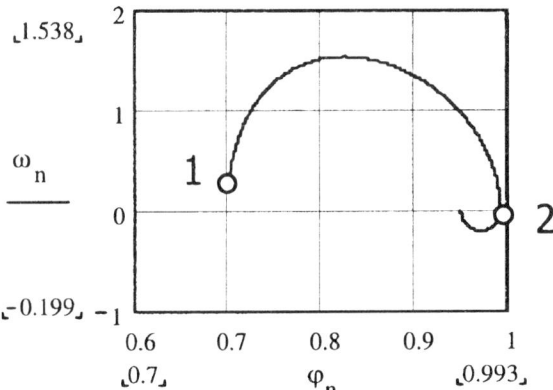

Figure 13. Control with two switching points.

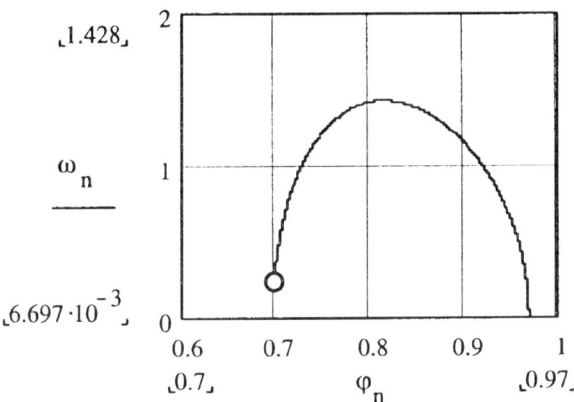

Figure 14. Control with one switching point.

From there a stable stopping motion for any initial conditions is found.

2.2 Simulation of a jumping motion upwards

In this case the control action may be chosen in a form as a dry friction:

$$u_{opt}(t) = -u_0 \, sign \, \dot\varphi \qquad (8)$$

The results of simulation of the equation (8) for the control with two switching points are shown in Fig. 15. - Fig. 18. Analysis of motion shows that the last and first step of control action must be corrected. It means that in the time moment when the normal reaction of foundation tends to zero, the system changes its structure (from one degree to two degrees of freedom with different differential equations). Therefore by changing the time intervals control action may be improved.

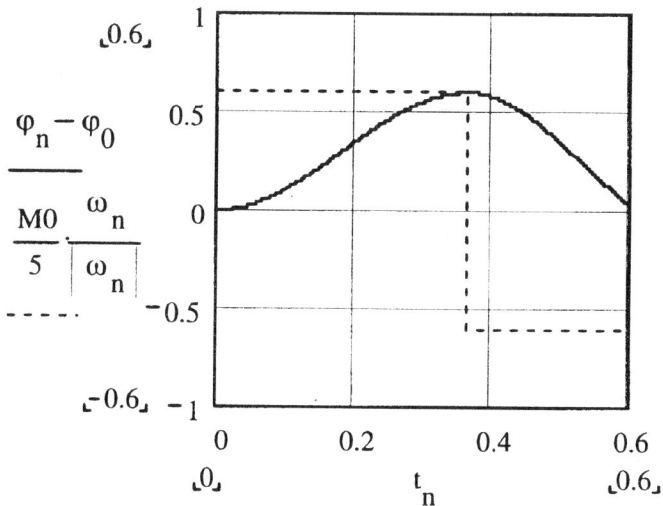

Figure 15. Control action and angle in the time domain.

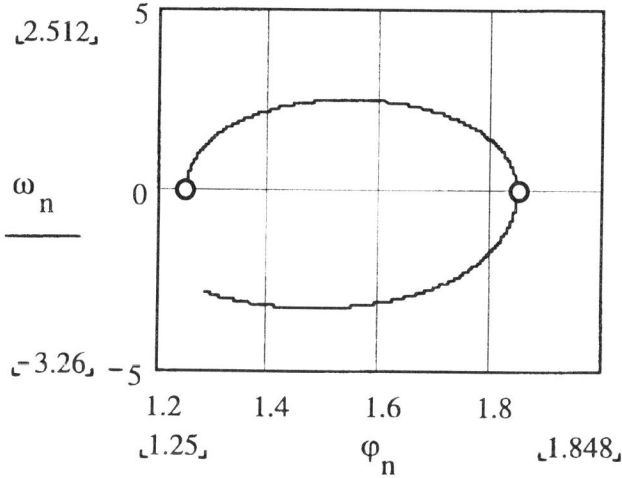

Figure 16. Motion in the phase plane in case of the control with two switching points.

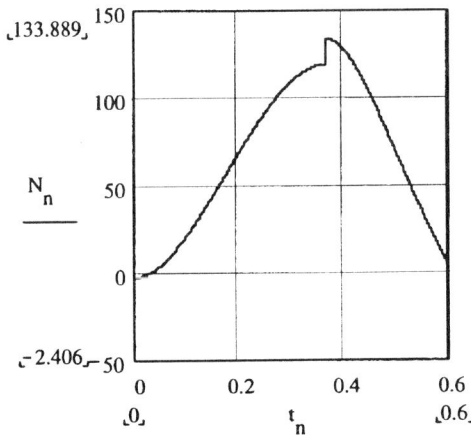

Figure 17. Normal reaction of the foundation.

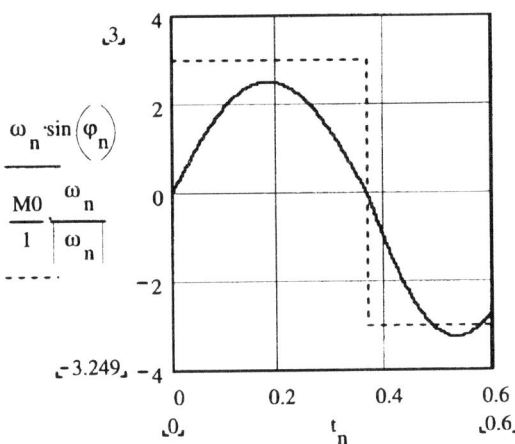

Figure 18. Velocity and control in time domain.

2.3 Calculation of a collision

Very important question for the system investigated above is the collision process when system sticks to a foundation. As it is shown by Viba et al.(1998,1999), in that time moment the collision of the object (bodies systems connected with pins and rough surfaces) must be determined by analysis of motion. To do that, the impact process in simultaneously more than in one point must be studied by using the main principles of theoretical mechanics (impulse, momentum and coefficient of restitution) and the law of dry friction.

2.4 Conclusion

In this paper we have shown that for adaptive control of mechanical systems we can use motion control, imitating muscle contraction process. To synthesize new control laws for start - stop systems the control action in the form of positive or negative dry friction may be applied. The use of advanced models of muscle contraction process is prospective for further investigations of human and animal behaviour during physical exercises.

3 References

Woledge R.C., Curtin N.A., Homsher E. (1985). *Energetic aspects of muscular contraction.* Academic Press.
Goubel F.(1998). *Éléments de méchanique musculaire*, Paris, Masson.
Tipans I.(1998). Mathematical model of a coupling of excitation and contraction in muscle fiber. In *Proceedings of the International Symposium. Analysis and synthesis of nonlinear dynamical systems in mechanics*, Riga Technical University, Latvia. 70 - 71.
Viba J. *(1988), Optimisation and synthesis of vibro-impact machines*, Riga, Zinatne.
Lavendelis E., Viba J., Sulcs A. (1996). Optimisation and synthesis of adaptive vibroimpact systems. In *Proceedings of the International Symposium. Analysis and synthesis of nonlinear dynamical systems in mechanics*, Riga Technical University, Latvia. 81 - 88.
Viba J., Pax C. (1996). Self - controlled hammer with a unilateral spring. In *Proceedings of International Seminar, Wave mechanics systems.* Technologija, Kaunas, Lithuania. 5-7.
Lavendelis E., Viba J., Grasmanis B., Fontaine J.-G. (1998). Calculation of collision in connected body systems to generate the start - stop motion. In *Twelfth CISM-IFToMM Symposium Ro.Man.Sy'98,*.Paris. Book of abstracts. 38.
Viba J., Grasmanis B., Fontaine J.-G. (1999). Simultaneous collisions in connected bodies systems. Solid mechanics and its applications. In *Proceedings of IUTAM / IFToMM Symposium,* Riga, Latvia, 24 - 28 August 1998, Kluwer academic publishers. Dordrecht/Boston/London, Vol.73, 267 - 274.

Inverse Simulation Study of Trampoline-Performed Somersaults*

Wojciech Blajer[1] and Adam Czaplicki[2]

[1] Department of Mechanics, Technical University of Radom, Poland
[2] Department of Biomechanics, Institute of Sport and Physical Education, Biała Podlaska, Poland

Abstract. Front and back somersaults on the trampoline are modeled. The developed mathematical models are used to solve the inverse dynamics problem, in which the applied moments of muscle forces at the joints that result in an actual (recorded) motion are determined. The nature of the stunts, the way the human body is maneuvered and controlled can be studied. The calculated torques are then used as control signals for the dynamic simulation. This provides a way to check the inverse dynamics procedures. The influence of typical control errors on somersault performance can be studied as well. Some examples of numerical calculations are reported.

1 Introduction

Front and back somersaults are common acrobatic stunts performed on the trampoline. Nevertheless, even the easiest somersaults are usually recognized only qualitatively as concerns some general guiding rules for their correct performance. The nature of the stunts, the way the human body is maneuvered and controlled, has not so far been well understood. The quantitative description may be useful for both cognitive and practical reasons, leading to better understanding the athlete movements and making a basis for more aware mastering of the somersault evolutions. The present contribution is an attempt towards an extensive study on the trampoline-performed human maneuvers, beginning with the kinematic analysis based on the recorded actual somersault performances, and getting through the inverse dynamics problem solution and direct dynamics simulations, which involve a nonlinear dynamical model of the trampolinist and recurrently interacting trampoline bad.

We define the *inverse dynamics problem* (see Garciá de Jalón and Bayo, 1994) as finding the applied torques (moments of muscle forces) at the joints that result in a given (recorded) motion of a biomechanical system. By comparing different inverse dynamics solutions, resulted from motion characteristics ranging from correct to incorrect somersaults, more strict rules of performance and control of the stunts can be understood. By definition then, if we use the calculated moments and forces as control signals for a dynamic simulation (*direct dynamics problem*) with the same biomechanical model, the outputs of the simulation should approximately match the data that served as input to the inverse dynamics problem. In this way, the direct simulation will provide a way to check the inverse dynamics procedures. Then,

---

* The research was supported by the State Committee for Scientific Research (KBN), Poland, under grant 9 T12C 060 17.

assumed the use of inverse dynamics solutions in dynamic simulations (*inverse dynamics simulation*) faithfully replicates the data used to produce the inverse dynamics solutions, the influence of typical control errors on somersault performance can be studied (by comparing outputs to control inputs slightly modified as compared to those obtained from inverse dynamics solution).

Reliability of solutions to the inverse dynamics problem and the direct dynamics simulation requires an adequate mathematical model of trampolinist and of interacting trampoline bad. The model characteristics need then to be identified (the bad stiffness, human body mass and geometric parameters, ...). Then, actual somersault performances must be recorded (filmed), and the obtained kinematic characteristics must be numerically recalculated to a required form of input data for the inverse dynamics problem. Algorithms for the numerical solutions of the inverse dynamics and the direct dynamics problems must finally been proposed. The computations may give a basis for some conclusions as well as the model and the methods verifications. All these issues are addressed in the sequel of this contribution.

2 The Model

Trampoline somersaults are difficult to model for many reasons: in general a space nature of motion, a very complicated description of human body, and a complex and recurrent interaction from the bad. The following limitations have thus been assumed in the paper.

- ☐ *Plane motion* is considered. This limits our analysis to the front and back somersaults (in tuck, pike and straight positions) without twisting. According to the nature of the stunts, the motions of both legs and of both arms are identical.

- ☐ The *trampolinist* is modeled as a *9-degree-of-freedom rigid multibody system*, whose position is described by $\mathbf{q} = [x_H \ y_H \ \varphi_1 \ \varphi_2 \ \varphi_3 \ \varphi_4 \ \varphi_5 \ \varphi_6 \ \varphi_7]^T$, where x_H and y_H are the hip coordinates, and φ_i ($i = 1, ..., 7$) are shown in Figure 1. The system is controlled by 6 torques $\boldsymbol{\tau} = [\tau_1 \ \tau_2 \ \tau_3 \ \tau_4 \ \tau_5 \ \tau_6]^T$ that model the moments of muscle forces at the joints.

- ☐ The *trampoline bed* is modeled as *weightless canvas* of known stiffness characteristics - the reactions from the bed, R_x, R_y and M_A (Figure 2), are assumed to depend linearly respectively on the bed deflections x_A and y_A, and the feet rotation angle θ_A.

A somersault begins/ends at the moment the athlete drops on the trampoline (touches the bed with his feet), and the jump can then be divided into two main phases:

- *a support phase*, when the feet push on the bed, and
- *a flying phase*, when the trampolinist does not touch the bed.

During the support phase, the downward motion of the body is first decelerated, and then the athlete rebounds upward. In the same time for front somersaults the upper body bends forward and downward (relative to the center of gravity) and the hips travel upward and backward. This allows the force acting on the feet from the trampoline to effectively push the body parts around its center of gravity resulting in a front somersault. A back somersault is done the same way except the upper body goes backward and downward while the hips go forward and upward. The quality of the somersault and amount of torque generated depends on the efficiency of this upper and lower body segmentation during the support phase. There are at least three parameters that must be considered by the athlete. They are how early the segmentation begins, how quickly it is completed and how much force it is done with. The most efficient somersault is achieved when each of the above parameters is maximized. Then, during the flying phase the total angular momentum of the body relative to the center of gravity remains constant, and the rotation of the body is regulated by changing from straight to tuck/pike position, and then back to straight position at the end of the phase.

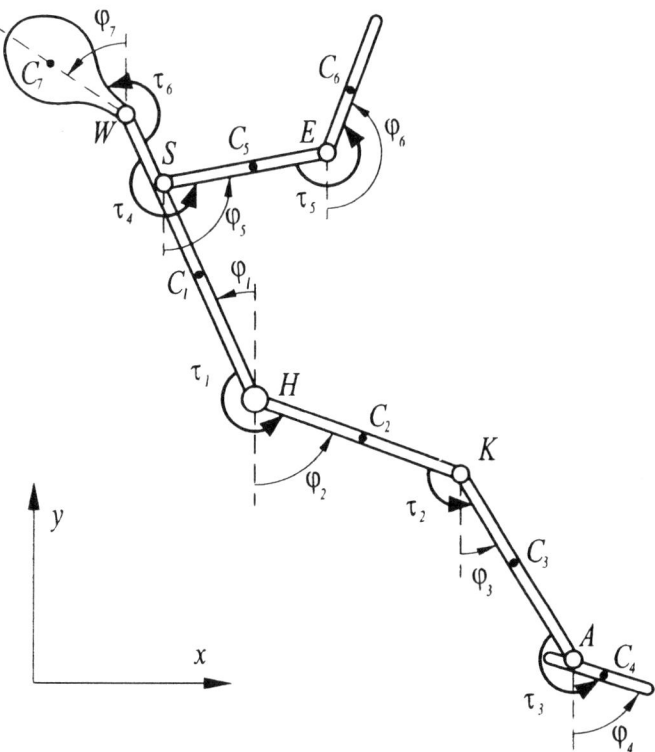

Figure 1. The trampolinist model.

3 The Dynamic Equations

The dynamic equations for the modeled trampolinist can be described by the following generic form (valid for both somersault phases) (see e.g. Blajer, 1997, Blajer and Schiehlen, 1992, Garciá de Jalón and Bayo, 1994, Risher, 1997):

$$\mathbf{M}(\mathbf{q})\ddot{\mathbf{q}} + \mathbf{h}(\mathbf{q},\dot{\mathbf{q}}) = \mathbf{f}(\mathbf{q}) + \mathbf{r}(\mathbf{q}) + \mathbf{B}^T \boldsymbol{\tau} \qquad (1)$$

where $\mathbf{q} = [x_H \; y_H \; \varphi_1 \; \varphi_2 \; \varphi_3 \; \varphi_4 \; \varphi_5 \; \varphi_6 \; \varphi_7]^T$ and $\boldsymbol{\tau} = [\tau_1 \; \tau_2 \; \tau_3 \; \tau_4 \; \tau_5 \; \tau_6]^T$ have already been defined, $\mathbf{M}$ is the 9×9-dimensional generalized mass matrix, $\mathbf{h}$ contains the dynamic generalized forces due to the centrifugal accelerations, $\mathbf{f}$ are the projections of the gravity forces onto the directions of **q**, **r** are the generalized forces arising from the trampoline reactions (**r** = **0** during the flying phase), and $\mathbf{B}^T$ is the 9×6-dimensional matrix of control distribution. For shortness we will not report here the explicit dynamic equations $\mathbf{M(q)\ddot{q}} + \mathbf{h(q,\dot{q})} = \mathbf{f(q)}$ for the free flying and uncontrolled human body, which can be obtained by using any standard method (see Blajer, 1997, and Garciá de Jalón and Bayo, 1994). The other components of equation (1) are:

$$\mathbf{r} = \begin{bmatrix} -R_x \\ R_y \\ 0 \\ -R_x l_2 \cos\varphi_2 + R_y l_2 \sin\varphi_2 \\ -R_x l_3 \cos\varphi_3 + R_y l_3 \sin\varphi_3 \\ M_A \\ 0 \\ 0 \\ 0 \end{bmatrix} ; \quad \mathbf{B}^T \tau = \begin{bmatrix} 0 & 0 & 0 & 0 & 0 & 0 \\ 0 & 0 & 0 & 0 & 0 & 0 \\ -1 & 0 & 0 & -1 & 0 & -1 \\ 1 & -1 & 0 & 0 & 0 & 0 \\ 0 & 1 & -1 & 0 & 0 & 0 \\ 0 & 0 & 1 & 0 & 0 & 0 \\ 0 & 0 & 0 & 1 & -1 & 0 \\ 0 & 0 & 0 & 0 & 1 & 0 \\ 0 & 0 & 0 & 0 & 0 & 1 \end{bmatrix} \begin{bmatrix} \tau_1 \\ \tau_2 \\ \tau_3 \\ \tau_4 \\ \tau_5 \\ \tau_6 \end{bmatrix} \quad (2),(3)$$

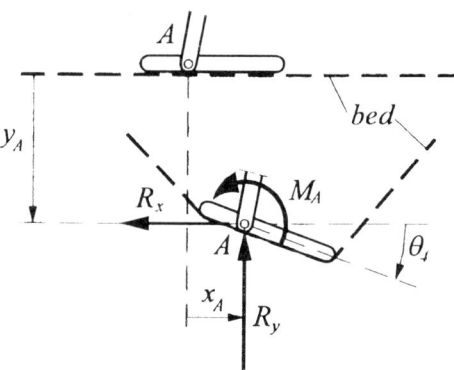

Figure 2. Interaction between the feet and the trampoline bed.

For rough preliminary calculations the reactions R_x, R_y and M_A from the bad have been taken as linear functions of displacements x_A, y_A and θ_A (see Figure 2), i.e.:

$$R_x = -k_x x_A ; \qquad R_y = -k_y y_A ; \qquad M_A = -k_\theta \theta_A \qquad (4)$$

where the stiffness coefficients k_x, k_y and k_θ have been estimated experimentally. Some more detailed measurements show that the characteristics are not linear, R_x and M_A depend on y_A as well, and all three R_x, R_y and M_A reactions should involve the damping effects. The improved mathematical model of the trampolinist-trampoline interaction will be considered in the further studies.

4 The Inverse Dynamics Problem

The solution of inverse dynamics problem needs motion characteristics $\mathbf{q}_d(t)$, $\dot{\mathbf{q}}_d(t)$ and $\ddot{\mathbf{q}}_d(t)$ to be determined first. These are measured from actual somersault performances using a camera with 120 Hz shutter frequency. The positions of markers placed on the athlete's leg, hip, trunk, head and arms (shown in Figure 3) are tracked. From these measurements, using a specialized program on PC computer, discrete trajectories $\mathbf{q}_d^*(t)$ are obtained. Cubic spline interpolation is then used to generate continuous $\mathbf{q}_d(t)$ consistent with $\mathbf{q}_d^*(t)$. The analytically given functions $\mathbf{q}_d(t)$ are finally differentiated to obtain $\dot{\mathbf{q}}_d(t)$ and $\ddot{\mathbf{q}}_d(t)$.

Figure 3. Back somersault in tuck position.

During the flying phase the precision of determination of $\mathbf{q}_d(t)$ can be verified using the condition that the center of gravity of the trampolinist must move vertically (or along a parabola), and the total angular momentum of the body must remain constant. The conditions are not valid during the support phase, however. Apart from determination of $\mathbf{q}_d(t)$, during this phase $x_{Ad}(t)$, $y_{Ad}(t)$ and $\theta_{Ad}(t)$ ($\theta_A = \pi/2 - \varphi_4$) should be found as well in order to estimate, according to equations (4), the reactions from the bed. High precision of determination of $x_{Ad}(t)$, $y_{Ad}(t)$ and $\theta_{Ad}(t)$ is required, and this is achieved by using special equipment.

Equation (1) represents 9 dynamic equations dependent on 6 controls τ. In other words, in the 9-dimensional configuration space of the system, only 6 directions are controlled while the remaining 3 ones are uncontrolled. The configuration space can then be split into a controlled and uncontrolled subspaces of dimensions 6 and 3, respectively. While the controlled subspace is defined by vectors represented in $\mathbf{B}$ as rows (columns of $\mathbf{B}^T$), the uncontrolled subspace can be spanned by vectors represented as columns of 9×3-dimensional matrix $\mathbf{D}$, an orthogonal complement matrix to $\mathbf{B}$, i.e. $\mathbf{D}^T \mathbf{B}^T = \mathbf{0}$ (see Blajer, 1997). In the case at hand we propose

$$\mathbf{D}^T = \begin{bmatrix} 1 & 0 & 0 & 0 & 0 & 0 & 0 & 0 & 0 \\ 0 & 1 & 0 & 0 & 0 & 0 & 0 & 0 & 0 \\ 0 & 0 & 1 & 1 & 1 & 1 & 1 & 1 & 1 \end{bmatrix} \qquad (5)$$

According to Blajer (1997), the premultiplication of equation (1) by $[\mathbf{M}^{-1}\mathbf{B}^T \ \mathbf{D}]^T$ yields projections of the dynamic equations into the controlled and uncontrolled subspaces

$$\mathbf{B}\ddot{\mathbf{q}} + \mathbf{B}\mathbf{M}^{-1}\mathbf{h} = \mathbf{B}\mathbf{M}^{-1}\mathbf{f} + \mathbf{B}\mathbf{M}^{-1}\mathbf{r} + \mathbf{B}\mathbf{M}^{-1}\mathbf{B}^T \tau \qquad (6)$$

$$\mathbf{D}^T\mathbf{M}\ddot{\mathbf{q}} + \mathbf{D}^T\mathbf{h} = \mathbf{D}^T\mathbf{f} + \mathbf{D}^T\mathbf{r} \qquad (7)$$

Using $\mathbf{q}_d(t)$, $\dot{\mathbf{q}}_d(t)$ and $\ddot{\mathbf{q}}_d(t)$, the control characteristics $\tau_d(t)$ that assure the realization of the specified motion can then be determined from equation (6), manipulated to

$$\tau_d(t) = (\mathbf{B}\hat{\mathbf{M}}^{-1}\mathbf{B}^T)^{-1}\left[\hat{\mathbf{M}}\ddot{\mathbf{q}} + \mathbf{B}\hat{\mathbf{M}}^{-1}(\hat{\mathbf{h}} - \hat{\mathbf{f}} - \hat{\mathbf{r}})\right] \qquad (8)$$

where $\hat{\mathbf{M}} = \mathbf{M}(\mathbf{q}_d)$, $\hat{\mathbf{h}} = \mathbf{h}(\mathbf{q}_d, \dot{\mathbf{q}}_d)$, $\hat{\mathbf{f}} = \mathbf{f}(\mathbf{q}_d)$, and $\hat{\mathbf{r}} = \mathbf{r}(\mathbf{q}_d, x_{Ad}, y_{Ad}, \theta_{Ad})$ ($\mathbf{r} = \mathbf{0}$ during the flying phase). Equation (7), rewritten to the form

$$\mathbf{D}^T\left(\hat{\mathbf{M}}\ddot{\mathbf{q}} + \hat{\mathbf{h}} - \hat{\mathbf{f}} - \hat{\mathbf{r}}\right) = \mathbf{0} \qquad (9)$$

can then be used to verify the input data. During the flying phase ($\hat{\mathbf{r}} = \mathbf{0}$), equation (9) is equivalent to the condition that the body center of mass moves along a parabola (or vertically with acceleration g pointed downward) and that the total angular momentum of the body parts remains constant. During the support phase ($\hat{\mathbf{r}} \neq \mathbf{0}$), equation (9) expresses the condition that the actual time-derivatives of linear momentum of the center of mass and of angular momentum with respect to the center of mass are due to the actual (measured) external forces $\hat{\mathbf{f}}$ and reactions from the bed $\hat{\mathbf{r}}$. Equation (9) serves thus to verify the input (measured) data.

Some representative results of determined control variations obtained as a solution to the inverse dynamics problem for a back somersault in tuck position are shown in Figure 4. The biggest variations in control torques can be observed during the support phase (0÷0.4 sec), while during the flying phase (0.4÷1.6 sec) the most important for the somersault performance are variations in hip and shoulder control torques. The variations of control torques in elbow and neck (not shown here) play a negligible role.

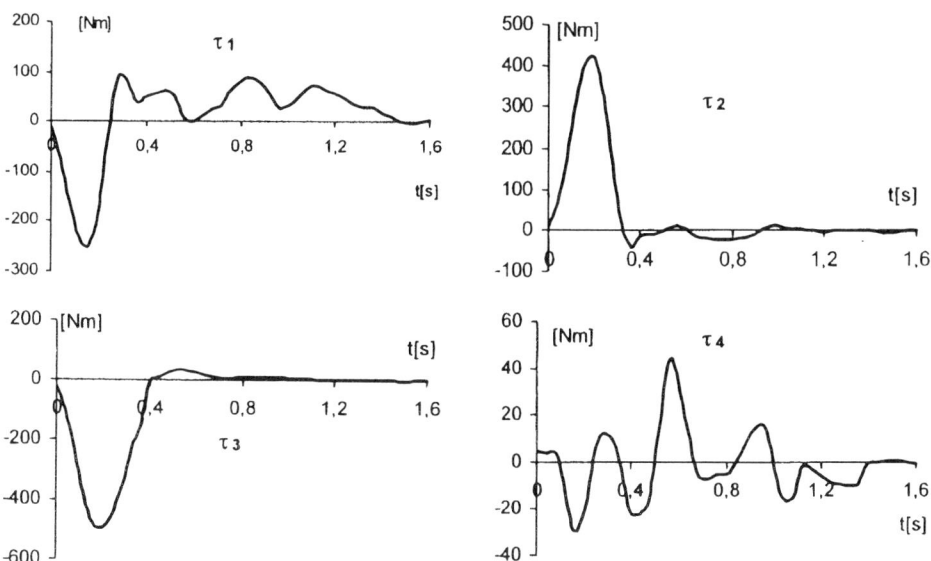

Figure 4. Back somersault in tuck position - solution of the inverse dynamics problem. The torques versus time in hip (τ_1), knee (τ_2), ankle (τ_3), and shoulder (τ_4) joints.

5 The Direct Dynamics Problem

To solve the direct dynamics problem equation (1) needs to be rearranged to

$$\ddot{\mathbf{q}} = \mathbf{M}^{-1}(\mathbf{q})\Big(\mathbf{f}(\mathbf{q}) - \mathbf{h}(\mathbf{q},\dot{\mathbf{q}}) + \mathbf{r}(\mathbf{q}) + \mathbf{B}^T \boldsymbol{\tau}_d(t) \Big) \qquad (10)$$

where $\tau_d(t)$ is a continuous time function calculated from the inverse dynamics. The initial state values $\mathbf{q}_0 = \mathbf{q}(t_0)$ and $\dot{\mathbf{q}}_0 = \dot{\mathbf{q}}(t_0)$ should match those used as input data to the inverse dynamics solution at $t = t_0$. A fourth-order adaptive step size Runge-Kutta integrator was used to produce the dynamic simulation with $\tau_d(t)$ as the control.

By definition, if $\tau_d(t)$ are calculated precisely enough, the solution $\mathbf{q}(t)$ and $\dot{\mathbf{q}}(t)$ to equation (10) should be consistent with $\mathbf{q}_d(t)$ and $\dot{\mathbf{q}}_d(t)$. In practice the numerically exact consistency can be achieved only at the beginning of simulation, assured that $\mathbf{q}_0 = \mathbf{q}_d(t_0)$ and $\dot{\mathbf{q}}_0 = \dot{\mathbf{q}}_d(t_0)$. For $t > t_0$ a difference between the simulated and measured state variables is always observed, and the difference tend to increase in simulation time. There are many reasons for this *inverse dynamics simulation failure* phenomenon. The most obvious are the integration truncation errors, which, in authors' opinion, are of minor importance, however. The other source of the simulation failure is an inaccuracy of the used mapping from the discrete $\tau_d^*(t)$ obtained from the inverse dynamics problem to the continuous $\tau_d(t)$ used in the direct dynamics simulation. In our studies, we used linear interpolation of sampled solution $\tau_d^*(t)$ to produce $\tau_d(t)$. The inverse dynamics simulation failure is finally caused by inaccuracy in determination of $\mathbf{q}_d(t)$, $\dot{\mathbf{q}}_d(t)$ and $\ddot{\mathbf{q}}_d(t)$, and, during the support phase, $x_{Ad}(t)$, $y_{Ad}(t)$ and $\theta_{Ad}(t)$. The inaccuracy is involved in the mapping from the measured (*) to continuous (d) data, and in the method of determination of $\dot{\mathbf{q}}_d(t)$ and $\ddot{\mathbf{q}}_d(t)$ from $\mathbf{q}_d(t)$. The calculated control variations $\tau_d^*(t) \to \tau_d(t)$ are influenced by these errors. To produce a direct dynamics simulation that well matches the measured movement pattern used as input to the inverse dynamics problem, much effort usually must be made in adequate preparation of the measured data.

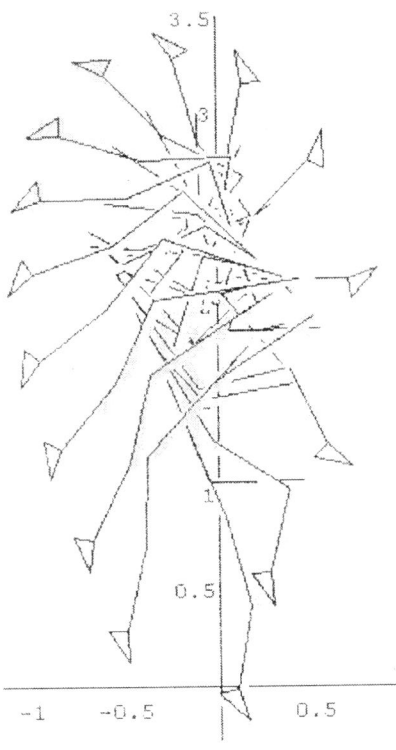

Figure 5. Back somersault in tuck position - sampled body configurations. Thick gray lines - the direct dynamics problem, solid black lines - measured data.

As concerns the flying phase, the authors achieved a good consistence between the inverse dynamics simulation and the data used as input in the inverse dynamics problem. The result obtained for a back somersault in pike position is illustrated in Figure 5, where the animations of the motions are demonstrated. The thick gray line shows the simulated positions in time, and the thin solid black line stands for the measured positions at the same sampled times. The difference is negligible.

The conformability was not so good for the support phase, however (the case is not demonstrated here). The inverse dynamics simulation failure was mainly due to the limited accuracy of used characteristics $x_{Ad}(t)$, $y_{Ad}(t)$ and $\theta_{Ad}(t)$. The current research is focussed on improving the measurements.

6 Concluding Remarks

A mathematical model for the analysis of front and back somersaults on the trampoline was developed. It can be applied to the solution of inverse dynamics problem as well as to the direct dynamics simulation. The same governing equations are used for the flying phase of somersault when the body rotates above the trampoline bed, and for the support phase when the athlete touches the bed. The only exception is that in the support phase the reactions from the bed on the feet are involved. The method of mathematical modeling includes the fact that the number of degrees of freedom of the system exceeds the number of controls.

The inverse dynamics problem is solved using motion characteristics measured (filmed) from actual somersault performances - the body position time variations (in the both phases) and the bed deflections (in the support phase). The applied moments of muscle forces at the joints that result in the given motions are then calculated. By analyzing the results the nature of the stunts, the way the human body is maneuvered and controlled, can be studied. This may be a valuable means for supporting the teaching and mastering of trampoline somersaults.

The torques calculated from the inverse dynamics are used as control signals for the dynamic simulations. In the flying phase, the obtained results faithfully replicate the data that served as input to the inverse dynamics problem, while, in the support phase, the conformability was not so good. To reduce the inverse dynamics simulation failure, the model of interactions between feet and bed need to be improved, and better accuracy of measurements of the bed deflections during the support phase must be achieved. The simulation results provide the way not only to check the inverse dynamics procedures - searching among different modifications of the control signals that result in coordinated (optimized) movement is possible, and the influence of typical control errors on somersault performance can be studied.

The future research will focus on more strict rules of body maneuvering and control in correctly performed somersaults. The fact that the performance will vary greatly with different individuals should also be considered.

References

Blajer, W. (1997). A geometric unification of constrained system dynamics. *Multibody System Dynamics* 1, 3-21

Blajer, W. and Schiehlen, W. (1992). Walking without impacts as a motion/force control problem. *Journal of Dynamic Systems, Measurement, and Control* 114, 660-665.

García de Jalón, J. and Bayo, E. (1994). *Kinematic and dynamic simulation of multibody systems: the real-time challenge*, Springer-Verlag, New York.

Risher, D.W., Schutte, L.M., and Runge, C.F. (1997). The use of inverse dynamics solutions in direct dynamics simulations. *Journal of Biomechanical Engineering* 119, 417-422.

Functional Biomechanics of Human Grasping and Requirements for Simple Robotic End-Effectors

Robert B. Addis and Bahram Ravani[1]

[1] Department of Mechanical and Aeronautical Engineering,
University of California-Davis, Davis, CA USA

Abstract. This paper explores functional biomechanics of human grasping with the aim of developing requirements for simple robotic end effectors. It uses a biomechanical evaluation to determine the degrees of freedom of the hand and enumerate all kinematically possible grasps. All grasps are then grouped into five grasp categories based on the type of oppositions formed between hand's elements. This is then used for developing the requirements for robot end effectors that can replace the hand's functions.

Introduction

Proper integration and use of an appropriate end-effector is one of the factors limiting the usefulness of a robot manipulator in effecting its work environment. Presently industrial end effectors are designed specifically for the application or simple two finger end-effectors are utilized for a variety of limited applications. In contrast, the human hand has adaptability unequaled by any mechanical manipulator or end effector developed to date. In an attempt to match some of the capabilities of the human hand, robotic researchers have developed dexterous robotic end- effectors (see for example, Crossley and Umholtz, 1977; Okada, 1979; Salisbury and Craig, 1982; Salisbury, 1986; Jacobsen, et al., 1986; Loucks, et. al., 1987; Salisbury, Brock, and O'Donnell, 1989; Blechschmidt, and Fessler, 1991). Similar attempts have been made in reconstructive hand surgery and the design of prosthetics for amputees in order to preserve and/or restore a patient's maximum prehensile abilities (see for example, Schlesinger, 1919; McBride, 1942; Taylor and Schwarz, 1948; Tubiana, 1981; Scott, 1990; Patterson and Katz, 1992). None of these works have resulted in either an end effector or a prosthetics hand design that has the combined ranges of strength, stability, compliance, fine motion, speed, and sensor variety as the human hand.

Initial studies, on human prehension, by Schlesinger (1919) through Taylor and Schwartz (1955) categorized human grasps by the geometry of the item gripped, e.g., spherical, cylindrical, etc. This was later refined by Napier (1956) to incorporate the reason for the grip, specifically "power" (for stability and strength) or "precision" handling (describing dexterous manipulation of an object by the fingers of the hand). This approach culminated in a taxonomy by Cutkosky and Wright (1986b) which mapped power and precision grasps as well as object geometry and size, in addition to applied forces and relative manipulation. Despite these efforts, there exists no practical categorization of human grasps that can define a canonical set useful in selection of robotic end effectors. This paper studies the functional biomechanics of human grasping to develop a canonical set of grasps that can provide a basis for selection of simple robotic end-effectors in various applications.

Functional Anatomy of Human Hand

The human pentadactyl form has an osseous skeleton consisting of 27 bones. It is composed of five rays that articulate at the wrist (carpus), which is formed by two rows of bones. The first row begins with the scaphoid (or navicular) located immediately proximal to the radius; next to this is an intermediate bone the lunate (or semilunar). The neighboring bone, the triquetrum (or pyramidal), is associated with the ulna and is adjacent to the pisiform, a sesamoid. The second or distal row is composed of the first through fifth carpals: the trapezium, the trapezoid, the capitate, and the hamate, which actually composes the union of the fourth and fifth carpals. Distal to the carpals are the 19 long bones that make up the five rays or digits of the hand (the thumb and four fingers). These five polyarticular chains are each made up of a metacarpal and three phalanges, with the exception of the thumb, which only has two phalanges resulting in it being the shortest ray. It is the position of the metacarpotrapezial joint, which accounts for the gap in the palm between the first and second rays. The short thumb column extends from the trapezium setting it in front of the carpal plane so that the first metacarpal makes a 45degree angle with the second metacarpal (with respect to the sagittal plane) when the hand is in its natural and relaxed position. This seller type (saddle shape) synovial joint provides exceptional mobility and is responsible for the thumb's ability to oppose the other rays.

Attached to the osseous skeleton is the fibrous skeleton composed of the aponeuroses (broad tendons), ligamentous system, and fibrous sheaths attached to and tying together the bones and dermis of the hand. This complex fibrous system has many functions but plays the essential role in stabilizing the carpals and rays of the hand, establishing the hand's degrees-of-freedom (DOF), and the range-of-motion at each joint. The fibrous skeleton also forms the complex pulley system, which guides the tendons of the intrinsic and extrinsic muscles to their fixations. It is through this pulley system that the voluntary degrees-of-freedom are activated and forces transmitted to an object. In this way the fibrous skeleton contributes directly to the measured strength of the hand.

Biomechanics of Human Hand

The osseous and fibrous skeletons combine to create two groups of elements in the hand: the fixed and the mobile (Littler, 1960). The fixed elements include the distal row of carpals and the second and third metacarpals. While some compliance exists within these fixed elements, due to the elastic nature of the fibrous skeleton, this group is the basis or framework from which the entire hand establishes its stability and rigidity. The mobile elements include two groups: the distal elements or phalanges, and the peripheral metacarpals, specifically, the thumb, the fourth, and the fifth metacarpals. The DOF contained within the mobile elements are what establishes the hand's functionality.

Examining the distal elements, the phalanges of all five fingers are capable of various degrees of flexion-extension at their individual phalangeal joints, with movement limited to bending towards the palm. The fibrous skeleton restrains the fingers from extending much beyond the frontal plane and entirely prevents the fingers from contacting the hand's dorsal surface. The four fingers (less the thumb) are also capable of adduction-abduction at their metacarpophalangeal joint which allows the lateral surfaces of neighboring fingers to oppose each other.

Of the peripheral metacarpals, the thumb is most unusual in that it is independent and highly mobile. It articulates with the trapezium and is capable of both adduction-abduction and anteposition-retroposition at the trapezometacarpal joint. This allows the thumb to be brought into opposition with the palmar surface, the anterior finger surfaces, the radial-lateral finger surfaces, and dorsal surfaces of the intermediate and distal phalanges. While the two ulnar metacarpals are considered mobile elements, they have only slight mobility which is limited to a flexion-rotation movement. Specifically, the fourth or ring metacarpal (a transitional element to the fifth metacarpal) has approximately 10 degrees of mobility in flexion-extension at the carpometacarpal joint, while the semi-independent fifth metacarpal, which articulates with the hamate and the base of the fourth metacarpal, displays 20° of flexion-extension at the carpometacarpal joint. Finally, axial rotation is present at all metacarpophalangeal joints but is a part of the hand's involuntary compliance mechanism, allowing the hand to conform to a grasped object more readily.

Table 1 summarizes the rays of the hand and their major DOF. The involuntary DOF are defined as the slight carpal deviations, axial rotations, and other joint compliance associated with the fibrous skeleton's elastic properties. Finally, the hand's grasping ability is further enhanced by being covered with a malleable dermis capable of conforming to innumerable shapes and objects. The dermis is also capable of varying the coefficient of friction between itself and an object being grasped via the sweat glands, another consideration when examining the grasp mechanics.

Table 1. Summary of Hand's Degrees-of-Freedom (DOF)

Ray	Involuntary DOF	Voluntary DOF	Total DOF
I – Thumb	1	4	5
II – Index	1	4	5
III – Middle	1	4	5
IV – Ring	3	4	7
V – Little	3	4	7
Totals	9	20	29

The hand's principal functional surface is its concave shaped palmar surface (the anterior volar), made-up of the carpals and finger metacarpals (second through fifth rays). Its broad surface may contact objects alone in many non-prehensile tasks or more typically work in opposition with the fingers and/or thumb to secure objects. In contrast, the hand's dorsal aspect or posterior surface (back of hand) has a convex shape and is limited to only nonprehensile functions since the fibrous skeleton prevents the fingers or thumb from interacting with its surface.

When the fingers are fully extended and adducted, each digit of the hand has a distinct length, with the third finger (functional axis) being the longest and the peripheral digits succeedingly shorter. The varying lengths of the individual finger phalanges results in the interphalangeal joints being staggered between adjoining fingers which allows nearly complete lateral opposition between the finger surfaces and minimizes the hand's profile. Further adduction can be realized by allowing the dorsal surface of the peripheral fingers to overlap the anterior surfaces of the third and fourth fingers phalanges. Lateral opposition between the rays of the hand is limited to neighboring fingers' except for the thumb, which can oppose the radial-lateral surface of all the fingers. The thumb is also unusual in that its phalange actually opposes the lateral side of the second metacarpal when fully extended and adducted due to its shorter length and its fixation at the base of the carpals.

Finger flexion-extension of the second through fifth rays is limited to the phalangeal joints. Finger flexion-extension occurs in the sagittal plane with the fingers capable of opposing the palmar surface but not the dorsum of the hand. In contrast, the thumb is capable of articulating at both its carpometacarpal and phalangeal joints for the combined motions of abduction-adduction and anteposition-retroposition. These motions allow the thumb to oppose the palm, the fingertips, the finger pads, the dorsal surfaces of the fingers' intermediate and distal phalanges, or the radial-lateral surfaces of the fingers. This mobility makes the thumb the master digit of the hand giving function to all others.

Besides absolute length, the relative length of each ray varies throughout flexion. The distal pads of each finger tip (or all the fingers as a whole) converge with the thumb pulp in thumb-finger opposition grips. When the digits are flexed to the palm (without the thumb) in palmar opposition grips, they converge in complete adduction to the base of the thenar eminence muscle with the axes of the rays converging towards the scaphoid tubercle. Dubousset (1971) and Kuczynski (1968) showed this convergence originates with the metocarpophalangeal and proximal interphalangeal articulations. The more ulnar the digit, the more obliquely it deviates as it approaches the palm, with no finger abduction evident or possible in the normal hand when fully flexed. As the fingers are extended though, the digits naturally diverge to their fully abducted position, but can be voluntarily adducted if desired. Unlike flexion-extension, abduction-adduction is a semi-active degree of freedom dependent on the degree of flexion.

Enumeration of Hand's Articulations

Ignoring the joints' range-of-motion for a moment and simply considering the hand's combined 29 DOF, there exists a maximum of 2^{29} variations or 536,870,912 unique joint articulation combinations. While far from infinite, this set is very large and unmanageable without some simplification. Nine DOF's are involuntary and result from the compliance in the fibrous skeleton, leaving 20 DOF or 1,048,576 (2^{20}) possible joint combinations to consider. A further simplification described in the previous section was involuntary finger adduction in response to finger flexion, resulting from the biomechanics of the metacarpophalangeal joint (Kuczynski, 1968; Dubousset, 1971). If one considers adduction-abduction of the fingers (rays II thru V) a semi-voluntary action, having to do more with the finger's flexion position (which is related to an object's geometry) than with a cognizant prehensile choice, the voluntary DOF can be reduced to 16, leaving 65,536 (2^{16}) joint combinations, or hand postures to study.

The remaining 16 DOF are grouped within the five rays of the hand: four DOF for the thumb (ray I) and three DOF per finger (rays II thru V). Specifically, the thumb's four DOF include anteposition-retroposition and abduction-adduction of the trapezometacarpal joint, and flexion-extension of its two phalangeal joints (metacarpo- and inter-). The three DOF for each finger are flexion-extension of their three phalangeal joints (metacarpo-, proximal, and distal phalangeal joints). Ignoring individual joint angles, the surfaces of these five polyarticulating chains act with or without the palmar surface to secure an object. In essence, all possible grasps are formed by some combination of surface interaction between these six major hand structures: the palm, the thumb, and four fingers, resulting in a minimum set of 2^6 or 64 fundamental combinations (Table 2) from which many hundred grasps originate.

The 64 combinations summarized in Table 2 are not to be confused with actual grasps. The table summarizes only the possible combinations between the hand's six major elements, not the various interaction(s) between their surfaces, which form the actual grasps. This is examined next.

Grasp Identification

The actual contact between an object and the major elements of the hand is made by some combination of the palmar surface, finger tips or finger pulp, thumb tip or thumb pulp, the lateral finger or thumb surfaces, and on occasion, the dorsal surface of the hand and/or fingers for non-prehensile interaction. These surfaces are represented by four major groups: 1) the palmar surface, 2) the tip or pad of each ray, 3) the ray's lateral surfaces, and 4) the hand's dorsal surfaces. Thus, the final grasp formed by the hand's elements is composed of one or more of these surfaces with up to 2^4 surface interactions possible (Table 3), depending on the type and number of hand elements involved. Table 3 summarizes the 16 surface interactions for the hand's elements. The complete set of static human grasps (see Addis 1998) can now be derived by combining hand elements (Table 2) and surface interactions in table 3.

We have performed an analysis on each combination in Table 2 to derive more than 334 individual grasp combinations (for ≥773 individual grasps) (see Addis 1998). This is fewer than half the 1024 (64 x 16) grasps based on the total number of combinations, 64, and surface variations, 16. It is fewer than half the 945 (63 x 15) grasps after accounting for the "no elements" and "no surface contact" conditions. The number of valid surface interactions is dependent on the biomechanics of the type and number of elements utilized to form the grasp.

A close examination of these 334 static grasps shows that their basic forms can be categorized into five general groups. The first group consists of the *"no grasp"* condition, in which the hand makes no contact with an object. This is a nonprehensile group of grasps which includes all of the hand postures associated with "hand gestures", the poses used for sign language, and involuntary reflex reactions.

Table 2. All possible interaction combinations between the hand's six major elements.

Combination No.	Palm	Thumb	Index Finger	Middle Finger	Ring Finger	Little Finger	Combination No.	Palm	Thumb	Index Finger	Middle Finger	Ring Finger	Little Finger	Combination No.	Palm	Thumb	Index Finger	Middle Finger	Ring Finger	Little Finger
No Elements							Three Elements							Four Elements						
1	O	O	O	O	O	O	23	X	X	X				43	X	X	X	X		
							24	X	X		X			44	X	X	X		X	
Single Element							25	X	X			X		45	X	X	X			X
2	X						26	X	X				X	46	X	X		X	X	
3		X					27	X		X	X			47	X	X		X		X
4			X				28	X		X		X		48	X	X			X	X
5				X			29	X		X			X	49	X		X	X	X	
6					X		30	X			X	X		50	X		X	X		X
7						X	31	X			X		X	51	X		X		X	X
							32	X				X	X	52	X			X	X	X
Two Elements							33		X	X	X			53		X	X	X	X	
8	X	X					34		X	X		X		54		X	X	X		X
9	X		X				35		X	X			X	55		X	X		X	X
10	X			X			36		X		X	X		56		X		X	X	X
11	X				X		37		X		X		X	57			X	X	X	X
12	X					X	38		X			X	X							
13		X	X				39			X	X	X		Five Elements						
14		X		X			40			X	X		X	58	X	X	X	X	X	
15		X			X		41			X		X	X	59	X	X	X	X		X
16		X				X	42				X	X	X	60	X	X	X		X	X
17			X	X										61	X	X		X	X	X
18			X		X									62	X		X	X	X	X
19			X			X								63		X	X	X	X	X
20				X	X															
21				X		X								All Six Elements						
22					X	X								64	X	X	X	X	X	X

Table 3. Possible variations in the surface contact between the hand's major elements (thumb, palm, fingers).

Variation No.	Surfaces				Variation No.	Surfaces				Variation No.	Surfaces			
	Palm	Finger Tip / Pad	Lateral	Dorsal		Palm	Finger Tip / Pad	Lateral	Dorsal		Palm	Finger Tip / Pad	Lateral	Dorsal
No Surfaces					Two Surfaces					Three Surfaces				
1	O	O	O	O	6	X	X			12	X	X	X	
					7	X		X		13	X	X		X
Single Surface					8	X			X	14	X		X	X
2	X				9		X	X		15		X	X	X
3		X			10		X		X	Four Surfaces				
4			X		11			X	X					
5				X						16	X	X	X	X

The second group consists of *unilateral grasps* (Malek, 1981). These non-prehensile grasps are capable of impeding a single degree-of-freedom (or direction) of an object's movement. This group includes grasps associated with opposing or applying a task-related uniaxial force or torque to an object (Iberall, 1997), e.g., *flat hand* supporting a heavy object (Lister, 1977), a *hook* shaped hand carrying a suitcase (Schlesinger, 1919), or using one or more of the finger tips to depress the keys on a type-writer or piano. The dorsal surface of the hand is capable of performing unilateral grasps. This group of nonprehensile unilateral grasps is the largest set identified in Addis 1998.

The third group is defined by the palmar surface and the various grasps formed in conjunction with the fingers and thumb. These prehensile grasps are capable of forming either bilateral or multilateral contacts with an object (Malek, 1981), immobilizing two or more of the object's degrees-of-freedom. This group contains all the digitopalmar and whole hand grasps generally associated with the fingers and/or thumb wrapping around an object and securing it against the palmar surface, e.g., cylindrical grasp (with or without the thumb and the object lying on the transverse or oblique palmar axis depending on its geometry and the task) as when holding a hammer or screwdriver, spherical grasp when holding a baseball or palmar grasp when holding a book (Schlesinger, 1919), etc. These grasps can be very powerful and stable depending on the number of fingers utilized. The relative importance of these grasps is readily apparent when you consider the fact that they constitute the second largest group identified in Addis 1998.

The fourth group consists of the grasps formed using only the thumb and finger(s). Like the third group, these prehensile grasps are capable of forming numerous bilateral and multilateral contacts with an object using either the tips or volar surfaces of the hand's rays. This group contains all the grasps associated with *thumb-finger pinch*, e.g., *tip prehension* as when holding a pearl, *flat/thin (two-finger) pincer* when holding a stick of gum, *three-jaw chuck* typical of holding a ball with the thumb and index and middle fingers, and *large (five finger) pincer* as in turning the lid on a jar (Schlesinger, 1919). These grasps constitute the third largest set of grasps identified in Addis 1998.

The fifth group consists of the *lateral oppositions* and *interdigit* contacts formed by the fingers and thumb. These prehensile grasps are also capable of forming various bilateral and multilateral contacts with an object depending on the number of fingers involved and the object's geometry. This group contains the grasps associated with objects being secured between the lateral surfaces of the fingers or between the thumb pad and lateral finger surfaces, e.g., holding a cigarette between the fingers or using the thumb and index finger to hold and turn a key. This group consists of approximately twenty grasps (see Addis 1998) and is relatively small and specialized due to the hand's biomechanics.

Table 4 summarizes the hand's elements, its degrees-of-freedom, and its contact surface points for the canonical set of grasps. Furthermore, it identifies the general class of robotic end-effectors, which can be substituted for the category of grasps under each group. The nonprehensile unilateral grasps would use a simple end-effector requiring zero DOF. This type of end-effector would use a fixed geometry to support, push, or engage an object. The palm and side opposition grasps emulate simple parallel end-effectors. The apparent difference between grasps within each group is the surface area brought to bear against the object and the object orientation with respect to the palm which is a function of the task intent, i.e., a screwdriver placed along the oblique palmar axis to take advantage of the upper limb's ability to provide a pronate-suponate motion to tighten a screw. A mechanical substitute can be tailored to the task and to the object, taking into account material properties and stability issues. Finally, the pad oppositions are the very definition of dynamic grasping, requiring one to three fingers working in unison with the thumb to provide various motions with respect to the palm. The third finger is utilized when spinning objects perpendicular to the palmar axis; the third finger provides the third contact point for stability while the thumb or other finger(s) reorient themselves. While all five rays can be utilized to form the dynamic grasps of pad opposition, the principle components include the thumb, index and middle finger. Capener (1956) recognized the specialized dynamic action of the two radial rays working in unison with the thumb naming it the *dynamic tripod*. He also summarized the apparent specialization of the two ulnar digits which customarily act together, especially in palmar grips, to provide the stability and control required of power grasps. The dynamic grips afforded by the pad opposition grasps would require a truly anthropomorphic-like end-effector to mimic *all* the hand's movements. This end effector would require hand-like configuration and geometry with an immobile palmar skeleton, a thumb, and three fingers (rays two through four) at a minimum. The thumb of the end-effector must be capable of four DOF consisting of anteposition-retroposition and flexion-extension of the trapezometacarpal joint, and flexion-extension of its metacarpophalangeal and interphalangeal joints. The end-effector's index and middle fingers require adduction-abduction and flexion-extension at their metacarpophalangeal joints, and flexion-extension of their two interphalangeal joints each. The end-effectors third finger would be equivalent to combining the hand's fourth or fifth ray into a singlewide finger

Table 4 Hand's elements, DOF, surface contact points, and equivalent robot end-effector for the canonical grasp oppositions.

Grasp	Unilateral	Palmar	Side	Pad
Elements	Any one or combination	Palm, finger(s), w/ or w/o thumb	Thumb & Index Finger or Inter-finger	Finger and Thumb Tips
Fingers	X	X	X	X
Palm	X	X		
Thumb	X	X	X	X
DOF / Finger-Thumb Motion	0 (fixed pose)	1 (for a fixed thumb pose)	1 (for a fixed finger pose)	Multiple DOF
Finger Flexion-Extension		X	X	X
Finger Adduction-Abduction				X
Thumb Flexion-Extension		X	X	X
Thumb Ante/Retroposition		X	X	X
Surface Contact Points	Any one or combination of hand surfaces	Anterior Surfaces	Thumb Tip and/or Finger Lateral Surfaces	Thumb & two to three finger tips minimum
Finger's Anterior Distal Volar	X	X		X
Finger's Anterior Intermediate and Proximal Volar	X	X		
Finger Lateral			X	
Thenar Eminence	X	X		
Hypothenar Eminence	X	X		
Thumb Anterior Distal	X	X	X	X
Thumb Proximal Volar		X		
Thumb Radial-Lateral	X			
Hand's Dorsal	X			
Equivalent Robot End Effector	Specialized tool, i.e., hook, platter, pointer	Parallel end effector	Parallel end effector	Anthropomorphic end effector

having three DOF in flexion-extension of its metcarpophalangeal and two interphalangeal joints. The Utah/MIT dexterous hand (McCammon and Jacobsen, 1990) is one of the closest approximations to such an end effector having demonstrated the ability to reproduce a number of grasps via telerobotic control.

Conclusions

In this paper, we have studied the biomechanics of human grasping to determine the mobility of human hand in grasping and enumerating all kinematic grasps. We have used the results to identify robotic end-effectors that can replace the human hand in performing various functions.

References

Addis, Robert B., (1998). Development and Application of a Human Grasp Taxonomy to Specify The Robotic End Effector Design for the LLNL DOR Process, M.S. Thesis, University of California-Davis, p. 209.

Blechschmidt, J.L. and Fessler, M.J. (1991). Motion comparision of the ASU finger and the human finger. Dept. of Mech. Engr. and Aero. Engr., Arizona State University.

Capener, N. (1956). The hand in surgery. *Journal of Bone and Joint Surgery* 38B:128.

Crossley, F.P.E. and Umholtz, F.G. (1977). Design for Three-Fingered Hand. *Mechanism and Machine Theory* 12:855-93.

Cutkosky, M.R. (1989). On grasp choice, grasp models and the design of hands for manufacturing tasks. *IEEE Transactions on Robotics and Automation* 5(3):269-279.

Cutkosky, M.R. and Wright, P.K. (1986b). Modeling Manufacturing Grips and Correlation with the Design of Robotic Hands. *Proceedings of the 1986 IEEE Int'l Conference on Robotics and Automation,* Washington DC: IEEE Computer Society Press, pp. 1533-1539.

Dubousset, J. (1971). Anatomie fonctionnelle de l'appareil capsulo-ligamentaire des articulations des doigts. In, R. Vilain (eds.), *Traumatismes Ostéo-Articulaires de la Main.* Paris: L'Expansion.

Iberall, T. (1997). Human Prehension and Dexterous Robot Hands. *The International Journal of Robotics Research* 16(3):285-299.

Jacobsen, S.C., Wood, J.E., Knutti, D.F., Giggers, K.B. (1986). The Utah/MIT Dexterous Hand: Work In Progress. In, D.T. Pham and W.B. Heginbotham, *International Trends in Manufacturing Technology, Robot Grippers.* UK: IFS Ltd, pp.341-389.

Kuczynski, K. (1968). The upper limb. In, R. Passmore and J.S. Robson (eds), *A Companion to Medical Studies*, Vol. 1. New York: Blackwell Scientific Publications.

Littler, J.W. (1960). The physiology and dynamic function of the hand. *Surgery Clin. North Am.* 40:259.

Loucks, C. S., Johnson, V.J., Boisiere, P.T., Starr, G.P., and Steele, J.P.H. (1987). Modeling and controll of the Stanford/JPL hand. *Proceedings of the 1987 IEEE International Conference on Robotics and Automation, Raleigh, North Carolina.* pp. 573-578.

MacKenzie, C.L., and Iberall, T. (1994). *The Grasping Hand.* Amsterdam: North-Holland.

Malek, R. (1981). Prehension and gesture: Ch 45 The grip and its modalities. In, R. Tubiana (ed.), *The Hand,* Vol 1. Philadelphia: W.B. Saunders, pp. 469-476.

McBride, E.D. 1942. *Disability Evaluation*, 3rd ed. Philidelphia, PA: J.B. Lippincott.

McCammon, I.D., and Jacobsen, S.C. (1990). Tactile sensing and control for the Utah/MIT hand. In, S.T. Venkataraman and T. Iberall (eds.), *Dexterous Robot Hands*. New York: Springer-Verlag, pp. 239-266.

Napier, J.R. (1956). The Prehensile Movments of the Human Hand. *Journal of Bone and Joint Surgery* 38B(4): 902-913.

Okada, T. (1979). Computer control of multi-jointed finger system. In, *6th Intrnational Joint Conference on Artificial Intelligence*, Tokyo.

Patterson, P. E. and Katz, J. A. (1992). Design and evaluation of a sensory feedback system that provides grasping pressure in a myoelectric hand. *Journal of Rehabilitation Research and Development* 29(1):1-8.

Salisbury, K. (1986). Teleoperator Hand Design Issues. *Proceedings IEEE International Conference on Robotics and Automation, San Francisco, CA*. 3:1355-1360.

Salisbury, K., Brock, D. and O'Donnell, P. (1989). Using an articulated hand to manipulate objects. In, M. Brady (eds.), *Robotic Science*. Cambridge, Mass.: MIT Press, pp. 540-562.

Salisbury, K. J. and Craig, J. J. (1982). Articulated hands: Force control and kinematics issues. *International Journal of Robotics Research* 1(1):4-17.

Schlesinger, G. (1919). Der Mechanische Aufbau der Kunstlischen Glieder (the Mechanical Structure of Artificial Limbs). In, M. Borchardt, et al. (eds.), *Ersatzglieder und Arbeitshilfen fur Kriegsbeschadigte und Unfallverletzte*. Berlin: Springer, pp. 321-699.

Scott, R. N. 1990. Feedback in myoelectric prostheses. *Clinical Orthopaedics and Related Research* 7(256):58-63.

Taylor, C.L. (1948). Patterns of hand prehension in common activities. *Engineering Prosthetics Research*, Special Technical Report 3. Los Angeles, CA: University of California, Department of Engineering.

Taylor, C.L. and Schwarz, R.J. (1955). The Anatomy and Mechanics of the Human Hand. *Artificial Limbs* 2:22-35.

Tubiana, R., (1981). Section 1: Functional anatomy; chapter 4, architecture and functions of the hand. In, R. Tubiana (eds.), *The Hand*, Vol. 1. Philidelphia: W. B. Saunders Co. pp. 19-93.

Appendix A

Programme and Organizing Committee

Chairmen:

Prof. A. Morecki
   Warsaw University of Technology
   Institute of Aeronautics and Applied Mechanics
   Nowowiejska 24
   00665 Warsaw
   POLAND

Prof. G. Bianchi
   CISM, Palazzo del Torso
   Piazza Garibaldi 18
   33100 Udine
   ITALY

Members:

Prof. J. Angeles
   Mc Gill University
   Mechancal Engineering Dept.
   817 Sherbrooke St. W.
   Montreal Quebec H3A 2K6
   CANADA

Prof. A.P. Bessonov
   Russian Academy of Sciences
   Mech. Engineering Research Inst.
   Griboedova 4
   Moscow-Centre, 101000
   RUSSIA

Prof. J-C. Guinot
   Laboratoire de Robotique de Paris
   Centre Universitaire
   Technologique de Vélizy
   10-12 avenue de l'Europe
   78140 Vélizy
   FRANCE

Prof. B. Heimann
   Universität Hannover
   Institut für Mechanik
   Appelstrasse 11
   30167 Hannover 1
   GERMANY

Prof. O. Khatib
   Stanford University
   Computer Science Department
   Stanford, CA 94305 2085
   USA

Prof. A.I. Korendyasev
   Russian Academy of Sciences
   Mech. Engineering Research Inst.
   Griboedova 4
   Moscow-Centre, 101000
   RUSSIA

Dr. J–C. Piedboeuf
Canadian Space Agency
6767 Route de L'Airoport
St. Hubert, Quibeec J34 8Y9
CANADA

Prof. W.O. Schiehlen
Universität Stuttgart
Institut B für Mechanik
Pfaffenwaldring 9
70550 Stuttgart 80
GERMANY

Prof. Atsuo Takanishi
Waseda University 59-308
Shinjuku-ku, Okubo 3-4-1
Tokyo 169
JAPAN

Dr. Eng. K. Tanie
Biorobotics Division
Robotics Department
Mechanical Eng. Laboratory
1-2 Namiki, Tsukuba
Ibaraki, 305
JAPAN

Prof. M. Vukobratovič
Institute "Mihajlo Pupin"
Volgina 15 - P.O. Box 906
11000 Beograd
YUGOSLAVIA

Prof. K.J. Waldron
Design Division, MC4021
Dept. of Mechanical Engineering
Stanford University
Stanford, CA 94305

Scientific Secretary:

Dr. C. Rzymkowski
Warsaw University of Technology
Institute of Aeronautics and Applied Mechanics
Nowowiejska 24,
00-665 Warszawa
POLAND

Tel.: (+48) 22 660 7992, Fax: (+42) 22 628 2587
E-mail: czarek@meil.pw.edu.pl

Secretary:

Dr. P. Agnola
CISM, Palazzo del Torso
Piazza Garibaldi 18
33100 Udine
ITALY

Tel.: (+39) 0432 248522, Fax: (+39) 0432 248550
E-mail: p.agnola@cism.it

Appendix B

List of Participants

AUSTRALIA

Karol Miller
The University of Western Australia
Department of Mechanical
and Materials Eng.
Nedlands, Perth, WA, 6907

CANADA

Jorge Angeles
Department of Mechanical Engineering
& McGill Centre for Intelligent Machines
McGill University
817 Sherbrooke St. W.
Montreal, Quebec, Canada H3A 2K6

Marek R. Kujath
Department of Mechanical Engineering
P.O. Box 1000
Halifax N3 B3J 2X4

Alexei Morozov
Department of Mechanical Engineering
& McGill Centre for Intelligent Machines
McGill University
3480 University str.2A7
Montreal, Quebec Canada H3A

Jean-Claude Piedboeuf
Canadian Space Agency
6767 Route de L'Airoport
St. Hubert, Quebec J34 8Y9

FINLAND

Tatu Leinonen
University of Oulu
Dept. of Mechanical Engineering
PO Box 4200
FIN –90014

FRANCE

Dahan Marc
Institut de Productique
L. Marc, 25 Av Alain Savary
F-25030 Besancon

Antoine Ferreira
Ecole Nationale Superieure d' Ingenieurs de
Bourges
10, boulevard Lahitolle
18020 – Bourges

Franck Geffard
Commissariat á l'Energie Atomique
Service de Téléopération et de Robotique
DTA/DPSA/STR - BP n°6
92265 Fontenay-Aux-Roses, CEDEX

Jean-Claude Guinot
Laboratoire de Robotique de Paris (LRP)
10-12 avenue de l'Europe
78140 Vélizy

Fethi Ben Ouezdou
Laboratoire de Robotique de Paris (LRP)
10-12 avenue de l'Europe
78140 Velizy Villacoublay

Gabriel Ramirez
Laboratoire de Mécanique des Solides
Faculté des Sciences SP2MI
Bd. 3, Téléport 2, BP. 179
86960 Futuroscope CEDEX

Jerome Szewczyk
Laboratoire de Robotique de Paris (LRP)
10-12 avenue de l'Europe
78140 Vélizy

GERMANY

Axel Buschman
Gerhard-Mercator-Universität
FB Maschinenbau/Mechanik
Lotharstraße 1
D-47048 Duisburg

Karsten Berns
Forschungszentrum Informatik
Haid-und-Neu-Str. 10-14
D-76131 Karlsruhe

Stefan Gruber
Institute B of Mechanics
University of Stuttgart
Pfaffenwald Ring 9
D-70550 Stuttgart

Bodo Heimann
Institut für Mechanik
Universität Hannover
Appelstr. 11
D-30167 Hannover

Manfred Hiller
Gerhard-Mecator-Universität Duisburg
Fachgebiet Mechatronik
Lotharstr. 1
D-47057 Duisburg

Klaus Löffler
Lehrstuhl B für Mechanik
TU München
Boltzmannstr. 15
D-85748 Garching

Jörg Müler
Gerhard-Mecator-Universität Duisburg
Fechgebiet Mechatronik
Lotharstr. 1
D-47057 Duisburg

Walter Schaaf
Fraunhofer Institute
Manufacturing Engineering and Automation
Nobelstr. 12
D-70569 Stuttgart

Igor Zeidis
TU Ilmenau
Fakultät für Maschinenbau
Max-Planc-Ring 12
D-98693 Ilmenau

Klaus Zimmerman
TU Ilmenau
Fakultät für Maschinenbau
Max-Planc-Ring 12
D-98693 Ilmenau

HUNGARY

Elisabeth Filemon
Technical University of Budapest
Department of Applied Mechanics
Muegyetem rkp 3
H-1111 Budapest

Tamás Insperger
Department of Applied Mechanics
Technical University of Budapest
H-1521 Budapest

Gábor Stépán
Department of Applied Mechanics
Technical University of Budapest
H-1521 Budapest

ITALY

Giovanni Bianchi
CISM, Palazzo del Torso
Piazza Garibaldi 18
33-100 Udine

List of Participants

Marco Ceccarelli
Dipartimento di Maccanica,
Strutture, Ambiente e Territorio
Lab. of Robotics and Mechatronics
University of Cassino, Via Di Biasio 43
03043 Cassino

Vincenzo Parenti-Castelli
DIEM – Facolta Di Ingegneria
University of Bologne
Via Risorgimento 2,
40136 Bologna

Giovanni Fabris
CISM, Palazzo del Torso
Piazza Garibaldi 18
33-100 Udine

Vinicio Turello
CISM, Palazzo del Torso
Piazza Garibaldi 18
33-100 Udine

JAPAN

Teru Hayashi
Toin University of Yokohama
Kurogane-cho 1614, Midori-ku
Yokohama 225-8502

Miura Hirofumi
Dept. of Mechanical Systems Engr.
Kogakuin University
1-24-2 Nishi Shinjuku
Shinjuku-ku, Tokyo, 163-8766

Mikio Horie
Tokyo Institute of Technology,
4259 Nagatsuta-cho, Midori-ku,
Yokohama, 226-8503

Makoto Kaneko
Dept. of Industrial and System Engineering
Hiroshima University
1-4-1 Kagamiyama
Higashi – Hiroshima, 739-8527

Hun-ok Lim
Kanagawa Institute of Technology
Humanoid Research Institute
Waseda University
1030 Shimoohgino, Atsugi
Kanagawa, 243-092

Atsuo Takanishi
Department of Mechanical Engineering
School of Science and Engineering
Waseda University
3-4-1 Okubo, Shinjuku-ku
Tokyo 169-8555

Hideaki Takanobu
Department of Mechanical Engineering
School of Science and Engineering
Waseda University
3-4-1 Okubo, Shinjuku-ku,
Tokyo 169-8555

LATVIA

Igor Tipans
Riga Technical University
1, Kalku Str.
Riga LV 1658

POLAND

Wojciech Blajer
Politechnika Radomska
ul. Malczewskiego 29
26-600 Radom

Adam Czaplicki
Institute of Sport and Psychical Education
ul. Akademicka 3
21-500 Biała Podlaska

Przemysław Herman
Politechnika Poznańska
Control, Robotics and Computer Science
ul. Piotrowo 3A
60-965 Poznań

Józef Knapczyk
Cracow University of Technology
ul. Warszawska 24
31-155 Kraków

Sławomir Łuszczak
Inst. of Aeronautics and Applied Mechanics
Warsaw University of Technology
Nowowiejska 24
00-665 Warsaw

Alicja Mazur
Institute of Engineering Cybernetics
Wrocław University of Technology
ul. Janiszewskiego 11/17,
50-372 Wrocław

Krzysztof Mianowski
Inst. of Aeronautics and Applied Mechanics
Warsaw University of Technology
Nowowiejska 24
00-665 Warsaw

Adam Morecki
Inst. of Aeronautics and Applied Mechanics
Warsaw University of Technology
Nowowiejska 24
00-665 Warszawa

Kazimierz Nazarczuk
Inst. of Aeronautics and Applied Mechanics
Warsaw University of Technology
Nowowiejska 24, 00-665 Warsaw

Andrzej Osyczka
Instytut Maszyn Roboczych
Politechniki Krakowskiej
Al. Jana Pawła II 37
31-864 Kraków

Cezary Rzymkowski
Inst. of Aeronautics and Applied Mechanics
Warsaw University of Technology
Nowowiejska 24
00-665 Warszawa

Krzysztof Tchoń
Instytut Cybernetyki Technicznej
Politechniki Wrocławskiej
ul. Janiszewskiego 11/17
50-372 Wrocław

RUSSIA

Eugene S. Briskin
Volgograd State Technical University
28 Lenin av.
Volgograd, 400066

Victor A. Glazunov
Mechanical Engineering Research Institute, RAS
M. Kharitonjevsky, 4, Moscow, 101830

Vladimir Pavlovsky
Keldysh Inst. of Appl. Math of RAS
Miusskaya squaze, 4
125047, Moscow, Russia

SPAIN

Francisco Valero
Department of Mechanical Engineering
Universidad Politécnica de Valencia
Camino de Vera s/n 46022 Valencia

Vincente Mata
Departmento de Ingenería Mecánica
Universidad Politécnica de Valencia
Camino de Vera s/n 46022 Valencia

UNITED KINGDOM

Robert Bicker
Department of Mechanical, Materials
and Manufacturing Engineering
University of Newcastle upon Tyne
Newcastle upon Tyne NE1 7RU

List of Participants

USA

Antal K. Bejczy
Autonomy and Control Section
Jet Propulsion Laboratory, MS-198-219
California Institute of Technology
4800 Oak Grove Drive
Pasadena, CA 91109-8099

Oussama Khatib
Robotics Laboratory
Department of Computer Science
Stanford University
Stanford, CA, 94305

Bahram Ravani
Department of Mechanical
and Aeronautical Engineering
University of California
One Shields Avenue
Davis, CA 95616

Bernard Roth
Department of Mechanical Engineering
Stanford University
Stanford, CA 94305-4021

James P. Schmiedeler
The Ohio State University
Department of Mechanical Engineering
206, West 18th Ave
Columbus, OH 43210

Steven A. Velinsky
Department of Mechanical
and Aeronautical Engineering
University of California, Davis
One Shields Avenue
Davis, CA 95616-5294

Kenneth J. Waldron
Design Division, MC4021
Department of Mechanical Engineering
Stanford University
Stanford, CA 94305

YUGOSLAVIA

Miomir Vukobratovič
Institute "Mihajlo Pupin"
Volgina 15 - P.O. Box 906
11000 Beograd